# Contemporary Issues in Animal Agriculture

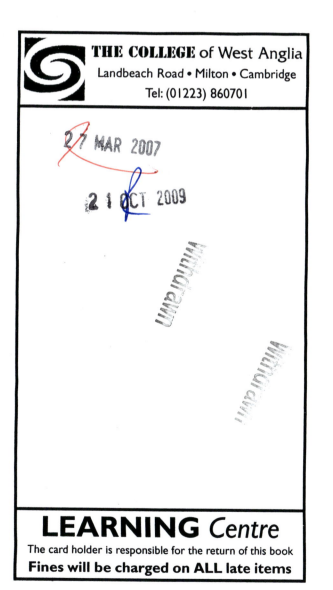

THE COLLEGE OF WEST ANGLIA

**Library of Congress Cataloging-in-Publication Data**

Cheeke, Peter R.
  Contemporary issues in animal agriculture / Peter R. Cheeke.—3rd ed.
    p. cm
  Includes bibliographical references and index.
  ISBN 0-13-112586-9 (alk. paper)
    1. Livestock—Social aspects. 2. Food of animal origin—Health aspects. 3. Animal
  nutrition. 4. Livestock—Ecology. 5. Animal industry—Environmental aspects. I. Title.

SF140.S62C48 2003
338.1'76—dc21                                                                    2002192997

**Editor in Chief:** Stephen Helba
**Executive Editor:** Debbie Yarnell
**Editorial Assistant:** Jonathan Tenthoff
**Managing Editor:** Mary Carnis
**Production Editor:** Emily Bush, Carlisle Publishers Services
**Production Liaison:** Janice Stangel
**Director of Manufacturing and Production:** Bruce Johnson
**Manufacturing Buyer:** Cathleen Petersen
**Creative Director:** Cheryl Asherman
**Cover Design Coordinator:** Miguel Ortiz
**Marketing Manager:** Jimmy Stephens
**Cover Photos:** Bison (upper left) courtesy of W. B. Kessler;
Naked chicken (upper middle), courtesy of A. Cahaner;
Broiler chickens (upper right), courtesy of R.A. Swick;
Sheep feedlot (middle left), courtesy of Peter Cheeke;
Horse (bottom left), courtesy of L. A. Lawrence;
Home on the range (bottom right), courtesy of Peter Cheeke

Pearson Education LTD.
Pearson Education Australia PTY, Limited
Pearson Education Singapore, Pte. Ltd.
Pearson Education North Asia Ltd.
Pearson Education Canada, Ltd.
Pearson Educación de Mexico, S.A. de C.V.
Pearson Education—Japan
Pearson Education Malaysia, Pte. Ltd.

636 C

0822189

10 9 8 7 6 5 4 3 2 1
ISBN 0-13-112586-9

# Contents

# About the Authors

Peter R. Cheeke is Professor Emeritus of Animal Nutrition at Oregon State University. He grew up on a small family farm in British Columbia, Canada. He graduated from the University of British Columbia with a BS in Agriculture (1963) and MS in animal nutrition (1965). He completed his academic training at Oregon State University with his PhD in animal nutrition (1968), under the supervision of Dr. James Oldfield. His PhD research was on interrelationships between vitamin E and selenium. Since 1969, he has been assistant associate and full professor at Oregon State University, with a teaching and research program in animal nutrition. He retired from OSU in 2000. One of his major research interests has been the study of natural toxicants in feeds and poisonous plants, particularly alkaloids in poisonous pasture weeds and toxins in potential new feedstuffs. He has written a book *Natural Toxicants in Feeds, Forages, and Poisonous Plants* (Prentice-Hall, 1998). He has worked with a variety of animal species, including ruminants, rabbits and poultry. He was a founder of the Oregon State University Rabbit Research Center, and is coauthor of *Rabbit Production* (Prentice Hall, 2000) and *Rabbit Feeding and Nutrition,* (Academic Press, 1987). He was honored by the western section of American Society of Animal Science with the Young Scientist Award in 1979 and the Distinguished Service Award in 2001. In 1990, he was named Distinguished Professor of Agriculture at Oregon State University. He has served on the editorial boards of the Journal of Animal Science, and Animal Feed Science and Technology. He has operated a small farm near Corvallis, Oregon—raising various types of livestock and poultry, including a small herd of beef cattle. He and his wife Karen now live on a small farm, with an equine facility for breeding and therapeutic riding instruction. He has had a life-long interest in the interrelationships of plants and animals, and in the survival and strengthening of the family farm.

# Preface

The objective of this book is to discuss some of the major controversial issues impacting animal production. Animal agriculture and animal scientists face many controversies and challenges. A growing segment of society views livestock production as unethical and inhumane. Until recently, animal scientists either ignored these concerns or attributed them to an inconsequential fringe element of society. It is beginning to sink in that societal concerns are legitimate, are here to stay, and will increase. Concerns about human exploitation of animals for food production and biomedical research have become mainstream. Food safety issues such as "mad cow disease" and *E. coli* contamination of meat make the headlines regularly. There is a steadily growing vegetarian movement based on the viewpoint that meat consumption is bad for one's health. The animal rights movement is opposed to the raising of domestic animals for any purpose. In the western United States, grazing of livestock on public lands is a major controversial issue. Air and water pollution, global warming, riparian zones, endangered species—the list goes on and on. Animal science graduates will be dealing with these types of issues, even to the extent of justifying to friends and relatives what they do for a living. In my opinion, it is critical that students have exposure to these contemporary issues during their education, rather than being hit with them as they enter "the real world."

There is currently great interest by society at large in the environment. More benign systems of resource utilization in agriculture, forestry, and other primary extractive industries are being sought. My intention has been to discuss these issues, particularly as they relate to animal production, in a global context. I have discussed these issues in a holistic manner, showing how animal production and other aspects of agriculture are integrated into the larger society as a whole.

Contemporary issues exist because there is no right answer. There is more than one valid point of view; otherwise there wouldn't be an issue. I have endeavored to critically evaluate societal concerns about various aspects of modern animal production and to present several points of view. One of my motivations for writing this book is my belief that the resolution of contentious issues is one of the major challenges facing the discipline of animal science. Ecological and environmental issues, as well as nutritional involvement in degenerative disease and aging, emerged as significant issues in the latter part of the twentieth century and will continue at the forefront of public concern in the twenty-first century. Agricultural students must be armed with thinking skills and a broad perspective to meet these ever-increasing challenges. We must be able to justify to society what we are doing, why we're doing it, and what consequences for the environment and food safety our actions produce. The sustainability of animal science and livestock production as careers will depend on how well the next generation meets these challenges. My hope is that this book will help point the way toward an acceptable middle ground that can accommodate the diverse viewpoints of society, including those of the livestock industry.

Some animal scientists would prefer to ignore controversial issues and just hope they will go away. They think it is a disservice to "our side" to give publicity to the ideas of "the enemy" such as environmentalists. I have to admit somewhat hesitantly that as a result of my research and contemplation in preparation of this book, my own views and sympathies have moved "to the left" a bit. I think that animal agriculture has a lot of room for improvement.

Another motivation for writing this book has been my perception that many animal scientists have given the knee-jerk response to societal concerns. I have noted colleagues dismissing concerns about cholesterol, meat consumption, and coronary heart disease, with comments about "so-called medical research" and "so-called medical authorities," while never having read a scientific paper on the subject themselves! My intention is that this book will provide a framework on which to build a reasoned and rational discussion of contemporary issues. If there are some things wrong with how we raise animals or with animal products, let's face up to it and deal with it.

I have endeavored to be reasonably open-minded and noninflammatory, without being bland. It is typical when dealing with controversial subjects to soften the blow by using words like "perhaps, maybe, might be, could be," and so on. I've tried to keep this to a minimum. I hope that a bit of fire and brimstone comes through on some topics. Some terms such as "factory farming" are inflammatory and "value-laden" to some people. As the old saying goes, if it looks like a duck, walks like a duck, and quacks like a duck, then it's probably a duck!

I have tried to write in clear and easily understood prose. There sometimes seems to be a perception among students and professors that a text should be challenging and difficult to read, as if to imply that a text is more valid and profound if incomprehensible than one that is easily understood. On the other hand, a book should expand the reader's horizons, including his or her vocabulary. I have attempted to accomplish both goals, to be mind-expanding and lucid.

Thus I hope that this book will contribute to expanding the horizons of ag students, and to instill an appreciation for the validity of other points of view. I've tried to avoid falling into the trap of *us* vs. *them*. There is a reasonable middle ground. There is no point in catering to either extreme by just "preaching to the choir." It does no good to reinforce existing agrarian prejudices. I hope I have challenged ag students to appreciate the validity of some of society's concerns about livestock production. Their future role will be to address these concerns, rather than just attempt to debunk them. I hope to do justice to the following mission statement from my university: "The highest aspiration of a university is to free people's minds from ignorance, prejudice, and provincialism and to stimulate a lasting attitude of inquiry."

I have cited recent literature extensively. These citations provide an entry into the scientific literature for those who want further information and will expose undergraduate students to scientific literature as contrasted with the popular press. They will gain an appreciation for the importance of grey, as opposed to black and white. There are no magic answers. Judgment and interpretation are important components of rational inquiry.

I have noted that colleagues who have complimented the first two editions of this text tend to be young, while often the "old guard" tends to regard my ideas as a touch ludicrous. My perception is that many students today are "ahead of the curve" in their appreciation of natural resource and environmental issues, and that animal science professors have some catching up to do. A "don't give an inch" defense of modern animal production techniques won't resonate with many students. They know better.

Each chapter has been written so that it can more or less stand alone. Chapter 4 is included not only because my primary area of interest is nutrition, but also because some knowledge of nutritional principles is desirable to deal effectively with issues such as "chemicals and hormones as feed additives," why beef cattle have saturated fat, why cattle and not chickens have been implicated in global warming, and so on.

Some issues, such as the relationships of dietary fatty acids and cholesterol to coronary heart disease and cancer, are very complex and difficult to present to a diverse audience without the perils of oversimplification, or, conversely, being overly technical and incomprehensible. I have cited sufficient recent publications to provide an entry to the literature for those who wish further information. The fairly extensive citation of recent research may reduce readability somewhat but is essential to validate statements that deal with controversial issues.

I hope the book is interesting. I've tried to pack in all sorts of interesting things, such as the recent discovery of a previously unrecorded species of wild cow in Vietnam (Chapter 1) to the role of the white cliffs of Dover in regulation of the global environment (Chapter 11).

I thank colleagues who have contributed photos and ideas. I am especially indebted to Helen Chesbrough for her Herculean efforts in typing and preparing the manuscript. Colleagues Steve Davis and Candace Croney reviewed Chapter 10; I should also acknowledge Steve Davis for critiquing the entire text during preparation of the second edition. I also thank those who have commented favorably to me on the first two editions; a little bit of positive reinforcement goes a long way with authors!

A final thought regarding this book is stimulated by the words of Swinnerton-Dyer (*Nature* 373:186, 1995):

> Most scientists . . . reach a moment at which they know they will never have another worthwhile idea, and the only result of their continuing to do research after that point is to clog up the learned journals with papers that no one should be expected to read. That may be the moment to write a book pulling one's subject together in a lucid way.

I have reached the former condition, and I hope I have succeeded in the latter!

My experience in writing books is that the job is never done. There comes a time to wrap things up, even if further improvement is possible. I guess that's why there will probably be a fourth edition!

# CHAPTER 1

# Domestication of Animals and Their Contributions to Human Welfare

**CHAPTER OBJECTIVES**

1. To briefly discuss the evolution of humans, emphasizing the roles of animals in the development of human society.

2. To discuss the domestication of the major livestock species, on a global basis, to illustrate the diversity of animals that contribute to human welfare.

3. To discuss the diverse contributions of domestic animals to humans, particularly emphasizing some of the roles that might be less familiar to North Americans.

Modern humans (*Homo sapiens*) evolved from earlier primate ancestors, such as *Homo erectus.* The prevailing consensus of anthropologists is that early humans evolved in Africa in the Pliocene epoch (2–5 million years ago) and radiated from Africa to Europe and Asia. The divergence between the lineages giving rise to modern apes and humans occurred in Africa between 4 and 10 million years ago (Fleagle, 1988). Humans and apes belong to a primate grouping known as hominoids. There are five genera of existing **hominoids:** the gibbons, gorillas, orangutans, chimpanzees, and humans.

Humans have been the hominoid success story (so far); all other hominoids are endangered species, largely because of the activities of their human relatives. Humans are the only surviving **hominids,** which are bipedal primates (bipedalism means walking upright on the hind limbs). The earliest hominids were *Australopithecus* species, of which Lucy, the fossil remains of a female *Australopithecus* found in Hadar in Ethiopia in 1974 by the American anthropologist Donald Johanson, is the most famous (Johanson and Edey, 1974). The earliest *Homo* species, *H. habilis* appeared in the fossil record about 2 million years ago. *Homo erectus* dates to about 1.6 million years ago. Artifacts found with fossil remains indicate that *H. erectus* were hunters, who killed animals as large as elephants and the large ungulates. They used fire, developed a wide range of tools, and developed social organizations of families and tribes. The transition from *H. erectus* to *H. sapiens* (wise man) was a gradual one, with increases in brain size and reduction in bone mass. The most recent intermediate species was *H. neanderthalensis* (Neanderthals). There were probably at least 20 hominid species (Tattersall, 2000).

Neanderthals survived in Europe until about 30,000 years ago. They rapidly disappeared from the fossil record, following the arrival of *H. sapiens* from Africa (Tattersall, 2000). A "recent out of Africa" model for human distribution argues that *H. sapiens* emerged in Africa about 100,000 years ago and spread globally, displacing and eliminating other *Homo* species encountered during their expansion (Adcock et al., 2001). Thus began the assault of the "wise man" on other species, which continues today. Our propensity for wreaking environmental havoc is apparently deep-seated in an evolutionary sense. The "out of Africa" hypothesis contends that all modern humans are descended from a single founder population of about 10,000 breeding adults, who remained isolated for thousands of years before expanding into Europe and Asia in a migrational wave beginning about 200,000 years ago (Winter et al., 2001).[1]

------

[1] The expansion out of Africa, perhaps over a million years ago (*Nature* 418:145–151, 2002), may have been motivated by the search for meat. Early *Homo* species were likely carnivores. As a population of carnivores increases, hunters must roam long distances to find prey. A few miles a year, over several thousand years, could account for dispersal of early humans into Europe and Asia. New fossil evidence pushes the emergence of *Homo* species in Africa back to 1–2 million years ago (*Nature* 148:145–151, 2002).

Our early *H. sapiens* ancestors used fire for cooking and developed increasingly sophisticated tools. Their body proportions became less apelike as technological skill in hunting and food gathering replaced brute strength. The use of fire for cooking was very significant, because heat treatment detoxifies many of the toxins found in plants. This allowed early humans to use many plant resources that were poisonous to other animals. *Homo sapiens* rapidly (in geological time) colonized all continents and habitats ranging from steamy equatorial jungles to frigid arctic tundra. Some of the most recent "cave men" were the Neanderthal and Cro-Magnon people. The Neanderthals of Europe appear to have been an evolutionary dead end, who were displaced by the earliest modern humans, the Cro-Magnon people. Based on DNA evidence from museum samples of Neanderthal tissues, Neanderthals went extinct without contributing to the gene pool of modern humans (Krings et al., 1997).

This background is relevant to this book in terms of dietary patterns of modern humans. Are we by evolutionary history vegetarians, carnivores, or omnivores? Undoubtedly, we are omnivores. Early hominids were **hunter-gatherers.** Our dental morphology is not that of herbivores or carnivores. There are no particular adaptations of the digestive tract typical of herbivores, such as an enlarged stomach or cecum for microbial fermentation of fiber. In contrast, there are leaf-eating monkeys (e.g., Colobus monkeys) with an enlarged, subdivided stomach with a microbial population, somewhat similar in function to the rumen of cattle. The cecum, an area of the hindgut that is enlarged in many herbivores (e.g., rabbit, horse), has in humans regressed to a vestigial form, the appendix. Humans are opportunistic feeders and basically eat anything edible, plant or animal, with a dietary preference for low fiber plant products. This is why people intrinsically prefer white bread to brown whole wheat bread, and oatmeal without oat hulls. Meat has been an important component of the human diet throughout our entire evolutionary history (Eaton and Konner, 1985). Evolution and diet are considered in more detail in Chapter 2.

The reporting of the complete human genome was achieved in February, 2001 (*Science* 291:1304, 2001). Molecular biology has added a powerful tool to paleontology (molecular paleontology). For example, Krings et al. (1997) sequenced Neanderthal mitochondrial DNA, and by comparing the number of base changes with the DNA of modern humans, concluded that an African common ancestry existed about 600,000 years ago. The genetic relationship between humans and the other four surviving hominoids (gibbons, gorillas, orangutans and chimpanzees) is extremely close. We share over 99 percent DNA in common with them (Goodman et al., 1990; Ruvulo, 1997). A single mutation at a functionally important site in the DNA sequence can have a very large phenotypic effect. For example, we have a "tail bone" but the genes that stimulate development of the tail are "turned off" in humans. Based on DNA evidence, chimpanzees (*Pan* spp.) and humans (*Homo*) are the most closely related hominoids. Only a few changes in gene expression or function are responsible for the morphological and functional differences between humans and chimpanzees (Chou et al., 1998). The differences in gene regulation and function explain variations between us and chimpanzees in susceptibility and response to diseases such as cancer, hepatitis, AIDS, malaria, and intestinal infections (Chou et al., 1998). We would be covered in body hair, like apes, except that all humans have the homozygous presence of a gene that "turns off" body hair growth (Winter et al., 2001). Perhaps it is a bit of a come-down to have to admit that there is very little genetic difference between humans, chimpanzees and gorillas.

# THE DOMESTICATION OF ANIMALS

Agriculture, the deliberate culture of crops and animals for food and other purposes, began about 10,000 years ago in the Middle East.

The change from a hunter-gatherer system to settled society resulted in an increased quantity of food available and the ability to store surpluses for winter and other lean times. The production of food did not require the labor of the entire social group, freeing some individuals to follow other pursuits. This division of labor and allocation of resources led to the rudiments of a market economy and formed the basis of our present society. The increased quantity of food produced by agriculture allowed the world's human population to dramatically increase in number, a process that continues unabated to this day.

A **domestic animal** is one that has been bred in captivity for purposes of economic profit to a human community, with complete human control over its breeding, organization of territory and food supply (Clutton-Brock, 1989a, b, 1999). Young animals of virtually all species can be tamed, but many species will not thrive or reproduce as adults. Thus there are only a few animal species that possess the innate characteristics necessary for domestication, and most of these now are, in fact, domesticated.

Characteristics of animals that increase their likelihood of being domesticated will be discussed briefly, particularly with reference to domestication of the dog, which was likely the first animal to be domesticated and is the only domesticated organism, plant or animal, characteristic of all hunter-gatherer cultures. Animals susceptible to domestication are usually social animals with a dominance hierarchical structure similar to that of humans. Animals that were domesticated were those that were hardy and able to adapt to varying diets and environments. Many animals, such as concentrate-selector ruminants (e.g., dik dik antelopes—see Chapter 4), are difficult to raise even with our best current knowledge; such animals are raised in zoos with difficulty, showing high mortality rates and poor reproduction. Animals that were domesticated had attributes of economic benefit to humans. Dogs, for example, had many benefits—for hunting, herding, guarding, and so on. Our ancestors didn't usually have the luxury of keeping animals for companionship. Until recently, dogs and cats were kept mainly for functional purposes. Dogs guarded the farm, kept strangers and predators away, helped herd the livestock, and so on. A farm without cats was soon overrun with rats and mice. Dogs and cats (and children!) until recently had to "earn their keep." Was animal domestication an intentional human act? Did our ancestors undertake domestication as a rational decision, in recognition of the potential benefits of bringing plants and animals under their control? We can never know. A contrasting explanation is the evolutionary view. In this view, the behavior, diets, and morphology of certain animals changed from that of their wild counterparts in response to the selection pressures of a new ecological niche—a domestic association with humans (Morey, 1994). Common sense would suggest that both scenarios were involved: humans recognized the benefits of domestication, and certain species were biologically equipped to take advantage of the opportunity, or as Budiansky (1992) proposes, animals chose domestication.

The first domestic animal was the **dog** (*Canis familiaris*), derived from the European grey wolf (Fig. 1-1), 12,000–14,000 years ago in Eurasia (Savolainen et al, 2002). While there is not complete agreement on the origins (Fig. 1-2) of all breeds of dogs, all **dogs** originated from one or more species of wild canines (*Canis* spp.). Most breeds were developed from the European and Asian wolf (*C. lupis*), with some infusion of the golden jackal (*C. aureus*). All canines have the same chromosome number; the dog, wolf, coyote, and golden jackal can all interbreed with fertile offspring. Body size of animals within a species tends to be higher with increasing distance from the equator (Bergmann's Rule); thus North American wolves are larger than coyotes, which are larger than jackals. Hence a wide range in body size of dog breeds reflects both human selection and the source of the original founder population. Mitochondrial DNA analysis suggests that all breeds of dogs are descended from the wolf, although there may have been more than one time and place of domestication (Vila et

**FIGURE 1-1** The wolf is the ancestor of domestic dogs. Dogs display many of the characteristics of juvenile wolves, including submissiveness, which makes them subservient to their human masters (surrogate pack leaders). The North American wolf, shown here, may be the ancestor of the Eskimo dog, the Husky although recent evidence (Leonard et al., 2002) indicates that Native American dogs originated from European wolves, and accompanied late Pleistocene humans across the Bering Strait from Siberia to Alaska. (Courtesy of Charles E. Kay.)

al., 1997). The association of wolves with humans was a mutually beneficial one in terms of hunting and food availability. Wolves have a hierarchical social structure, with a leader and submissive pack members, as does human society. This has allowed humans and dogs to interact on a behavioral basis, with the human master serving the role of the dog pack leader. Behavioral studies have shown that the patterns of behavior of dogs in human society are the same as those of wolves in wolf society (Clutton-Brock, 1989a). The animals most

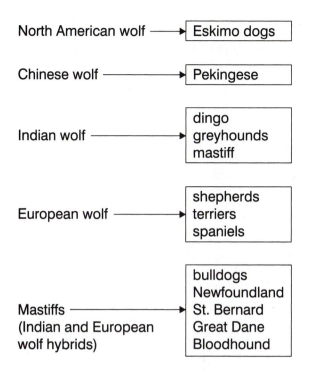

**FIGURE 1-2** Origin of modern dog breeds.

easy to domesticate are those with this type of social structure. The Canadian scientist Valerius Geist studied bighorn mountain sheep and was able to associate with them in the wild as literally a member of the flock (Geist, 1971). This type of interaction is not possible with deer, bison, or most other wild ungulates. In some societies dependent upon a particular species, domestication never did occur. The American Plains Indians, for example, were dependent on wild bison, but domestication did not occur, undoubtedly because of the refractory nature of the bison's behavior. Budiansky (1992) argues, not entirely convincingly, that certain species such as the wolf *chose* to be domesticated as an evolutionary survival strategy. He also claims, more convincingly, that domesticated species such as the dog exhibit **neoteny,** a phenomenon in which animals retain juvenile features of their ancestors into adulthood. Domestic dogs display the docility, submissiveness, and anatomical features of the juvenile wolf. These aspects will be discussed in more detail.

Characteristic changes in morphology, physiology, and behavior occur when animals are domesticated. In the initiation of the domestication process, only a small number of individuals from the wild population are selected and constitute the "**founder population.**" These selected animals will not have the full genetic diversity of the wild population. Mutations that would be deleterious in the wild might not be deleterious or may even be desirable in the newly domesticated population, so the domesticated strain will rapidly diverge from the wild population. With dogs, for example, it is likely that more easily tamed, submissive individuals would have been selected, and aggressive animals that attacked their human caretakers would have been killed.

Dogs have an exceptionally good ability to interpret visual cues of humans (social-cognitive abilities), allowing them to interact with humans in unique ways. Hare et al (2002) found that dogs were superior to our nearest relatives, the chimpanzees, in utilizing visual cues to locate hidden food. They concluded that the social-cognitive abilities of dogs converged with those of humans during domestication.

The dog has several neotonic features, including submissive behavior to the parent (the human master is a surrogate parent). The shape of the head and jaw and relative sizes of the teeth and other parts of the skeletal system in the dog are those of the juvenile wolf. Adult wolves rarely bark, but the juveniles, like dogs, do. Interestingly, there are similar resemblances between adult humans and juvenile apes (McKinney, 1998).

A fascinating aspect of domestication is the relationship of **coat color** to temperament and susceptibility to stress. Mutations of coat color different from the wild form are often associated with lessening of the fear response and other behavioral changes characteristic of taming. Hemmer (1990) uses the term "**environmental appreciation**" to describe the behavior of domestic animals. Domesticated animals have lower environmental appreciation than their wild counterparts, meaning that their alertness, fear responses, motor activity,

## HUMAN FOUNDER POPULATIONS

*(Review of R. Lewis. 2001. "Founder Populations Fuel Gene Discovery." The Scientist, April 16, 2001, pp. 8–9.)*

Founder populations of humans are modern groups that descend from a few individuals who left one area to settle in another, for political, religious, or social reasons. Currently there is much interest in these populations to search for genes responsible for various diseases such as breast cancer, cystic fibrosis, cardiovascular disease, schizophrenia, and so on. If these conditions were present in the founder populations, they tend to become "genetically concentrated" as the population becomes increasingly homozygous for particular traits. The main founder populations in the world are those of Quebec, Finland, Iceland, Sardinia, Costa Rica, the northern Netherlands, Newfoundland, and a few discrete ethnic groups such as Ashkenzai (eastern European) Jews.

The French Canadians of Quebec are a good example of a founder population. They have several monogenic disorders at a higher incidence than for the general population, including cystic fibrosis, familial hypercholesterolemia, and pseudo vitamin D-deficiency rickets. The original founding population of fewer than 5,000 individuals was established from 1608 until 1660, when immigration from France ceased. The Quebec population grew, by natural increase, from 2,000–4,000 founding genotypes to over 6 million today, in 14 generations. This unique population provides a laboratory for identifying genes that lie behind the major diseases of humans. The population of Iceland was similarly derived from a small founder group, and detailed health records for Icelandic families are available for the past several hundred years. Founder groups in the United States include Utah Mormons and Old Order Amish. Studies of gene interactions in founder group populations will likely have huge impacts on the knowledge and treatment of human disease.

and so on are less. In other words, they have less need to "rely on their wits" for survival and so can relax their guard when under the care of humans. There is an intricate biochemical relationship between the hormones associated with stress responses (e.g., adrenalin, noradrenalin) and melanin pigments of hair (Fig. 1-3). Hemmer (1990) reviewed studies showing relationships between **coat color and behavior**, explainable on the metabolic basis that up to a certain point (DOPA) the pigments that determine coat color (melanins) and the catecholamine neurotransmitters that control behavioral responses (flight or fright hormones) share a common biosynthetic pathway. Thus a mutation for a new coat color, by enhancing melanin pigment synthesis, could divert DOPA from catecholamine synthesis and thus alter behavior to a more placid state. Selection for a nonwild color has been a central feature of the domestication process (Hemmer, 1990).

Selection for behavioral change in domestic animals entered a new phase in the twenti-

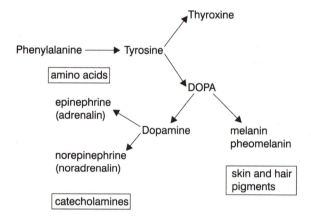

**FIGURE 1-3** Relationships between hair pigmentation and behavior may involve a common biosynthetic pathway of hair pigments (melanins) and neurotransmitters (catecholamines).

eth century, with the development of factory farming of animals. There is evidence of selection of livestock and poultry for tolerance of increased crowding under confinement conditions (see Chapter 8). Ethical aspects of this selection are discussed in Chapter 10.

## THE RUSSIAN FARM FOX EXPERIMENT

*(Review of L. N. Trut. 1999. "Early Canid Domestication: The Farm-Fox Experiment." Amer. Scientist 87:160–169.)*

A Russian geneticist, Dmitry Belyaev, undertook a long-term experiment to determine the genetic changes in wild foxes selected for a single factor, tameness. The experiment began in 1959, with 30 male and 100 female wild foxes. Each resulting pup was evaluated and assigned to a group: pups that flee or bite when handled; pups that can be petted but show no friendly response; and pups that exhibit friendly behavior: tail wagging and whining. By the sixth generation, it was necessary to add another (Elite) group: pups that were eager for human contact, sniffing and licking people as dogs do. By the time of this report (1999), selection had occurred for 30–35 years. The Elite foxes were unmistakenly domesticated; they were docile, eager to please, and craved human attention. Remarkably, with succeeding generations selected only for tameness, pronounced physical changes occurred. By the eighth generation, coat color changes were observed, with loss of pigment on some parts of the body, and a star-shaped pattern on the face. Next came floppy ears and rolled tails, turned up over the back. They began to look remarkably similar to several breeds of dogs.

Russian scientists have interpreted these findings as a shift in the rates of timing of developmental processes. Newborn fox pups have floppy ears. In the Elite line, these were carried over to adulthood. Even novel coat colors may be attributed to changes in the timing of embryonic development, during which embryonic pigment cells (melanoblasts), formed in the neural crest, migrate to the embryo's epidermis. A delay in the migration of some melanoblasts results in their deaths, thus altering skin pigmentation.

The changes observed in these foxes, which were selected for the single trait of tameness, were characteristic of domestication processes in general. Most domestic animals show the same basic and physiological changes. As Charles Darwin noted in Chapter 1 of *On the Origin of Species*, "not a single domestic animal can be named which has not in some country droopy ears"—a feature not found in any wild animal except the elephant.

The Russian experiment demonstrated that in just a few generations of selection for tameness, a wild animal was transformed into a domesticated one. With the collapse of the Soviet Union, the institute was faced with severe financial limitations. To generate funds, some of the foxes have been sold to Scandinavian fur farmers, who have been pressured to develop animals that do not suffer stress in captivity. They also plan to market pups as house pets.

The significance of this experiment is that it demonstrates that a simple selection process for animals that respond favorably to humans could account for the development of domestic animals.

---

The development of the human-dog bond is of considerable interest and of practical importance today, especially to veterinarians. The **human-canine bond** has applications in human medicine. Ownership of dogs has been shown to favorably influence several medical conditions, including preventing loneliness and depression, reducing high blood pressure and obesity, and enhancing physical fitness, all in a very cost-effective manner (Allen, 1997). Better knowledge of the human-canine bond can not only facilitate these medical benefits, but also aid veterinarians and others dealing with undesirable canine behavior, such as attacks and bites. There are primitive dogs still in existence (e.g., New Guinea singing dog); Brisbin and Risch (1997) suggest that the study of these animals could provide clues to the solution of current problems such as canine aggression, public health issues, genetic diseases of modern dogs, and pet overpopulation.

Domestication of animals has had a tremendous influence on **human history**. The huge empire of Genghis Khan and his mounted warriors, the conquering of the Native Americans in North and South America, and innumerable other victories and defeats owe their existence to the domestication of the horse. The development of agriculture and modern society would not have been possible without the domestication of draft animals (horses, oxen, yaks, buffalo, camels, etc.). The development of nomadic societies, herding cattle, sheep, reindeer, and so on would not have been possible without animal domestication. Development of trade between distant cultures, such as the Silk Road linking Europe and China, was possible only with camels and other "beasts of burden." Such trade led to exchange of culture, ideas, language, and genes. The only cultures that developed the wheel were those with draft animals. Thus the history of modern humans is intertwined with that of domestic animals. No human society lacking domestic animals has evolved beyond the primitive hunter-gatherer stage. Domestication of most of the major domestic animals began 8,000–10,000 years ago as agriculture developed. Their origins will be briefly discussed. A timeline for domestication of some of the common domestic animals is shown in Fig. 1-4, with the most recent being laboratory rodents (rats and mice).

Modern breeds of **cattle** belong to two species, *Bos taurus* and *Bos indicus*. Both are derived from the aurochs, or giant wild ox (*Bos primigenius*). The **aurochs** were a very successful species widespread throughout Europe and Asia. They survived in the wild in central Europe until the seventeenth century, with the last individual killed in Poland in 1627 (Clutton-Brock, 1989a). Dozens of breeds were developed in Europe, as farmers in different localities selected for the traits most important to their needs. There are minor groups of cattle in southeast Asia that are derived from different **wild bovids**. The Bali cattle of Indonesia are domesticates of the wild **banteng,** a wild bovid now found in very small numbers in the wild. The wild **gaur** of India and southeast Asia is the progenitor of the mithan (*Bos frontalis*). The **kouprey** (*Bos*

*sauveli*) was another wild bovid in the forests of southeast Asia. It may be extinct, with the last individuals seen in 1982 in war-torn Cambodia (NRC, 1983). A more recent report (Anonymous, 1996) indicates that kouprey were seen in the wild in 1988, and the species may persist in Cambodia and Thailand. Incredibly, a previously unknown (to scientists) wild bovid (*Pseudoryx nghetinhensis*) was discovered in Vietnam in 1992 (Dung et al., 1993, 1994). It is estimated that a few hundred animals of the new species, known as the **spindle-horned ox** (Fig. 1-5), may exist in remote forests in northern Vietnam (Kemp et al., 1997). Unfortunately, the animal has not been seen alive; its existence is known from skulls and skins sold in markets by hunters.[2] Likewise, the continued existence of kouprey is indicated by the occasional skulls offered for sale in local markets (Anonymous, 1996).

The **water buffalo** (Fig. 1-6) is a very important work, milk and meat animal in many parts of Asia, as well as to a lesser extent in some other areas including parts of Europe such as Hungary and southern Italy. Buffaloes are of two species, the Asian (*Bubalus bubalus*) and the African (*B. syncerus*). All domestic buffaloes are derived from the Asian species, with domestication occurring about 4,000 years ago. The water buffalo is widely used as a work animal in much of Asia. It is an efficient utilizer of low quality roughages, and is commonly fed on rice straw and coarse vegetation growing on ditch banks. In spite of its tropical origin, the buffalo is not heat tolerant. It compensates for this by wallowing in water and mud.

Cattle, water buffalo, and yaks are the only bovids domesticated in significant numbers. **Yaks** (*Bos granniens*) are native to the high mountains of Asia (e.g., Tibet, Nepal and Himalayan areas of China), where both domestic

---

[2] In 1996, an adult female was captured, but it died in captivity. Two captured juveniles also died.

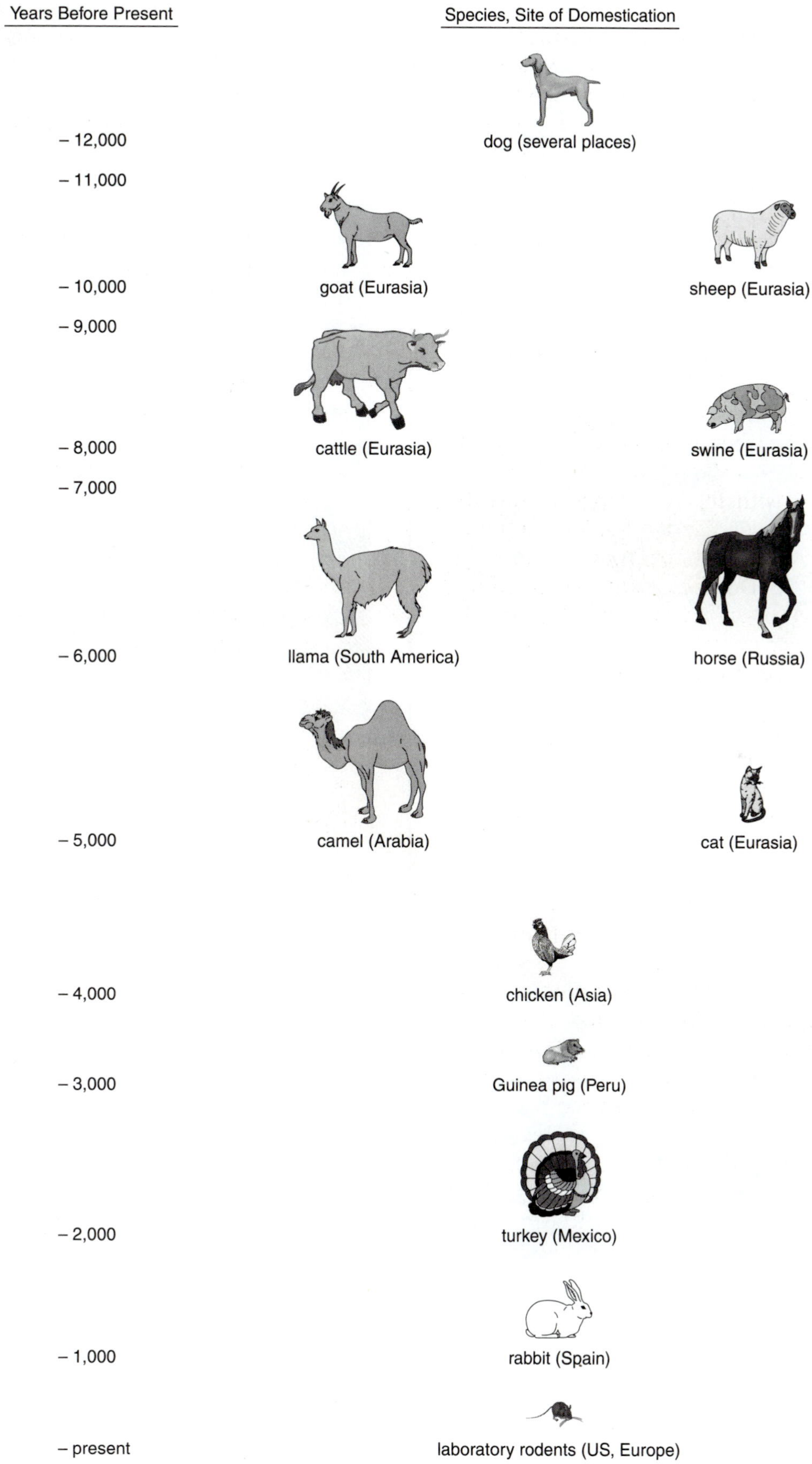

| Years Before Present | Species, Site of Domestication |
|---|---|

dog (several places)

– 12,000

– 11,000

goat (Eurasia)

sheep (Eurasia)

– 10,000

– 9,000

cattle (Eurasia)

swine (Eurasia)

– 8,000

– 7,000

llama (South America)

horse (Russia)

– 6,000

camel (Arabia)

cat (Eurasia)

– 5,000

chicken (Asia)

– 4,000

Guinea pig (Peru)

– 3,000

turkey (Mexico)

– 2,000

rabbit (Spain)

– 1,000

laboratory rodents (US, Europe)

– present

**FIGURE 1-4** A timeline for domestication of some of the common domestic animals.

**FIGURE 1-5** The spindle-horned ox (*top*) is a new species of wild bovid discovered by scientists in 1992. It occurs in remote forests in Vietnam. The gaur (*bottom*) is a wild bovid found in the forests of India. (Courtesy of D. Kenny and the Denver Zoo.)

**FIGURE 1-7** Yaks are native to the high mountains of Asia. They are used for milk (*top*) and transportation (*bottom*). (Courtesy of D. Miller).

**FIGURE 1-6** Water buffalo in many Asian countries convert rice straw and other crop residues into milk and animal power. They are especially useful in the cultivation of flooded rice paddies.

(Fig. 1-7) and wild yaks are found. Yaks are used to provide milk, hides for warm clothing, work, and meat. Efforts have been made to domesticate the **American bison,** or more usually, to create hybrids of bison and domestic cattle. The Canada Department of Agriculture had a program to develop "cattalo" (Peters and Slen, 1966), while American ranchers developed "beefalo." Neither hybrid proved successful. The cattalo program was terminated, and genetic evidence indicates that beefalo contain little or no bison genes (Lenoir and Lichtenberger, 1978). Although crossing bison (*Bison bison*) with cattle has not proven successful, there is increasing interest in bison ranching in North America. Bison

**FIGURE 1-8** The musk ox is native to the tundra of Greenland and northern Canada and is well-adapted to harsh, arctic conditions. It produces a valuable, fine-textured wool called qiviut. Several efforts to domesticate the musk ox have been made, with little success so far because of very poor reproduction and calf survival. The animals pictured are part of a breeding program at the University of Saskatchewan in Canada.

can survive more harsh winter conditions than can cattle, and the meat commands a premium price. Bison are enjoying a period of being viewed as noble and harmonious with the land, while beef cattle and cattle ranchers are often viewed with disdain and suspicion. Bison also are viewed sympathetically as "survivors" that were almost completely exterminated in the buffalo hunting orgies of the late 1800s. The **musk ox** (Fig. 1-8) (*Ovibos moschatus*) is of interest as a candidate for domestication. It is native to the arctic tundra. It produces a high quality wool (qiviut) which has potential as a luxury fiber (Rowell et al., 2001). Efforts to domesticate musk oxen have not been very successful, mainly because of very poor reproduction (Flood, 1989).

Sheep and goats are closely related and are members of the family *Bovidae*. Domestic **sheep** (*Ovis aries*) are derived from the European wild sheep (mouflon). The North American bighorn sheep are also descendants of the European wild sheep, which migrated across the land bridge linking Siberia and Alaska in

the Pleistocene period. The sheep and goat were the first animals to be domesticated after the dog. The flocking instinct of sheep facilitated their domestication. As with the dog, the process involved gradually increasing social interaction between two social species (humans and sheep). Mutual benefit from the association would strengthen the bond. Sheep have been modified by human selection into animals that in many environments are totally dependent upon human protection from predators for their survival. The domestic **goat** (*Capra hircus*) is derived from the wild goat or bezoar that is native to the Eurasian mountains of Iran, Turkey, and adjacent areas. Both sheep and goats have been extensively modified by human selection. There are dairy, meat, and fiber breeds of both species. Sheep are largely social animals that associate in herds or flocks, while goats are often less gregarious. Both species are intermediate feeders (see Chapter 4), but goats are more browsers than grazers. They are capable of making efficient use of feed resources but may also cause great ecological damage if raised with poor grazing management. Archeological evidence (bones in caves and at cooking sites) indicates that the initial domestication of goats occurred with a shift from pure hunting to a symbiotic pastoralism with rudiments of herd management. Tribesmen began to control movement of herds, and by selectively hunting young male goats, altered the sex ratio, age structure, and breeding season of the herded animals, ultimately leading to sufficient changes in the managed goats to create a "genetic firewall" separating them from wild populations (Zeder and Hesse, 2000).

The **horse** (*Equus caballus*) was domesticated in eastern Europe about 4000–3000 B.C. Although it has been used as a meat animal in various cultures, its primary role has been for riding and draft purposes. In Mongolia, mares are milked and horse milk is an important dietary item. Horses were important work animals over the entire globe until the first decades of the twentieth century and are still very important in many developing countries.

## EVOLUTION OF HORSES

*(Review of B. J. MacFadden and T. E. Cerling. 1994. "Fossil Horses, Carbon Isotopes and Global Change."*
*Trends in Ecology 9:481–486.)*

The 55 million year fossil record of horses (Equidae) provides a classic example of evolutionary history. The evolution of the horse is a complex story, which can be related to adaptation to changing environments. There was explosive diversification and formation of new Equidae species during the middle Miocene (20–10 million years ago), mainly in North America. In this paper, the authors used the measurement of radioactive carbon isotopes in fossil horse teeth to relate anatomical changes in the horse to environmental changes in the same period. Plants use two main types of photosynthesis: in $C_3$ photosynthesis, the carbon from $CO_2$ is incorporated into compounds containing 3 carbons, while in $C_4$ photosynthesis, compounds with 4 carbons are formed. The teeth of fossil horses can be analyzed for their content of the isotopes $^{13}C$ and $^{14}C$. The $C_3$ and $C_4$ plants incorporate different proportions of $^{13}C$ and $^{14}C$; thus the $^{13}C/^{14}C$ ratio in teeth can be used to determine the diet type of a fossilized animal. In the Miocene period, climate change resulted in a change from forests and savannahs to grasslands. During the forest period, plants had mainly $C_3$ photosynthesis, while during the grassland stage, $C_4$ plants predominated. During this time, Miocene horses had rapid dental changes, with low-crowned teeth (brachydont) changing to high-crowned teeth (hypsodont). Hypsodonty is an adaptive shift from browsing to grazing, correlated with the advent of grasslands. Grasses contain abrasive particles of silica (silica phytoliths) that cause increased tooth wear. Thus the increased height of the tooth crown in modern horses is an adaptation to a diet high in grasses. This is corroborated by the $^{13}C/^{14}C$ ratios in the teeth of fossil horses; the changes correlate with changes in dental anatomy and in vegetation. Thus the measurement of carbon isotopes in fossilized teeth provides a powerful tool for interpreting evolutionary changes related to diet.

Another technique involving teeth can give clues to the diet of extinct animals. During the Pleistocene era, elephants inhabited North America; these were the forest-dwelling mastodons and the plains-dwelling mammoths. Silica phytoliths vary between plants and can be used to differentiate between them. Gobetz and Bozarth (2001) found that silica phytoliths of grasses were identifiable in the calculus or plaque on the teeth of mastodons. Phytoliths of diatoms were also present. These findings suggest that mastodons grazed on grass (rather than browsing on trees) in a cool, moist Pleistocene environment, possibly near water. Interesting, eh?

Much of eastern Europe, the former Soviet Union, China, and several Latin American countries still depend heavily on horse power, although this will undoubtedly decrease as modernization occurs. Although in terms of conserving fossil fuels, providing employment, and so on, a case can be made for continued use of animal power in many countries, most of the people involved would rather use a tractor or truck and will do so, as soon as economics allow. The other domestic equid species besides the horse is the **donkey** (*Equus asinus*). The donkey is the only major domestic animal that was domesticated in Africa, in ancient Egypt. Wild asses persist in small numbers in Asia and Africa. Donkeys have many advantages over horses and other work animals in arid areas. They survive on low quality feed, and are efficient in conserving and utilizing water. They have an exceptional tolerance to dehydration and are much more efficient in digestion of low quality, fibrous feeds than is the horse (Izraely et al., 1989). Donkeys will continue to serve as patient beasts of burden for peasant farmers in many countries for years to come. For example, there are over 300,000 working equids (mainly donkeys) in South Africa (Wells and Krecek, 2001). In parts of South Africa (the homelands) in the 1980s, there was a misguided government-sponsored donkey extermination program.

Budiansky (1997) suggests that the domestication of the horse saved it from extinction. The modern horse evolved in North America. About 11,000 years ago, at the end of the Pleistocene, many large herbivores in North America became extinct (e.g., horses, mastadons, mammoths, etc.). Climatic changes, causing the conversion of grasslands to forests, contributed to this extinction. The European wild horse faced similar environmental changes, and many species of large herbivores in Europe did become extinct at that time. Budiansky (1997) presents evidence indicating domestic horses arose from a remnant population of surviving horses in the Ukraine, about 6,000 years ago. The "wild horses" of North America are actually feral animals, descended from escaped or released domestic horses. A few true wild horses persist in Asia. **Przewalski's horse** (*E. przewalskii*) (Fig. 1-9) was last recorded in the wild in 1968, in Mongolia. A new wild population has been started with zoo-bred animals, in Uzbekistan (Pereladova et al., 1999). Another herd of these horses, also known as takhi, has been established in Mongolia (Boyd, 1998). Another wild Asian horse is the kulan (*E. hemionus kulan*), still found in the wild in small numbers.

The domestic **pig** (*Sus scrofa*) is derived from the wild European pig, which is still found in the forests of Europe (wild boar). Remains of domesticated pigs have been found in archaeological sites throughout Europe and Asia, dating back to 7000 B.C. Until recently, pigs were raised on table scraps, waste products, and food that they could find for themselves. Pigs were very effective in the hardwood forests of Europe in harvesting acorns, nuts, and other food. In parts of Europe, pigs were herded by a swineherd through the forests to gather their own food (Brownlow, 1992). In other areas, such as China, most families kept their own pig and used it as a living garbage disposal unit. Only in the twentieth century has intensive swine production based on grains and other concentrates developed. In many parts of the world, pigs, along with chickens, play a very important role as scavengers on small farms and even in cities. Chinese breeds of pigs are currently of interest to animal breeders because of traits such as very high prolificacy and the ability to survive on forage and other fibrous foodstuffs (McLaren, 1990).

The domestic **chicken** (*Gallus domesticus*) is descended from one or more species of southeast Asian jungle fowl. It is believed that fowl were first used for cock-fighting, and later were recognized as a source of meat and eggs. The oldest archaeological records of chicken bones indicate domestication about 3000–2000 B.C. in India, from where they spread throughout Europe and Asia (West and Zhou, 1989). It is not conclusively known whether chickens were introduced into South America by Pacific Islanders before the time of Columbus or whether their arrival in South America was with Europeans. No pre-Columbus skeletal remains have been found in South American archaeological sites. Chickens are very suitable for backyard production on small-holder (peasant) farms. They can scavenge most of their food, and are supplemented with kitchen wastes. Large-scale intensive poultry production based on concentrate feeds has developed only in the second half of the twentieth century, coinciding with intensification and mechanization of grain production. In many countries, indigenous chickens are still very important as scavengers (Aini, 1990).

The **turkey** (*Meleagris gallopavo*) is native to North and Central America. Turkeys were domesticated by American Indians in Central America and transported to Europe by the

**FIGURE 1-9** Przewalski's horses in the wild in Mongolia. (Courtesy of Lee Boyd.)

Spaniards. In North America, turkeys have until recently been produced mainly for the Christmas and Thanksgiving markets. They are now raised intensively year-round. Turkeys have been rapidly altered by genetic selection, with increases in meatiness and size. Their conformation has been so altered that natural mating is virtually impossible; the birds are now mated almost exclusively by artificial insemination.

**Waterfowl** (ducks and geese) are also important poultry species in many countries. Ducks (*Anas platyrhynchos*) are raised in large numbers in Asia for meat and eggs and are also widely used in rice paddies for insect control. **Ducks** are herded from field to field in Asian countries (Fig. 1-10), resulting in the development of breeds such as the Indian Run-

ner, which walks with an upright posture rather than the usual waddle. Ducks were domesticated in Asia, with Chinese records dating back about 3,000 years. Ducks raised in Asian countries such as Indonesia are comparable to Leghorn chickens in egg production. The Muscovy duck is native to South America and was domesticated by American Indians. It is important in several countries, such as France, as a meat bird. Domestic **geese** (*Anser anser*) are derived from the European and Asian wild geese and have been domesticated for centuries. They are raised for meat and feathers (down). Geese, along with the ostrich, are unique among the common domestic poultry in being herbivorous, and thus having the capability of being raised on pasture. However, adequate growth rates of geese for intensive meat production can be achieved only with concentrate feeding. **Ostriches** (*Struthio camelus*) have been domesticated in South Africa for production of plumes (feathers), meat, and leather (Fig. 1-11). Ostrich production has recently increased in importance in the United States. Whether or not a viable industry will develop remains to be seen.

**FIGURE 1-10** Ducks are important sources of meat and eggs in Asian countries. Ducks in Indonesia (*top*) are used for insect control and gleaning of grain in rice paddies (*bottom*). (Courtesy of Yono C. Raharjo.)

**FIGURE 1-11** Ostriches have been domesticated in South Africa, where ostrich farming has been an important activity for over 100 years. The original interest in the ostrich was mainly as a source of feathers (plumes), but the emphasis has shifted to the skins for leather and the carcass for meat. A small ostrich industry has developed in the United States.

The **emu,** a large ratite (flightless bird) native to Australia, is being raised in the United States and Australia for meat, leather, and oil (Fig. 1-12).

Other avian species raised domestically include guinea fowl and pigeons. **Guinea fowl** (*Numida meleagris*), which occur wild in Africa, are raised in Europe for meat production (Fig. 1-13). The young birds are called keets. **Pigeons** (*Columba livia*) are derived from the rock dove of Europe and Asia. Pigeon raising is common in Europe, where the young (squabs) are a gourmet food item. Other birds domesticated for meat production include various species of **gamebirds** such as quail, partridges, and pheasants.

Camels and llamas are important livestock in some countries. The fossil record indicates that **camelids** evolved in North America and spread into South America and into Asia across the Alaska-Siberian land bridge during an ice age. They, along with horses, elephants (mammoths), and many other large animals, later became extinct in North America. Wild camelids still exist in Asia (wild two-humped camel in Mongolia) and South America (guanaco and vicuna). **Camels** (*Camelus* spp.) were domesticated in Asia and the Middle East, while llamas and alpacas were domesticated in the Andes by 4000 B.C. (Mason, 1984). The camel (Fig. 1-14) is widely used as a pack and

**FIGURE 1-12** The emu is a ratite bird native to Australia. Emu farming is of emerging interest in many parts of the world. The birds are raised for meat, leather, and oil.

**FIGURE 1-13** Guinea fowl, native to Africa, are raised in Europe for meat production.

**FIGURE 1-14** The camel is widely used as a draft animal in many parts of Africa and Asia (*top*), and is famous as the "ship of the desert" (*bottom*). (Bottom photo courtesy of the Food and Agriculture Organization (FAO) of the United Nations.)

**FIGURE 1-15**  The llama is an important domestic animal in the Andean region of South America. Interest in llama production is growing rapidly in the United States, where they are raised as pack animals, as sources of wool, and as exotic pets. (Courtesy of J. A. Pfister.)

work animal in many arid regions of the world, such as North Africa, the Middle East, India, and Pakistan. In some African countries, camels are raised as meat animals. Large numbers of camels are found in deserts in Australia. These feral animals are the descendants of escaped pack animals. They are not used for commercial purposes in Australia but are considered pests and cause great damage to native trees and shrubs, which have not evolved defenses against large mammalian herbivores. **Llamas** (*Lama glama*) and **alpacas** (*Lama pacos*) are used in the Andean countries of South America as pack animals and sources of wool. The llama (Fig. 1-15) is primarily a pack animal, while the alpaca produces a high quality wool. Llamas can survive on low quality roughage and are better adapted to the harsh conditions of the high altitude Andean regions than are cattle, sheep, or goats. Llama numbers in the United States are increasing rapidly; although these animals are used to some extent as pack animals by hikers, they are primarily kept as exotic pets.

There is a great diversity in animal species that have been domesticated by humans. Many of these animals are important in localized geographical areas or with certain ethnic groups but are of minor importance on a global scale. They run the gamut of insects (honey bees), molluscs (snails), and reptiles (alligators and crocodiles) to large herbivores (reindeer). A brief discussion of some of these will be given to indicate the diverse roles of domesticated animals in contributing to human welfare.

The **reindeer** (*Rangifer tarandus*), a member of the deer (Cervidae) family, is an important livestock species to the people of northern Scandinavia, Russia, and Siberia. Reindeer herding (Fig. 1-16) is particularly associated with the Lap people (Laplanders) of Finland, Norway, and Sweden. Reindeer are used as work animals, and as sources of meat, fur, and milk. Reindeer are uniquely suited to their environment and survive the winter by utilizing lichens as their main food supply. Their dependence on lichens makes reindeer production very susceptible to the vagaries of air pollution, since lichens absorb their nutrients directly from the air. The Chernobyl nuclear accident, for example, decimated part of the reindeer industry when the lichens were contaminated with radioactive elements. Efforts to establish reindeer industries in Alaska and Canada have met with only modest success in spite of valiant efforts, such as a five

**FIGURE 1-16** Animals have been important sources of food and clothing for humans throughout our evolutionary history, predating most other innovations. In northern Europe, reindeer are herded to provide food, skins, and animal power. The animals feed on the arctic lichens, converting this meager resource into products useful to humans. (Courtesy of M. Nieminen.)

**FIGURE 1-17** Deer farming is an expanding activity in New Zealand and several western European countries. Deer are raised for the meat (venison), while the antlers are valuable for sale in China and other Asian countries as medicinal products, including supposed value as an aphrodisiac. There is considerable interest in deer farming in North America, although it is illegal in many areas, for reasons discussed in Chapter 6. (Courtesy of R. D. Buckmaster.)

year trek of breeding stock across the Canadian arctic.

**Deer** farming (Fig. 1-17) is increasing rapidly in importance and has become a major livestock industry in New Zealand (Barry and Wilson, 1994). Deer farming is also increasing rapidly in Canada and the United States and is quite important in Europe. Several species are

raised commercially, including the fallow deer (*Cervus dama*) and the red deer (*Cervus elaphus*), also known as the wapiti or elk. What is called an elk in North America is called a red deer in Europe, while what is called an elk in Europe is a moose in North America. The terminology can get quite confusing! Fallow deer are grazers rather than browsers and so can be raised on pasture like cattle. They are particularly raised for their meat (venison). The red deer is raised extensively in New Zealand for its antlers, which are highly valued in the Orient for medicinal purposes (including, but not exclusively, use as a supposed aphrodisiac). Deer farming is increasing rapidly in importance in many countries, particularly for venison production. The raising of deer in parks and farms has saved at least one species, the Pere David deer of China, from extinction. It has not existed in the wild for 2,000–3,000 years. The priest Father David (Pere David) sent breeding stock to Europe, and shortly thereafter, the only herd surviving in China was killed in a flood and civil unrest. Pere David's deer have been reintroduced successfully into the wild in China (Jiang et al., 2000). Deer farming in North America has been vigorously opposed by some (Geist, 1985) on the grounds that escapees may destroy, by competition or

**FIGURE 1-18** The eland is an African antelope with potential for domestication. However, efforts to accomplish this have failed, because the eland does not offer any significant advantages over current domesticated animals such as cattle.

crossbreeding, indigenous species of wild deer. An additional potential problem with elk farming is the existence of **chronic wasting disease (CWD)** in some elk populations. This is an encephalopathy similar to mad cow disease (see Chapter 9).

Efforts have been made to domesticate various types of antelope such as the eland (*Taurotragus oryx*), an African antelope (Fig. 1-18). According to Kyle (1994), these "efforts came to naught because they did not ask sufficiently critically: what can an eland do which a cow cannot?" (p. 2)

**Rabbits** (*Oryctalogus cuniculus*) have been domesticated for hundreds of years and are an important meat animal in France, Italy, and Spain. There is a worldwide rabbit industry, but with the exception of the preceding countries, it is only marginally successful. Rabbits are raised for meat, fur, and Angora wool—a luxury fiber. Some of the potential advantages of rabbits as a small livestock species have been reviewed by Cheeke (1986). They can be fed diets based on forages and agricultural by-products and thus are not competitive with humans for grain. They have a high reproductive rate and a rapid growth rate. However, rabbit production is difficult to automate, so it cannot effectively compete with industrial poultry production. The main potential for rabbit production is on a backyard scale for subsistence farmers, particularly in developing countries.

**Guinea pigs** (*Cavia porcellus*) are native to the Andean region of South America and were domesticated several thousand years ago. They are a significant meat source for the Indians of Peru, Colombia, Ecuador, and Bolivia. Because they cannot jump or climb, they are raised in simple enclosures, and even in cardboard boxes. In many ways, they are an ideal meat animal for subsistence farmers. However, except in areas where they are already eaten, it is doubtful that people in other cultures would find it acceptable to consume guinea pigs.

The **capybara** (*Hydrochoersus hydrochaeris*) is the world's largest rodent, reaching an adult weight of over 100 pounds (Fig. 1-19). It is native to South America. Although the animal is not truly domesticated at this time, capybara farming and ranching is occurring on a limited scale. The capybara has an enlarged hindgut with cecal fermentation and digests forages almost as efficiently as do ruminants. It is a semiaquatic animal and can utilize aquatic weeds and forages that are not suit-

**FIGURE 1-19** The capybara, native to the floodplains of the Amazon and other large South American rivers, is the world's largest rodent, with adults exceeding 100 lbs in weight. It is a semiaquatic animal with potential for meat and leather production. The animals pictured are part of a breeding program at the University of Sao Paulo, Piracicaba, Brazil.

able for grazing by cattle. Capybara ranching is complementary to cattle ranching in the floodplains of South America. The capybara is likely to increase in importance as a meat animal and source of hides for leather.

Many other animals are farmed or ranched, usually for meat production. In many cases, it is questionable as to whether the animals can truly be considered domesticated. Such examples would include snails, crocodiles, and oysters. Animals raised for commercial purposes, besides mammals and birds, include reptiles (e.g., crocodiles), amphibians (e.g., frogs), insects (e.g., bees), crustaceans (e.g., shrimp), and molluscs (e.g., snails, oysters). **Aquaculture,** involving the farming of fish such as catfish, salmon, trout, carp, and many other species, is increasing rapidly in importance. Again, it is questionable whether or not these fish can truly be considered domestic animals.

Many wild animals have potential for agriculture. Many are small animals that would be useful for subsistence farmers in developing countries. The term **microlivestock** has been coined to describe these animals (NRC, 1991), as well as small domestic species such as rabbits and guinea pigs. They include such diverse species as miniature antelope (duikers, dik dik, mouse deer), various rodents (giant rat, grass-cutter), and lizards (iguana). Advantages of microlivestock are their low feed requirements (i.e., the amount of forage needed per day for five guinea pigs versus one cow), low space requirements and inexpensive, easily constructed facilities and equipment, ease of handling by women and children (who do much of the work on small farms), ease of transport to market (on a bicycle, in a box carried on the head, etc.), and a small carcass that does not require refrigeration. Although the domestication and production of microlivestock has a lot to recommend it, it should not be lost sight of that people in developing countries have no desire to remain forever poor and at a subsistence level of living as peasant farmers.

Another system with a lot of potential is **game farming** or ranching. This is rapidly gaining momentum in southern Africa. A combination of native wild herbivores may utilize forage more efficiently than do the domestic species. For example, giraffe browse on tree foliage out of reach of other herbivores, zebra graze on coarse, mature forage too low in digestibility to support ruminants, dik dik antelope utilize browse in dense forest, and so on. Managing the wild species as a sustainable resource can save animals from extinction. In arid, desert areas, animals like the oryx have potential, because they have low water requirements. Cattle, in contrast, are notoriously inefficient in water utilization (see Chapter 6). Game animals can also generate income via photographic and/or hunting safaris and as trophy animals. The income from these endeavors can exceed that of domestic livestock raised on the same property.

# ROLES OF ANIMALS AND ANIMAL PRODUCTS IN HUMAN WELFARE

Humans derive many benefits from animal production (Bradford, 1999). Those widely familiar to most people will be mentioned only briefly; many people, however, may be unaware of the importance of some animal products such as dung. It is worth noting again that while humans derive benefit from animals, animals are also beneficiaries of the relationship. They are provided with food, shelter, freedom from predators, affection, and it is hoped, a humane, painless death rather than the alternative faced by wild animals. Very few wild animals die peacefully of old age or live lives free of fear and stress.

Animals are important sources of **animal protein,** in the form of meat, eggs, milk, and—in some cultures—blood. Animal proteins generally have a higher nutritional value than plant proteins, both in terms of amino acid composition and amounts of other nutrients such as minerals and vitamins. Most plant proteins contain toxic factors (e.g., phytates, trypsin inhibitors, lectins, alkaloids, cyanogens, etc.), which may cause toxicities (e.g., cyanide in cassava) or reduced nutrient availability (bound niacin in corn, low iron and phosphorus availability in grains). There is a common adage that "you are what you eat." A human is considerably more like a pig or a cow in body composition than like soybeans or lettuce leaves, so it should not be surprising that meat and other animal products are excellent sources of nutrients for humans. Milk and egg protein are considered the highest quality sources of protein for humans, because of their content and quality of essential amino acids. Again, upon reflection, the reason for this is obvious. Milk provides the sole source of nutrients for baby animals in their period of most rapid growth and development, so obviously it must contain all of the nutrients needed to accomplish this. Similarly, an egg must contain an adequate amount of all nutrients to convert a raw egg into a baby bird.

In many societies, particularly nomadic ones, livestock are a major source of **wealth** and are a means of accumulating capital (Fig. 1-20). They are used for barter, exchanges with in-laws in marriage arrangements, and so on. In times of food scarcity, they are a mobile form of **food storage** (milk, blood, and meat). Nomadic tribes of the Sahelian region of North Africa appear to have been in reasonable ecological balance with the environment until recently. Modern innovations such as international borders and fences that restrict transhumanence (migrations), well-intentioned efforts to provide more wells and water sources, use of imported European breeding stock, and so on have destroyed this balance. Desertification in the Sahelian region of Africa is probably more attributable to politicians, outside experts, and aid programs than to livestock producers!

Domestic animals have long had important symbolic or other roles in human cultures. In some cases, their involvement in traditional cultural activities may increasingly be controversial, particularly in the context of western sensibilities. Animal sacrifice (e.g., voodoo in Haiti), bullfighting and cockfighting are examples of activities that are regarded as cultural essentials in some countries, and barbaric rituals in others. In an increasingly complex and interconnected world, environmentalists, animal rights advocates, and defenders of indigenous cultures sometimes find themselves opposing cultural practices (e.g., animal sacrifice) that are considered imperative or sacred by the cultures that they are supporting.

In many countries, such as India, animal manure (**dung**) is a major source of **fuel** (see Chapter 6). The dung is collected and formed into "dung patties," which can be easily stacked and transported. They are a major source of fuel for cooking. Fuel for this purpose is in short supply in many countries, and the search for cooking fuel by impoverished people has led to extensive deforestation, soil erosion, flooding, and loss of wildlife habitat. In India, cattle and buffalo consume forage such as rice straw and ditch-bank or roadside weeds, which are unsuitable as fuel for cooking fires and convert them to dung (as well as using these resources to support work and milk production), which can be used as a cooking fuel in simple stoves. Thus the cattle of India, which are much maligned by many experts from industrialized countries as being sacred and nonproductive, actually have important roles in resource utilization. Without the conversion of straw to dung, deforestation in India would be much

**FIGURE 1-20** In many countries, domestic animals are important as sources of wealth, serving as "living banking institutions." This is true of many African cultures, such as the Masai of East Africa, shown here. (Courtesy of W. D. Hohenboken.)

**FIGURE 1-21** Cattle in India are considered sacred, and are allowed "free rein" to wander at will, even into restaurants! Their sacred status arose because of the many important contributions cattle have traditionally made to Indian society. (Courtesy of Robert A. Swick.)

worse than it already is. In treeless areas, peasants would have no other source of cooking fuel. In the early days of the American west, "buffalo chips" played a similar role as the homesteaders' fuel source. Cattle have had such important roles as sources of milk, work, and cooking fuel in India that they have come to be viewed as sacred. Although they are allowed free, undisturbed movement in cities (Fig. 1-21), they are trained to leave crops alone. Contrary to common western belief, these "sacred cows" are useful and play important roles in the economy of India (Jacobson, 1999).

While use of dung as fuel is desirable, the process can be made more efficient by introducing a modest amount of modern technology. The manure can be anaerobically digested in thick plastic bags to produce methane, which can be stored in tanks and used as a source of natural gas for cooking. The remaining sludge retains nitrogen and other plant nutrients that are lost when the dung is directly burned, so with **biogas production** the dung can be used as both fuel and fertilizer. Simple biogas generation units are a reality in many developing countries.

Animal products have in the past been very important as sources of clothing. **Wool** from sheep, camels, and llamas, cashmere and mohair from goats, and other animal fibers have been important in producing clothing. Synthetic fibers have largely replaced animal fibers, except for luxury fibers (cashmere, mohair, angora rabbit hair) and sheep wool for higher quality suits and sweaters. Similarly, animal **fur and skins** for hats, gloves, and coats were once virtually essential in cold climates (the garments of Eskimos being an extreme example). Synthetic fibers can now be used instead. Animal furs are mainly used now for luxury garments such as mink coats, and their use for this purpose seems to be rapidly declining, in part because of animal rights activities. At one time, horse **hair** was an important commodity, used for stuffing cushions, automobile seats, and so on. Hair removed from cattle hides was important for making felt, while pig hair was used as bristles for brushes. These specialized uses of animal fibers are rapidly declining, and even wool, which is an important commodity, will undoubtedly continue to decline in importance.

**Leather** is prepared from hides, mainly of cattle. As with wool and hair, it is still widely used but has to a considerable degree been replaced by synthetics. This trend will continue. As with wool fiber, there is an upscale market for expensive, exotic types of leather for boots and jackets. Ostrich, alligator, and crocodile farming (Fig. 1-22) for exotic leathers is increasing. **Feathers** have also been of importance. Ostrich plumes have been used for garments, dusters, and so on, while goose down was widely used for pillows and vests.

In summary, except for luxury applications, synthetic materials can substitute for wool, fur, hair, feathers, skin, and leather, so these animal products will likely decline in importance. They are also viewed by many as frivolous products and increasingly are the target of animal rights activists and eco-terrorists. Even use of wool is opposed by animal rights groups, on the grounds that sheep shearing is cruel and inhumane.

Meat processors like to say that they use everything but the squeal, and this statement is basically true. Many **pharmaceutical and medical products** are prepared from animal

**FIGURE 1-22** Crocodiles, such as these on a crocodile farm in South Africa, are a source of meat and leather. Crocodiles and their American cousins, the alligators, are reptile species with potential for farming (I hesitate to say potential for domestication!). The animals pictured here are fed mainly on ostrich wings and other waste products from ostrich farms in South Africa. Alligator farming in conjunction with swine production in the United States has been proposed as a means of disposing of dead pigs from large swine operations.

sources. For example, drugs such as insulin are derived from slaughterhouse material, while sausage casings and condoms are prepared from intestines. The importance of slaughterhouse material (e.g., pancreas glands) as sources of drugs and medicines will decrease as these products become increasingly available through **biotechnology.** However, some of the biotechnology techniques involve livestock. Fertilized eggs of dairy cattle can be injected with genes to cause the resulting transgenic animal to synthesize a desired drug or hormone in the mammary gland to be secreted in the milk. Thus there is the likelihood of lactating animals such as transgenic cattle and goats being used as biological factories to produce drugs. Several drugs have already been experimentally produced in this manner. These include tissue plasminogen activator (TPA), a blood clot dissolving drug, and human lactoferrin, an iron-binding protein. The researchers euphemistically refer to these gene-splicing techniques as "breakthroughs in the pharm-yard."

It often comes as a surprise to Americans who travel to other parts of the world to observe that a lot of the rest of the world still operates to varying degrees on **animal power.** This includes not only the so-called Third World countries of Asia, Africa, and South America, but considerable areas of Europe as well. In European countries such as Poland and much of Russia, horses and oxen are still very important sources of farm power. In some rapidly developing countries that are deficient in petroleum such as Brazil and Chile serious efforts are being made to retain the use of animal power. In Chile, for example, both horses and oxen are extensively used on large, sophisticated farms producing fruit and vegetables for export to North America. There are research projects in Chile to design more efficient and modern horse-drawn equipment for use in cultivating and other activities. Besides reducing dependence on petroleum, use of work animals provides employment opportunities in countries with large numbers of underemployed people. It is being increasingly recognized that animal power will remain important in many tropical countries for years to come, and that the continued use of work animals has beneficial effects on the environment and social structure. Oxen and buffalo used for work eat mainly straw and other low quality feeds, so they are not competing with humans for food. They do not require expensive inputs that automation would require such as fuel, repairs, and spare parts. Buffalo can be used to cultivate flooded rice paddies, in which use of tractors would be difficult. Teleni and Murray (1991) cite evidence that draft animals provide the power for the cultivation of nearly 50 percent of the world's cultivated land and the hauling of over 25 million carts. More than 240 million cattle and 60 million buffalo are kept as work animals. In many Asian countries, elephants are important draft animals (Fig. 1-23). They are usually captured from the wild and tamed, rather than being raised from breeding herds, so are not truly domesticated animals.

**FIGURE 1-23** Elephants are important draft animals in numerous Asian countries such as Thailand. (Courtesy of Robert A. Swick.)

Some of the contributions of livestock to human welfare, in a global context, are summarized in Table 1-1.

There are many ways that the efficiency of work animals could be increased. Instead of raising steers for use as oxen, cows can be used. In addition to work, they will provide milk and calves for meat production. The breeding season can be adjusted so that calving and early lactation do not coincide with

**TABLE 1-1** Contributions of livestock to human well-being in a global context (Adapted from Hodges, 1999)

| |
|---|
| - provision of work-draft animals |
| - transportation of goods and people |
| - animal fat and protein for improved nutrition |
| - milk to enable human infants to survive |
| - leather, wool, and horn for clothing and shelter |
| - animal fat for burning for illumination |
| - dung fuel for cooking and heating |
| - animal power for pumping water to irrigate crops |
| - benefits on crops in integrated farming systems |
| - conversion of fibrous vegetation to human-edible food |
| - influence on human culture and values, traditions and rituals. |

the periods of heaviest farm work. Plows and cultivation equipment can be modified so less power is required; hence, one milk cow could replace two oxen. Awadhwal and Yule (1991) have described the development of new animal-drawn tillage implements that can be used by small farmers to put into place new crop production systems such as the broadbed and furrow system that increases crop production in the semiarid tropics (Fig. 1-24). These types of developments are promising, because they reflect a direction of research towards **appropriate technology.** In the past, there was often a tendency to regard animal power as an archaic concept, to be replaced with fossil fuel-powered devices as soon as possible.

# ROLE OF LIVESTOCK PRODUCTION IN SUSTAINABLE AGRICULTURE

Virtually all societies that have progressed beyond the hunter-gatherer stage have developed agricultural systems in which livestock are an integral part. Agriculture has been sustainable for thousands of years in Europe, Asia, Africa, and South America. Relatively little agricultural development of North America and Australia occurred until the arrival of Europeans.

The terms **sustainable agriculture** and **alternative agriculture** have become fashionable because of concerns that the current technology and petroleum-based agriculture of the developed countries may not be sustainable on a long-term basis (NRC, 1989). **Monoculture** production of corn and soybeans in the United States, for example, has resulted in very high yields due to high inputs of fertilizer and agricultural chemicals, at the expense of massive soil erosion and groundwater contamination with nitrogen and herbicides. About 10 tons of soil per acre are lost annually from Iowa crop land (Little, 1987). Pimentel et al. (1995) estimate a U.S. loss of

**FIGURE 1-24** (*Left*): Hundreds of millions of people in developing countries depend upon animal power for cultivation and planting of crops. For many years, aid programs and foreign specialists virtually ignored the roles of buffalo, oxen, donkeys, camels, and other animals in food production. Fortunately, that attitude is changing, and recent research on increasing the efficiency of animal power has led to improvements in simple agricultural implements. In this photo, an "agribar" is illustrated. This is a simple, light tool bar to which a variety of implements can be attached. The implement shown here is a broadbed former, which is used in a new broadbed and furrow system of planting, which increases crop yields through better moisture conditions and weed control. (*Right*): Groundnuts (peanuts) being cultivated and weeded, using a simple cultivator attached to the agribar. The oxen walk in the furrows between the broadbed. Further descriptions of these animal-powered implements are provided by Awadhwal and Yule, 1991. (Courtesy of N. K. Awadhwal and D. F. Yule.)

17 tons of soil per hectare per year (6.9 tons per acre) from croplands, and 6 tons per hectare per year from pastures. These figures have been disputed; Trimble and Crosson (2000) in a detailed analysis of soil erosion in the United States concluded: "It is questionable whether there has ever been another perceived public problem for which so much time, effort, and money were spent in light of so little scientific evidence" (p. 248). Further, "we do not seem to have a truly informed idea of how much soil erosion is occurring in this country" (p. 248). At any rate, livestock production would appear to have both positive and negative effects on soil erosion. Intensive cultivation of corn and soybeans to produce feed for livestock increases soil erosion, while production of forages for pasture and hay reduces soil erosion.

Livestock production as part of a sustainable agriculture program can have a major impact on reducing **soil erosion.** Production of forage for ruminant animals usually re-

quires crop rotation of grains with grass and legume forages. Grasses with their fibrous root systems increase soil organic matter, improve soil texture, increase water retention, and reduce runoff. Deep-rooted legumes such as alfalfa bring up minerals from the subsoil and open up subsoil channels to improve water percolation to replenish groundwater. Livestock manure, when spread on croplands, increases organic matter and improves soil texture. Crop rotations involving hay and pasture production have a major impact on a long-term basis in sustaining soil fertility and reducing erosion. Of course, most of the major grain-producing areas of the world were originally grasslands, and their high productivity when first cultivated was a reflection of the favorable effects of grasses on the soil.

Sustainable agriculture implies optimizing the utilization of resources while minimizing damage to the environment. Integration of crop and animal production is necessary for **optimal resource utilization.** Livestock can utilize crop residues for feed and bedding, while the manure produced has a greater fertilizer and soil amendment value than crop residues left on the field.

**Crop-livestock integration** optimizes utilization of human resources. Raising livestock ties farmers and their families to the land. The stewardship of the land is likely to be greater than when only crops are produced. It is much easier to mine the soil as an absentee owner. Children raised on farms with livestock learn invaluable lessons on life, death, and responsibility. Many of the concerns that modern agriculture has with the "animal rights" movement can be traced to the lack of a rural upbringing of almost the entire U.S. population (less than 1 percent of the population lives on real farms). Urban people are brought up with the notion that animals are little people. (If you disagree with this, select and read some children's books dealing with animals and form your own opinion.) Farm children learn early that animals don't behave in the idealistic, bucolic manner that they do in children's books, and that they can be (and often are) incredibly cruel to each other, as they establish their social order (pecking order) without the niceties of human civilization. Animals die and are killed (baby chicks eaten by a skunk, sheep killed by coyotes, a deformed calf "put down" by the farmer). People raised under rural conditions may have a better understanding of the rhythms of nature than children have who are brought up in the city. They know where food comes from and how it gets from the ground or farmyard to the table.

Many people who were brought up on farms couldn't wait until they were old enough to escape the farm and head for the city. They may not even be aware of how much the farm experience shaped their psyche and gave them a work ethic, a sense of responsibility, and a good dose of common sense. Millions of these resourceful people grew up on farms in the 1930s, 40s, and 50s in North America and are now approaching retirement. My personal opinion is that many of their children and grandchildren (the infamous present generation) have not ac-

quired these valuable personal characteristics to as great a degree, because of their urban upbringing. The challenge of sustainable agriculture in North America is to somehow make production agriculture both appealing and profitable to the next generation in order to preserve these positive social attributes. Farmwork is hard, even with modern equipment and automation, and increasingly unprofitable.

The previous comments on the sustainability of integrated crop-livestock production need to be evaluated within the context of modern **industrial agriculture.** Raising cattle in huge feedlots, consolidating dairy farms into confinement units with 1,000–10,000 cows, consolidating swine and poultry production into huge confinement units as the trend is now are not necessarily examples of sustainable agriculture. They may not be examples of crop-livestock integration. They are in some cases a frontal assault on the environment, with massive groundwater and air pollution problems and legitimate animal welfare concerns. Thus I don't claim that livestock production is inherently "good" or "bad," beyond reproach, or conversely, beyond repair. I do suggest that industrial crop and livestock production, based on high inputs of petroleum energy and agricultural chemicals, may not be the optimal route to follow on the road to sustainable agriculture.

In several European countries, farmers are considered the **guardians of the countryside.** The English cherish their countryside, with hedgerows, cottages, rock walls and farm animals (Fig. 1-25). Legislation exists to preserve this safety valve for urban dwellers, by preventing the amalgamation of small farms into large units, by conserving hedgerows, and so on. British farmers are paid grants by the government to restore and maintain hedges. In England, the Council for the Protection of Rural England (CPRE) exists to promote "the beauty, tranquility and diversity of rural England by encouraging the sustainable use of land and other resources in town and country" (*www.cpre.org.uk*). In Switzerland, the alpine scenery with small dairy farms and

**FIGURE 1-25**  The countryside of England, with rock walls and hedges enclosing small fields. These attractive vistas are part of the English tradition and culture. Farmers are considered "guardians of the countryside," preserving this heritage for urban citizens to enjoy. English farmers receive government assistance to compensate for their preservation of the countryside. Cattle, sheep and other domestic animals are an essential component of this living cultural heritage. (Courtesy of J.R. Carlson.)

bell-wearing cows is preserved by government subsidies that make these small farms profitable and allow farmers to make a decent living. The European Union provides payments to farmers in designated stewardship areas for using livestock grazing to promote species diversity (Mearns, 1997a). The frontier mentality of the United States, with its emphasis on individual property rights, has not encouraged this kind of legislative activity. The current mood of the general U.S. populace seems to be moving in the direction of conservation, environmental concerns, green-belts, multiple use of public lands, and concern for the survival of the small family farm. Farm animals play a role in this process. Farmers will increasingly

be viewed as stewards of the land, preserving the environment for future generations. The ethic "it's my land and I'll do what I want with it" is increasingly viewed as being at variance with stewardship. Sustainable agriculture will be discussed in more detail in Chapter 11.

# LIVESTOCK IN DEVELOPMENT PROGRAMS

It is a perception among many animal scientists that livestock-related research plays "second fiddle" to plant science research, both in industrialized and developing countries. This is particularly true in aid programs in developing countries. Livestock are often viewed as the "bad guys" by expatriate "experts." It is my perception that many of the people involved in international development programs have an idealistic or altruistic viewpoint, and have incorporated ideas into their thinking that livestock compete with humans for food (Chapter 3) or are the cause of desertification and other forms of environmental degradation (Chapters 6 and 7), and that people in developing countries should be discouraged from raising animals. In fact, livestock play exceedingly important roles in the culture and economy of people in developing countries, and efforts to dissuade them from raising animals will be both bewildering and unrealistic to the people involved—the foreign experts are talking to deaf ears and lose credibility in the process. Agriculture in tropical countries often involves very intricate interrelationships, and tinkering with one facet often overturns the cart somewhere else in the system. For example, development of higher-yielding rice with short straw resulted in inferior quantity and quality of straw available for feeding to buffalo, the draft animal needed to grow the rice.

In two papers, Mearns (1997a, b) described many instances of traditional livestock production systems that provide positive environmental benefits that are not fully recognized by policy makers. Just as industrial animal production has environmental costs that are not fully reflected in the market value of its products, traditional livestock systems have environmental benefits that are not well captured by market mechanisms (Mearns, 1997a). Some of these benefits include favorable effects on biodiversity, nutrient recycling, and diverse outputs of products (blood, hides, dung fuel, transportation, flexible household capital reserves, risk management insurance, cultural

and ceremonial values, etc.). These multiple values often don't hit the radar screens of ag economists and politicians.

One organization with specific activities involving livestock and small animals is **Heifer Project International (HPI)**. HPI is a nongovernmental organization (NGO) with a mission to alleviate poverty and hunger. It is the major NGO that specializes in animal agriculture as the vehicle for development of people and communities, helping families become self-sufficient in food and income.

The philosophy of HPI is summarized by Aaker (1994):

> We believe a healthy environment is a prerequisite for people to have hope for the future. People are not isolated from an ecosystem. It is not possible to help people help themselves through livestock programs without considering their relationship to the various ecosystems in which they live. Such an approach would only be another example of what humans have done too much of already—tinkering with the parts instead of working with the whole. Humans are but one part of this whole, as are animals.

A cornerstone of the HPI program is that people receive training and motivation before they are given an appropriate animal. They then are responsible for "passing on the gift" by helping others in their community with offspring from their animal and by sharing knowledge they have obtained. The organization's first project was a shipment of 18 heifers to Puerto Rico in 1944. Since that time, countless numbers of cattle, sheep, goats, rabbits, poultry, beehives, and fish have been provided to families all over the world, including impoverished communities in the United States. HPI is based in Little Rock, Arkansas (HPI, P.O. Box 808, Little Rock, AR 72203; *www.heifer.org*), and, as a livestock-oriented aid organization, is very worthy of support by animal scientists, livestock producers, and other animal enthusiasts.

# STUDY QUESTIONS

1.  What is our genus and species? What do these words mean?

2.  What characteristics of wolves and sheep made them relatively easy to domesticate, as compared to deer and bison?

3.  What is the origin of domestic breeds of cattle? Do any wild cattle still exist?

4.  What are the major domesticated bovids?

5.  What were the first three species of animals to be domesticated? When did this occur?

6.  What is the only common animal that was domesticated in Africa?

7.  Which of our common livestock and poultry species were domesticated in North and South America?

8.  Deer are raised commercially in New Zealand. What are the major commercial products of deer farming?

9.  What are some advantages of rabbits and guinea pigs as meat animals?

10.  Under what conditions might capybara be suitable meat-producing animals?

11.  What are microlivestock?

12.  Why are milk and eggs sources of excellent quality protein and other nutrients?

13.  What are some of the other useful products obtained from livestock production besides meat and milk?

14.  How does animal production contribute to sustainable agriculture?

15.  Are American ranchers and cowboys "guardians of the countryside and western cultural traditions"? Are subsidized grazing fees justifiable in terms of protecting agricultural land from urbanization and "development"?

# REFERENCES

AAKER, J. 1994. *Livestock for a small earth. The role of animals in a just and sustainable world.* Seven Locks Press, Washington, D.C.

ADCOCK, G. J., E. S. DENNIS, S. EASTEAL, G. A. HUTTLEY, L. S. JERMIIN, W. J. PEACOCK, and A. THORNE. 2001. Mitochondrial DNA sequences in ancient Australians: Implications for modern human origins. *Proc. Nat. Acad. Sci.* 98:537–542.

AINI, I. 1990. Indigenous chicken production in South-east Asia. *World's Poult. Sci. J.* 46:51–57.

ALLEN, D. T. 1997. Effects of dogs on human health. *J. Amer. Vet. Med. Assn.* 210:1136–1139.

ANONYMOUS. 1996. Status and protection of Asian wild cattle and buffalo. *Conservation Biology* 10:931–935.

AWADHWAL, N. K., and D. F. YULE. 1991. New implements for small farmers of the semi-arid tropics. *World Anim. Rev.* 68:73–77.

BARRY, T. N., and P. R. WILSON. 1994. Venison production from farmed deer. *J. Agric. Sci.* 123:159–165.

BOYD, L. 1998. The 24-h time budget of a takh harem stallion (*Equus ferus przewalskii*) pre- and post-reintroduction. *Appl. Anim. Behav. Sci.* 60:291–299.

BRADFORD, G. E. 1999. Contributions of animal agriculture to meeting global human food demand. *Livestock Prod. Sci.* 59:95–112.

BRISBIN, I. L., Jr., and T. S. RISCH. 1997. Primitive dogs, their ecology and behavior: Unique opportunities to study the early development of the human-canine bond. *J. Amer. Vet. Med. Assn.* 210:1122–1126.

BROWNLOW, M. J. C. 1992. Acorns and swine: Historical lessons for modern agroforestry. *Quart. J. Forestry* 86:181–190.

BUDIANSKY, S. 1992. *The covenant of the wild. Why animals chose domestication.* William Morrow and Company, Inc., New York.

BUDIANSKY, S. 1997. *The nature of horses. Exploring equine evolution, intelligence, and behavior.* The Free Press (A division of Simon and Schuster, Inc.), New York.

CHEEKE, P. R. 1986. Potentials of rabbit production in tropical and subtropical agricultural systems. *J. Anim. Sci.* 63:1581–1586.

CHOU, H.-H., H. TAKEMATSU, S. DIAZ, J. IBER, E. NICKERSON, K. L. WRIGHT, E. A. MUCHMORE, D. L. NELSON, S. T. WARREN, and A. VARKI. 1998. A mutation in human CMP-sialic acid hydroxylase occurred after the *Homo-Pan* divergence. *Proc. Natl. Acad. Sci.* USA 95:11751–11756.

CLUTTON-BROCK, J. 1989a. *A natural history of domesticated mammals.* Univ. of Texas Press, Austin.

CLUTTON-BROCK, J. 1989b. *The walking larder. Patterns of domestication, pastoralism and predation.* Unwin Hyman, Inc., London.

CLUTTON-BROCK, J. 1999. *A natural history of domesticated mammals.* 2nd ed., Cambridge University Press, Cambridge.

DUNG, V. V, P. M. GIAO, N. N. CHINH, D. TUOC, P. ARCTANDER, and J. MACKINNON. 1993. A new species of living bovid from Vietnam. *Nature* 363:443–445.

DUNG, V. V., P. M. GIAO, N. N. CHINH, D. TUOC, and J. MACKINNON. 1994. Discovery and conservation of the Vu Quang ox in Vietnam. *Oryx* 28:16–21.

EATON, S.B., and M. KONNER. 1985. Paleolithic nutrition: A consideration of its nature and current implications. *New Engl. J. Med.* 312:283–289.

FLEAGLE, J.G. 1988. *Primate adaptation and evolution.* Academic Press, San Diego.

FLOOD, P. F. (Ed.). 1989. Proceedings of the Second International Muskox Symposium. *Can. J. Zoo.* 67:1092–1166.

GEIST, V. 1971. *Mountain sheep.* University of Chicago Press, Chicago.

GEIST, V. 1985. Game ranching: Threat to wildlife conservation in North America. *Wildl. Soc. Bull.* 13:594–598.

GOBETZ, K. E., and S. R. BOZARTH. 2001. Implications for late Pleistocene mastodon diet from opal phytoliths in tooth calculus. *Quarternary Res.* 55:115–122.

GOODMAN, M., D. A. TAGLE, D. H. A. FITCH, W. BAILEY, J. CZELUSNIAK, B. F. KOOP, P. BENSON, and J. L. SLIGHTOM. 1990. Primate evolution at the DNA level and a classification of hominoids. *J. Mol. Evol.* 30:260–266.

HARE, B., M. BROWN, C. WILLIAMSON and M. TOMASELLO. 2002. The domestication of social cognition in dogs. *Science* 298: 1634–1636.

HEMMER, H. 1990. *Domestication. The decline of environmental appreciation.* Cambridge University Press, Cambridge, UK.

HODGES, J. 1999. Animals and values in society. *Livestock Prod. Sci.* 58:159–194.

IZRAELY, H., I. CHOSHNIAK, C. E. STEVENS, M. W. DEMMENT, and A. SHKOLNIK. 1989. Factors determining the digestive efficiency of the domesticated donkey (*Equus asinus asinus*). *Quart. J. Exp. Physiol.* 74:1–6.

JACOBSON, D. 1999. A reverence for cows. *Nat. Hist.* 108(6):58–63.

JIANG, Z., C. YU, Z. FENG, L. ZHANG, J. XIA, Y. DING, and N. LINDSAY. 2000. Reintroduction and recovery of Pere David's deer in China. *Wildlife Soc. Bull.* 28:681–687.

JOHANSON, D. C., and M. E. EDEY. 1974. *Lucy. The beginnings of humankind.* Warner Books, Inc., New York.

KEMP, N., M. DILGER, N. BURGESS, and C. V. DUNG. 1997. The saola *Pseudoryx nghetinhensis* in Vietnam—new information on distribution and habitat preferences, and conservation needs. *Oryx* 31:37–44.

KRINGS, M., A. STONE, R. W. SCHMITZ, H. KRAINITZKI, M. STONEKING, and S. PAABO. 1997. Neanderthal DNA sequences and the origin of modern humans. *Cell* 90:19–30.

KYLE, R. 1994. New species for meat production. *J. Agric. Sci.* 123:1–8.

LENOIR, F., and M. J. LICHTENBERGER. 1978. The Y chromosome of the Basolo hybrid beefalo is a Y of *Bos taurus. Vet. Rec.* 102:422–423.

LEONARD, J. A., R. K. WAYNE, J. WHEELER, R. VALADEZ, S. GUILLEN, and C. VILA. 2002. Ancient DNA evidence for Old World origin of New World dogs. *Science* 298:1613–1616.

LITTLE, C. E. 1987. *Green fields forever. The conservation tillage revolution in America.* Island Press, Washington, D.C.

MACFADDEN, B. J., and T. E. CERLING. 1994. Fossil horses, carbon isotopes and global change. *Trends in Ecology* 9:481–486.

MASON, I. L. 1984. *Evolution of domesticated animals.* Longman Group Ltd., London.

MCKINNEY, M. L. 1998. The juvenilized ape myth—our "overdeveloped" brain. *BioScience* 48:109–116.

MCLAREN, D. G. 1990. Potential of Chinese pig breeds to improve pork production efficiency in the USA. *Anim. Breed. Abstr.* 58:347–369.

MEARNS, R. 1997a. Livestock and environment: potential for complementarity. *World Anim. Rev.* 88(1):1–14.

MEARNS, R. 1997b. Balancing livestock production and environmental goals. *World Anim. Rev.* 89(2):24–33.

MOREY, D. F. 1994. The early evolution of the domestic dog. *Amer. Scientist* 82:336–347.

NATIONAL RESEARCH COUNCIL. 1983. *Little-known Asian animals with a promising economic future.* National Academy Press, Washington, D.C.

NATIONAL RESEARCH COUNCIL. 1989. *Alternative agriculture.* National Academy Press, Washington, D.C.

NATIONAL RESEARCH COUNCIL. 1991. *Microlivestock. Little-known small animals with a promising economic future.* National Academy Press, Washington, D.C.

PERELADOVA, O. B., A. J. SEMPERE, N. V. SOLDATOVA, V. U. DUTOV, G. FISENKO, and V. E. FLINT. 1999. Przewalski's horse—adaptation to semi-wild life in desert conditions. *Oryx* 33:47–58.

PETERS, H. F., and S. B. SLEN. 1966. Range calf production of cattle x bison, cattalo and Hereford cows. *Can. J. Anim. Sci.* 46:157–164.

PIMENTEL, D., C. HARVEY, P. RESOSUDARMO, K. SINCLAIR, D. KURZ, M. MCNAIR, S. CRIST, L. SHPRITZ, L. FITTON, R. SAFFOURI, and R. BLAIR. 1995. Environmental and economic costs of soil erosion and conservation benefits. *Science* 267:1117–1123.

ROWELL, J. E., C. J. LUPTON, M. A. ROBERTSON, F. A. PFEIFFER, J. A. NAGY, and R. G. WHITE. 2001. Fiber characteristics of qiviut and guard hair from wild muskoxen (*Ovibos moschatus*). *J. Anim. Sci.* 79:1670–1674.

RUVOLO, M. 1997. Molecular phylogeny of the hominoids: Inferences from multiple independent DNA sequence data sets. *Mol. Biol. Evol.* 14:248–265.

SAVOLAINEN, P., Y. ZHANG, J. LUO, J. LUNDEBERG and T. LEITNER. 2002. Genetic evidence for an East Asian origin of domestic dogs. *Science* 298:1610–1613.

TATTERSALL, I. 2000. Once we were not alone. *Sci. Amer.* January:56–62.

TELENI, E., and R. M. MURRAY. 1991. Nutrient requirements of draft cattle and buffaloes. pp. 113–119 In: Ho, Y. W., H. K. Wong, N. Abdullah, and Z. A. Tajuddin (Eds.). *Recent advances on the nutrition of herbivores.* Malaysian Soc. Anim. Prod., Serdang, Malaysia.

TRIMBLE, S. W., and P. CROSSON. 2000. U.S. Soil erosion rates—myth and reality. *Science* 289:248–250.

TRUT, L. N. 1999. Early canid domestication: The farm-fox experiment. *Amer. Scientist* 87:160–169.

VILA, C., P. SAVOLAINEN, J. E. MALDONADO, I. R. AMORIM, J. E. RICE, R. L. HONEYCUTT, K. A. CRANDELL, J. LUNDEBERG and R. K. WAYNE. 1997. Multiple and ancient origins of the domestic dog. *Science* 276: 1687–1689.

WELLS, D., and R.C. KRECEK. 2001. Socioeconomic, health and management aspects of working donkeys in Moretele 1, North West Province, South Africa. *J. S. Afr. Vet. Ass.* 72:37–43.

WEST, B., and B.-X. ZHOU. 1989. Did chickens go north? New evidence of domestication. *World's Poult. Sci. J.* 45:205–218.

WINTER, H., L. LANGBEIN, M. KRAWCZAK, D. N. COOPER, L. F. JAVE-SUAREZ, M. A. ROGERS, S. PRAETZEL, P. J. HEIDT, and J. SCHWEIZER. 2001. Human type I hair keratin pseudogene φHaA has functional orthologs in the chimpanzee and gorilla: Evidence for recent inactivation of the human gene after the *Pan-Homo* divergence. *Hum. Genet.* 108:37–42.

ZEDER, M. A., and B. HESSE. 2000. The initial domestication of goats (*Capra hircus*) in the Zagros mountains 10,000 years ago. *Science* 287:2254–2257.

# CHAPTER 2

# Animal Products in the Human Diet

**CHAPTER OBJECTIVES**

1. To discuss the contributions of animal protein sources (milk, meat, and eggs) to the human diet.

2. To discuss the causes of the major diseases of Western civilization that are linked to diet and consumption of animal products.

3. To discuss methods of improving animal products to reduce fat, saturated fat, and cholesterol contents of meat, milk, and eggs.

4. To discuss methods of improving animal products, by increasing the content of healthful components.

5. To discuss the impacts of ongoing changes in the Western diet on the livestock industries and possibilities for minimizing animal industry disruptions.

Interest in vegetarianism in North America and Europe seems to be increasing. My perception is that "meat eaters" are often somewhat apologetic for indulging in the practice and think that "someday I'll quit eating so much meat." Presumably, these attitudes have developed as a result of numerous reports in the popular press of adverse effects of meat and eggs on cancer, heart disease, blood cholesterol levels, and so on and the beneficial effects of fruits, vegetables, and fiber on these same health problems. Rarely is there much publicity on the "good news"

about animal products: that they are highly palatable, rich in most nutrients, and contain nutrients in a balance similar to the nutritional needs of humans.

## WHAT KIND OF ANIMAL ARE WE?

The dietary habits of an animal are closely influenced by its taste perception, teeth (dentition), and digestive tract. The fact that a cow likes to eat grass, has teeth well adapted for grazing and masticating grass, and has an enlarged digestive tract with a symbiotic population of microbes that digest grass suggests that a cow is an animal whose basic diet is grass or other forage—that is, it is an herbivore. Equally obviously, a cow does not find flesh palatable; does not have canine teeth, claws, speed, stealth and other features necessary for stalking and killing prey; and does not have a digestive tract well adapted to the digestion of flesh (meat). Without belaboring the obvious, it is totally apparent that a cow is an herbivore and not a carnivore. Equally obvious, by the same type of reasoning, is that a tiger (and other cats) is a carnivore and not an herbivore. (In defiance of this obvious fact, some devout vegetarians and animal rights advocates attempt to feed their pet cats on a vegetarian diet, much to the annoyance of the cats!)

Using the same type of logic, perhaps we can find clues from our taste preferences, dentition, and nature of our digestive tract, as well as from the dietary habits of our ancestors, as to what kind of animal we are. Are humans carnivores or herbivores (vegetarians)? We are neither. We, along with rats, pigs, chickens, and numerous other animals, are omnivores. We prefer, and are best adapted to, fairly nutrient-rich, low fiber materials such as seeds, meat, nuts and fruit. We are not well adapted to eating fibrous materials such as forage or isolated grain fiber (bran). A horse, rabbit, or cow will avidly consume oat or wheat bran. Try eating a cup of bran or oat hulls and discover for yourself that our palatability responses are not for high fiber materials. A few hours later, your digestive tract will tell you the same thing!

In contrast to the preceding view, Popovich et al. (1997) assert that humans evolved on high fiber, low fat diets high in fruit and foliage, with a virtual absence of animal products. They based this assertion on the feeding behavior of other great apes (e.g., gorilla), which consume a vegetarian diet of fruit and foliage. The sacculated colon of humans and gorillas is closer to that of an herbivore than to an omnivore (Popovich et al., 1997). These authors further suggest that the

short chain fatty acids (e.g., butyrate) produced by colonic fermentation have beneficial effects in preventing colitis and colon cancer. The natural gorilla feeding pattern of foraging throughout the day has health benefits in reducing serum cholesterol levels. Captive gorillas fed diets containing meat and eggs develop high serum cholesterol levels, premature cardiovascular disease, and ulcerative colitis (Popovich et al., 1997). Sound familiar? Consumption of animal products as a significant part of the diet is certainly justified by our anatomy and digestive physiology and our evolutionary heritage. However, the good health of some of the millions of vegetarians around the world indicates that animal products are not dietary essentials for humans. This is a modern development; it was not until the identification in 1948 of vitamin $B_{12}$ as the "animal protein factor" that it was possible for nonruminants such as humans, poultry, and swine to be healthy on diets with no animal products. The development of the modern swine and poultry industries based on corn-soy diets, with no animal protein source, occurred after the commercial availability of synthetic vitamin $B_{12}$. There are no plant sources of vitamin $B_{12}$. It is synthesized only by bacteria, by which means it is obtained by herbivores. Until the availability of synthetic vitamin $B_{12}$, the only way for omnivores to obtain it was by eating animal tissue, or, in the case of some human populations, by consuming fermented products in which microbes synthesized the vitamin.

Meat and other animal products as components of the human diet cannot now be justified on the grounds that they are nutritionally essential. It is possible for humans to obtain adequate amounts of all nutrients from consumption of plant products only (with a supplement of vitamin $B_{12}$). This is made easier and gastronomically more exciting if milk and eggs are included in the diet (i.e., lacto-, ovo-vegetarianism). Equally apparent is that the decision as to whether or not to eat meat and other animal products is a personal one, just as whether you will or won't eat tomatoes or avocados is your own business and nobody else's. Livestock producers should not operate from the viewpoint that the general public has an obligation to buy their products. Vegetarians should not be regarded as the enemy.

The ethical question of whether it is morally right or wrong to raise animals for food can be discussed endlessly, with no single right answer. It is a topic like religion, abortion, and certain environmental issues. Some people fervently and passionately believe that slaughtering a cow is murder. Some people equally fervently believe that this viewpoint is absurd and that animals are part of the food chain. Some live to die so others can eat. In view of the fact that this discussion can go on interminably, I will present my opinion and the objectives for the rest of this chapter, in the hope that this will provide a framework for others to use in their formulation of ideas on this subject.

The role of herbivores in the global ecosystem is to convert vegetation to a more nutritious form (flesh) for animals higher up the food chain. To compensate for the fact that they will be killed and eaten, herbivores have a high reproduction rate to replenish and maintain their populations. The smaller the animals are, the more likely they will be caught and eaten, so the higher the reproductive rate (otherwise they would have become extinct eons ago). For example, rabbits and meadow voles have several large litters each year. In spite of the fact that most of the offspring will serve as food for carnivores, the species survives because enough individuals reach breeding age to carry on the species. No wild rabbit dies non-violently of old age in a normal ecosystem. Larger herbivores, such as deer, cattle, horses, and so on do not reproduce as rapidly as rabbits, because many predators that eat rabbits (hawks, eagles, bobcats, cats, foxes, etc.) are not normally likely or able to kill and eat a deer or a horse. Nevertheless, if these large herbivores are artificially protected from predation, their populations will increase incredibly rapidly. There are only two ways other than predation

or slaughter that their populations can be controlled: either by artificial birth control or by massive starvation. In the case of deer, wild horses, elk, and other large wild herbivores, either we reintroduce and protect their natural predators such as wolves, grizzly bears, and mountain lions, or we substitute human predation—that is, hunting. If we wish to reintroduce the large predators, we find that they generally cannot coexist with large human populations, whereas deer and other wild herbivores can. The rapid and largely out-of-control increase in human populations has resulted in massive disruptions of ecological balances between predators and prey. We can either remove ourselves (by creating very large wildlife preserves in which most or all human activity is excluded, letting nature take its course) or attempt to intelligently manage wild herbivore populations by hunting. A clean kill by a bullet is likely to be less traumatic to the victim than a kill by a pack of wolves.

In summary, my belief is that all plant and animal life is interrelated, and artificially trying to eliminate predation (killing for meat) is unnatural and ultimately leads to massive ecological disruption. Vegetation needs herbivores for survival, and herbivore numbers are regulated by predation. Regulation of herbivores by starvation in the absence of predation disrupts the plant balance. If plants or animals are artificially protected from being eaten, the system will eventually collapse. For the complete ecological cycle, plants must be eaten by herbivores, and the herbivores must be eaten by carnivores or omnivores. Thus cattle, sheep, and other herbivorous animals have evolved to serve as prey and now serve the ecological role formerly served by wild herbivores such as bison. The one animal totally out of balance with the ecosystem is the human animal. Methods of reintroducing a balance between the human population and the rest of life on this planet are beyond the scope of this book but nevertheless are urgently needed. If necessary, Nature (or Gaia; see Chapter 11) will probably accomplish this, with famines, new virulent diseases (e.g., AIDS), climate changes, or other calamities.

Human nutritional requirements reflect our evolutionary history extending millions of years into the past (Eaton and Konner, 1985). Genetically, today's humans are the same biological organisms as our ancestors when they took up agriculture a scant 3000–10,000 years ago (Eaton et al., 1996). Thus our ancestral dietary pattern has continuing relevance to the nutritional needs of contemporary humans. Small populations of hunter-gatherers still exist. These populations have a low intake of saturated fat, because of the leanness of wild game. However, their cholesterol intake is very high, about 480 mg per day (Eaton et al., 1996). Despite this, they maintain very low serum cholesterol levels (about 125 mg/dl) (Eaton, 1992). Their high ratio of polyunsaturated to saturated (P:S) intakes (1.4 for hunter-gatherers, 0.4 for contemporary Americans), low saturated fatty acid intake, and low total fat intake apparently offsets the high cholesterol intake (Eaton et al., 1996). Hunter-gatherers do not consume cereal grains but have a high intake of fruits and vegetables, providing a high intake of phytochemicals, vitamins, and minerals (Johns, 1996). Ancestral human diets contained much more protein and fiber than do contemporary diets. Thus meat consumption is consistent with good health, which, based on archeological evidence, hunter-gatherers apparently enjoyed. However, the fatty acid composition of contemporary intensively fattened meat animals deviates markedly, in an undesirable manner, from the fat content and fatty acid composition of wild game animals (Crawford, 1968). This aspect will be discussed further under "Modification of Animal Products to Improve Human Health" later in this chapter.

Humans, like many other mammals, have a dietary preference for high-energy foods. Throughout our evolutionary history, a principal endeavor of humans has been to obtain enough to eat. Sweet foods are indicative of substances containing a high concentration of sugars and thus are high in energy; similarly, fatty substances are high in energy. Therefore, we are attracted to sweet substances (sweet tooth) and fat-rich foods (Birch, 1992). Unfortunately, our current sedentary lifestyle and abundant food supply are served too well by our evolutionary history. Humans are better equipped metabolically to cope with scarcity than with abundance (Birch, 1992). (Perhaps this is true for other aspects of our lives—e.g., affluence—as well?)

A reasonable assumption is that the diets of the few remaining populations of hunter-gatherers reflect the diets of pre-agricultural humans. In other words, the diets of modern-day hunter-gatherers may represent a standard reference for modern human nutrition and a model for defense against diseases of affluence (Cordain et al., 2000). In general, pre-industrial societies show a high dependence on animal-origin foods (Brand-Miller and Holt, 1998; Cordain et al., 2000; O'Dea, 1991). Ratios of stable isotopes in bones can be used to assess the nature of the diet. Carbon-13-enriched bones of early hominids suggest that hominids consumed high quality animal foods before the development of stone tools and the origin of the genus *Homo* (Sponheimer and Lee-Thorp, 1999). Extensive discussion on the importance of meat in the evolution of the human diet is presented in proceedings of a symposium, "Foraging Strategies and Natural Diet of Monkeys, Apes and Humans," edited by Widdowson and Whiten (1991), in a book, *The Hunting Apes: Meat Eating and the Origins of Human Behavior*, reviewed by Boesch (1999), and in a book, *Meat Eating and Human Evolution*, edited by Stanford and Bunn (2001). Cordain (2002) has popularized his work on the Paleolithic diet as "the Paleo diet" for modern humans.

# NUTRITIONAL PROPERTIES OF ANIMAL PRODUCTS

## *Milk and Eggs*

As mentioned in Chapter 1, milk and eggs by biological necessity have a very high nutritional value. They serve as the sole source of nutrients for young, rapidly growing animals when their nutritional requirements are at their highest. An egg must contain adequate amounts of all required nutrients, or it will be incapable of supporting the development of a live chick. Milk must be nutritionally adequate to feed a rapidly growing suckling animal. Therefore, it should come as no surprise that milk and eggs are widely recognized as two of the most nutritious products humans can consume. On the other hand, it should also be recognized that because most humans are not rapidly growing young animals (children have a much slower growth rate than other animals, and adult humans have even lower requirements), milk and eggs may be consumed at levels that provide higher levels of nutrient intake than required. This may account for some of the health concerns with animal products. For example, an embryonic chick needs a lot of cholesterol to synthesize hormones, other steroid compounds and cell membranes during its development within the egg. Eggs are high in cholesterol to provide the embryo with this essential metabolite. However, this level of **cholesterol** may exceed that needed by an adult human. The adverse effects of cholesterol in eggs can be counteracted by omega-3 fatty acids (defined later in this chapter). Eggs enriched in omega-3 fatty acids are commercially available (see Modification of Animal Products to Improve Human Health—this chapter). By modifications of the diets of laying hens, eggs can be enriched in various nutrients.

**Milk** has enjoyed the reputation of being "nature's most nearly perfect food." Recently, several vociferous critics of dairy products have attracted attention. Even the almost universal acceptance of dairy products being a healthful source of dietary calcium has been subject to criticism. A fundamental argument for most of the dairy food critics is the claim that humans are the only animals that aren't weaned; no other species consumes milk in adulthood. While true, this may have little bearing on the health effects of milk consumption. Maijala (2000) provided a good review of the positive aspects of the dairy industry. Historically,

many of today's nutritional food beliefs turn into tomorrow's food fads or quackery, emphasizing that caution is needed in evaluating the changing whims of consumers. In the following discussion, an effort has been made to critically evaluate human health aspects of consumption of dairy foods.

Milk contains the sugar **lactose,** which requires digestion by the enzyme lactase. Most animals, except for those during the suckling period, have very low levels of intestinal lactase activity, because milk or other sources of lactose are not a normal component of the post-weaning diet. Many humans have low lactase secretion as adults and may suffer from **lactose intolerance.** This condition results when the undigested lactose is fermented by bacteria in the colon, causing gas production and distress in the lower tract, including diarrhea. Those ethnic groups that have had a long history of using dairy products, such as Europeans and the people of India, have adequate levels of lactase secretion. People of many other ethnic groups without an evolutionary history of milk consumption have low levels of lactase secretion and are intolerant of dietary lactose. This is one of the few recent biochemical adaptations to a dietary condition known in the human population. Lactose intolerance need not be a problem today, because sources of the lactase enzyme are readily available and sold in supermarkets and pharmacies. Despite this, activists have claimed that inclusion of dairy products in the USDA Food Guide Pyramid is a type of "**nutritional racism**" because lactose intolerance disproportionately affects people of African, Asian, and Hispanic origin.

Cow's milk has traditionally been used in preparation of infant formulas. It is recognized that it is not a perfect substitute for human milk, which contains various bioactive substances (growth factors, hormones, cytokines, glycoconjugates, nucleotides, long-chain polyunsaturated fatty acids, etc.) that cow's milk may not contain or contain in the same proportion as human milk (Hamosh, 1997; Weaver, 1997; Zinn, 1997). However, cow's milk comes much closer to duplicating human milk than do other sources such as soy milk (see below). While somewhat speculative, it may be possible through genetic engineering techniques to genetically modify cows to cause them to produce various bioactive factors found in human milk (Meisel, 1997). Herds of such genetically modified cows could be maintained by cloning. The biology of bioactive peptides in milk has been reviewed by Schanbacher et al. (1997).

Among the nutritional benefits of milk are high quality protein and an abundance of calcium. **Protein quality** refers to the balance of essential amino acids and is discussed further in Chapter 4. Dairy products are excellent sources of **calcium** and are useful in maintaining adequate calcium intake to aid in preventing osteoporosis. **Osteoporosis** is a reduction in bone mass, ultimately causing a loss of structural integrity of the bone followed by fractures caused by slight trauma. It occurs most frequently in postmenopausal white women and the elderly of both sexes (NRC, 1989). Approximately 20 percent of U.S. women suffer one or more osteoporotic fractures by age 65, and up to 40 percent experience fractures after age 65. Men and black women have a greater bone mass than white women so take longer to develop sufficient bone depletion to cause fractures.

Bone is a metabolically active tissue; the bone mineral is in a dynamic state of mobilization and deposition. Osteoporosis occurs when mobilization of bone mineral exceeds redeposition (remodelling). Although all dietary factors involved in the onset of osteoporosis are not understood, there is a relationship between dietary calcium intake and bone mass (NRC, 1989). A daily intake of 800 mg calcium is appropriate for women between the ages of 25 and 45. Intakes of 1,200 mg per day are recommended for girls and women between ages 10 and 25 to achieve maximum bone mass (NRC, 1989). Dietary phosphorus, protein and fiber may adversely affect calcium. Excess dietary phosphorus promotes calcium mobilization from bone and increased urinary calcium excretion; excess dietary protein may increase urinary calcium ex-

cretion, and dietary fiber and phytates (see Chapter 5) depress calcium absorption.

Dairy products are some of the best dietary sources of calcium. The widespread practice in the United States of girls and women consuming large amounts of diet soda and having low intakes of milk is undesirable. **Soda drinks** contain phosphorus and virtually no calcium, thus promoting poor bone mineralization and high urinary calcium excretion. The long-term effects of this selection of beverages does not bode well for the skeletal health of older women in the future. In addition, soft drinks contain fructose as a sweetening agent. Fructose bypasses the metabolic controls on the conversion of glucose to activated acetate, the precursor of cholesterol. Thus dietary fructose can increase serum cholesterol levels, as well as promote dental caries. The tendency of young people to avoid dairy products and to consume soft drinks may lead to a future population of toothless osteoporotics with heart disease!

In a retrospective study, Renner (1994) found that the calcium intake via milk and dairy products of patients with osteoporosis had been markedly lower during childhood and adolescence than in age-matched controls. Inadequate calcium intake during early life is a decisive risk factor for the development of osteoporosis, which cannot be reversed by increased calcium intake later (Renner, 1994).

While it is common knowledge and generally accepted as fact that consumption of **dairy products** is beneficial in preventing osteoporosis, another point of view is that dairy products have a negative effect (Hegsted, 1986). Osteoporosis, assessed by incidence of hip fracture, actually shows a positive association with dietary calcium intake (Fig. 2-1). In a U.S. study, Feskanich et al. (1997) found that women consuming greater amounts of calcium from dairy foods had significantly increased risk of hip fracture while no increase in risk was observed for the same levels of calcium from nondairy sources. They concluded, "These data do not support the hypothesis that higher consumption of milk or other food

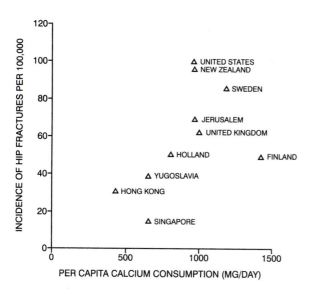

**FIGURE 2-1** A comparison of estimated calcium consumption in the food supply with the incidence of hip fractures in women, in countries differing in consumption of dairy products. It is noteworthy that the incidence of hip fractures (an indication of bone health) is highest in those countries with high consumption of dairy products. (From Hegsted, 1986; Courtesy of the *Journal of Nutrition*.)

sources of calcium by adult women protects against hip or forearm fractures" (p. 992). Hegsted (1986) suggests that the high protein intake associated with consumption of dairy products may have a negative effect, especially in diets high in other protein sources such as meat, by increasing urinary calcium excretion. A lifelong adaptation to a high calcium diet may suppress formation of calcitriol (activated vitamin D) and eventually impair the body's ability to utilize dietary calcium and maintain bone mass. Osteoporosis, like several other major chronic diseases, is largely a disease of affluent western cultures, characterized by diets high in animal products. As Hegsted (1986) notes, "If women in general required 800 mg calcium per day or more, calcium deficiency and osteoporosis would be rampant throughout the world" (p. 2317). However, studies in China do suggest that rural Chinese women who consumed dairy products and

had a relatively high calcium intake had higher bone density than urban women (Hu et al., 1993). However, other factors such as exercise could also have played a role in these differences. Chinese women in Beijing have very low rates of hip fracture compared to Caucasian U.S. women of similar age (Ling et al., 1996). Differences between Asian and Caucasian women in susceptibility to osteoporosis have a genetic component, including differences in allelic variations in the vitamin D receptor gene, associated with indices of bone turnover and density (Eisman, 1998).

In a symposium "Required versus Optimal Intakes: A Look at Calcium" (*J. Nutr.* 124:1404S–1430S, 1994), it was concluded that the calcium requirements for peak bone development and reduced age-related bone loss are higher than the current NRC recommendations (Miller and Weaver, 1994). The paper of Hegsted (1986), suggesting a possible negative effect of dairy products on calcium, received little mention. Perhaps it is disingenuous to mention that the symposium was partially sponsored by the National Dairy Council and Land O'Lakes, Inc., and that the guest editor was from the National Dairy Council. In another current paper (Prentice, 1997) entitled "Is Nutrition Important in Osteoporosis?" it was concluded, "It is, however, worth reflecting that, on a global basis, osteoporotic fracture incidence is highest in populations with high Ca intakes, such as in Northern Europe, suggesting that low Ca intake *per se* is not a major predisposing factor for osteoporosis" (p. 364). Prentice (1997) concluded, "The issue of whether nutrition is important in the aetiology of osteoporosis is complex and unresolved" (p. 365).

In summary, it must be concluded that the jury is still out as to whether consumption of dairy products helps protect against osteoporosis, or conversely, helps promote it.

**Soy milk** products are increasingly being used by vegetarians as substitutes for "real" milk. The phytates in soy products may reduce the absorption of calcium by as much as

20 percent (Heaney et al., 1991). Typically, diets used by soy milk consumers will also be high in other soy products, cereals, and other plant products—all of which are high in phytate and fiber, which substantially impair calcium absorption (Renner, 1994).

In 1992, the U.S. dairy industry was influenced by reports that the development of **insulin-dependent diabetes** in humans might be associated with the consumption of cow's milk by babies. Researchers at the University of Toronto reported that cow's milk may trigger an autoimmune response that destroys the insulin-secreting β-cells of the pancreas in genetically susceptible individuals (Karjalainen et al., 1992; Martin et al., 1991). A 17 amino acid peptide fraction of bovine serum albumin (BSA) may be the active component of milk causing these effects. These studies suggest that in susceptible individuals, antibodies to BSA may react with a membrane protein of the β-cells of the pancreas, causing an autoimmune response leading to destruction of β-cells and decreased insulin secretion. Inadequate insulin secretion is the cause of insulin-dependent diabetes mellitus, characterized by elevated blood sugar levels (hyperglycemia). Diabetic individuals had higher levels of antibodies to BSA than did healthy subjects in these studies. If these studies are confirmed and consumption of cow's milk by infants is found to be a significant factor in the etiology of diabetes, several means of circumventing the problem are evident. The most obvious would be to encourage breast-feeding of infants, particularly in families with a history of diabetes. A technological fix would be genetic engineering of dairy cows to alter the structure of BSA sufficiently to cause it to lose its ability to react with β-cell membrane proteins.

Consumption of dairy products has beneficial effects against **hypertension** or high blood pressure (Hamet, 1995). Populations consuming at least the RDA of calcium (800 mg per day, and 1,200 mg per day for pregnant women) have a lower incidence of sodium sensitivity and a lower prevalence of pregnancy-induced hypertension than those whose calcium intake is lower (Hamet, 1995). Thus consumption of dairy products is a useful dietary strategy to help reduce the risk of high blood pressure.

Dairy products have protective effects against the development of gastrointestinal tract tumors (**colon cancer**) (McIntosh et al., 1995; Bounous et al., 1991). The whey proteins appear to be the active protective component (McIntosh et al., 1995). The mechanism of action is not known but may be related to calcium and phosphorus in milk. Van der Meer et al. (1997) reviewed the effects of dietary fats and milk products on colon cancer. Dietary fat may promote colon cancer by increasing free fatty acids and **secondary bile acids**[1] in the colon. These substances damage the colon epithelial cells, causing an increased rate of colon cell turnover (hyperproliferation). Hyperproliferation of the colon epithelial cells increases the risk of mutation in the oncogenes and tumor-suppressor genes, as well as promoting the growth of premalignant polyp cells into tumors. A high intake of fat promotes colon cell hyperproliferation. Colon cancer results from about six mutations in tumor suppressor and oncogenes in the colonic epithelial cells, resulting in a progressive sequence of the following:

normal epithelium → hyperproliferative epithelium → adenoma → carcinoma

Van der Meer et al. (1997) presented evidence that the hyperproliferative effects of meat and milk fat are dependent on the calcium content of the diet. Milk fat, accompanied by the high calcium content of milk, does not promote hyperproliferation. Milk calcium increases the fecal excretion of bile acids and fatty acids and lowers their concentration in the colon. The active form of calcium in this precipitation is calcium phosphate. Milk con-

---

[1] Primary bile acids are produced in the liver, stored in the gall bladder, and secreted into the intestine via the bile duct. Secondary bile acids are produced by microbial metabolism of the primary bile acids in the colon.

tains both these minerals. Van der Meer et al. (1997) suggest that this interaction with calcium may explain why meat consumption may be associated with an increased risk of colon cancer: meat contains high amounts of fat and phosphate but very low calcium. Interestingly, addition of milk to tea increases the effectiveness of tea in preventing mammary and colon cancer (Weisburger et al., 1997). Tea contains phenolic compounds that have antioxidant and anticancer effects (Mukhtar and Ahmad, 2000; Yang et al., 2000).

Consumption of fermented milk products (e.g., yogurt) has protective effects against breast cancer (Biffi et al., 1997). The mode of action has not been identified. Conversely, high intakes of calcium and dairy products promote prostate cancer (Giovannucci, 1999).

Milk and dairy products contain **conjugated linoleic acid (CLA),** which has anticancer activity (Ha et al., 1989; Ip et al., 1991). CLA is also found in cooked beef (Shantha et al., 1994). It occurs in dairy products and red meat because it is formed from linoleic acid in the rumen. Ip and Scimeca (1997) hypothesized that CLA inhibits the formation of cancer-promoting metabolites of arachidonic acid (a long chain fatty acid) in the tissues. The cancer-preventing properties of milk components have been reviewed by Parodi (1997, 1999). Besides CLA, milk fat contains a phospholipid, **sphingomyelin,** which has cancer-inhibitory properties. Milk fat contains **butyric acid,** which inhibits proliferation of cancer cells. Finally, Parodi (1997) indicates that dairy cattle have the ability to extract anticancer agents from their diet and excrete them in milk. These include β-carotene and β-ionone from forages such as alfalfa, and gossypol from cottonseed. Low levels of gossypol inhibit proliferation of cancer cells (Hu et al., 1994). The occurrence of cancer-preventing substances in foods of animal origin is significant, and counter to conventional wisdom that animal-origin foods are cancer-promoting.

The finding that CLA in ruminant fat (meat and milk) is apparently a potent inhibitor of

cancer has been great news to the dairy and beef industries. The potential health benefits of CLA probably exceed any negative effects of ruminant fat on human health. The CLA is actually a mixture of unsaturated fatty acids, composed of various cis-trans and positional isomers of linoleic acid (18:2 ω-6). In the rumen, the bacterium *Butyrivibrio fibrisolvens* hydrogenates linoleic acid to stearic acid. The CLA (Fig. 2-2) are intermediates in this conversion (Fig. 2-3).[2] In meat and milk fat, the main CLA is **rumenic acid** (Fig. 2-3); it is considered the most biologically active CLA isomer (Bessa et al., 2000). Feeding sources of linoleic acid (e.g., oilseeds, corn—see Table 2-1) is an effective way of increasing CLA content of ruminant products.

Because trans fatty acids have potential harmful effects on human health, rumen manipulation of biohydrogenation should be designed to increase CLA production while reducing the trans C 18:1/CLA ratio (Bessa et al., 2000), which is the vaccenic acid/ rumenic acid ratio. The use of ionophores (e.g., rumensin) in conjunction with lipid supplementation may be a means of accomplishing this. The trans fatty acids in ruminant fat do not have the same negative effects on human health as those in hydrogenated margarine (Demeyer and Doreau, 1999).

A postulated mode of action of CLA in anticancer activity is as an antioxidant, a seemingly incongruous role of an unsaturated fatty acid. Oxidation of lipids, proteins, and nucleic acids and alteration of DNA are involved in carcinogenesis. The CLA may also have beneficial effects against atherosclerosis (Rudel, 1999), although present reports are not definitive. Of interest in this regard is that consumption of dairy products often lowers serum cholesterol in humans (Eichholzer and Stahelin, 1993). Perhaps CLA is the "hypocholesterolemic factor" they proposed for dairy products. Major dietary sources of CLA

---

[2] Fatty acid terminology is explained in more detail later in this chapter—see Diet and Coronary Heart Disease.

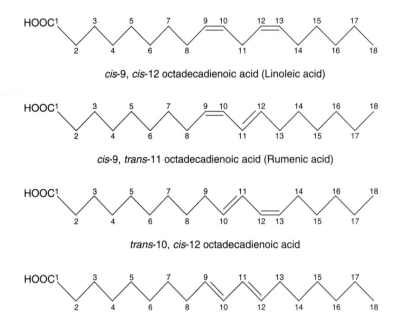

*cis*-9, *cis*-12 octadecadienoic acid (Linoleic acid)

*cis*-9, *trans*-11 octadecadienoic acid (Rumenic acid)

*trans*-10, *cis*-12 octadecadienoic acid

*trans*-9, *trans*-11 octadecadienoic acid

**FIGURE 2-2**   Linoleic acid and its main conjugated linoleic acid (CLA) isomers produced in the rumen.

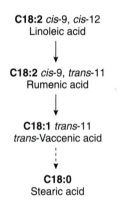

**C18:2** *cis*-9, *cis*-12
Linoleic acid

**C18:2** *cis*-9, *trans*-11
Rumenic acid

**C18:1** *trans*-11
*trans*-Vaccenic acid

**C18:0**
Stearic acid

**FIGURE 2-3**  Pathway of biohydrogenation of linoleic acid in the rumen.

are whole milk, cheeses, butter, yogurt, and beef (Ma et al., 1999). Synthetic CLA can be fed to swine and poultry to increase CLA content of pork and poultry products (Du et al., 1999). The richest CLA source, in amount per serving, in the compilation of Ma et al. (1999) was fast-food hamburger! (Table 2-1).

**Eggs** are good sources of all nutrients. As previously mentioned, it would be remarkable if they were not, because they have to provide all the nutrients needed to convert egg yolk and albumin into a baby chick! The main concern with eggs in the human diet has been with their high cholesterol content. Eggs are rich in cholesterol for the same reason they are rich in nutrients; cholesterol is needed by the developing embryo for a variety of metabolic functions (Fig. 2-4). Dietary cholesterol and human health are discussed later in this chapter. Eating eggs does not appear to increase the risk of coronary heart disease (CHD). In a dietary study involving 37,851 men and 80,082 women, Hu et al. (1999) found that the risk of CHD was no higher in the men and women who ate one egg every day than in those who ate only one egg per week.

## Meat

Meat consists primarily of skeletal muscle and connective tissue. Nutritionally, it is mainly protein and water, with variable quantities of fat. It is a good source of some minerals and vitamins but deficient in others.

Meat protein has a good balance of **essential amino acids.** Because the tissues of other animals are very similar to those of humans, it would be surprising indeed if meat

**TABLE 2-1** Levels of CLA in beef and dairy products.
(Adapted from Ma et al., 1999).

| Product | % Fat | mg CLA/g | mg CLA/serving |
|---|---|---|---|
| Sirloin tip roast[a] | 9.2 | 0.3 | 28.7 |
| Extra lean ground beef[a] | 10.0 | 0.1 | 11.5 |
| Ground beef[a] | 20.4 | 0.4 | 36.2 |
| Rib roast[a] | 27.8 | 0.8 | 77.6 |
| Fast-food hamburger[a] | 25.6 | 0.8 | 81.0 |
| Whole milk | 3.2 | 0.1[b] | 25.6 |
| 1% milk | 1.0 | 0.04[b] | 10.5 |
| 2% milk | 2.1 | 0.1[b] | 25.8 |
| Goat cheese | 28.5 | 0.7 | 34.3 |
| Brie cheese | 27.9 | 1.0 | 52.1 |
| Parmesan cheese | 28.3 | 1.2 | 12.0 |
| Mozzarella cheese | 24.9 | 1.1 | 57.1 |
| Cheddar cheese | 34.6 | 1.4 | 71.7 |
| Cream cheese | 33.8 | 0.9 | 13.8 |
| Yogurt | 5.4 | 0.2 | 42.7 |
| Butter | 91.1 | 4.3 | 64.1 |
| Sour cream | 12.6 | 0.6 | 9.2 |
| Cottage cheese | 3.1 | 0.2 | 26.7 |

[a]cooked

[b]mg CLA/l of sample

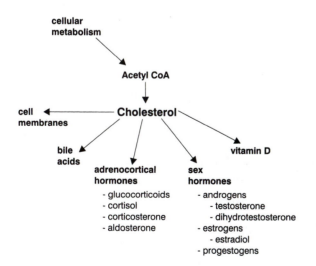

**FIGURE 2-4** Cholesterol is an essential metabolite in animals. Some of its most important metabolic roles are illustrated.

did not have a high protein quality. Meat protein has very high digestibility (92–100%), whereas plant proteins have a much lower digestibility (usually less than 80%). This is because some of the plant protein is associated with indigestible cell wall material. The plant cell wall structure does not allow full access of digestive enzymes to the cell contents, and plants contain phenolic compounds (tannins) that complex with proteins and reduce their digestibility. Thus, compared to most plant proteins, meat is a high quality, highly digestible source of protein in the human diet. However, enthusiasm for the high protein quality of meat has to be tempered by acknowledging that humans do not require high quality dietary protein. The growth rate of humans is extremely slow. We take 18–20 years to reach mature size. In contrast, a baby pig is much smaller than a human baby when born, but in less than two years a pig can reach a body weight of over 400 pounds. The laboratory rat, pig, and chicken, which are often used to evaluate dietary proteins for their quality, have extremely rapid growth rates. The amino acid content of wheat meets the needs of humans, whereas it does not for rats, pigs, and chicks. Humans can meet their amino acid requirements from bread and other plant sources such as beans. The concept that animal protein is required by humans is incorrect (Shorland, 1988).

Most people in the United States and other industrialized countries consume more protein than necessary for good health. Besides wasting protein, this overconsumption of protein has environmental effects—the excess nitrogen ends up in the sewer! The average American consumes between 130 and 200 percent of the **recommended daily allowance (RDA)** of protein (NRC, 1989). This is true for both vegetarians and nonvegetarians. The amount of meat needed to satisfy our daily protein requirement is quite small (no more than six ounces of lean meat per day).

One of the most common concerns about meat is its fat content. The fat content is variable, depending upon how long the animal was "finished." Body fat, with the exception of a small quantity found in cell membranes and nerve tissue, is the form in which excess energy consumed is stored. When an animal consumes more calories than necessary to provide the chemical energy (ATP) to meet its energy needs, the excess cannot be easily excreted but rather is stored as body fat. A major concern of human nutritionists with some types of **fat in meat** is with the content of saturated fatty acids. These will be defined and discussed in more detail later in this chapter as will the content and significance of cholesterol in animal products.

Meat contains almost no carbohydrates. Except for the lactose in milk, all the important carbohydrates in the human diet are of plant origin. These include the highly digestible starches and sugars and the poorly digested fiber components of the plant cell wall material, such as cellulose, hemicellulose, and lignin. Thus true dietary fiber can come only from plant sources. Meat does contain connective tissue (gristle), which may perform some of the roles of fiber in the gut, such as physical stimulation of the gut lining.

Meat is a good source of all minerals except calcium. Some of the minerals in meat are of special significance, because their availability (bioavailability) is much higher from meat than from plant sources. Meat has a high content of absorbable iron. **Iron** exists in foods in the ionic form (ferric and ferrous iron salts) and as organic iron associated with organic molecules such as heme, the iron-binding porphyrin in hemoglobin and myoglobin. Myoglobin is an important oxygen-carrying pigment in muscle tissue, storing oxygen for aerobic metabolism when muscular work is required. **Heme iron,** which makes up 40–60 percent of the iron in meat, is several times more absorbable than the non-heme iron found in plant sources. Meat in the diet improves the absorption of iron in plant sources as well (NRC, 1989). Both the fat and lean components of meat are involved in the enhancement of iron absorption. Fatty acids facilitate iron uptake by intestinal mucosal cells, while meat protein may promote the reduction of dietary iron to the more absorbable ferrous form or may form soluble iron chelates that are taken up by the intestinal mucosal cells (Kapsokefalou and Miller, 1993).

A study in Finland (Salonen et al., 1992) has implicated high blood iron level as an important risk factor for heart disease and heart attacks. The high content and bioavailability of iron in meat thus might be viewed as negative attributes under some circumstances. The possible relationship between iron and coronary heart disease will be discussed later in this chapter.

Zinc from meat has a higher bioavailability than zinc from plant sources (Hortin et al., 1993). **Zinc** is nutritionally very important as a cofactor in the fundamental process of growth involving DNA synthesis and is essential for normal growth, wound healing, normal immune responses, and taste responses.

Zinc from plant sources has a low bioavailability because of the presence of phytates in most plant materials. Phytates are organic molecules in plants which bind phosphorus, zinc, copper and numerous other minerals, and interfere with their absorption.

Most other minerals needed by humans are found in significant amounts in meat. **Selenium** in meat is of high bioavailability (Shi and Spallholz, 1994). Meat from animals raised in areas with high-selenium soils (e.g., North Dakota) is enriched with selenium (Hintze et al., 2001).

Meat is an excellent source of the **B-complex vitamins,** especially thiamin ($B_1$), riboflavin ($B_2$), pyridoxine ($B_6$), niacin and cyanocobalamin ($B_{12}$). Meat is the main natural source of **vitamin $B_{12}$** in the human diet. Meat (except liver) is a poor source of the fat soluble vitamins, vitamins A, D, E and K. Liver is an excellent source of vitamin A, because it is the major storage site in the body for this vitamin. All meat products are poor sources of vitamin C. The vitamins present in meat have a high bioavailability. This is not always true with plant sources; niacin and biotin, for example, often exist in plants in a poorly available bound form.

Thus meat is an excellent source of many important nutrients, is very digestible, and is palatable to humans. Per capita meat consumption in virtually all countries shows a direct correlation with per capita income. When they can afford it, people of all cultures upgrade their diets by consuming more animal protein.

# CONCERNS ABOUT RED MEAT AND HUMAN HEALTH

**Red meat** has an image problem. In North America at least, red meat is often regarded by health-conscious consumers as less desirable than **white meat** (poultry) and fish. What are the nutritional differences, if any, between red and white meat?

The color of meat is due largely to two iron-containing pigments, hemoglobin and myoglobin. Myoglobin makes up 80–90 percent of the total pigment in meat. Red meat (beef and lamb, leg muscles of poultry) has more **myoglobin** than pork, poultry, or fish. Veal is a white meat because the muscles of immature animals raised on milk have a low myoglobin content. The breast muscles of chickens and turkeys have a low myoglobin content and so are white. White muscle fibers have less oxygen storage (because of low myoglobin) so rely on anaerobic metabolism of glucose for energy, producing lactic acid as a waste product. White fibers contract rapidly in short bursts but are easily fatigued. Red fibers have more myoglobin, use aerobic metabolism (citric acid cycle), and contract more slowly but for a longer duration. All muscles contain a mixture of red and white fibers, with the red fibers predominating in so-called red meat. This difference in myoglobin content does not sound ominous. So why does red meat have an image problem?

Red meat, especially beef, has an image problem for several reasons. With medical authorities, it is because the body fat of ruminants tends to have a higher proportion of saturated fatty acids than does the fat of nonruminants, for reasons discussed under Diet and Coronary Heart Disease. Saturated fats have been linked to elevated blood cholesterol levels, which in turn have been associated with increased incidence of cardiovascular disease. People have incorrectly thought of "red meat," "ruminant meat" and "saturated fat" as being synonymous. Ruminant body fat tends to have a higher content of saturated fatty acids than meat from nonruminants. This has absolutely nothing whatsoever to do with whether it is red or white meat but is a result of rumen fermentation.

A new theory for explaining heart disease involves blood iron levels (Salonen et al., 1992). People with levels of more than 200 micrograms (µg) of ferritin per liter of blood appear to be at greater risk for heart disease than those with lower levels of less than 200 µg ferritin per liter. **Ferritin** is an iron-containing blood and tissue protein that stores excess body iron. Because red meat has a higher iron content than white meat, due to its higher myoglobin content, this **iron overload theory** could specifically implicate red meat as a risk factor in heart disease, because meat is the major source of absorbable iron in the diet.

With the general public, several other issues contribute to an image problem with red meat, particularly with beef. Per capita consumption of beef in the 1950s and 1960s increased rapidly in North America, as people enjoyed "the good life" after the economic Depression of the 1930s and the trauma of World War II. Their children, influenced by the Vietnam War, civil rights concerns, and other social issues, were perhaps the best known anti-establishment generation of all time. Beef consumption was a symbol of the affluent establishment. According to many gurus of the anti-establishment movements, beef production was a waste of valuable resources, including grains and tropical rain forests (see The Hamburger Connection—Chapter 6). It was claimed that beef cattle are fed diets "laced with chemicals and hormones." Who wants to eat chemicals and hormones? Beef became a symbol of the extravagant, resource-consuming American who was destroying the global environment to live a life of luxury, while most of the rest of the world suffered pestilence and famine. It made good press! Numerous other controversial issues, such as grazing on public lands, have contributed to a negative perception of beef. As with most issues, there is another side; "red meat" is not the villain that many have painted it.

The major human diseases for which meat consumption has been linked as a causative factor are coronary heart disease and cancer. These will be discussed individually in the following sections. Meat consumption has also been linked to **rheumatoid arthritis** (Grant, 2000). In a fairly nonspecific analysis of the relationship between prevalence of rheumatoid arthritis and national diet for several European countries (Fig. 2-5), Grant (2000) found meat fat to have the highest statistically significant correlation. An association with meat fat could relate to production of series 2 prostaglandins, which have inflammatory activity. Other possible factors in meat discussed by Grant (2000) include nitrite and iron, which increase production of free radicals. Free radicals may be involved in inflammatory effects noted in rheumatoid arthritis.

# DIET AND CORONARY HEART DISEASE

Animal fats in meat, milk, cheese, and eggs usually have a higher proportion of saturated fatty acids and a lower proportion of unsaturated fatty acids, especially polyunsaturates, than vegetable and fish oils. Animal products are the only sources of cholesterol in the human diet, since plants don't contain significant amounts of cholesterol. These properties are nutritionally significant, because blood cholesterol levels tend to be increased by dietary cholesterol and saturated fat intake and lowered by consumption of a higher proportion of unsaturated fatty acids. This has led to numerous suggestions by various groups and agencies that it would be nutritionally prudent for the average American to make dietary changes that would reduce total fat, saturated fat, and cholesterol intakes (NRC 1989, 1991, 1992). The following terms are frequently used in the discussion of dietary fat, cholesterol, animal and plant products, and cardiovascular disease, so it is appropriate to define them here.

**Lipids** are substances in plant and animal tissue that are soluble in organic solvents such as diethyl ether. The majority of dietary lipids are triglycerides (fats and oils), but there are many other types (e.g., phospholipids, cholesterol, carotenoids, chlorophyll pigments, waxes, etc.). In discussions of animal and human metabolism, the term lipid is used more or less synonymously with fat. Technically, this is incorrect; for example, cholesterol and beef tallow are both lipids, but only one of them (beef tallow) is a fat. The "basic units" of fat structure are the **fatty acids,** which consist

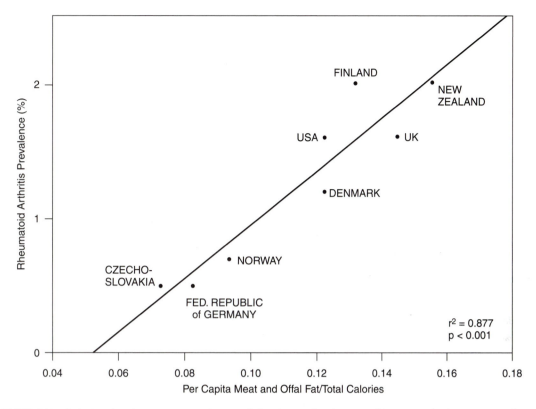

**FIGURE 2-5** Relationship between prevalence of rheumatoid arthritis and meat intake (as portion of total calories). (Courtesy of W. B. Grant.)

of a backbone or chain of carbon atoms, with a carboxyl group −COOH at one end of the chain and a methyl group −CH$_3$ at the other end. The carbons may be linked together by single bonds −CH$_2$−CH$_2$− or in some cases by double bonds −CH=CH−. Most fatty acids important in dietary and body fats have an even number of carbon atoms, and are "long chain," with 16, 18 or 20 carbons. (This is because they are synthesized by the addition of two carbons at a time, from acetyl Coenzyme A.)They are conveniently identified by reference to their numbers of carbons and double bonds: for example, 18:0 (stearic acid), 18:1 (oleic acid), 18:2 (linoleic acid), 18:3 (linolenic acid). Thus, for example, linoleic acid has 18 carbons and two double bonds.

A **saturated fatty acid** is one that is completely saturated with hydrogen. It consists of a chain of carbon atoms joined together, with a carboxyl group −COOH at one end of the

molecule. A carbon atom can form four chemical bonds. In a saturated fatty acid, each carbon atom is saturated with hydrogen; it is chemically incapable of taking up any more hydrogen. The most common dietary sources of saturated fatty acids are the body fat of ruminants and the tropical oils such as palm oil. The principal saturated fatty acids in meat are stearic (18:0), palmitic (16:0), and myristic (14:0) acids.

As the name suggests, an **unsaturated fatty acid** is not saturated with hydrogen, so it is capable of taking up more. This is because it has one or more double bonds between carbon atoms:

$$-CH=CH-$$

An unsaturated fatty acid can be made partially or completely saturated by the addition of hydrogen (**hydrogenation**) to the double bond(s). This may be done chemically (as in making margarine) or biologically (in the rumen). Corn oil is a liquid; it is converted to margarine by making it more saturated by adding hydrogen:

$$-CH = CH- + H_2 \longrightarrow -CH_2-CH_2-$$
unsaturated FA                    saturated FA

**Monounsaturated fatty acids (MUFA)** contain one double bond (e.g., oleic acid [18:1]). They are important components of some vegetable oils such as olive and canola oils (Table 2-2).

**Polyunsaturated fatty acids (PUFA)** contain two or more double bonds in the carbon chain. The more double bonds, the greater the degree of unsaturation. Many fatty acids in plant oils have two or three double bonds, while fish oils are often highly unsaturated with four or more double bonds. The most common PUFA in vegetable oils are linoleic (18:2) and linolenic (18:3) acids.

Fats and oils are **triglycerides** (the currently preferred term is **triacylglycerol**), meaning that they contain three fatty acids combined with glycerol:

$CH_2 - O -$ Fatty acid$_1$
|
$CH - O -$ Fatty acid$_2$
|
$CH_2 - O -$ Fatty acid$_3$

**Fats** are higher in saturated fatty acids than most oils and are usually solid at room temperature (e.g., lard, tallow). **Oils** are high in unsaturated fatty acids and are liquid at room temperature. The more the degree of unsaturation of the fatty acids, the lower the melting point (the lower the temperature required to change the oil from liquid to solid). The melting point is also influenced by the number of carbon atoms in the fatty acids. The shorter the carbon chain, the lower the melting point. For example, coconut oil is highly saturated but is a liquid at room temperature because it has a high content of short chain fatty acids. The fatty acid composition of a number of common fats and oils is given in Table 2-2.

**Cholesterol** is a lipid but not a fat. It has a totally different chemical structure than a triglyceride (fat). It is a type of compound called a steroid:

Cholesterol is synthesized in animals by complex chemical reactions, starting from a simple compound, acetyl coenzyme A (acetyl CoA). Cholesterol is the central compound in the synthesis of many other steroids, such as bile acids, adrenal cortex hormones, sex hormones, and vitamin D (Fig. 2-4). It is also an essential component of all cell membranes.

Cholesterol is an essential compound in metabolism. All tissues of the human body synthesize cholesterol. It is also absorbed from the diet when animal products are consumed. Excess cholesterol is excreted from the body in the feces, as bile salts and neutral sterols (both are formed in the liver and excreted in the bile). Thus while the average person may think that cholesterol is "bad," it is essential for human life. However, our tissues can synthesize all of the cholesterol needed, so it is not necessary in the diet.

Although plant foods do not contain significant amounts of cholesterol, it is synthesized in plants and converted to various plant sterols, such as ergosterol (pro-vitamin D), saponins, steroid alkaloids and glycosides, and so on. The cholesterol molecule has a fleeting existence in plants, as it is just a step in the synthesis of other sterols.

Cholesterol is transported in the blood as a component of plasma proteins called **lipoproteins.** These lipoproteins can be separated on the basis of their density. Those with the highest lipid content are the least dense

**TABLE 2-2** Fatty acid composition of some common fats and oils.

| Fat or Oil | Fatty Acids, % of Total* | | | | | |
|---|---|---|---|---|---|---|
| | 16:0** | 18:0 | 18:1 | 18:2 | 18:3 | 20:4 |
| Beef fat | 27.7 | 12.9 | **44.4** | 1.2 | 1.5 | - |
| Canola oil | 4.3 | 1.7 | **59.1** | 22.8 | 8.2 | 0.5 |
| Chicken fat | 15.1 | 7.2 | **36.9** | 31.8 | 3.0 | - |
| Coconut oil | **8.0** | 2.8 | 5.6 | 1.6 | - | - |
| Corn oil | 12.0 | 2.7 | 30.1 | **54.7** | 1.4 | 0.2 |
| Cottonseed oil | 20.9 | 1.9 | 16.0 | **59.6** | 0.1 | - |
| Lamb fat | 24.4 | 22.3 | **37.3** | 1.9 | 3.7 | - |
| Linseed oil | 6.4 | 3.3 | 17.0 | 15.6 | **57.7** | - |
| Olive oil | 14.0 | 2.6 | **74.0** | 8.1 | 1.0 | 0.4 |
| Palm oil | **42.0** | 5.4 | 39.1 | 10.6 | - | 0.2 |
| Peanut oil | 11.1 | 3.0 | **52.1** | 27.8 | 0.5 | 0.7 |
| Pork fat | 26.3 | 14.5 | **43.3** | 9.6 | 0.7 | - |
| Rapeseed oil | 1.7 | 0.1 | **14.3** | 13.4 | 8.9 | 0.9 |
| Safflower oil | 12.3 | 1.8 | 11.2 | **74.3** | - | 0.5 |
| Soybean oil | 11.5 | 4.3 | 27.3 | **49.7** | 6.9 | 0.2 |
| Sunflower oil | 6.8 | 3.9 | 15.7 | **73.5** | - | - |

*The major fatty acids in each fat and oil are highlighted in bold type. For example, canola, olive and peanut oils are major sources of monounsaturated fatty acid (18:1). Linseed oil is very high in highly unsaturated fatty acid (18:3). The major fatty acids in coconut oil are short chain (e.g., 12:0, 14:0) and do not appear in the table. Similarly, the major fatty acid in rapeseed oil is erucic acid (22:1), a poisonous fatty acid used for industrial purposes. Note that the most abundant fatty acid in beef, pork, and lamb fat is unsaturated (18:1). The composition of animal fats is much more variable than that of vegetable oils; the values given are representative of typical analyses.

**Common names are 16:0, palmitic acid; 18:0, stearic acid; 18:0, oleic acid; 18:2, linoleic acid; 18:3, linolenic acid; 20:4, arachidonic acid.

(fat floats on water) while those with the least lipid and most protein have the highest density. Thus by ultracentrifugation they can be separated into various groups. The major ones are the following:

1. chylomicrons
2. very low density lipoproteins (VLDL)
3. low density lipoproteins (LDL)
4. high density lipoproteins (HDL)
5. very high density lipoproteins (VHDL)

Chylomicrons are very high in fat and consist of fat droplets surrounded by a pro-

teinaceous membrane. This is the form in which fat droplets are taken from the site of fat absorption (intestinal mucosa) to the liver.

Cholesterol undergoes continual deposition and mobilization from the tissues, including deposits in artery walls. Free cholesterol is removed from the tissues by the **HDL** and taken to the liver, where it is removed from circulation and converted to bile acids. This is the so-called **good cholesterol**, because it is lowering tissue cholesterol levels. The HDL can be regarded as a cholesterol scavenger.

The **LDL** fraction is the **bad cholesterol**. It transports cholesterol from the liver for deposition in the tissues. The blood LDL/HDL cholesterol ratio is a good predictor of the potential for coronary heart disease; the higher the ratio (the more bad cholesterol relative to good cholesterol), the greater the risk. A LDL/HDL ratio below 2.5 is considered good; a ratio greater than 5.0 signifies high risk. The metabolism of LDL and HDL and their roles in coronary heart disease have been reviewed by Steinberg and Witztum (1990). The roles of the liver in synthesis of cholesterol and in the maintenance of cholesterol and HDL homeostasis have been reviewed by Dietschy et al. (1993). The liver is extremely important in regulating cholesterol status and the activities of blood lipoproteins.

**Arteriosclerosis** refers to lesions of the arteries (Fig. 2-6) characterized by thickening, calcification, and lipid accumulation (hardening of the arteries). The term **atherosclerosis** refers specifically to the type of arteriosclerotic lesions with a high lipid content. The lesions begin in young people as fatty streaks and progress to elevated lesions called fibrous **plaques.** This occurs particularly in the coronary arteries and the abdominal aorta. The plaques become increasingly larger and protrude into the lumen of the arteries. They become calcified and lipid-enriched, with triglyceride and cholesterol deposits. The plaques may hemorrhage, causing ulceration and thrombosis. **Thrombosis** is an advanced lesion that so severely occludes the artery that blood supply to the heart muscle is threatened. This leads to the death (necrosis) of part of the heart muscle, causing coronary heart disease and **heart attack.** A heart attack is caused by sudden loss of heart function (arrhythmia). With immediate emergency treatment, the patient may be revived, with the final outcome determined by how much of the heart muscle has died.

The initial lesions in atherosclerosis involve oxidation of the lipids in the LDL. This oxidation occurs in the lining of the arteries. The oxidized LDL stimulates local macrophages to engulf the oxidized LDL, further oxidize it, and produce enlarged fat-filled macrophages called **foam cells.** The foam cells undergo necrosis,

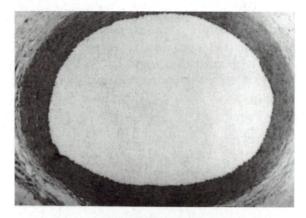

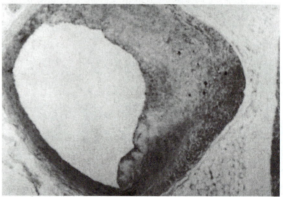

**FIGURE 2-6**   The appearance of a healthy artery *(top)* contrasted with an atherosclerotic coronary artery *(bottom),* showing the constriction of the blood vessel caused by the buildup of plaque. (Courtesy of J. E. Fincham, SA Medical Research Council.)

and in the process, the oxidized LDL causes damage to the endothelial layer of the artery wall. The buildup of these lesions causes atherosclerosis. Antioxidants such as vitamin E protect against atherosclerosis by preventing the formation of oxidized LDL (Diaz et al., 1997). Vitamin E ($\alpha$-tocopherol) is a constituent of the LDL particles, preventing the LDL from oxidative damage (Meydani, 2001). Vitamin E also appears to have non-antioxidant roles in preventing atherosclerosis, such as inhibiting production of proinflammatory substances such as cytokines (Meydani, 2001) and prostaglandins (Wu et al., 2001). Inflammatory effects are reflected in a blood protein, C-reactive protein (CRP); elevated CRP may be a stronger predictor of cardiovascular events than LDL-cholesterol (Ridker et al., 2002).

**Coronary heart disease (CHD)** is the usual clinical sign of atherosclerosis and is the major cause of death among adults in the United States and other industrialized countries (NRC, 1989). The major risk factors for CHD are the following:

1. high serum cholesterol and high serum LDL cholesterol
2. high blood pressure (hypertension)
3. cigarette smoking
4. being a male or a postmenopausal female
5. low serum HDL cholesterol
6. diabetes mellitus
7. family history of premature CHD
8. other factors such as obesity, lack of physical activity, personality type (e.g., Type A), high serum ferritin

The important relationship between serum cholesterol and CHD is shown in Fig. 2-7, which includes data generated from a study involving over 360,000 men aged 35–57 years (Martin et al., 1986). Elevated serum cholesterol (**hypercholesterolemia**) is an important risk factor in CHD. Every 1 percent increase in serum cholesterol increases the risk of CHD by about 2 percent (Hopkins, 1992), or, putting it more positively, for every 1 percent decrease in serum cholesterol, there is a 2 percent decrease in risk of CHD. The adverse effects of hypercholesterolemia are not controversial. What has been controversial is the significance of dietary cholesterol and type of dietary fat in hypercholesterolemia.

**Dietary cholesterol** in the majority of people makes a small contribution to the metabolic pool of cholesterol in the human body (NRC, 1989). However, it is now generally agreed by medical authorities that reduction of dietary cholesterol has very little impact on blood cholesterol levels and CHD incidence in most people (Grundy, 1990; Grundy and Denke, 1990; McNamara, 1990; Rifkind and Grouse, 1990; Steinberg and Witztum, 1990). Lowering blood

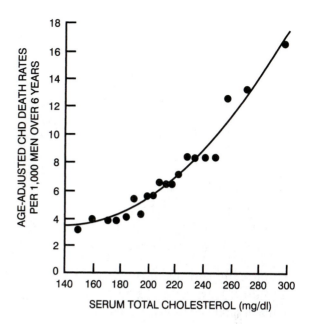

**FIGURE 2-7** Relationship of serum cholesterol to CHD mortality in 361,662 men ages 35–57 during an average follow-up of 6 years. (Modified from the National Research Council (1989) based on data of Martin et al (1986).)

cholesterol levels lowers the risk of CHD, but reducing dietary cholesterol in the majority of people will not appreciably lower blood cholesterol. Other dietary factors (i.e., certain unsaturated fatty acids) are involved in lowering blood cholesterol. The exception to this is those individuals who have a genetic predisposition to hypercholesterolemia. These individuals have abnormal activities of enzymes involved in cholesterol homeostasis, resulting in high blood cholesterol levels and an inability to homeostatically buffer additional cholesterol obtained from the diet. These people, identified clinically by their very high blood cholesterol levels, may benefit by avoiding dietary cholesterol, but usually require medication in addition to diet modification in order to control their high blood cholesterol levels. Humans in general respond less to dietary cholesterol than do other primates (Grundy and Denke, 1990) and most laboratory animals. With some animals such as rabbits the blood cholesterol level is easily elevated by feeding sources of cholesterol. Research results with these species are, therefore, not usually very relevant to humans.

It is known that the relationship between dietary and plasma cholesterol is curvilinear and is related to the **LDL receptor activity** and its ability to modulate absorbed cholesterol (Hopkins, 1992). The response to increased dietary cholesterol depends on the current (baseline) cholesterol intake. People who typically consume little or no cholesterol in their diet are sensitive to added cholesterol, whereas those with moderate to high cholesterol intakes (typical Western diet) are insensitive to further challenge (Hopkins, 1992), because their liver LDL receptors have been saturated (down-regulated).

It has been known for many years that fats with a fairly high content of certain saturated fatty acids cause elevated serum cholesterol. The major saturated fatty acids that have cholesterol-increasing effects are palmitic acid (16:0), myristic acid (14:0), and lauric acid (12:0). Stearic acid (18:0) and saturated fatty acids with 10 carbons or less are not hypercholesterolemic. The fat of ruminant animals has a higher content of saturated fatty acids than do vegetable oils, which, with the exception of coconut and palm oils, contain almost entirely unsaturated fatty acids. Reiser et al. (1985) showed that beef fat having about 20 percent stearic acid was not hypercholesterolemic (did not increase serum cholesterol). The body fat of ruminants has a predominance of C16 and C18 fatty acids, with about 80 percent of the fatty acids consisting of 14:0, 16:0, 18:0, and 18:1 (Byers and Schelling, 1988).

In development of a National Cholesterol Education Program, it was noted that patients with elevated blood cholesterol (hypercholesterolemia) showed similar reductions in blood cholesterol with either lean red meat or lean white meat in the diet (Davidson et al., 1999).

While dietary cholesterol is almost universally viewed negatively, cholesterol is important in many metabolic functions, particularly in young individuals. The reason that eggs are high in cholesterol is because this substance is needed for the growth and development of the embryo to produce a baby chick. Schoknecht et al. (1994) demonstrated that the growth

and behavioral responses of baby pigs were improved by supplementation with cholesterol. They concluded, "These data suggest that these neonatal pigs are unable to produce sufficient cholesterol to meet requirements for normal growth and brain development and are dependent on dietary cholesterol in milk" (p. 305). They further suggested that, because of the close resemblance of brain development in humans to that of the pig, it would be prudent to include dietary sources of cholesterol (i.e., animal products such as milk, eggs, and meat) in **diets for infants and young children.** It may well be that the tendency of some enthusiastic vegetarians today to ensure that their children never consume animal products may have long-term detrimental consequences on their growth and mental development.

## Effect of Species and Diet on Body Fat Composition of Livestock

Body fat (**adipose tissue**) is a dynamic substance, which is not simply deposited to remain there indefinitely. It undergoes constant mobilization into the blood, metabolism in the liver and mixing with recently absorbed fat, and redeposition. In nonruminants, such as swine and poultry, the end result of this process is that the composition of the body fat will resemble that of the dietary fat consumed recently. If a pig is fed tallow as a fat source, its body fat will be hard and resemble tallow. If it is fed peanut oil, its body fat will be soft and oily, like peanut oil (Fig. 2-8). What if it is fed no fat in the diet? Animals fed a fat-free diet still may deposit body fat, because fat is synthesized in the body from carbohydrate. Most of the fatty acids synthesized in the body are saturated, although there is some conversion of saturated to unsaturated fatty acids by desaturase enzymes. Thus swine and poultry fat will contain a mixture of saturated and unsaturated fatty acids, with the composition heavily dependent on the composition of the dietary fat currently being consumed.

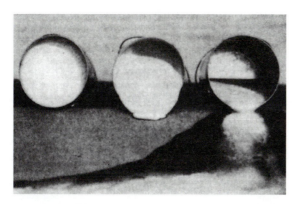

**FIGURE 2-8** The effect of dietary fatty acid composition on the carcass fat of nonruminant animals. These are lard samples from pigs fed saturated *(left)*, unsaturated *(middle)*, and polyunsaturated *(right)* fats in the diet. The carcass fat of pigs and poultry closely resembles the dietary fat. (Taken from USDA Bulletin 1492 [1928].)

The situation in ruminants is quite different. The rumen is, in chemical terms, a reducing environment, meaning that it is rich in hydrogen (**biological reduction** involves gain of hydrogen). Thus, unsaturated fatty acids in the diet of ruminants are converted to the saturated form by the process of hydrogenation (addition of hydrogen) in the rumen. Also, the dietary fat level is usually quite low in ruminant diets, because at levels much above 5 percent, fats disrupt rumen function due to a lack of emulsifying agent. Fats are emulsified post-ruminally in the small intestine by bile. Thus in ruminants, much of the body fat is synthesized from dietary carbohydrate, by way of acetate produced from rumen fermentation of carbohydrate. Saturated fatty acids are the end product of fatty acid synthesis from acetate. Regardless of the type or level of dietary fat, the carcass fat of ruminants will be fairly high in saturated fatty acids.

Rumen microbes synthesize a wide variety of **odd-carbon chain and branched-chain fatty acids,** many of which are of the trans configuration (Byers and Schelling, 1988). These fatty acids (see Figs. 2-2 and 2-3) are present primarily in the membrane phospholipid and as free fatty acids. Ruminant body fat contains a small amount of these odd-carbon and branched-chain fatty acids. Unsaturated fatty acids with a trans configuration tend to behave as saturated fatty acids with respect to their cholesterolemic effects. The cis and trans configuration of fatty acids refers to the orientation of the carbon chain (acyl chain) about the double bond. With the **cis** configuration, the acyl groups are on the same side of the double bond, whereas with the **trans** configuration, they are on opposite sides (*trans* means across):

$$\underset{\text{cis}}{-\overset{\overset{\displaystyle H}{|}}{C}=\overset{\overset{\displaystyle H}{|}}{C}-} \qquad\qquad \underset{\text{trans}}{-\overset{\overset{\displaystyle H}{|}}{C}=\underset{\underset{\displaystyle H}{|}}{C}-}$$

Naturally occurring unsaturated fatty acids are usually of the cis configuration. The configuration affects the shape: cis fatty acids are L shaped, whereas trans fatty acids are more straight. This difference in shape has a marked effect on membrane structure and function. The primary source of **trans fatty acids** is from margarine, because industrial hydrogenation of vegetable oils produces trans fatty acids. Ruminant fat, including milk fat, is also a source of trans fatty acids. The trans analogs of unsaturated fatty acids (e.g., elaidic acid [18:1] is the trans analog of oleic acid) do not have cholesterol-lowering properties and may increase LDL and lower HDL (Mensink and Katan, 1990).

Another means by which trans fatty acids may have a role in heart disease is by increasing **serum lipoprotein(a)** (Nestel et al., 1992). This is a large molecule in the blood made up of apolipoprotein B, cholesterol, other lipids, and a protein called apo(a). High serum lipoprotein(a) levels are a risk factor for CHD (Mensink et al., 1992). Diets high in trans-monounsaturated fatty acids (e.g., elaidic acid in margarine) may thus increase the risk of CHD by increasing serum lipoprotein(a). This finding attracted media attention in late 1992, becoming the current nutritional bombshell of the day. Newspaper articles suggested that margarine might be hazardous to one's health and that perhaps butter was not so bad after all. To further complicate matters, Hornstra et al. (1991) reported that a palm oil-enriched diet lowers serum lipoprotein(a) in humans. Palm oil, of course, has received a lot of bad publicity in the popular media because of its high content of saturated fatty acids.

The roles and significance of apolipoproteins in CHD are complicated. For those who wish further information, apolipoproteins and their effects on cholesterol metabolism have been reviewed by Rothblat et al. (1992).

## Saturated Fatty Acids and Cholesterol Metabolism

Some saturated fatty acids elevate blood cholesterol. Most do not. It is important not to make a blanket statement that saturated fatty acids cause hypercholesterolemia.

Three saturated fatty acids have been shown in various studies to be hypercholesterolemic. They are lauric (12:0), myristic (14:0), and palmitic (16:0) acids. Research suggests that **palmitic acid** has less effect than formerly believed (Hayes et al., 1991). The effect of palmitic acid (16:0) depends upon the blood cholesterol status of the subject. In individuals with normal plasma cholesterol levels (less than 225 mg/dl) and moderate cholesterol intakes (less than 300 mg/day), 16:0 has no effect on plasma cholesterol (Hayes and Khosla, 1992), but in hypercholesterolemic subjects (> > 225 mg/dl), and especially those consuming more than 400 mg cholesterol per day, dietary 16:0 elevates blood cholesterol. The effect seems to involve LDL receptor activity. This area has been reviewed by Hayes and Khosla (1992). Lauric and myristic acids are the two major hypercholesterolemic fatty acids. These shorter chain fatty acids are found mainly in butterfat and tropical oils. Beef fat is low in both of these fatty acids. **Stearic acid** (18:0), one of the main saturated fatty acids in beef fat, is not hypercholesterolemic (Bonanome and Grundy, 1988; Grundy and Denke, 1990;

Reiser et al., 1985). The only fatty acid in beef fat with significant cholesterol-raising effects in beef fat is palmitic acid (Table 2-1), which, according to Hayes et al. (1991), may actually have little hypercholesterolemic activity. This is supported by the finding of Reiser et al. (1985) that beef fat was not hypercholesterolemic. Lean beef is equivalent to chicken in its effects on plasma cholesterol and triglycerides (Denke, 1994; Scott et al., 1994). Lean beef can be used successfully in cholesterol-lowering diets (Davidson et al., 1999).

Stearic acid, a non-hypercholesterolemic fatty acid in beef, might nevertheless have a negative influence on coronary heart disease. Stearic acid causes hypercoagulability of the blood by activation of factor XII and by aggregation of blood platelets (Hoak, 1994). It has not yet been determined if consumption of stearic acid-rich foods such as beef can cause a thrombogenic effect *in vivo*.

Grundy and Denke (1990) reviewed the mechanisms by which certain saturated fatty acids cause hypercholesterolemia. They hypothesize that saturated fatty acids increase LDL-cholesterol by inhibiting the LDL receptors in the liver for removing LDL-cholesterol from the blood. Stearic acid (18:0) is rapidly converted to oleic acid (18:1) by desaturases. Palmitic acid (16:0) is chain elongated to stearic and then desaturated to oleic. **Chain elongation** is slow, whereas **desaturation** is rapid. Thus stearic acid is rapidly converted to the monosaturated oleic acid, which does not interfere with LDL receptors. The 12:0, 14:0, and 16:0 must be chain-elongated to 18:0. This accounts for 12:0 and 14:0 having greater hypercholesterolemic effects than 16:0. There is evidence that some people (high responders) are more sensitive to hypercholesterolemic effects of saturated fatty acids than are others (Grundy and Denke, 1990). Shorland (1988) proposed a different mechanism to explain the hypercholesterolemic effect of palmitic acid. According to his hypothesis, palmitic acid inhibits synthesis of long chain fatty acids from acetyl CoA by inhibiting acetyl CoA carboxylase. This redirects more acetyl CoA to cholesterol biosynthesis. According to this hypothesis, palmitic acid (and 12:0 and 14:0) elevate serum cholesterol by promoting cholesterol biosynthesis.

## *Unsaturated Fatty Acids and Cholesterol Metabolism*

Unsaturated fatty acids, especially the polyunsaturated fatty acids (PUFA) with more than one double bond, have a serum cholesterol-lowering effect. The mechanisms are not completely known (NRC, 1989). The association of PUFA with HDL may facilitate removal of cholesterol from the serum lipoproteins by the liver and its excretion in bile. The monounsaturated fatty acids, such as oleic acid (18:0), have a lesser but still physiologically significant cholesterol lowering effect.

The PUFA in vegetable oils are mostly of a type known as **omega-6 ($\omega$-6)**, while the fatty acids of fish oils are of the **$\omega$-3** type. The meaning of these terms is shown by looking at the location of the double bonds:

an omega-6 fatty acid (vegetable oil type):

$$CH_3-CH_2-CH_2-CH_2-CH_2-CH$$
$$=CH-(CH_2)_n-COOH$$

an omega-3 fatty acid (fish oil):

$$CH_3-CH_2-CH=CH-(CH_2)_n-COOH$$

The omega designation refers to the location of the first double bond from the terminal end (methyl end) of the fatty acid. The double bonds in fatty acids are "methylene interrupted," meaning that each double bond is separated by a $-CH_2$. Thus if you know where the first double bond is, you then know where the rest of them are. For example:

18:2($\omega$-6):linoleic acid:

$$CH_3-(CH_2)_4 CH=CH-CH_2-CH$$
$$=CH-(CH_2)_7-COOH$$

18:3($\omega$-3):linolenic acid:

$$CH_3-CH_2-CH=CH-CH_2-CH=CH-CH_2-CH$$
$$=CH-(CH_2)_7-COOH$$

The $\omega$-3 fatty acids are significant as dietary factors in preventing CHD. It was observed that Greenland Eskimos had very low rates of CHD, in spite of a diet that was very high in fat and animal products and very low in carbohydrates and plant-origin foods (Budowski, 1988). Fish oils are highly unsaturated, and the fatty acids are of the $\omega$-3 group. They cause a drop in serum levels of LDL-cholesterol ("bad" cholesterol) and are particularly effective in reducing serum triglyceride concentrations. In addition to their favorable effects on serum lipids, $\omega$-3 fatty acids are also involved in modulating the synthesis of prostaglandins, prostacyclins, thromboxanes, and leukotrienes. These substances have hormone or hormone-like roles in vasoconstriction and **blood platelet aggregation** and may be useful in "thinning the blood" in terms of preventing blood clot formation and thrombosis. However, these effects also have a negative side, in that prolonged bleeding times, reduced platelet counts, and reduced platelet adhesiveness, associated with a change in the platelet fatty acid composition from $\omega$-6 to $\omega$-3 fatty acids, are also seen in Greenlanders and Japanese with a high fish intake. They have spontaneous bleeding episodes and bruise easily. With moderate intakes of fish oils, overall beneficial effects on CHD result, leading to the current recommendations that Americans should increase their consumption of fish. Cold water marine fish (e.g., salmon, mackerel) are particularly good sources of $\omega$-3 fatty acids. There are several extensive reviews of the biological effects of $\omega$-3 fatty acids and their roles in reducing CHD (Budowski, 1988; Kinsella et al., 1990; Weaver and Holub, 1988).

The $\omega$-3 fatty acids have also been observed to have favorable effects on preventing fatal cardiac arrhythmia (fibrillation), as reviewed by Nair et al. (1997). Animal products such as eggs can be enriched in $\omega$-3 fatty acids by modifications of the hen's diet (see Modification of

Animal Products to Improve Human Health—this chapter).

A prime cause of heart attacks is the occurrence of a thrombus or clot in the coronary artery. This process is accelerated by the aggregation and adhesion of a type of blood cell, the platelets. The aggregation of platelets is controlled by **prostaglandin** derivatives synthesized from long chain PUFA. **Thromboxane** is a clotting factor synthesized in the platelets; it is derived from arachidonic acid ($20:4\omega6$). **Prostacyclin,** synthesized in the walls of blood vessels, inhibits platelet aggregation. The balance between thromboxane and prostacyclin regulates thrombosis. **Eicosapentaenoic acid (EPA),** which is $202:5\omega3$, is found in fish oils. This fatty acid is also used to make thromboxane and prostacyclin. However, they differ slightly from those derived from arachidonic acid. The thromboxane derived from $22:5\omega3$ does not cause platelet aggregation. Therefore, the tendency for clot formation is reduced. The synthesis of prostanoids (prostaglandins and thromboxanes) involves two oxygens ($O_2$), catalyzed by **cyclooxygenase (COX).** There are two isoforms of COX, COX-1, and COX-2. The COX-2 enzyme has been implicated in stimulation of cancer development. Some foods and beverages (e.g., grapes, wine), contain small phenolic molecules that are COX-2 inhibitors and may play a role in treating or preventing cancer (Dannenberg, 2001).

Some of the major $\omega$-3 and $\omega$-6 fatty acids are the following:

| $\omega$-3 | | $\omega$-6 | |
|---|---|---|---|
| C18:3$\omega$3 | Linolenic acid (LNA) | C18:2$\omega$6 | Linoleic acid (LA) |
| C20:5$\omega$3 | Eicosapentaenoic acid (EPA) | C20:4$\omega$6 | Arachidonic acid (AA) |
| C22:5$\omega$3 | Docosapentaenoic acid (DPA) | | |
| C22:6$\omega$3 | Docosahexaenoic acid (DHA) | | |

The lipid content of the retina and grey matter in the mammalian brain is high in both AA and DHA. These long chain PUFA are derived from their respective dietary essential fatty acid precursors, LA and LNA, through desaturation and chain elongation. The AA and DHA accumulate rapidly in the human brain during the third timester of pregnancy and in the early postnatal period (Wainwright, 2000), when the rate of brain growth is maximal. Broadhurst et al. (1998) proposed that the evolution of the enlarged brain of *Homo sapiens,* in the East African Rift Valley widely considered the cradle of human origins, was due to the consumption of fish, which provided the fatty acid nutrients necessary for brain growth. They claim that the uniquely complex human neurological system could not have evolved in the absence of consumption of animal-based foods. The lipid profile of tropical freshwater fish has a DHA:AA ratio that is closer to human brain phospholipids than any other food source known (Broadhurst et al., 1998). Fish are also an excellent source of Zn, Cu, I, and other trace elements necessary for PUFA metabolism and for normal brain development and function. According to Broadhurst et al. (1998), "There is good evidence today that lack of abundant, balanced DHA and AA *in utero* and infancy leads to lower intelligence quotient and visual acuity, and in the longer-term contributes to clinical depression and attention-deficit hyperactivity disorder. We are not so far removed from our Paleolithic ancestors that we can expect our present agricultural, processed-food-based diet to provide indefinitely for our continued intellectual development" (p. 18). Food for thought!

## Cholesterol Oxides

Cholesterol in food products such as meat and eggs can be oxidized to produce **cholesterol oxidation products (COP).** Free radicals produced by peroxidation of unsaturated fatty acids stimulate COP production.

The COP have been shown to be cytotoxic, atherogenic, and mutagenic and may have a role in both atherosclerosis and cancer (Guardiola et al., 1996). An additional benefit of feeding high levels of vitamin E to animals prior to slaughter (besides increasing shelf life) may be a reduction in COP production.

## Fatty Acid Composition of Fat of Wild Ruminants

In a review of **Paleolithic nutrition** (the caveman diet), Eaton and Konner (1985) point out that while early humans consumed a diet high in meat, the actual nutrients consumed may have been quite different from those in today's Western-type diet. The importance of meat in the caveman diet is revealed by the abundance of animal bones and hunting tools in archaeological sites and the absence of much evidence of plant foods and tools for processing plant foods. The dawn of agriculture about 10,000 years ago marked a shift away from a meat diet to one with more plant foods (as indicated by seed-grinding tools [mortar and pestle] in archaeological sites).

There are considerable differences in the fat composition of wild ruminants compared with modern sheep and cattle. Wild ruminants are generally much leaner, having lower quantities of depot fat and intramuscular fat (marbling). A greater proportion of the tissue fat of wild ruminants consists of structural lipids such as phospholipids in cell membranes. Crawford and Gale (1969) and Gale et al. (1969) surveyed a number of wild African herbivores (eland, wildebeest, buffalo, etc.) and found much higher levels of long chain PUFA such as $C20:4\omega3$, $C22:5\omega3$, and $C22:6$ 3. The total percentage of $\omega$-6 PUFA of total fatty acids was about 30 percent in wild ruminants vs. 12 percent in domestic cattle, while for $\omega$-3 PUFA the averages were 9 and 4 percent respectively. Thus the fat of wild ruminants seems to contain a much higher proportion of the cholesterol-lowering $\omega$-3 and $\omega$-6 PUFA. This may reflect the greater proportion of

structural lipid, plus diet differences. The main fatty acid in grass is $C18:3\omega3$ (linolenic acid), whereas seeds (e.g., cereal grains) contain mainly $C18:2\omega6$ (linoleic acid). Crawford and Gale (1969) speculated that seed capsules of herbage consumed by wild ruminants might bypass hydrogenation in the rumen, and serve as a source of higher $\omega$-6 PUFA through desaturation and chain elongation.

These differences in fatty acid composition could be of importance in meat products produced by game farming (e.g., deer, eland). However, the high PUFA in structural phospholipids of wild ruminants contributes to undesirable flavors and odors of the meat. For example, in studies by Larick et al. (1989), meat from forage-finished American bison exhibited off-flavors including ammonia, gamey, bitter, liverish, old, rotten and sour perceptions by taste-panel participants.

Simopoulos (1998) notes that while the human genetic profile has not changed much in the past 40,000 years, our current Western diet has changed markedly over this time, with an increase in total fat, saturated fat, trans fatty acids and $\omega$-6 essential fatty acids but a decrease in $\omega$-3 fatty acids. The ratio of $\omega$-6 to $\omega$-3 fatty acids is 10–20:1 today, whereas in pre-agricultural times it was about 1:1. This change arose because of our use of vegetable oils (mainly $\omega$-6) and grain-fed meat animals, with carcass fat high in saturated and $\omega$-6 fatty acids. Simopoulos (1998) proposes that the high incidence of "diseases of Western civilization" such as coronary heart disease, hypertension, arthritis and diabetes is related to a diet out of step with our evolutionary heritage. Archeological evidence suggests that early humans cracked bones to eat bone marrow and ate brains and other high fat tissues (Eaton et al., 1998). These tissues are excellent sources of C-20 and $C$-$22\omega$-3 fatty acids (Cordain et al., 1998). In addition, the body fat of concentrate-selector wild ruminants (see Chapter 4), such as deer, has higher unsaturated fatty acid levels than for grazers such as cattle (Meyer et al., 1998). Small, concentrate-selector ruminants probably were important food sources for pre-industrial peoples.

## Negative Effects of Polyunsaturated Fatty Acids and Low Serum Cholesterol

The human diet contains a great variety of natural mutagens and potential carcinogens (Ames, 1983), and many of these exert their effects through the generation of oxygen radicals (Shorland, 1988). The **lipid oxidation** chain reaction (rancidity) produces a variety of mutagens and carcinogens, such as fatty acid hydroperoxides, cholesterol hydroperoxide, endoperoxides, fatty acid epoxides, reactive aldehydes, and alkoxy and hydroperoxy radicals (Shorland, 1988). The PUFA in cell membranes are susceptible to attack by these reactive substances. Diets high in PUFA may provoke greater cancer incidence (NRC, 1989). Oxidative damage is also implicated in the aging process. Thus the NRC (1989) recommendation to reduce total calories from lipids, regardless of source, is sound. Reducing intakes of saturated fatty acids and increasing intakes of PUFA may have some positive effects in terms of CHD but negative effects on cancer incidence. The PUFA should not be regarded as a nutritional panacea.

There is a U-shaped relationship between mortality and serum cholesterol levels. A zone of minimum mortality exists between about 188 and 225 mg cholesterol per dl. Incidence of CHD increases with higher cholesterol levels, while at levels below 188 mg/dl, mortality from cancer, hemorrhagic stroke, cirrhosis, and other nonspecific causes increases. A population-wide dietary change to lower blood cholesterol would likely actually increase human mortalities from cancer and other conditions associated with low blood cholesterol (Frank et al., 1992; McNamera, 1990; Muldoon et al., 1990).

## Diet and Stroke

Stroke is the third major cause of death among adults, following heart disease and cancer (NRC, 1989). A **stroke** is the abrupt onset of neurological disability due to loss of blood supply to an area of the brain served by an artery that has become blocked or clogged by atherosclerosis or other cause. There is a U-shaped relationship of stroke incidence with serum cholesterol—the highest incidence occurs with both high and low levels of serum cholesterol, and the lowest stroke incidence with intermediate cholesterol levels (NRC, 1989). There seems to be no consistent relationship of stroke with consumption of animal products or saturated fatty acids. Gillman et al. (1997) found that dietary fat and saturated fatty acids reduced the incidence of ischemic stroke in men. (Ischemic stroke is due to functional constriction or obstruction of a blood vessel.)

## Other Dietary Factors and Coronary Heart Disease

The interpretation of dietary effects on a medical condition is often difficult. It is rarely possible to modify one component of a diet without changing something else. The incidence of CHD is highest in populations with high intakes of fat and saturated fatty acids, often coming from animal sources. Does this indicate that fat is the causative factor, or could it be that diets lower in fat are usually higher in fruits and vegetables? Perhaps these foods contain protective substances. Certain plant constituents are well known for their effects on reducing serum cholesterol. These include **saponins** in legumes (e.g., beans) and **soluble fiber** in barley and oats. Soluble fiber refers to β-glucans and other nonstarch polysaccharides (NSP). Both barley bran and oat bran have cholesterol-lowering properties, whereas wheat bran has no effect or may even be hypercholesterolemic (Topping, 1991). The mode of action of NSP is probably to bind with bile acids, enhancing their excretion in the feces. The bile acids are sterols synthesized in the liver from cholesterol, and their emulsifying action increases the enterohepatic recycling of cholesterol by enhancing its absorption. Increasing bile acid excretion in the feces reduces the amount of cholesterol reabsorbed following its excretion in the bile.

Other dietary constituents from plant sources such as **antioxidants** may have favorable effects in reducing CHD. Duthie et al. (1989) reviewed evidence that free radicals produced from lipid autooxidation (rancidity) may play a role in atherosclerosis. Lipid hydroperoxides may cause free radical-mediated damage to the artery wall, initiating conditions leading to platelet aggregation around the site of injury. Lipid hydroperoxides inhibit synthesis of prostacyclins, which, as discussed earlier, inhibit platelet aggregation and adhesion. Duthie et al. (1989) hypothesized that antioxidants may have protective effects against CHD via their inhibition of lipid peroxidation. The beneficial effects of dietary monounsaturated fatty acids such as oleic acid (18:1) may be because they are incorporated into the LDL structure, making it resistant to autooxidation (Parthasarathy et al., 1990).

Surprising as it may initially seem, a new theory of CHD involving blood **iron** could provide a unifying theme in CHD incidence regarding the possible involvements of peroxides, red meat, fish oils, and aspirin, and regarding differences in men vs. premenopausal women and in premenopausal vs. postmenopausal women. Salonen et al. (1992) in a study in Finland (which has the highest incidence of CHD of any country) found that men with high (over 200 μg/l) blood levels of ferritin were at much greater risk for occurrence of CHD than were men with lower ferritin levels. Ferritin is a storage form of excess body iron, so high blood ferritin levels reflect an overabundance of iron in the body. It has long been known that certain minerals such as iron and copper stimulate the autooxidation of lipids to produce free radicals, peroxides and fat rancidity. Feed manufacturers are aware of this and do not mix trace minerals and vitamins in the same premix, because iron and copper will cause autooxidation and destruction of the vitamins. The **iron overload theory of CHD** proposes that excess iron causes oxidation of the LDL proteins in the blood (de Valk & Marx,

1999). Oxidative damage to the LDL proteins changes their physical structure, causing them to be recognized by white cells as cellular debris. The white cells engulf the oxidized LDL and are deposited on the artery wall. Fibrotic lesions around the deposits develop, causing a narrowing of the artery (arteriosclerosis), leading to CHD (Steinberg, 1997). Saturation of the antioxidant defenses of the blood by excessive iron promotes free radical-mediated tissue damage. Iron is the most potent promoter of free radical formation from peroxides (Halliwell and Gutteridge, 1986). Vitamin E has the most protective effect against development of these lesions (Diaz et al., 1997).

While confirmation of the iron hypothesis is needed, there has been evidence for several years (e.g., Sullivan, 1989, 1992) that high iron levels were correlated with high CHD incidence. The iron theory can be used to rationalize the involvement of several risk factors. Premenopausal women have a much lower CHD incidence than men of the same age. There are large losses of iron in menstruation. After menopause, loss of iron by menstruation ceases, and the CHD rate rises markedly in women to the same rate as for men. Oral contraceptives reduce menstruation and increase the rate of CHD in women. Both fish oils and aspirin, which reduce CHD incidence, prolong bleeding, causing iron depletion. Finally, the association of CHD with meat intake could be rationalized because meat is the major source of dietary iron, and the iron in meat is particularly well absorbed. Red meat is high in iron because of its high myoglobin content. This could be the "smoking gun" in the red meat-human health debate.

**Homocysteine** is a sulfur-containing amino acid produced by the demethylation of methionine. Blood levels of this amino acid are elevated in patients with arterial occlusive diseases (coronary, cerebral, and peripheral artery occlusion or blockage) and may be a risk factor for these diseases (Malinow, 1994a, b, 1996). A possible association of elevated homocysteine levels with meat consumption may be a consequence of excessive intake of protein and methionine, increasing the biosynthesis of homocysteine. However, meat is also a good source of vitamin $B_{12}$. Supplementation with folic acid and vitamin $B_{12}$ is very effective at reducing blood homocysteine levels in patients with elevated levels (Brattstrom, 1996). In a European study comparing serum homocysteine levels in vegetarians and meat eaters, Krajcovicova-Kudlackova et al. (2000) found a greater frequency of hyperhomocysteinemia in vegetarians than in meat eaters (29% vs. 5%), and significantly lower serum vitamin $B_{12}$ levels in vegetarians. They concluded that the elevated serum homocysteine levels in vegetarians were a consequence of subclinical vitamin $B_{12}$ deficiency. Probably other aspects of meat consumption, such as the hypercholesterolemic effects of saturated fatty acids, are of greater importance than homocysteine in arterial disease. A symposium proceedings with a number of papers presented on homocysteine is published in the *Journal of Nutrition* (126:1235S–1300S, 1996), and a review of nutritional effects is that of Weir and Scott (1998).

Atherosclerotic lesions are often calcified (Demer, 1995). A variety of bone matrix proteins, such as osteopontin, have been identified in atherosclerotic lesions. Of concern is that dietary calcium supplements, intended to counteract osteoporosis, may aggravate cardiovascular disease by promoting calcification of blood vessel walls. Critics of milk consumption have used this concern to link milk drinking with vascular disease.

## Dietary Recommendations for Prevention of CHD

The Committee on Diet and Health of the National Research Council (NRC, 1989) made the following recommendations:

> Reduce total fat intake to 30% or less of calories. Reduce saturated fatty acid intake to less than 10% of calories and the intake of cholesterol to less than 300 mg daily. The intake of fat and cholesterol can be reduced by substituting fish, poultry without skin, lean meats, and low- or non-fat dairy products for fatty meats and whole-milk dairy products; by choosing more vegetables, fruits, cereals, and legumes; and by limiting oils, fats, egg yolks, and fried and other fatty foods.

What are the potential impacts of these recommendations, if followed, on the livestock industries? Are the impacts good, bad, or indifferent? I will present my own views, as a starting point for discussion.

1. I believe that the NRC recommendations are based on sound science. Literally hundreds of experiments with humans and many species of experimental animals have shown adverse effects of saturated fatty acids, particularly palmitic (16:0), myristic (14:0), and lauric (12:0) acids, on serum cholesterol levels. Many studies, involving hundreds of thousands of people, have shown that at blood levels above 180 mg cholesterol per dl, there is a positive relationship between CHD and serum cholesterol. The higher the serum cholesterol, the higher the risk of CHD. Populations in which the average serum cholesterol level is less than 180 mg/dl are virtually free of atherosclerosis and CHD, while those groups of people in which the average serum cholesterol level is above 220 mg/dl have high rates of CHD (NRC, 1992). Populations in the former group are those on the lower end of the per capita meat consumption scale, while those in the latter group are populations with high intakes of animal products (NRC, 1989).

2. How much meat, milk, and eggs should people eat? Is it good if beef cattle, sheep, dairy, swine, and poultry production expands, and bad if these industries shrink in size?

In my opinion, the ideal size of each of the livestock industries in the United States is such that the amount of meat, milk, and eggs produced is approximately equal to the amount that the human population should consume for good or optimal health and at the same time provides a decent livelihood for the producers of these products. If that means that the poultry industry should be twice as large as now or half as large as now, fine. If it means that there should be two, three or five times as many beef cattle as now or 10 percent, 25 percent or 50 percent as many as now, fine. We have been conditioned, particularly in the industrialized countries, to think of growth of an industry as being good, and a retrenching, bad. For example, if sheep numbers are increasing, sheep producers and their industry representatives are excited—the sheep industry is expanding and that's great. When numbers decline (as they have more or less continuously since the 1930s), sheep producers and many animal scientists are glum. "What can we do to revive the sheep industry?" is the lament. In fact, does it matter if sheep numbers exceeded 50 million in the United States in 1940 and are now under 10 million? How many sheep in the United States is the optimal number? Perhaps it is only 2 million. Perhaps (although this stretches credulity) it is 100 million. The point I'm making is that if in the past the average American has consumed more meat and/or eggs than desirable for optimal health, and it is now recognized as being desirable to reduce this intake, which in turn results in a demand for fewer eggs and less beef and fewer numbers of laying hens and beef cattle, so what? Acceptance of this reality does not make a person "antilivestock."

3. It is my opinion that it is more useful to the livestock industries and animal scientists to come to grips with the demonstrated relationships among saturated fat and cholesterol intakes and CHD than to claim that there is no relationship or that there's some sort of conspiracy against animal products by the medical community. One of my colleagues uses the term "so-called medical authorities" to justify dismissing their research. My opinion is that the livestock industries and animal scientists, instead of always "crying foul," should take a serious look at their products and work to develop strategies that will make them better products. Meat, milk, and eggs are fine products, as discussed and stressed in the opening of this chapter. They can be made even better by modifying their fat, saturated and unsaturated fatty acid, and cholesterol compositions. How can we alter the lipid content of meat, milk, and eggs to overcome their adverse effects, in most people, of excesses of total fat, saturated fat, palmitic acid, and cholesterol? Methods for modifying the composition of animal products are discussed later in this chapter.

# MEAT, DIETARY FAT, AND CANCER

Cancer is the second most common cause of death in the United States. (CHD is the number one killer.) The incidence of many cancers increases with age, so that as the American population becomes increasingly older, cancer cases will increase, as will other age-related pathologies such as Alzheimer's disease. Although it is often claimed in the popular press and on TV that there is an epidemic of cancer in the United States, in fact the rate of many cancers, adjusted for age, is actually decreasing. Ironically, the major cancer type that is increasing in incidence is lung cancer, attributable largely to smoking. This is ironic because it is a risk factor that an individual can easily eliminate, whereas other perceived risk factors, such as pesticide residues on food, are in fact not significant risk factors at all. However, many people seem to be more concerned about them than things like smoking, which is under their personal control (but which also require effort and will power to change, whereas it takes little personal effort to complain about agricultural chemicals).

**Cancer** is a disease caused by unrestrained growth of populations of cells. Cancer cells have three main properties: (1) they have a faster rate of growth than normal cells, which is not restrained by normal controls on cell growth and division; (2) they invade neighboring tissues; and (3) they metastasize (spread) to nonadjacent tissues in other parts of the body. The initiation of cancer is the induction of cellular mutations by outside agents. This involves damage to cellular DNA. There are various DNA repair mechanisms that repair most damage, but occasionally a damaged cell will reproduce itself and propagate the mutation. This process is known as **tumor initiation.** The actual cancer may develop many years later and requires the action of a **tumor promoter.** A promoter is a chemical that doesn't initiate cancer but stimulates the growth of the mutated cells. Most carcinogens (cancer-causing substances) are both initiators and promoters. Cellular mutation and potential tumor development occur in individuals of all ages. In younger individuals, the immune system is usually capable of responding to and destroying such mutated cells, thus preventing the tumor development. As we age, however, our immune systems degenerate. Therefore, the older we get, the less able we are to destroy the cancerous cells as they develop and the greater the risk of cancer.

There is some evidence that the amount and type of dietary fat can promote the development of certain types of cancers. Saturated fatty acids and animal fats have been implicated in some cases, so will be discussed briefly here. There are reports, summarized by NRC (1989) suggesting a correlation of colorectal cancer with a dietary pattern of high fat intake, particularly of saturated fats, and a low vegetable intake. The association may be due to the fat intake, or alternatively, to a low dietary fiber level. Fiber has been hypothesized to have protective effects against colorectal cancer. A high fat, low fiber diet is typical of the United States and other industrialized countries, which have much higher colon cancer rates than do countries in which intake of animal protein sources is low. Breast cancer and prostate cancer show the same trend, with a high incidence associated with diets high in fat and animal protein. These three types, colon, breast and prostate cancer, are the major cancers that seem to be associated with high fat, low fiber diets. It is not currently known if the fat is specifically involved or if the critical factors are related to a low intake of vegetables. Plants are sources of fiber and β-carotene, and other substances, such as indole-3-carbinol and sulforaphane in vegetables of the cabbage family, which may

have **anticancer activity** (Zhang et al., 1992). An apparent inconsistency is that while there is a strong positive correlation of cancer incidence with meat consumption, there is no evidence that vegetarian diets reduce cancer risk (Bingham, 1999; Hill, 1999).

Both saturated and unsaturated fats have been implicated in carcinogenesis. Unsaturated fatty acids in vegetable oils have been shown to be promoters of certain types of cancer (NRC, 1989). This involves specifically the ω-6 series of PUFA; the ω-3 fatty acids in fish oils are not tumor promoting. In the case of colon cancer, dietary fats may have promoter activity by enhancing the concentrations of secondary bile acids in the gut. **Secondary bile acids** are produced by microbial metabolism of the primary bile acids that are secreted via the bile. High fat diets tend to increase bile secretion. Esterified secondary bile acids are less toxic than the free bile acids. People on high fiber diets and vegetarians have a higher percentage of esterified bile acids in the colon than do omnivores (Adlercreutz, 1998). Diets high in fiber produce high fecal bulk and increase speed of gut passage (laxative effect), both of which would tend to dilute the concentration of bile acids in the colon contents. Numerous studies have shown that the concentration of mutagens (potential carcinogens) in the colon is higher in meat-eaters than in vegetarians (NRC, 1989). Some of these may have their origin in the cooking process. In particular, smoked meat (e.g., in BBQ) may contain benzopyrene and other **mutagens in wood smoke.**

Consumption of red meat has shown a higher association with incidence of **colon cancer** than intakes of poultry and fish, while all sources of fiber (vegetables, fruits, and grains) were associated with the lowest risk of colon cancer in studies involving 7,284 men (Giovannucci et al., 1992) and 88,751 women (Willett et al., 1990). These authors concluded that diets high in red meat and animal fat and low in fiber increased the risk of colorectal cancer. The two major factors involved in colorectal cancer are inheritance and diet

(Winawer and Shike, 1992). The development of colon cancer proceeds through mucosal cell proliferation, adenoma formation, adenoma growth, and transformation to malignant tumors. Adenomas begin as small benign polyps, which predate the clinical diagnosis of colorectal cancer by about 10 years (Giovannucci et al., 1992). Dietary factors may have an effect on the earliest stage of mucosal cell proliferation, as well as on adenomas and tumor development. The carcinogenic effect of dietary factors may depend on an inherited disposition that makes the colonic mucosal cells susceptible to the carcinogenic effect of the diet (Winawer and Shike, 1992).

Another possible mechanism for a stimulatory effect of fat on colon cancer involves formation of **free radicals** in the gut. Erhardt et al. (1997) compared the production of reactive oxygen species (e.g., superoxide, hydroxyl radicals) in the feces of individuals consuming either a vegetarian diet or a diet rich in fat (50%) and meat and low in fiber. The reactive hydroxy free radical concentration was 13 times higher in subjects consuming the high fat diet than when the same individuals consumed a vegetarian diet. There was a 42 percent higher fecal iron concentration with the meat diet, which may be significant because iron is a very potent oxidant. It is postulated that in the oxygen-containing area near the colon mucosa, bacteria release superoxides, which are converted to highly reactive hydroxyl radicals through the catalytic action of $Fe^{2+}$. These free radicals may convert other substances to mutagens. A diet low in fiber and high in meat and fat may promote the formation of free radicals and the secondary bile acids that solubilize peroxidizable fatty acids. Phytates in fiber-rich plant materials may inhibit this process by binding iron, which promotes peroxidation.

There are several other possible mechanisms of action by which meat could affect colon cancer, besides the effects of saturated fat. A heavily browned meat surface increases the risk of colorectal cancer (Voskuil et al., 1997) by the formation of potentially carcinogenic **hete-** **rocyclic amines (HCA).** The HCAs are metabolized by **acetylator enzymes** (Minchin et al., 1993). There are genetic differences among humans in acetylator activity (Minchin et al., 1993). Mutations in Ras oncogenes are involved in colorectal cancer (Bos, 1989). Ras oncogene mutations are closely linked to a high activity of acetylator enzyme activity (Oda et al., 1994). Thus there is a complex interaction between genetic disposition to colorectal cancer, specific gene mutations, and meat consumption (Voskuil et al., 1997). Obviously, this is a complicated situation but one which tends to support an involvement of meat consumption with occurrence of colon cancer. The production of HCA during the cooking of meat is greatly influenced by cooking time (Knize et al., 1994). Microwave cooking reduces HCA formation (Felton et al., 1994). Consumption of well-done meat, and consequently, exposure to HCA, increases risk of breast cancer (Zheng et al., 1998). The nutritional and toxicological significance of HCA in cooked meat has been reviewed by Stavric (1994) and Skog et al. (1998).

Prolonged cooking of meat at high temperatures (e.g., roasting) and browning of meat may also lead to some interactions between sugars, proteins and fats, producing substances called **advanced glycation end products (AGEs).** These compounds produce inflammatory responses in blood vessels, increasing risk of heart disease.

As discussed in the section on dairy products, another proposed explanation for an effect of meat on colon cancer risk is related to its high phosphorus but very low calcium contents. **Calcium phosphate** may precipitate bile acids and fatty acids in the colon, reducing their oxidative damage to colon epithelial cells and thus reducing epithelial hyperproliferation (Van der Meer et al., 1997).

Another effect of meat fat on cancer may involve **protein kinase C (PKC)** activation (Pajari et al., 1998). PKC is a family of intracellular enzymes that mediate several hormonal, mitogenic, and tumor-promoting signals in cells. A sustained activation or inhibition of PKC has been proposed to play an important role in the causation of colon cancer (Pajari et al., 1998). Meat contains several bioactive lipids such as glycerophospholipids, sphingolipids, and other cell membrane components, which may escape digestion and reach the colon. These compounds may modify signal transduction processes by activating PKC.

Another proposed mechanism of action relating meat consumption to colon cancer involves formation of **nitrosamines** in the colon (Bingham, 1997). High levels of meat in the diet increase the formation of ammonia and nitrosamines in the colon. Some of the chromosomal mutations found in colorectal cancer, such as the Ras oncogene mentioned earlier, are consistent with nitrosamine-induced effects. Nitrosamines are formed by bacterial metabolism of amino acids. Bingham (1997) in an evaluation of published studies on meat consumption and colon cancer observed that red meat and processed meat tended to increase colon cancer risk, whereas white meat (chicken) was associated with no effect or a reduction in risk. He observed that red meat consumption increased fecal nitrosamines, but white meat did not, indicating a specific difference that might be linked to the higher iron content of red meat and a role of iron as a catalyst in enzymes involved in nitrosamine formation.

Bingham (1997) reviewed data showing a striking increase in colon cancer rates in Japan, over the time in which Japanese diets have become increasingly westernized and meat consumption has increased. Interestingly, Japanese people have a higher proportion of individuals with high acetylator activity; as discussed earlier, high meat-eating fast-acetylator phenotypes are at an increased risk of colon cancer (Roberts-Thompson et al., 1996). The acetylator enzyme, N-acetyltransferase, catalyzes the formation of mutagenic products from heterocyclic amines, via hydroxylation and acetylation (addition of an acetyl group) in the liver and colon (Roberts-Thompson et al., 1996).

It is very difficult in the studies of colon cancer in the human population, such as those cited, to differentiate between the effects of meat and fat consumption and fiber intake. Is the observed adverse effect due to the high fat or the low fiber intake; conversely, is the lower incidence with a high fiber diet due to protective effects of fiber and other plant components or to a lower fat intake? Regardless, rates of colorectal cancer in various countries are strongly correlated with per capita consumption of red meat and animal fat and inversely associated with fiber consumption (Giovannucci et al., 1992). Even the most dedicated animal scientist or meat supporter must be somewhat dismayed by the preponderance of evidence suggesting a role of meat consumption in the etiology of colon cancer.

The incidence of **hormone-dependent cancers** such as breast, ovary, endometrial, and prostate cancers is lower in people consuming low fat, high fiber vegetarian or semivegetarian diets than in those consuming the typical Western diet. Adlercreutz (1990) has proposed that the protective effects of high fiber diets may be partially due to the presence of compounds called lignans (not the same as lignin) in many plant foods. **Lignans** are phenolic compounds that are estrogenic (phytoestrogens) or may be converted to phytoestrogens by microbes in the gut. A Western-type diet elevates plasma levels of sex hormones and reduces serum steroid-binding globulin, increasing the bioavailability of the steroid sex hormones. Adlercreutz (1990) reviewed evidence that lignans in grains and pulses (peas and beans) are modified by gut bacteria to estrogenlike compounds. These compounds stimulate synthesis in the liver of steroid hormone binding globulin, reducing levels of bioavailable sex hormones in the blood. They may also interfere with the uptake of estrogens at the tissue level by binding with estrogen receptors. Thus plant food-based diets may have protective effects against hormone-dependent cancers by reducing the uptake of estrogens and other steroid hormones by ovarian, breast, and prostate tissue. This hypothesis, if valid, does not incriminate meat as a promoter of cancer but does support the widely held view and NRC recommendation that the Western diet could be improved by increasing the intake of whole grains, beans, and other plant-based foods.

A high intake of animal fat or red meat is statistically linked to an increased risk for **prostate cancer** (Giovannucci, 1995). Milk and other dairy product consumption is also correlated with prostate cancer (Giovannucci et al., 1998). The rationale behind this effect is somewhat complicated. Calcitriol, or 1,25 (OH)$_2$ vitamin D$_3$, inhibits cellular proliferation of prostate tumor cells. High calcium intakes depress production of calcitriol. Hence higher calcium intakes associated with dairy products could enhance tumor proliferation.

Diets high in fruits and vegetables lower the risk of prostate cancer. The red carotenoid pigment of tomatoes, **lycopene,** has been specifically implicated as having protective effects against prostate cancer (Giovannucci, 1999). The lycopene in cooked tomato products (e.g., ketchup, tomato paste) is more readily absorbed than from raw tomatoes.

Prostate cancer rates are low in most Asian countries, where consumption of soybean products is high (Habito et al., 2000). Prostate disease is dependent on sex hormones. In the prostate gland, testosterone is converted to dihydrotestosterone (DHT), which stimulates growth of prostate tissue. Soybeans contain estrogenic isoflavones that inhibit the growth of prostate tumor cells. In a study in which meat in the diet was replaced with tofu (Habito et al., 2000), changes in serum biologically active sex hormones in men were noted.

Meat consumption may be linked with cancer by way of association. Fast food meat products are usually consumed with french fried potatoes. Cooking french fries at high temperature causes a reaction between the amino acid asparagine and sugars, producing a substance called **acrylamide.** Acrylamide has been shown to be carcinogenic in laboratory animals.

It should be recognized that the incidence of many types of cancers is much lower in industrialized countries than in developing countries where the per capita intake of animal products is much lower and plant products much higher. For example, liver and stomach cancer rates are low and declining in North America and Europe and high in most other countries. The United States has one of the lowest rates of stomach cancer in the world, whereas in 1930 stomach cancer was the leading cause of death for U.S. men (NRC, 1989). High intakes of salted or smoked fish and pickled vegetables are correlated with high stomach cancer rates in Japan and many other countries. These foods are high in salt, nitrates, and nitrites. Nitrate or nitrite in salt-preserved foods may be converted to **nitrosamines,** which are potent carcinogens. Numerous plants that are eaten in Japan and other Asian countries are carcinogenic, including the bracken fern, which is a widely consumed delicacy in the Orient. Cycads that are used as foods in some countries also contain a potent carcinogen.

Liver cancer is rare in the United States but rates are high in Africa and Southeast Asia. This high incidence seems to be related to exposure to hepatitis B virus infection in early life, and the consumption of foods contaminated with aflatoxin. **Aflatoxin** is the most potent carcinogen known. It is produced by the mold *Aspergillus flavus* that grows on grains and other foods under warm, humid conditions. The mycotoxin **fumonisin,** produced by *Fusarium fungi* growing on maize, has been linked to a high incidence of human esophageal cancer in South Africa (Rheeder et al., 1992).

In summary, cancer incidence is definitely affected by diet. The incidence of some types of cancer is high in Western countries where diets high in fat and low in fiber are consumed. There is little direct evidence implicating meat consumption per se; the association is probably more a reflection of high fat intake regardless of source, and a low intake of plant-origin foods. On the other hand, the incidence of several other types of cancer is very low in Western countries and much higher in developing countries that have a low meat, high plant-origin food diet.

# OBESITY AND CONSUMPTION OF ANIMAL PRODUCTS

**Obesity** is basically caused by the consumption of excess energy (calories) from carbohydrates, proteins, and lipids. For almost the entire evolutionary history of humans, the principal preoccupation of our ancestors was to get enough to eat. It is only recently (and only in the so-called developed countries) that we have become exposed to unlimited food. Our ancestral instincts (sweet tooth, attraction to fatty foods with a high energy content consumed in a more active lifestyle) served us well until now. Now, the major nutritional preoccupation of many people is "counting calories" to try to limit intake to control obesity. Overconsumption has become our major nutritional disease. All other nutritional **diseases of Western civilization**—diabetes, CHD, some types of cancer, obesity—have overconsumption as their primary cause. In addition, food processing has increased some of the tendency to overconsume. On an evolutionary basis, a sweet tooth had survival value, attracting humans and other animals (except strict carnivores) to digestible low fiber plant parts highest in sugars and starch, such as ripe fruit. One would have difficulty in becoming obese when attracted to apples by a sweet tooth. The availability of refined table sugar completely overrides the survival value of a sweet tooth; never before in our evolutionary history could one have become obese because of an attraction to sweet taste. Similarly, we (and other animals) find energy-rich fatty foods highly palatable. The availability of table sugar and refined fats and oils has allowed for a tremendously unnatural consumption of concentrated sources of energy such as refined sugars and oils, both directly (sugar added to coffee and as a top-dressing of desserts, etc., oils used as salad dressing, butter and other spreads on bread and baked potatoes, etc.) and "hidden" in cookies and pastries, French fries, coffee creamer, and so on.

Thus the blame for obesity can't be placed on any particular food but rather on a general tendency to eat energy-rich foods because they best satisfy our basic instincts (Taubes, 2001). It takes real willpower to bypass these treats and dine on carrot sticks, boiled cabbage, low fat, stringy, tough meat from a three year old grass-fed beef steer, and so on. However, obesity does not seem to characterize "meat eaters." Reiser and Shorland (1990) noted, "It is a common observation that obesity is more prevalent among the less advantaged, high carbohydrate subcultures in the United States than among the affluent high meat consuming groups" (p. 25). The stereotypical obese Eastern European peasant, as another example, consumes a low meat diet based on carbohydrate sources such as bread and potatoes. Prior to the development of agriculture 10,000 years ago, meat-eating hunter-gatherers were taller and leaner than their agriculturally inclined descendents (Cordain, 1999).

It is a common observation in cancer studies with laboratory animals that tumor incidence is related to food intake. **Caloric restriction** of animals markedly reduces tumorigenesis and the induction of cancer by known carcinogens. Control of appetite and avoidance of tobacco are two seemingly simple behaviors that would probably have more effect on reducing the incidence of "diseases of Western civilization" than any other lifestyle modification.

Shorland (1988) uses the term **nutritional distortion** to indicate the addition to the diet of a component that has been selectively removed from plant or animal tissues, such as fats and oils or sugar, or the incorporation of the residue (e.g., white flour) after removal of a component (e.g., bran). The diseases of Western civilization, including obesity, are largely due to caloric concentration by **food processing and refining.** Table sugar and vegetable oils are two obvious examples. Refined flour is energy-concentrated, because the fiber (bran) has been removed. Especially in the case of vegetable oils, this processing

# MEAT CONSUMPTION, INSULIN RESISTANCE, AND DIABETES

*(Review of J. C. Brand-Miller and S. Colagiuri. 1999. "Evolutionary Aspects of Diet and Insulin Resistance." World Rev. Nutr. and Diet. 84:74–105.)*

**Insulin resistance** is when greater than normal insulin levels are required to elicit a normal glucose response or where physiological concentrations of insulin produce a less than normal response. The long-term result is increased risk of non-insulin-dependent diabetes mellitus and coronary heart disease. Insulin resistance occurs in epidemic proportions in some populations such as Australian aborigines and Pima Indians[1] of Arizona. Pima Indians in Mexico, who have retained their traditional diet, have a much lower incidence of diabetes and obesity than their relatives in Arizona, who have adopted the U.S. diet and lifestyle. It is proposed that insulin resistance had survival value for aboriginal peoples consuming a high meat diet. The **"thrifty genotype"** hypothesis proposes that cycles of feast and famine during human evolution selected for a genotype that promotes excessive weight gain during times of food abundance. The selective force for insulin resistance was the low-carbohydrate, high meat diet during the last 2 million years of human evolution. Certain ethnic groups were especially dependent on a high meat diet. The Mongoloid ancestors of Native Americans occupied the Siberian mammoth steppe during the final 20,000 years before the last Ice Age. Their diets were based solely on animal foods with almost no carbohydrate. The ancestors of the Australian aborigines who arrived in Australia 50,000 years ago had a hunting and shellfish-gathering life, with plant foods having a minor role. The Eskimos (Inuit) of the Arctic regions had almost no carbohydrate in their diet. Thus the ancestors of some of the world's present aboriginal peoples were primarily carnivores.

Carnivores have some unique metabolic characteristics. In this paper, the metabolic similarities of human carnivores with feline carnivores (cats) were outlined. In comparing the metabolic and biochemical dietary adaptations of cats (obligate carnivores) with those of humans (omnivores), it became apparent that evolution has shaped both feline and hominid metabolic machinery towards an animal-based diet. Some examples are the following:

1. Cats can neither synthesize vitamin $B_{12}$ nor absorb bacterially produced $B_{12}$ and are wholly dependent on animal flesh as their $B_{12}$ source. The same is true for humans. The human nutritional requirement for vitamin $B_{12}$ clearly indicates that a vegetarian diet did not shape the human genome.

2. Taurine is an amino acid not found in plants but is an essential nutrient for all mammalian cells. Cats have lost the ability to synthesize taurine. Since animal flesh is rich in taurine, selective pressure for taurine biosynthesis was relaxed in cats. Humans have a low ability to synthesize taurine. Vegan vegetarians have very low tissue taurine levels.

3. Plants are largely devoid of the $C_{20}$ and $C_{22}$ fatty acids that mammals require. Herbivores have evolved hepatic enzymes (desaturases and elongases) that allow $C_{18}$ plant-based fatty acids to be chain elongated and desaturated to the $C_{20}$ and $C_{22}$ products. Both cats and humans have very low levels of these enzymes, because their need to metabolize $C_{18}$ fatty acids has been reduced by diets high in flesh.

4. Vitamin A does not occur in plants; herbivores convert β-carotene to vitamin A. Cats have lost their ability to synthesize vitamin A from carotene and obtain their vitamin A needs from the organs (liver, kidney) of their prey. Humans also have a poor ability to convert β-carotene to vitamin A.

5. Human infants are susceptible to iron deficiency. In the past, a predominantly meat-based weaning diet would have relaxed the need for large iron stores. Iron in meat has a much higher bioavailability than plant iron.

Aboriginal peoples in many parts of the world have recently undergone marked lifestyle changes, including a change from traditional meat-based diets to processed foods high in fat and sugar. Traditional meat diets were often based on lean meat (except in the case of Eskimos). The genetically based insulin resistance of these people, arising from their traditional lifestyles, makes them extremely susceptible to disorders of carbohydrate metabolism, including obesity and diabetes. Educational intervention with dietary counseling would be an appropriate course of action for affected populations. The introduction of commercial dry foods for cats, containing large amounts of carbohydrate, will likely lead to similar metabolic problems for this carnivorous species.

An "insulin-index" for foods has been developed. Highly refined or concentrated carbohydrate sources (bread, potato, bakery products) have the highest insulin index, while protein foods and nonrefined carbohydrate sources have the lowest indices. The insulin index is described in detail by Holt et al., (1997).

---

[1]The terms "Indians" and "Native Americans" are sometimes contentious in the "political correctness" sense. Both are used here, with no intention of insensitivity. The use of "Native Americans" is not without critics; in Canada, it raises the specter of American imperialism (Native Canadians?). In fact, in Canada, the term "First Nations" peoples is used. Perhaps the term "Indigenous Peoples" should be used. I mention all this only because I have received student criticism for using the term "Indian" in class, with the (invalid) implication that it is a derogatory word.

## THE OBESITY BUG

(Review of Dhurandhar et al., 2000. "Increased Adiposity in Animals Due to a Human Virus." Int. J. Obesity 24:989–996; Holmes, B. 2000. "The Obesity Bug." New Scientist, Aug. 5, pp. 26–30.)

A medical doctor in India, Nikhil Dhurandhar, had a chance meeting with a family friend, who was a veterinary pathologist. The pathologist casually mentioned that he was studying a viral epidemic that was sweeping through Indian poultry flocks, killing hundreds of thousands of chickens. He was amazed that the dead birds had very large amounts of body fat. Dhurandhar was intrigued, because normally one would expect a sick bird to waste away, rather than get fat. They isolated the virus, SMAM-1, and injected it into healthy chickens. Six weeks later, the injected chickens had 50 percent more body fat than controls, and lower blood cholesterol. Dhurandhar moved to the United States, in an attempt to study the situation in more detail. He was not allowed to import the SMAM-1 virus, so instead he and his colleagues used a human adenovirus (Ad-36) available in the United States. Three experiments with chickens and one with mice were conducted. In agreement with the results from India, the viral-infected animals had increased body fat and low serum cholesterol. Blood samples from 313 obese people and 92 lean people were obtained. Just four of the lean people (5%) had antibodies to Ad-36, while 100 (32%) of the obese people had Ad-36 antibodies. Thus the intriguing question arises—could some human obesity be the result of a subclinical viral infection? Hence, "the obesity bug."

This research was controversial; the authors experienced considerable difficulty in getting it published in a peer-reviewed journal. This is often the case with research that doesn't mesh with "establishment expectations." The idea that a person could "catch" obesity is difficult for most medical authorities to accept.

may have other implications. Oils in plant tissues occur with natural antioxidants such as tocopherol (vitamin E) and certain phenolic compounds. Shorland (1988) expresses the concern, "The indications are that polyunsaturated oils taken out of context of the tissues in which they reside may have adverse nutritional effects" (p. 193). Perhaps the loading of ruminants with fat (marbling) by feedlot feeding on concentrates can be viewed as nutritional distortion. The ruminant evolved as a forage eater, not as a seed eater. As discussed earlier, the wild ruminant has a much higher content of $\omega$-3 PUFA in its body fat than the feedlot fed animal, which has been fed an unnatural diet to load its intramuscular tissue with triglycerides.

# MODIFICATION OF ANIMAL PRODUCTS TO IMPROVE HUMAN HEALTH

The main components of animal origin foods often linked to adverse effects on human health are total fat, saturated fatty acids, and cholesterol. Can meat, milk, and eggs be produced and/or processed to minimize or eliminate these possibly undesirable properties? Besides the health implications of meat fat, there are major economic losses. As Bergen and Merkel (1991) state, "The animal production industry (in the U.S.) spends about $2 billion yearly to deposit fat on animals and another $2 billion yearly to remove the fat from meat products" (p. 2953).

Upon reflection, it is apparent that a great many improvements have already been made. The level of coronary heart disease has been declining for many decades in the United States, while meat consumption has risen. In the first few decades of the twentieth century, the major sources of fat in the American diet were lard and butter. Lard-type pigs, depositing massive amounts of body fat, were the basis of the swine industry. Through genetic selection, the pig has been changed from the obese, lard-type pig to the very lean, meaty animal of today (Fig. 2-9). Similarly, the composition and conformation of beef cattle have been markedly altered by genetics. The ideal beef steer of 1950 was a blocky, overly fat animal and the ideal cut of beef was loaded with fat. Thus a tremendous change in the fat content of meat animals, in a positive direction, has already occurred. The swine industry did not collapse with the replacement of lard by margarine and vegetable shortening; the dairy industry didn't collapse when margarine replaced butter. Much further improvement in lowering the fat content of meat animals is possible through selection and breeding. The main hindrances are political, economic, and social. Meat grading standards have lagged far behind consumer desires. Producers are usu-

**FIGURE 2-9** The livestock industry has made great progress in reducing the fat content of meat animals. For example, hogs are leaner. (*Top*): 1940s hog. (*Bottom*): 1990s hog. (Courtesy of National Live Stock and Meat Board.)

ally paid on a live-weight basis rather than on a lean meat production basis so have little economic incentive to change. There is the striking anomaly of dairy scientists referring to low milk fat as a pathology or metabolic disease (milk fat depression), while the excess fat produced by the dairy industry winds up as billions of pounds of surplus butter stored by the U.S. government. Consumers want low fat milk, while the milk producer is penalized by receiving a lower price for low fat milk. Nevertheless, milk has retained a good share of the market place through the development of modified products. Low fat and fat-free milk, flavored milk, long shelf-life unrefrigerated milk, and low cholesterol milk are all either realities in the market or easily could be. Food scientists have developed processing procedures to remove cholesterol from milk. Milk products with a variety of flavors are or could be competitive with soft drinks. A major improvement in health, especially of girls and women, could be made by an increase in milk and a decrease in soft drink consumption. The dairy industry has the products needed to have a major favorable influence on osteoporosis. A new trend in the dairy industry is component pricing, whereby dairy producers receive payment on the basis of milk protein or solids nonfat. Multiple component pricing involves payment for butterfat, protein, and other solids per pound of each component, providing an incentive for producers to market milk with higher protein and less fat.

Genetic selection for leanness has been effective in lowering carcass fat content of meat animals but sometimes with a price. The swine industry has had substantial problems with **porcine stress syndrome** (PSS), a recessive genetic trait. Pigs with PSS are susceptible to malignant hyperthermia (elevated body temperature) and heart failure if exposed to stress (sorting, transportation, high environmental temperature, and general anesthesia). Pigs expressing the PSS gene are heavily muscled, lean, easily excited, and have a high incidence of pale, soft, exudative (PSE) meat. PSE pork is difficult to process and unappealing to consumers. Selection of pigs for lean carcasses has resulted in an increased incidence of the PSS gene and increasing problems with PSE pork. The heterozygous animals (carriers) may not exhibit PSE but are susceptible to stress, causing increased handling and transportation losses. The PSS/PSE gene can be eliminated by identification of susceptible animals and avoiding their use in breeding programs. An older method of identifying PSS individuals is the halothane test, which measures the pig's tolerance to halothane gas, a general anesthetic. Newer methods involve DNA testing, which can identify carriers as well as PSS homozygous individuals. Use of DNA probes for PSS should lead to the development of lean strains of PSS-free pigs.

A number of genetic variants of livestock are especially high in muscle content and/or low in fat. There are several European breeds of cattle (e.g., Belgian Blue, Piedmontese) with a gene that causes muscle hypertrophy (double muscling). This increases yield of lean tissue but results in a narrow pelvis that causes calving difficulty. Most of the calves must be delivered by caesarean operation. Thus double-muscled cattle probably have little commercial potential, although Wheeler et al. (2001) suggest that breeds such as the Piedmontese could be used as terminal sires to produce progeny with improved tenderness. The double-muscling gene codes for inactivated myostatin (Wheeler et al., 2001), while in the Belgian Blue breed there has been selection for other genes contributing to muscling independent of inactive myostatin. The double-muscled trait in cattle has been reviewed by Arthur (1995). A novel form of muscle hypertrophy in sheep is associated with extensive muscling of the loin and hindquarters (Carpenter et al., 1996) and has been given the descriptive name **callipyge** (calli, beautiful; pyge, buttocks). Sheep with the callipyge gene have an increased size of the most valuable muscles of the loin and leg, leaner carcasses, and increased feed efficiency (Fig. 2-10).

Australian scientists have identified genes associated with beef tenderness (*Feedstuffs*, Aug. 27, 2001, p. 3). A practical method of identifying "tender cattle" was developed. An animal with a fast flight time has a poor temperament, producing progeny with tough beef. Flight time is measured as the time taken by an animal to travel 2 m after it is released from a scale. Sires can be selected for slow flight time early in life, to improve the eating quality of their progeny's meat.

The most successful strategy to reduce the fat content of meat has been the use of large, later-maturing animals (Bergen and Merkel, 1991). In early-maturing animals, deposition of protein as muscle tissue (meat) plateaus at a lower body weight than for later-maturing, larger animals, with the result that much of the gain in the finishing period is body fat. Later-maturing animals are slaughtered at a younger physiological age, producing meat with a higher proportion of protein and lower amount of fat.

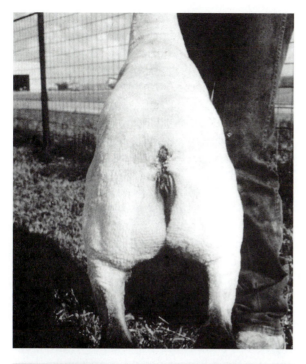

**FIGURE 2-10** A callipyge lamb *(top)*. These animals have enlarged loin muscles *(bottom left)* as compared with normal lambs *(bottom right)*. (Courtesy of Noelle Cockett.)

The fat content of meat animals can be readily controlled by diet. Body fat is the end product of the storage of excess energy consumed, in both livestock and humans. Body fat content is reduced if protein, amino acids, and other nutrients are balanced with respect to the energy content of the diet. Thus a rapidly growing animal can be kept lean by properly balancing the protein and energy levels of the diet. As mature size is approached, growth of lean tissue decreases and the extra energy intake is deposited as storage fat. The inevitable result of "finishing" animals is to increase the carcass fat content.

The simplest method to produce lean animals is to limit-feed (Bergen and Merkel, 1991), to ensure that there is not excess energy to be deposited as fat. Bergen and Merkel (1991) do not consider this to be practical under North American conditions, because it has traditionally been very labor intensive. However, with automated feeding systems, limit-feeding should be quite easy to accomplish.

A common justification for the finishing period is that fat is necessary to give meat its tenderness and flavor. This is particularly true for intramuscular fat (marbling) in beef. Research to find methods of maintaining these qualities while decreasing the fat content should be a priority. According to Bergen and Merkel (1991), levels above 3 percent intramuscular fat (marbling) have little additional effect on meat palatability.

Another approach to reducing fat content of meat is the use of substances that alter metabolism to reduce fat deposition. As long ago as the early 1960s, Canadian scientists recognized that body composition could be altered to produce less fat and more lean tissue by using dietary additives to promote mobilization of body fat (Cunningham and Friend, 1964). In the 1980s, the availability of biotechnology-produced growth hormone, as well as the development of compounds called β-**agonists,** stimulated great interest in the **repartitioning of nutrients** to promote protein and reduce fat accretion. Examples of β-agonists include cimaterol, clenbuterol, and ractopamine. Ractopamine is the only one cleared for commercial use by the FDA. The β-agonists stimulate protein synthesis, reduce protein degradation, enhance mobilization of body fat, and reduce the rate of fat synthesis. Growth hormone, which is produced by the pituitary gland, increases formation of lean tissue. **Growth hormone** (also known as somatotropin) is now being produced synthetically using biotechnology techniques. The potential commercial administration to livestock of growth hormone and repartitioning agents (β-agonists) will likely be limited by negative public perception of "chemicals, hormones, drugs, and additives." These substances are discussed in more detail in Chapter 5.

Passive immunization of animals with adipocyte plasma membranes to produce cytotoxic antibodies has been proposed as a means of reducing fat uptake by adipose tissue (Nassar and Hu, 1991). Cytotoxic antibodies cause destruction of fat cell membranes, reducing the number of fat cells and hence lowering the potential of tissues to store fat. Pigs treated in this manner have shown fat reductions of 25–30 percent (Moloney, 1995). Similar effects have been seen in sheep (Moloney et al., 1998). If economically successful, this approach could be useful because it does not involve administration of feed additives or growth promotants that elicit consumer concern.

A more difficult problem than fat content is the saturated fatty acid content of ruminant fat, because of the hydrogenation of unsaturated fats in the rumen. Australian researchers have developed **rumen protected fat.** Small droplets of vegetable oil can be microencapsulated with a protein such as casein, which is then chemically treated (e.g., with tannin or formaldehyde) to make it resistant to digestion in the rumen. Thus the fat is protected against digestion in the rumen, but the treated protein is digested in the intestine, allowing the unsaturated fatty acids to be absorbed and deposited in the adipose tissue and milk fat (Ashes et al., 1992). To date, the process has been too expensive for routine use. Another means of protecting unsaturated fatty acids from rumen biohydrogenation is by the addition of an amino acid to the carboxyl group, forming an acyl amide. The presence of a free carboxyl group is necessary for hydrogenation of unsaturated fatty acids in the rumen. Fotouhi and Jenkins (1992) reported that linoleoyl methionine (produced by reacting the amino group of methionine with the carboxyl of linoleic acid) when administered orally to sheep increased the linoleic acid content of the body fat. If such acyl amides can be economically produced, they could have potential as feed additives to increase the unsaturated fatty acid content of ruminant body and milk fat.

Development of methods to increase the unsaturated fatty acid content of ruminant fat has a down side. Unsaturated fats are suscep-tible to oxidation. Lipid oxidation in meat is associated with rancid flavor, objectionable odors, and increased drip loss, all of which reduce the palatability and nutritional qualities of the meat. Dietary supplementation of meat animals with higher than conventional levels of vitamin E (alpha-tocopherol) increases the **vitamin E** content of the muscle tissue, where it is incorporated into the cellular membranes. Vitamin E is an antioxidant which prevents free radical-induced oxidation of unsaturated fatty acids. Thus vitamin E supplementation of animals in the finisher phase can improve meat quality (Buckley et al., 1995) and extend **shelf life** (Liu et al., 1995). In beef, extra vitamin E protects the red pigment of red meat, myoglobin, from oxidation and discoloration. Thus beef from vitamin E-treated animals has a longer shelf life, retaining its red coloration in packages in the meat counter for a longer period than regular beef (Liu et al., 1995). Improved supermarket appearance of beef, as well as the higher nutritional value from vitamin E treatment, offers an opportunity for the beef industry to improve consumer acceptance of its products.

Because neither stearic nor oleic acids increase serum cholesterol, it would be desirable to modify the fatty acid composition of beef fat by increasing the conversion of palmitic acid, which is hypercholesterolemic, to stearic and oleic acids:

$$\text{palmitic (16:0)} \xrightarrow{\text{elongation}} \text{stearic (18:0)} \xrightarrow{\text{desaturase}} \text{oleic (18:1)}$$

St. John et al. (1991) studied elongase and desaturase activity of bovine adipose tissue. The desaturation of stearic acid was the limiting step in the conversion of palmitate to oleate. They suggested that selection of animals for increased desaturase activity is possible. Thus, there appear to be several potential means of altering the body fat of ruminants.

The rumen is an adaptation allowing the animal to digest cellulose and subsist on roughages. It serves these processes well. It is not an efficient type of gut for utilization of concentrate feeds, as will be discussed in more detail in Chapter 3. Despite this fact, beef cattle in the United States are fed high concentrate diets in feedlots, for a variety of currently valid reasons. One consequence of this feeding system is a high energy intake and deposition of surplus energy as fat (finishing). A shorter feeding period on concentrates, with a longer period on forages, could result in leaner beef. Reducing the length of time that feedlot cattle are fed high concentrate diets to 112 days can produce palatable beef without excessive fat (Duckett et al., 1993). The Japanese beef breed, the Wagyu, has a higher ratio of monounsaturated (MUFA) to saturated (SFA) fatty acids than does beef from European-origin breeds (Boylston et al., 1995). Consumption of meat with higher levels of MUFA and lower levels of SFA would have a favorable effect on blood cholesterol levels and risk of coronary heart disease. If consumer demand for lean beef translated into higher prices paid to producers, a forage-based system could offer a means of producing meat with a lower fat content. Another alternative might be to use a high concentrate diet for rapid growth, followed by a finishing period on a high roughage diet (Griebenow et al., 1997).

There are a number of problems with forage-finished beef. Grass-fattened cattle tend to have a lower dressing percentage (because of greater gut size), higher shrinkage, and lower quality grade. **Grass-fed beef** is more variable in flavor and color than that from grain-fed animals (Schroeder et al., 1980). Forage-finished beef often has a darker muscle color that deteriorates rapidly during retail display, lowering appeal to consumers. Grass-fed beef is less tender and of lower palatability to humans. The off-flavor is associated with phospholipids in the structural fat and with flavor constituents produced by rumen fermentation of forage lipids (Melton, 1990). A high concentration of PUFA in the phospholipids contributes to the undesirable flavor of grass-fed beef (Larick and Turner, 1989) and may lead to oxidative rancidity and off-flavors. The flavor of the fat may be adversely affected by

products of autooxidation of PUFA and by the storage of odiferous (odor-causing) compounds in the fat when the meat is cooked (Larick et al., 1987; Larick and Turner, 1989). There are a large number of volatile flavor compounds in meat, most of which are derivatives of fatty acids (Melton, 1990). Increased perceptions of ammonia, gamey, liver-tasting and rotten flavors may be noted with forage-finished beef (Larick et al., 1987). These are associated with PUFA in the phospholipids and with diterpenoids produced by microbial fermentation of chlorophyll (phytol) in the rumen of forage-fed animals (Larick et al., 1987). The fat of forage-finished animals may be yellow, because of the deposition of carotenoid pigments from green forage (Fig. 2-11). **Yellow fat** reduces consumer appeal and may provoke the perception that the meat is from a sick animal. Thus, as with most issues, the situation of forage- vs. grain-finished beef is more complicated than it first appears.

A major disadvantage of grass-fed beef is its lack of tenderness, or more frankly, its toughness. Americans accustomed to grain-fed beef do not accept tough meat. There are methods with potential for improving the tenderness of grass-fed beef. A high calcium content of muscle tissue activates enzymes called **calpains**, which promote post-mortem tenderness (Swanek et al., 1999; Karges et al., 2001). Calcium absorption is promoted by vitamin D. Preslaughter treament of cattle with high calcium diets and vitamin D may offer a practical means of increasing tenderness (Swanek et al., 1999; Karges et al., 2001; Scanga et al., 2001). Development of this technique offers potential as a means of increasing the acceptability of forage-fed beef.

Interest in the nutritional and healthful properties of **fish** in the human diet is largely due to the ω-3 fatty acids. These are of two general types: those with 20 carbons and five double bonds (20:5 ω-3) and those with 22 carbons and six double bonds (22:6 ω-3). The 20:5 fatty acid (eicosapentaenoic acid, EPA) is the more important nutritionally, as a precursor of prostaglandins, prostacyclins, thromboxanes,

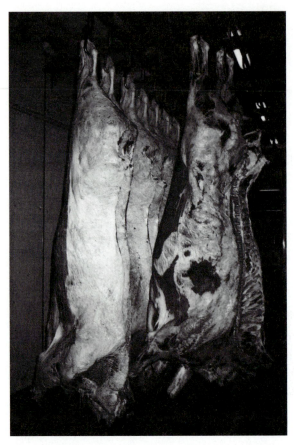

**FIGURE 2-11** Carcasses with yellow fat, from forage-fed animals *(right)* compared with normal carcasses *(left)*. (Courtesy of R. K. Tume.)

and leukotrienes. Meat, fish, fish oils, and eggs are the only significant sources of $\omega$-3$C_{20}$ PUFA for humans (Scollan et al., 2001). The **$\omega$-3 fatty acid content** of animal products can be increased by the use of fatty fish as a feed ingredient. Hulan et al. (1988, 1989) used redfish (*Sebastes* sp.) as a feed additive for broilers and enhanced the $\omega$-3 fatty acid content of the meat to values comparable to levels in fish. There were no fishy flavors or undesirable effects. For people who would rather eat chicken than fish, this feeding system might be useful. Some plant oils, such as soybean oil and linseed oil, also have appreciable quantities of $\omega$-3 fatty acids. Cunnane et al. (1990) increased the $\omega$-3 fatty acid content of pork by feeding 5 percent linseed (flax seed) to pigs. Ajuyah et al. (1991) and Chan-

mugam et al. (1992) observed a similar result with broilers. Flax seed has the highest $\omega$-3 fatty acid content of any vegetable oil. Thus, if consumer demand warranted, pork and poultry meat could be produced with an enhanced content of polyunsaturated and $\omega$-3 fatty acids (Leskanich and Noble, 1997). Rumen-protected $\omega$-3 PUFA could be fed to ruminants to increase tissue levels of $\omega$-3 fatty acids of meat and milk. Ruminant tissues can be enriched with $C_{20}$ $\omega$-3 PUFA because rumen microbes do not hydrogenate 20:5 $\omega$-3 and 22:6 $\omega$-3 fatty acids (Scollan et al., 2001). These authors found that feeding fish oil to beef steers doubled the proportions of these fatty acids in muscle phospholipids. The 18:1 trans fatty acid (trans-vaccenic acid) was also increased, but the trans fatty acids in ruminant fat, in contrast to those in margarine, are not a significant risk factor for cardiovascular disease (Scollan et al., 2001). Fish oil supplementation also increases the desirable CLA in beef fat (Enser et al., 1999).

Monounsaturated fatty acids such as oleic acid (18:1) have cholesterol-lowering properties but are less susceptible to autooxidation and development of rancid off-flavors than PUFA. Feeding of vegetable oils or full-fat oil seeds that are high in monounsaturated fatty acids but low in PUFA (e.g., high oleic acid peanuts, canola seed) may offer a means of increasing the unsaturated fatty acid content of pork or poultry fat without undesirable effects on meat quality.

**"Designer eggs"** are those enriched with nutrients or other biologically active compounds that have beneficial effects on human health. They are produced by feeding special dietary formulations to laying hens. Examples of designer eggs include those enriched in $\omega$-3 fatty acids (Farrell, 1998) and conjugated linoleic acid (Du et al., 1999). The benefits of enrichment in $\omega$-3 fatty acids have been reviewed by Simopoulos (2000). Western diets are deficient in these fatty acids relative to the diets on which humans evolved. Stadelman (1999) has reviewed the numerous beneficial effects of eggs on human health in a paper entitled "The Incredibly Functional Egg," a take-off on the American Egg Board's slogan, "the incredibly edible egg."

Because of the public perception of the adverse effects of cholesterol in the human diet and the fact that animal-origin foods are the only source of dietary cholesterol, there is interest and potential in reducing the cholesterol content of meat, milk, and eggs. The production of **low cholesterol meat** and eggs is a challenging task, because cholesterol is an essential metabolite in animal tissue. Some dietary factors can reduce serum cholesterol. For example, alfalfa contains **saponins** that have cholesterol-lowering properties. They combine with cholesterol in the gut and prevent it from being reabsorbed (they block the enterohepatic cycling of cholesterol). Further research is required to determine if alfalfa or saponins can be used to reduce levels of cholesterol in meat. An excellent source of saponin is the yucca plant. Yucca extract is currently used for its ammonia-binding properties (see Chapter 5). While alfalfa meal and/or alfalfa saponins in the diet lower the blood cholesterol level of chickens, they do not affect egg cholesterol levels (Nakaue et al., 1980). This is because egg cholesterol is synthesized from metabolic precursors in the ovary so does not arise directly from the blood cholesterol. There is little likelihood of being able to lower the cholesterol content of eggs by modification of the hen's diet (Griffin, 1992; Hargis, 1988). Likewise, there probably is not much potential for dramatically lowering the cholesterol content of meat. Contrary to popular belief, cholesterol in meat is associated with the lean and not with the fat (Reiser, 1975). This is because tissue cholesterol occurs mainly in cell membranes, where it is essential for membrane function. Production of lean meat may actually increase its cholesterol content, because the concentration of cell membranes will be higher than when the muscle tissue is diluted with fat. Fat cells are large, so the ratio of cell membrane to total weight is low, resulting in a low cholesterol content of adipose tissue. There is not much difference in the cholesterol contents of various types of meat (Chizzolini et al., 1999).

Processing methods have been developed to remove cholesterol from milk. If cost effective, this will permit the marketing of **low cholesterol dairy products,** including cheese and ice cream. Milk can be passed through a column of polymer-supported saponins to remove cholesterol (Micich et al., 1992). **Low cholesterol eggs** can also be produced from liquid eggs used industrially in the food industry for production of mayonnaise, ice cream, bakery items, and salad dressing. Awad et al. (1997) reviewed several methods of removing cholesterol, including solvent extraction, supercritical fluid extraction, conversion of cholesterol to an inactive product using cholesterol reductase, and complexing with β-cyclodextrin. β-cyclodextrins are cyclic oligosaccharides containing seven glucose units; they form complexes with cholesterol. Low cholesterol eggs produced by processing with β-cyclodextrin have similar taste and functional properties as regular eggs (Awad et al., 1997).

As pointed out earlier, medical authorities now generally agree that dietary cholesterol in most individuals has little effect on blood cholesterol levels (NRC, 1989). It is the type and amount of dietary fat that is more important. If and when the negative public perception of the word "cholesterol" diminishes, interest in producing low cholesterol food products will likely decline also.

# UNDERNUTRITION, SUPERNUTRITION, AND LONGEVITY

Many years ago, McKay et al. (1939) reported that if laboratory rats were deliberately limited in their caloric intake, their **longevity** was markedly increased. In fact, they could be fed on a very low plane of nutrition and kept in a stunted condition for longer than the normal life span of rats, and then when realimented (full fed) they grew and reached normal or almost normal adult weight. There is considerable evidence that there is genetic control of the aging process, with tissue cells capable of a finite number of cell divisions. This "biological clock" seems to operate by a shortening of the chromosomes each time a cell divides, until a critical length of the telomere (nongene end section of chromosomes) is reached (de Lange, 1998). When the rate of metabolism is slowed by undernutrition, the **aging process** is retarded because of the slower rate of cell division. In simple terms, we have a finite amount of metabolism possible before a genetic program causes the system to self-destruct. A fast rate of metabolism uses up the allotted metabolic capacity more quickly, whereas a slow metabolic rate prolongs it. In a review of this subject, Lynn and Wallwork (1992) stated, "Rate of living, i.e. metabolic rate, is proportional to rate of aging, i.e. rate of degenerative cell destruction" (p. 1917).

The Western-style low fiber diet, with a high density of protein and energy from meat, milk, vegetable oils, sugar, and so on, supports a very rapid growth rate. Children on a typical Western diet grow faster, reach puberty quicker, and are taller and heavier as adults than those in developing countries consuming a higher fiber, lower energy diet based on complex carbohydrates (rice, cassava) with less or no animal protein. Maximizing the growth rate of humans is not necessarily the best strategy for optimizing longevity and freedom from degenerative diseases. Should our eating habits be more like those of feedlot steers and broiler chickens or like those of range cattle and wild pigs? It is interesting that, in contrast to what the preceding discussion might suggest, longevity is usually greater for people in the developed countries than for those in the lesser developed countries. Factors other than diet, such as public hygiene, may also account for this difference.

The preceding comments should be qualified by a note of caution. These comments should not be interpreted as suggesting that children should be deliberately underfed to increase their longevity. Undernutrition of growing children can cause permanent stunting and is obviously not desirable. I believe, however, that the topic is worthy of brief discussion. There is considerable criticism of the Western-style diet, with its emphasis on nutrient-rich items such as meat and its low fiber content. There is even a hard-core environmentalist position that a reduction in size of humans would be desirable to reduce our ecological impacts.

**Supernutrition** (excessive intake of calories and protein) is characteristic of the Western diet and is linked to a high incidence of the diseases of affluence such as CHD and cancer. A unifying concept to explain this relationship is that in energy metabolism, free radicals are produced. In the metabolic pathways in which carbohydrates, fatty acids and amino acids are oxidized to yield ATP, carbon dioxide and water, oxygen ($O_2$) is converted to the oxygen free radical ($O_2^-$). While this is metabolically essential, the escape of trace amounts of $O_2^-$ can lead to formation of hydrogen peroxide by the enzyme superoxide dismutase, and the formation of other free radicals by reaction of $O_2^-$ with unsaturated fatty acids. These reactive products can cause membrane damage important in CHD, and DNA damage leading to the initiation of cancer. The more calories consumed, the more oxygen consumed, and the greater the likelihood of tissue damage caused by free radicals. This may explain why supernutrition promotes and undernutrition retards the aging process and the diseases of aging such as CHD and cancer.

**The Mediterranean diet,** characteristic of parts of Italy, Spain, Greece, and other countries of the Mediterranean region, is associated with a low incidence of CHD (Sanders, 1991). Cereal products, fish, legumes, olive oil, fruits, and vegetables are major components of the Mediterranean diet, while meat is used more as a condiment than a main dish. Olive oil has a high content of monounsaturated fatty acids, is neutral to serum cholesterol and CHD, and is stable to oxidative heat damage. Certain other vegetable oils, such as canola oil, share these beneficial properties. In the Mediterranean countries, olive oil is obtained from the whole fruit by mechanical means (pressing) rather than by solvent extraction. Thus in addition to the oil, other components with antioxidant activity are transferred to the oil (Trichopoulou and Vasilopoulou, 2000). Extra virgin olive oil is an especially good source of these antioxidants. A variety of wild plants are commonly included in salads in the Mediterranean diet; some of them are very high in antioxidant activity (Trichopoulou and Vasilopoulou, 2000).

# VEGETARIANISM

Some segments of American society seem to be becoming polarized into vegetarians vs. "meat eaters" or carnivores. Actually, omnivore is a more appropriate term, since few if any humans are strict carnivores (eating only flesh). Interestingly, those societies in which people are basically carnivores (e.g., Greenland Eskimos) with a very high intake of animal flesh and very low consumption of plant-origin food are characterized by very low incidence of diseases of civilization such as diabetes and CHD. Their incidence of these diseases increases markedly when they are exposed to a Western diet high in table sugar and refined oils. As previously discussed, the omega-3 fatty acids in fish oils probably account for the low incidence of heart disease in Greenland Eskimos. The Masai of Kenya are another ethnic group with very high intakes of animal-origin foods and a very low incidence of CHD. The Masai people tend to have low serum cholesterol levels, despite a high intake of animal fats. The hypocholesterolemic effects of saponins in various herbs and vegetables consumed by Masai people are believed to account for the low blood cholesterol levels and low incidence of heart disease (Johns and Chapman, 1995) (see p. 78).

There are other examples of cultures with diets high in either meat or saturated fats. In a comparison of two Polynesian island popula-tions, both with high intakes of saturated fat from consumption of coconuts, Prior et al. (1981) found these populations to have high saturated fat intakes, very high serum cholesterol levels, and a low incidence of coronary heart disease. These characteristics suggest that emphasis on saturated fat and total serum cholesterol as predictors of cardiac problems is simplistic. The Sami people of Finland (Laplanders) are reindeer herders. Although a major component of their diets is red meat (reindeer meat), they have an exceptionally low mortality from coronary heart disease (Luoma et al., 1995). They have high serum cholesterol levels. They have high serum vitamin E and selenium concentrations, derived from the consumption of lean meat (vitamin E) and fish (selenium). These nutrients provide a good antioxidant status, which is protective against heart disease (Diaz et al., 1997).

Is a vegetarian diet adequate for humans? Is it less healthy than a diet containing meat and other animal proteins? Perhaps our experience with livestock is relevant. In the last 50 years, poultry and swine in the United States have been raised almost exclusively on a vegetarian diet based on corn and soybean meal. If it were not nutritionally adequate, we certainly wouldn't expect the high levels of animal performance that are achieved with corn-soy diets, although, admittedly, longevity and aging are not usually important factors in animal production.

There are some nutrients of particular concern with vegetarian diets. These include iron, vitamin $B_{12}$, folic acid, calcium, zinc, and pyridoxine (vitamin $B_6$). When properly fortified with the appropriate minerals and vitamins (as we do for corn-soy swine and poultry diets), vegetarian diets are nutritionally adequate. Some comprehensive reviews on nutritional aspects of vegetarianism are those of Dwyer (1988, 1991).

There are some definite advantages of diets high in plant components. Vegetarian type diets usually result in a lower intake of total fat, saturated fat, and cholesterol, and higher intakes of

## MEDITERRANEAN DIETS

*(Review of A. Trichopoulou and P. Lagiou. 1997. "Healthy Traditional Mediterranean Diet: An Expression of Culture, History, and Lifestyle." Nutr. Rev. 55:383–389.)*

The term "Mediterranean diet" refers to dietary patterns found in olive-growing areas of the Mediterranean and includes areas such as southern France and Italy, Greece, Spain, the Arab Middle Eastern countries, and North Africa. The populations involved have different cultures, religions, economic prosperity, education, alcohol intake and diet, but a unifying factor is that olive oil figures prominently. Olive oil is important not only for its own health benefits, but also because it is associated with the consumption of large quantities of vegetables in salads and legumes in cooked foods. In all cases, the ratio of monounsaturated to saturated fats is much higher than in other parts of the world.

The traditional Mediterranean diet has eight components:

1. high monounsaturated/saturated fat ratio due to high intake of olive oil
2. moderate ethanol consumption, mostly from wine with meals
3. high consumption of legumes
4. high consumption of grains and cereals, including bread
5. high consumption of fruits
6. high consumption of vegetables
7. low consumption of meat and meat products
8. moderate consumption of milk and dairy products, mainly cheese and fish

The good health of Mediterranean people includes lower mortality rates from coronary heart disease, and cancers of the colon, breast, endometrium, and prostate. Besides diet, other factors possibly involved include a relaxing psychosocial environment, mild climatic conditions, preservation of an extended family structure, and possibly even the afternoon siesta.

Olive oil contains various phytochemicals, including vitamin E, and a phenolic compound that inhibits lipid peroxidation. It doesn't contribute to posteating hyperglycemia (i.e., diabetes). Olive oil fatty acids seem to increase bone density and reduce osteoporosis. High consumption of fruits and vegetables in Mediterranean diets provides various phytochemicals and vitamins in abundance.

It may be ironic that as the healthful benefits of Mediterranean diets are being scientifically recognized, the advent of transnational fast-food restaurants is rapidly changing traditional dietary patterns of people in the Mediterranean region to a less healthful fast-food-containing diet.

A symposium on "Mediterranean Diets: Science and Policy Implications" was published in *Am. J. Clin. Nutr.* 61(65):1313S, 1995.

fiber and health-promoting phytochemicals. Numerous **phytochemicals** are known to have protective effects against heart disease and cancer. The low incidence of CHD in the Masai, for example, has been attributed to their high consumption of **saponins** in plant products that they consume (Chapman et al., 1997; Johns and Chapman, 1995). Saponins are well known for their cholesterol-lowering properties (reviewed by Cheeke, 1996). **Chlorophyll,** the ubiquitous compound in green plants, has anticancer activity (Breinholt et al., 1995; Egner

et al., 2001). **Tannins** and phenolic compounds in many plants have antioxidant activity that has protective effects against CHD and cancer (reviewed by Cheeke, 1998).

Soy products are major components of vegetarian diets. **Soybeans** contain numerous phytochemicals and have other attributes that have favorable effects on human health. For example, in a series of symposia on the Role of Soy in Preventing and Treating Chronic Disease (Messina and Erdman, 1995; Setchell, 1998), beneficial effects of soy are documented for cholesterol reduction, coronary heart disease, and cancer. Active components of soy include saponins and isoflavones such as genistein, which favorably affect heart disease, and phytates and trypsin inhibitors, which have anticancer activity. While it has been fashionable for many people to ridicule "tofu eaters," they may get the last laugh!

Key et al. (1999) reviewed health benefits of vegetarian diets. Vegetarians on average are thinner than nonvegetarians. They have a lower risk of coronary heart disease. They tend to have a lower incidence of constipation, diabetes, diverticular disease, gallstones, hypertension, and emergency appendicectomy. Barnard et al. (1995) calculated medical costs attributable to meat consumption. These authors are associated with Physicians Committee for Responsible Medicine, an organization of vegetarian medical doctors with an anti-meat agenda. Despite that bias, they raise a legitimate point. Most people likely accept the idea that smoking causes an increased incidence of various diseases, increasing medical costs for smokers over those for nonsmokers. Similarly, if meat consumption increases the incidence of various disorders over that of nonmeat-eaters, it should be possible to calculate the extra health care costs attributable to meat consumption. Barnard et al. (1995) calculated the direct U.S. medical costs attributable to meat consumption at $28.4–61.4 billion annually. White and Havala (1995) even suggest that vegetarians should receive lower insurance rates. That would certainly provide an incentive for eating less meat.

Although various studies such as those just cited suggest that vegetarians have better health than the general population, interpretation of why is more complicated than simply comparing diets. For example, in a comparison of over 6,000 vegetarians with 5,000 meat eaters, Thorogood et al. (1994) found a lower mortality rate from heart disease and cancer in vegetarians. However, they concluded, "Our data do not provide justification for encouraging meat eaters to change to a vegetarian diet" (p. 70), because "vegetarians tend to be thinner, smoke less, and are generally of higher socioeconomic status than the general population, all factors which are major determinants of total, cardiovascular, and cancer mortality" (p. 67). Similarly, Beilin (1993) found that interpretation of health effects of a vegetarian diet were complicated by alcohol consumption; people who adhere to a vegetarian diet tend to drink little alcohol, while moderate to heavy drinkers are not known for their avoidance of meat.

The position on vegetarian diets of the American Dietetic Association is as follows: "It is the position of the American Dietetic Association (ADA) that appropriately planned vegetarian diets are healthful, are nutritionally adequate, and provide health benefits in the prevention and treatment of certain diseases" (Anonymous, 1997, p. 1317).

Some vegetarian diets seem a bit extreme or bizarre to nonadherents. For example, the Hallelujah diet (Donaldson, 2001) consists of raw fruits and vegetables, carrot juice, dehydrated barley grass juice, raw nuts and seeds, olive oil, flax seed oil, cooked vegetables, whole grain products, tubers, and a vitamin $B_{12}$ supplement (yum, yum!). All animal products except butter are eliminated. Also excluded are refined flour, refined sugars, refined, bleached vegetable oils, hydrogenated fats, and table salt. Although this diet apparently meets nutrient requirements (Donaldson, 2001), it is unlikely to attract a large following.

Ethical aspects of vegetarianism, including its relationship to feminism, are discussed in Chapter 10.

# CONCLUSIONS: ANIMAL PROTEINS AND HEALTH

Animal products are valuable sources of nutrients and eating enjoyment. They contain no fiber so are highly digestible and concentrated sources of protein and energy. The principal components of meat that are of potential concern in the human diet are the fat content, and specifically, some of the saturated fatty acids. Many possibilities exist for the improvement of animal products by modifying the fat type and amount. Inclusion of animal protein sources in balanced diets is consistent with good health, enjoyable eating, and longevity.

The National Research Council nutritional recommendations for reducing chronic disease risk (NRC, 1989) have been presented in a form useful for direct application (NRC, 1991, 1992). The nine dietary guidelines that are the bottom-line of these publications are the following:

1. Reduce total fat intake to 30 percent or less of your total calorie consumption. Reduce saturated fatty acid intake to less than 10 percent of calories. Reduce cholesterol intake to less than 300 milligrams (mg) daily.

2. Eat five or more servings of a combination of vegetables and fruits daily, especially green and yellow vegetables and citrus fruits. Also, increase your intake of starches and other complex carbohydrates by eating six or more daily servings of a combination of breads, cereals, and legumes.

3. Eat a reasonable amount of protein, maintaining your protein consumption at moderate levels.

4. Balance the amount of food you eat with the amount of exercise you get to maintain appropriate body weight.

5. It is not recommended that you drink alcohol. If you do drink alcoholic beverages, limit the amount you drink in a single day to no more than two cans of beer, two small glasses of wine, or two average cocktails. Pregnant women should avoid alcoholic beverages.

6. Limit the amount of salt (sodium chloride) that you eat to 6 grams (g) (slightly more than 1 teaspoon of salt) per day or less. Limit the use of salt in cooking and avoid adding it to food at the table. Salty foods, including highly processed salty foods, salt-preserved foods, and salt-pickled foods, should be eaten sparingly, if at all.

7. Maintain adequate calcium intake.

8. Avoid taking dietary supplements in ex-

cess of the U.S. Recommended Daily Allowance (U.S. RDAs) in any one day.

9. Maintain an optimal level of fluoride in your diet and particularly in the diets of your children when their baby and adult teeth are forming.

Items 1, 3, and 7 relate most closely to consumption of animal products. Consumption of moderate amounts of lean meat is consistent with items 1 and 3. For adults, NRC (1992) recommends consuming no more than 6 ounces of lean meat per day. A 3 ounce portion of meat is equal to about the size of a deck of cards. In order to maintain adequate calcium intake while limiting cholesterol and saturated fat, low fat or nonfat milk and other dairy products should be selected (NRC, 1992). Thus changes to more healthful dietary habits do not mean that animal products are eliminated. However, frequent consumption of a 16 ounce well-marbled steak along with a baked potato covered with sour cream and butter is not consistent with a healthy diet.

The husband and wife team of William and Sonja Connor of the Oregon Health Sciences University has been active in promotion of what they call "the new American diet" (Connor and Connor, 1986). Their program proposes a gradual, three-phase modification of the typical high fat (about 40% of total calories), low fiber American diet to a low fat, high fiber diet. The general goals of "the new American diet" are the following:

1.  to decrease cholesterol intake
2.  to reduce intake of saturated fat
3.  to reduce total fat intake
4.  to increase intake of sources of complex carbohydrates and fiber (whole grains, bread, potatoes, beans, pasta, cereals, rice, fruits and vegetables)
5.  to reduce salt intake
6.  to reduce intake of refined sugars
7.  to keep alcohol consumption low
8.  if overweight, to reduce total food intake

The specific goals of the Connor program, which call for greater reductions in total fat, saturated fat, and cholesterol intakes than the NRC recommendations, are the following:

1. to reduce daily cholesterol intake from the current average of 400–500 mg per day to less than 100 mg per day
2. to reduce fat intake by half, from 40% of total calories to 20%
3. to reduce saturated fat intake from the current 14% of total calories to 5–6%
4. to increase carbohydrate intake from the present 45% of total calories to 65%, with emphasis on complex carbohydrates and fiber
5. to reduce the intake of refined sugar from 20% of calories to 10%
6. to reduce salt intake by at least 50%

The Connor group has developed a **cholesterol-saturated fat index** (CSI) for foods to assist diet planners in selecting foods that contribute to meeting the preceding goals. The derivation of the CSI is described by Connor et al. (1986). It is calculated by an equation:

$$CSI^* = 1.01 \times g \text{ saturated fat}^* + 0.05 \times mg \text{ cholesterol}^*$$
$$^*\text{per } 100 \text{ g of the foodstuff.}$$

The use of the CSI allows the selection of foods that are low in both cholesterol and saturated fat. Some foods such as shellfish are high in cholesterol but very low in saturated fat. Others such as chocolate and palm oil products (e.g., coffee creamer) are high in saturated fat but lack cholesterol. The use of the CSI allows the selection of foods that help to reduce serum cholesterol and CHD. Examples of CSI values for some common food items are listed in Table 2-3. The lower the number, the greater the potential for a desirable effect in reducing serum cholesterol and CHD.

**TABLE 2-3**  The Cholesterol-Saturated Fat Index (CSI) of common dietary items (Adapted from Connor and Connor, 1986; Connor et al., 1986).

| Item | CSI |
|---|---|
| Protein sources (per 100 g): | |
| White fish | 4 |
| Salmon | 5 |
| Shellfish (shrimp, crab, lobster) | 6 |
| Poultry (no skin) | 6 |
| Beef, pork, and lamb: | |
| 10% fat (ground sirloin, flank steak) | 9 |
| 15% fat (ground round) | 10 |
| 20% fat (ground chuck, pot roasts) | 13 |
| 30% fat (ground beef, pork and lamb steaks, ribs, pork and lamb chops, roasts) | 18 |
| Cottage cheese | 6 |
| Most cheeses (e.g., cheddar, Roquefort, Swiss, Brie, Jack, American, cream cheese, Velveeta, cheese spreads) | 26 |
| Whole eggs (2) | 29 |

*(Continued)*

**TABLE 2-3**   (Continued)

| Item | CSI |
|------|-----|
| Fats (per 4 tablespoons): | |
| Peanut butter | 5 |
| Most vegetable oils | 8 |
| Mayonnaise | 10 |
| Soft margarines | 10 |
| Hard stick margarines | 15 |
| Butter | 37 |
| Coconut and palm oils, chocolate | 47 |
| Frozen desserts and milk products (1 cup): | |
| Sherbet, low fat frozen yogurt | 2 |
| Ice milk | 6 |
| Ice cream, 10% fat | 13 |
| Rich ice cream, 16% fat | 18 |
| Skim milk | 0 |
| 1% milk | 2 |
| 2% milk, plain low fat yogurt | 4 |
| 3.5% whole milk | 7 |
| Liquid nondairy creamers | 22 |
| Sour cream | 37 |
| Imitation sour cream | 43 |

# STUDY QUESTIONS

1. What is the relevance, if any, that humans apparently evolved as omnivores? Do humans require specific nutrients in animal products or can we meet our nutritional needs from a vegetarian diet?

2. How does the body fat composition differ between domesticated ungulates (e.g., cattle fattened in a feedlot) and wild ungulates such as deer and bison? Do these differences have any implications as far as the health of people consuming the meat of these animals? In terms of health, did our meat-eating ancestors eat meat with the same nutritional properties as the meat we eat today?

3. Why do humans have a "sweet tooth" and like fatty foods?

4. What is osteoporosis? What relationship, if any, does drinking soft drinks instead of milk have on the likelihood of later development of osteoporosis?

5. Is soy milk equivalent to cow's milk in its effect on bone health? Why or why not?

6. How does consumption of dairy products influence hypertension?

7. What is the significance of the occurrence of conjugated linoleic acid in dairy products?

8. "I don't eat pizza because I am lactose-intolerant." Discuss. (Hint: Does cheese contain lactose?)

9. How does the bioavailability of iron in meat compare to that of iron in plant-based products?

10. What are the nutritional differences between red and white meat? Which is higher in cholesterol? Why? Why is red meat red? Are beef cattle the only source of red meat?

11. What is the mechanism by which high intakes of iron might influence incidence of coronary heart disease?

12. What is the difference between cholesterol and fat? Is cholesterol a type of fat? Is it a lipid?

13. What are some good sources of monounsaturated fatty acids?

14. What are the roles of HDL and LDL in cholesterol metabolism?

15. Explain the differences in how saturated and unsaturated fats affect cholesterol metabolism and influence incidence of coronary heart disease.

16. What are some of the potential nutritional hazards of not allowing infants and children to consume milk, meat, or any other foods of animal origin?

17. What are the main fatty acids in beef fat? Which ones, if any, might have undesirable effects on blood cholesterol levels of beef consumers? Can the fatty acid composition of beef tissue be altered by feeding, genetics, and so on?

18. Often, people who are promoting "new" meat animals such as bison, water buffalo, rabbits, ostrich, emu, and so on make the claim that the meat from their particular favorite animal is "low in cholesterol." Is there such a thing as low cholesterol meat? Why does meat contain cholesterol? Which has more cholesterol, fatty meat or lean meat? Why?

19. Why is palmitic acid hypercholesterolemic while stearic acid is not?

20. The indigenous people of Greenland and other high Arctic areas consume a diet made up almost exclusively of meat and fat, with virtually no plant foods eaten. What positive and negative effects on their health are attributable to this type of diet? On their traditional diet, do they have a high incidence of coronary heart disease?

21. Discuss the iron overload theory of coronary heart disease. Is there a difference between "red meat" and "white meat" with respect to this theory?

22. What is the optimal level of meat in the human diet?

23. Discuss the effects of animal products and plant foods in the human diet on cancer.

24. What are some of the underlying causes of the "diseases of Western civilization"?

25. What are processed and refined foods? How could these products cause nutritional distortion?

26. What are the positive and negative aspects of grass-fattened beef?

27. How does nutrition affect longevity and aging?

28. Discuss the NRC nutritional recommendations for the human diet. What would be the implications of the widespread adoption of these recommendations on the livestock industries of the United States?

29. Is meat an essential part of the well-balanced human diet? Critique the eating of meat, from the viewpoint of a vegetarian. Critique the adequacy of a vegetarian diet, from the viewpoint of a meat consumer or producer.

30. Speculate on the role of meat in the human diet 100 years from now.

31. A nominee for the position of surgeon general of the United States is quoted as stating that the elimination of animal products from the American diet would reduce health care costs markedly and virtually eliminate heart disease and other degenerative diseases. Discuss the validity of this perception.

# REFERENCES

ADLERCREUTZ, H. 1990. Western diet and Western diseases: Some hormonal and biochemical mechanisms and associations. *Scand. J. Clin. Lab Invest.* Suppl. 201:3–23.

ADLERCREUTZ, H. 1998. Evolution, nutrition, intestinal microflora, and prevention of cancer: A hypothesis. *Proc. Soc. Exp. Biol Med.* 217:241–246.

AJUYAH, A. O., K. H. LEE, R. T. HARDIN, and J. S. SIM. 1991. Influence of dietary full-fat seeds and oils on total lipid cholesterol and fatty acid composition of broiler meats. *Can. J. Anim. Sci.* 71:1011–1019.

AMES, B. N. 1983. Dietary carcinogens and anticarcinogens. *Science* 221:1256–1264.

ANONYMOUS. 1997. Position of the American Dietetic Association: Vegetarian diets. *J. Amer. Dietetic Assoc.* 97:1317–1321.

ARTHUR, P. F. 1995. Double muscling in cattle: A review. *Aust. J. Agric. Res.* 46:1493–1515.

ASHES, J. R., P. ST. VINCENT WELCH, S. K. GULATI, T. W. SCOTT, G. H. BROWN, and S. BLAKELEY. 1992. Manipulation of the fatty acid composition of milk by feeding protected canola seeds. *J. Dairy Sci.* 75:1090–1096.

AWAD, A. C., M. R. BENNICK, and D. M. SMITH. 1997. Composition and functional properties of cholesterol reduced egg yolk. *Poult. Sci.* 76:649–653.

BARNARD, N. D., A. NICHOLSON, and J. L. HOWARD. 1995. The medical costs attributable to meat consumption. *Preventive Med.* 24:646–655.

BEILIN, L. J. 1993. Vegetarian diets, alcohol consumption, and hypertension. *Ann. NY Acad. Sci.* 676:83–91.

BERGEN, W. G., and R. A. MERKEL. 1991. Body composition of animals treated with partitioning agents: Implications for human health. *FASEB J.* 5:2951–2957.

BESSA, R. J. B., J. SANTOS-SILVA, J. M. R. RIBEIRO, and A. V. PORTUGAL. 2000. Reticulo-rumen biohydrogenation and the enrichment of ruminant edible products with linoleic acid conjugated isomers. *Livestock Production Sci.* 63:201–211.

BIFFI, A., D. CORADINI, R. LARSEN, L. RIVA, and G. DE FRONZO. 1997. Antiproliferative effect of fermented milk on the growth of a human breast cancer cell line. *Nutrition and Cancer* 28:93–99.

BINGHAM, S. 1997. Meat, starch and non-starch polysaccharides. Are epidemiological and experimental findings consistent with acquired genetic alterations in sporadic colorectal cancer? *Cancer Letters* 114:25–34.

BINGHAM, S. A. 1999. Meat or wheat for the next millennium? Plenary Lecture. High-meat diets and cancer risk. *Proc. Nutr. Soc.* 58:243–248.

BIRCH, L. L. 1992. Children's preferences for high-fat foods. *Nutr. Rev.* 50:249–254.

BOESCH, C. 1999. A theory that's hard to digest. A review: Stanford, C. The Hunting Apes: Meat Eating and the Origins of Human Behavior. Princeton University Press, 1999. 245 pp. Nature 399:653.

BONANOME, A., and S. M. GRUNDY. 1988. Effect of dietary stearic acid on plasma cholesterol and lipoprotein levels. *New Eng. J. Med.* 318:1244–1248.

BOS, J. L. 1989. Ras oncogenes in human cancer: A review. *Cancer Res.* 49:4682–4689.

BOUNOUS, G., G. BATIST, and P. GOLD. 1991. Whey proteins in cancer prevention. *Cancer Lett.* 57:91–94.

BOYLSTON, T. D., S. A. MORGAN, K. A. JOHNSON, J. R. BUSBOOM, R. W. WRIGHT, JR., and J. J. REEVES. 1995. Lipid content and composition of Wagyu and domestic breeds of beef. *J. Agric. Food Chem.* 43:1202–1207.

BRAND-MILLER, J. C., and S. COLAGIURI. 1999. Evolutionary aspects of diet and insulin resistance. In: Simopoulos, A.P. (Ed.). *Evolutionary aspects of nutrition and health. World Rev. Nutr. and Diet.* 84:74–105.

BRAND-MILLER, J. C., and S. H. A. HOLT. 1998. Australian aboriginal plant foods: A consideration of their nutritional composition and health implications. *Nutr. Res. Rev.* 11:5–23.

BRATTSTROM, L. 1996. Vitamins as homocysteine-lowering agents. *J. Nutr.* 126:1276S–1280S.

BREINHOLT, V., J. HENDRICKS, C. PEREIRA, D. ARBOGAST, and G. BAILEY. 1995. Dietary chlorophyllin is a potent inhibitor of aflatoxin $B_1$ hepatocarcinogenesis in rainbow trout. *Cancer Res.* 55:57–62.

BROADHURST, C. L., S. C. CUNNANE, and M. A. CRAWFORD. 1998. Rift Valley lake fish and shellfish provided brain-specific nutrition for early Homo. *British J. Nutr.* 79:3–21.

BUCKLEY, D. J., P. A. MORRISSEY, and J. I. GRAY. 1995. Influence of dietary vitamin E on the oxidative stability and quality of pig meat. *J. Anim. Sci.* 73:3122–3130.

BUDOWSKI, P. 1988. ω-3 Fatty acids in health and disease. *World Rev. Nutr. and Diet.* 57:214–274.

BYERS, F. M., and G. T. SCHELLING. 1988. Lipids in ruminant nutrition. pp. 298–312. In: Church, D.C. (Ed.). *The ruminant animal.* Prentice Hall, Englewood Cliffs, N.J.

CARPENTER, C. E., O. D. RICE, N. E. COCKETT, and G. D. SNOWDER. 1996. Histology and composition of muscles from normal and callipyge lambs. *J. Anim. Sci.* 74:388–393.

CHANMUGAM P., M. BOUBREAU, T. BOUTTE, R. S. PARK, J. HEBERT, L. BERRIO, and D. H. HWANG. 1992. Incorporation of different types of ω-3 fatty acids into tissue lipids of poultry. *Poult. Sci.* 71:516–521.

CHAPMAN, L., T. JOHNS, and R. L. A. MAHUNNAH. 1997. Saponin-like in vitro characteristics of extracts from selected non-nutrient wild plant food additives used by Maasai in meat and milk based soups. *Ecol. Food Nutr.* 36:1–22.

CHEEKE, P. R. 1996. Biological effects of feed and forage saponins and their impacts on animal production. pp. 377–385. In: Waller, G.R., and K. Yamasaki (Eds.). *Saponins used in food and agriculture.* Plenum Press, New York.

CHEEKE, P. R. 1998. *Natural toxicants in feeds, forages and poisonous plants.* Interstate Publishers, Danville, IL.

CHIZZOLINI, R., E. ZANARDI, V. DORIGONI, and S. GHIDINI. 1999. Calorific value and cholesterol content of normal and low-fat meat and meat products. *Trends in Food Sci. & Tech.* 10:119–128.

CONNOR, S. L., and W. E. CONNOR. 1986. *The new American diet.* Simon and Schuster, Inc. New York.

CONNOR, S. L., J. R. GUSTAFSON, S. M. ARTAUD-WILD, D. P. FLAVELL, C. J. CLASSICK-KOHN, L. E. HATCHER, and W. E. CONNOR. 1986. The cholesterol/saturated fat index: An indication of the hyper-cholesterolaemic and atherogenic potential of food. *Lancet* 1:1229–1232.

CORDAIN, L. 1999. Cereal grains: Humanity's double-edged sword. In: Simopoulos, A. P. (Ed.). *Evolutionary aspects of nutrition and health. World Rev. Nutr. and Diet.* 84:19–73.

CORDAIN, L. 2002. *The Paleo diet.* John Wiley and Sons, New York.

CORDAIN, L., C. MARTIN, G. FLORANT, and B. A. WATKINS. 1998. The fatty acid composition of muscle, brain, marrow and adipose tissue in elk: Evolutionary implications for human dietary lipid requirements. *World Rev. Nutr. and Diet.* 83:225–226.

CORDAIN, L., J. B. MILLER, S. B. EATON, N. MANN, S. H. A. HOLT, and J. D. SPETH. 2000. Plant-animal subsistence ratios and macronutrient energy estimations in worldwide hunter-gatherer diets. *Am. J. Clin. Nutr.* 71:682–692.

CRAWFORD, M. A. 1968. Fatty-acid ratios in free-living and domestic animals. *Lancet* 1:1329–1333.

CRAWFORD, M. A., and M. M. GALE. 1969. Linoleic acid and linolenic acid elongation products in muscle tissue of *Syncerus caffer* and other ruminant species. *Biochem. J.* 115:25–27.

CUNNANE, S. C., P. A. STITT, S. GANGULI, and J. K. ARMSTRONG. 1990. Raised omega-3 fatty acid levels in pigs fed flax. *Can. J. Anim. Sci.* 70:251–254.

CUNNINGHAM, H. M., and D. W. FRIEND. 1964. Effect of nicotine on nitrogen retention and fat deposition in pigs. *J. Anim. Sci.* 23:717–722.

DANNENBERG, A. J. 2001. Inhibition of COX-2. A novel approach to breast cancer prevention. *J. Nutr.* 130 (Suppl. 1):190S(Abst.).

DAVIDSON, M. H., D. HUNNINGHAKE, K. C. MAKI, P. O. KWITEROVICH, JR., and S. KAFONEK. 1999. Comparison of the effects of lean red meat vs lean white meat on serum lipid levels among free-living persons with hypercholesterolemia. *Arch. Intern. Med.* 159:1331–1338.

DE LANGE, T. 1998, Telomeres and senescence: ending the debate. *Science* 279:334–335.

DEMER, L. L. 1995. Editorial. A skeleton in the atherosclerosis closet. *Circulation* 92:2029–2032.

DEMEYER, D., and M. DOREAU. 1999. Targets and procedures for altering ruminant meat and milk lipids. *Proc. Nutr. Soc.* 58:593–607.

DENKE, M. A. 1994. Role of beef and beef tallow, an enriched source of stearic acid, in a cholesterol-lowering diet. *Am. J. Clin. Nutr.* 60:1044S–1049S.

DE VALK, B., and J. J. M. MARX. 1999. Review Article. Iron, atherosclerosis, and ischemic heart disease. *Arch. Intern. Med.* 159:1542–1548.

DHURANDHAR, N. V., B. A. ISRAEL, J. M. KOLESAR, G. F. MAYHEW, M. E. COOK, and R. L. ATKINSON. 2000. Increased adiposity in animals due to a human virus. *Int. J. Obesity* 24:989–996.

DIAZ, M. N., B. FREI, J. A. VITA, and J. F. KEANEY, JR. 1997. Antioxidants and atherosclerotic heart disease. *New Engl. J. Med.* 337:408–416.

DIETSCHY, J. M., S. D. TURLEY, and D. K. SPADY. 1993. Role of liver in the maintenance of cholesterol and low density lipoprotein homeostasis in different animal species, including humans. *J. Lipid Res.* 34:1637–1659.

DONALDSON, M. S. 2001. Food and nutrient intake of Hallelujah vegetarians. *Nutr. Food Sci.* 31:293–303.

DU, M., D. U. AHN, and J. L. SELL. 1999. Effect of dietary conjugated linoleic acid on the composition of egg yolk lipids. *Poult. Sci.* 78:1639–1645.

DUCKETT, S. K., D. G. WAGNER, L. D. YATES, H. G. DOLEZAL, and S. G. MAY. 1993. Effects of time on feed on beef nutrient composition. *J. Anim. Sci.* 71:2079–2088.

DUTHIE, G. G., K. W. J. WAHLE, and W. P. T. JAMES. 1989. Oxidants, antioxidants and cardiovascular disease. *Nutr. Res. Rev.* 2:51–62.

DWYER, J. T. 1988. Health aspects of vegetarian diets. *Am. J. Clin. Nutr.* 48:712–738.

DWYER, J. T. 1991. Nutritional consequences of vegetarianism. *Annu. Rev. Nutr.* 11:61–91.

EATON, S. B. 1992. Humans, lipids, and evolution. *Lipids* 27:814–820.

EATON, S. B., and M. KONNER. 1985. Paleolithic nutrition: A consideration of its nature and current implications. *New Eng. J. Med.* 312:283–289.

EATON, S. B., S. B. EATON III, A. J. SINCLAIR, L. CORDAIN, and N. J. MANN. 1998. Dietary intake of long-chain polyunsaturated fatty acids during the Paleolithic. In: Simopoulos, A. P. (Ed.). *The return of ω 3 fatty acids into the food supply. I. Land-based animal food products and their health effects. World Rev. Nutr. and Diet.* 83:12–23.

EATON, S. B., S. B. EATON III, M. J. KONNER, and M. SHOSTAK. 1996. An evolutionary perspective enhances understanding of human nutritional requirements. *J. Nutr.* 126:1732–1740.

EGNER, P. A., J-B. WANG, Y.-R. ZHU, B.-C. ZHANG, Y. WU, Q.-N. ZHANG, G.-S. QIAN, S. Y. KUANG, S. J. GANGE, L. P. JACOBSON, K. J. HELZLSOUER, G. S. BAILEY, J. D. GROOPMAN, and T. W. KENSLER. 2001. Chlorophyllin intervention reduces aflatoxin-DNA adducts in individuals at high risk for liver cancer. *PNAS* 98:14601–14606.

EICHHOLZER, M., and H. STAHELIN. 1993. A Review. Is there a hypocholesterolemic factor in milk and milk products? *Internat. J. Vit. Nutr. Res.* 63:159–167.

EISMAN, J. A. 1998. Plenary Lecture. Genetics, calcium intake and osteoporosis. *Proc. Nutr. Soc.* 57:187–193.

ENSER, M., N. D. SCOLLAN, N. J. CHOI, E. KURT, K. HALLETT and J. D. WOOD. 1999. Effect of dietary lipid on the content of conjugated linoleic acid (CLA) in beef muscle. *Anim. Sci.* 69:143–146.

ERHARDT, J. G., S. S. LIM, J. C. BODE, and C. BODE. 1997. A diet rich in fat and poor in dietary fiber increases the in vitro formation of reactive oxygen species in human feces. *J. Nutr.* 127:706–709.

FARRELL, D. J. 1998. Enrichment of hen eggs with ω-3 long-chain fatty acids and evaluation of enriched eggs in humans. *Am. J. Clin. Nutr.* 68:538–544.

FELTON, J. S., E. FULTZ, F. A. DOLBEARE, and M. G. KNIZE. 1994. Effect of microwave pretreatment on heterocyclic aromatic amine mutagens/carcinogens in fried beef patties. *Fd. Chem. Toxic.* 32:897–903.

FESKANICH, D., W. C. WILLETT, M. J. STAMPFER, and G. A. COLDITZ. 1997. Milk, dietary calcium, and bone fractures in women: A 12-year prospective study. *Am. J. Public Health* 87:992–997.

FOTOUHI, N., and T. C. JENKINS. 1992. Ruminal biohydrogenation of linoleoyl methionine and calcium linoleate in sheep. *J. Anim. Sci.* 70:3607–3614.

FRANK, J. W., D. M. REED, J. S. GROVE, and R. BENFANTE. 1992. Will lowering population levels of serum cholesterol affect total mortality? Expectations from the Honolulu Heart Program. *J. Clin. Epidemiol.* 45:333–346.

GALE, M. M., M. A. CRAWFORD, and M. WOODFORD. 1969. The fatty acid composition of adipose and muscle tissue in domestic and free-living ruminants. *Biochem. J.* 113:6p.

GILLMAN, M. W., L. A. CUPPLES, B. E. MILLEN, R. C. ELLISON, and P. A. WOLF. 1997. Inverse association of dietary fat with development of ischemic stroke in men. *JAMA* 278:2145–2150.

GIOVANNUCCI, E. 1995. Epidemiologic characteristics of prostate cancer. *Cancer* 75:1766–1777.

GIOVANNUCCI, E. 1999. Tomatoes, tomato-based products, lycopene, and cancer: Review of the epidemiologic literature. *J. Natl. Cancer Inst.* 91:317–331.

GIOVANNUCCI, E., E. B. RIMM, A. WOLK, A. ASCHERIO, M. J. STAMPFER, G. A. COLDITZ, and W. C. WILLETT. 1998. Calcium and fructose intake in relation to risk of prostate cancer. *Cancer Res.* 58:442–447.

GIOVANNUCCI, E., M. J. STAMPFER, G. COLDITZ, E. B. RIMM, and W. C. WILLETT. 1992. Relationship of diet to risk of colorectal adenoma in men. *J. Natl. Cancer Inst.* 84:91–98.

GRANT, W. B. 2000. The role of meat in the expression of rheumatoid arthritis. *Brit. J. Nutr.* 84:589–595.

GRIEBENOW, R. L., F. A. MARTZ, and R. E. MORROW. 1997. Forage-based beef finishing systems: A review. *J. Prod. Agric.* 10:84–91.

GRIFFIN, H. D. 1992. Manipulation of egg yolk cholesterol: A physiologist's view. *World's Poult. Sci. J.* 48:101–112.

GRUNDY, S. M. 1990. Cholesterol and coronary heart disease. *JAMA* 264:3053–3059.

GRUNDY, S. M., and M. A. DENKE. 1990. Dietary influences on serum lipids and lipoproteins. *J. Lipid Res.* 31:1149–1172.

GUARDIOLA, F., R. CODONY, P. B. ADDIS, M. RAFECAS, and J. BOATELLA. 1996. Biological effects of oxysterols: Current status. *Fd. Chem. Toxic.* 34:193–211.

HA, Y. L., N. K. GRIMM, and M. W. PARIZA. 1989. Newly recognized anticarcinogenic fatty acids: Identification and quantification in natural and processed cheeses. *J. Agric. Food Chem.* 37:75–81.

HABITO, R. C., J. MONTALTO, E. LESLIE, and M. J. BALL. 2000. Effects of replacing meat with soybean in the diet on sex hormone concentrations in healthy adult males. *Brit. J. Nutr.* 84:557–563.

HALLIWELL, B., and J. M. C. GUTTERIDGE. 1986. Oxygen free radicals and iron in relation to biology and medicine: Some problems and concepts. *Arch. Biochem. Biophys.* 246:501–514.

HAMET, P. 1995. The evaluation of the scientific evidence for a relationship between calcium and hypertension. *J. Nutr.* 125:311S–400S.

HAMOSH, M. 1997. Should infant formulas be supplemented with bioactive components and conditionally essential nutrients present in human milk? *J. Nutr.* 127:971S–974S.

HARGISS, P. S. 1988. Modifying egg yolk cholesterol in the domestic fowl—A review. *World's Poult. Sci. J.* 44:17–29.

HAYES, K. C., and P. KHOSLA. 1992. Dietary fatty acid thresholds and cholesterolemia. *FASEB J.* 6:2600–2607.

HAYES, K. C., A. PRONCZUK, S. LINDSEY, and D. DIERSEN-SCHADE. 1991. Dietary saturated fatty acids (12:0, 14;0, 16:0) differ in their impact on plasma cholesterol and lipoproteins in nonhuman primates. *Am. J. Clin. Nutr.* 53:491–498.

HEANEY, R. P., C. M. WEAVER, and M. L. FITZSIMMONS. 1991. Soybean phytate content: Effect on calcium absorption. *Am. J. Clin. Nutr.* 53:745–747.

HEGSTED, D. M. 1986. Calcium and homeostasis. *J. Nutr.* 116:2316–2319.

HILL, M. J. 1999. Meat or wheat for the next millennium? A debate pro meat. Meat and colorectal cancer. 1999. *Proc. Nutr. Soc.* 58:261–264.

HINTZE, K. J., G. P. LARDY, M. J. MARCHELLO, and J. W. FINLEY. 2001. Areas with high concentrations of selenium in the soil and forage produce beef with enhanced concentrations of selenium. *J. Agric. Food Chem.* 49:1062–1067.

HOAK, J. C. 1994. Stearic acid, clotting, and thrombosis. *Am. J. Clin. Nutr.* 60:1050S–1053S.

HOLMES, B. 2000. The obesity bug. *New Scientist* Aug. 5:26–30.

HOLT, S. H. A., J. C. BRAND MILLER and P. PETOCZ. 1997. An insulin index of foods: the insulin demand generated by 1000-kf portions of common foods. *Am. J. Clin. Nutr.* 66:1264–1276.

HOPKINS, P. N. 1992. Effects of dietary cholesterol on serum cholesterol: A meta-analysis and review. *Am. J. Clin. Nutr.* 55:1060–70.

HORNSTRA, G., A. C. VAN HOUWELINGEN, A. D. M. KESTER, and K. SUNDRAM. 1991. A palm oil-enriched diet lowers serum lipoprotein(a) in normocholesterolemic volunteers. *Atherosclerosis* 90:91–93.

HORTIN, A. E., G. ODUHO, Y. HAN, P. J. BECHTEL, and D. H. BAKER. 1993. Bioavailability of zinc in ground beef. *J. Anim. Sci.* 71:119–123.

HU, F. B., M. J. STAMPFER, E. B. RIMM, J. E. MANSON, A. ASCHERIO, G. A. COLDITZ, B. A. ROSNER, D. SPIEGELINAN, F. E. SPEIZER, F. M. SACKS, C. H. HENNEKENS, and W. C. WILLETT. 1999. A prospective study of egg consumption and risk of cardiovascular disease in men and women. *JAMA* 281:1387–1405.

HU, J. F., X. H. ZHAO, J. B. JIA, B. PARPIA, and T. C. CAMPBELL. 1993. Dietary calcium and bone density among middle-aged and elderly women in China. *Am. J. Clin. Nutr.* 58:219–227.

HU, Y. -F., C. -J. CHANG, R. W. BRUEGGEMEIER, and Y. C. LIN. 1994. Presence of antitumor activities in the milk collected from gossypol-treated cows. *Cancer Lett.* 87:17–23.

HULAN, H. W., R. G. ACKMAN, W. M. N. RATNAYAKE, and F. G. PROUDFOOT. 1988. Omega-3 fatty acid levels and performance of broiler chickens fed redfish meal or redfish oil. *Can. J. Anim. Sci.* 68:533–547.

HULAN, H. W., R. G. ACKMAN, W. M. N. RATNAYAKE, and F. G. PROUDFOOT. 1989. Omega-3 fatty acid levels and general performance of commercial broilers fed practical levels of redfish meal. *Poult. Sci.* 68:153–162.

IP, C., and J. A. SCIMECA. 1997. Conjugated linoleic acid and linoleic acid are distinctive modulators of mammary carcinogenesis. *Nutrition and Cancer* 27:131–135.

IP, C., S. F. CHIN, J. A. SCIMECA, and M. W. PARIZA. 1991. Mammary cancer prevention by conjugated dienoic derivative of linoleic acid. *Cancer Res.* 51:6118–6124.

JOHNS, T. 1996. *The origins of human diet and medicine. Chemical ecology.* The University of Arizona Press, Tucson.

JOHNS, T., and L. CHAPMAN. 1995. Phytochemicals ingested in traditional diets and medicines as modulators of energy metabolism. pp. 161–188. In: Arnason, J.T., et al. (Eds.). *Phytochemistry of medicinal plants.* Plenum Press, New York.

KAPSOKEFALOU, M., and D. D. MILLER. 1993. Lean beef and beef fat interact to enhance nonheme iron absorption in rats. *J. Nutr.* 123:1429–1434.

KARGES, K., J. C. BROOKS, D. R. GILL, J. E. BREAZILE, F. N. OWENS, and J. B. MORGAN. 2001. Effects of supplemental vitamin D$_3$ on feed intake, carcass characteristics, tenderness, and muscle properties of beef steers. *J. Anim. Sci.* 79:2844–2850.

KARJALAINEN, J., J. M. MARTIN, M. KNIP, J. ILONEN, B. H. ROBINSON, E. SAVILAHTI, H. K. AKERBLOM, and H. M. DOSCH. 1992. A bovine albumin peptide as a possible trigger of insulin-dependent diabetes mellitus. *New Engl. J. Med.* 327:302–307.

KEY, T. J., G. K. DAVEY, and P. N. APPLEBY. 1999. Health benefits of a vegetarian diet. *Proc. Nutr. Soc.* 58:271–275.

KINSELLA, J. E., B. LOKESH, and R. A. STONE. 1990. Dietary ω-3 polyunsaturated fatty acids and amelioration of cardiovascular disease: Possible mechanisms. *Am. J. Clin. Nutr.* 52:1–28.

KNIZE, M. G., F. A. DOLBEARE, K. L. CARROLL, D. H. MOORE, II, and J. S. FELTON. 1994. Effect of cooking time and temperature on the heterocyclic amine content of fried beef patties. *Fd. Chem. Toxic.* 32:595–603.

KRAJCOVICOVA-KUDLACKOVA, M., P. BLAZICEK, J. KOPCOVA, A. BEDEROVA, and K. BABINSKA. 2000. Homocysteine levels in vegetarians versus omnivores. *Ann. Nutr. Metab.* 44:135–138.

LARICK, D. K., and B. E. TURNER. 1989. Influence of finishing diet on the phospholipid composition and fatty acid profile of individual phospholipids in lean muscle of beef cattle. *J. Anim. Sci.* 67:2282–2293.

LARICK, D. K., H. B. HEDRICK, M. E. BAILEY, J. E. WILLIAMS, D. L. HANCOCK, G. B. GARNER, and R. E. MORROW. 1987. Flavor constituents of beef as influenced by forage- and grain-feeding. *J. Food Sci.* 52:245–251.

LARICK, D. K., B. E. TURNER, R. M. KOCH, and J. D. CROUSE. 1989. Influence of phospholipid content and fatty acid composition of individual phospholipids in muscle from Bison, Hereford and Brahman steers on flavor. *J. Food Sci.* 54:521–526.

LESKANICH, C. O., and R. C. NOBLE. 1997. Manipulation of the ω-3 polyunsaturated fatty acid composition of avian eggs and meat. *World's Poult. Sci. J.* 53:155–183.

LING, X., L. AIMIN, Z. XIHE, C. XIAOSHU, and S. R. CUMMINGS. 1996. Very low rates of hip fracture in Beijing, People's Republic of China. *Am. J. Epidemiol.* 144:901–907.

LIU, Q., M. D. LANARI, and D. M. SCHAEFER. 1995. A review of dietary vitamin E supplementation for improvement of beef quality. *J. Anim. Sci.* 73:3131–3140.

LUOMA, P. V., S. NAYHA, K. SIKKILA, and J. HASSI. 1995. High serum alpha tocopherol, albumin, selenium and cholesterol, and low mortality from coronary heart disease in northern Finland. *J. Int. Med.* 237:49–54.

LYNN, W. S., and J. C. WALLWORK. 1992. Does food restriction retard aging by reducing metabolic rate? *J. Nutr.* 122:1917–1918.

MA, D. W. L., A. A. WIERZBICKI, C. J. FIELD, and M. T. CLANDININ. 1999. Conjugated linoleic acid in Canadian dairy and beef products. *J. Agric. Food Chem.* 47:1956–1960.

MAIJALA, K. 2000. Position paper. Cow milk and human development and well-being. *Livestock Prod. Sci.* 65:1–18.

MALINOW, M. R. 1994a. Homocyst(e)ine and arterial occlusive diseases. *J. Int. Med.* 236:603–617.

MALINOW, M. R. 1994b. Plasma homocyst(e)ine and arterial occlusive diseases: A mini-review. *Clin. Chem.* 40:173–176.

MALINOW, M. R. 1996. Plasma homocyst(e)ine: A risk factor for arterial occlusive diseases. *J. Nutr.* 126:1238S–1243S.

MARTIN, J. H., B. TRINK, D. DANEMAN, H. M. DOSCH, and B. H. ROBINSON. 1991. Milk proteins in the etiology of insulin-dependent diabetes mellitus (IDDM). *Ann. Med.* 23:447–452.

MARTIN, M. J., S. B. HULLEY, W. S. BROWNER, L. H. KULLER, and D. WENTWORTH. 1986. Serum cholesterol, blood pressure, and mortality: Implications from a cohort of 361 662 men. *Lancet* 2 (Oct. 25):933–936.

MCINTOSH, G. H., G. O. REGESTER, R. K. LELEU, P. J. ROYLE, and G. W. SMITHERS. 1995. Dairy proteins protect against dimethylhydrazine-induced intestinal cancers in rats. *J. Nutr.* 125:809–816.

MCKAY, C. M., L. A. MAYNARD, G. SPERLING, and L. L. BARNES. 1939. Restricted growth, lifespan, ultimate body size, and age changes in the albino rat after feeding diets restricted in calories. *J. Nutr.* 18:1–13.

MCNAMARA, D. J. 1990. Relationship between blood and dietary cholesterol. pp. 63–87. In: Pearson, A.M., and T.R. Dutson (Eds.). *Meat and health, Advances in meat research,* Vol. 6. Elsevier Applied Science, New York.

MEISEL, H. 1997. Biochemical properties of bioactive peptides derived from milk proteins: potential nutraceuticals for food and pharmaceutical applications. *Livestock Prod. Sci.* 50:125–138.

MELTON, S. L. 1990. Effects of feeds on flavor of red meat: A review. *J. Anim. Sci.* 68:4421–4435.

MENSINK, R. P., and M. B. KATAN. 1990. Effect of dietary trans fatty acids on high-density and low-density lipoprotein cholesterol levels in healthy subjects. *New Eng. J. Med.* 323:439–445.

MENSINK, R. P., P. L. ZOCK, M. B. KATAN, and G. HORNSTRA. 1992. Effect of dietary cis and trans fatty acids on serum lipoprotein(a) levels in humans. *J. Lipid Res.* 33:1493–1501.

MESSINA, M., and J. W. ERDMAN, JR. 1995. First international symposium on the role of soy in preventing and treating chronic disease. *J. Nutr.* 125:567S–808S.

MEYDANI, M. 2001. Symposium: Molecular mechanisms of protective effects of vitamin E in atherosclerosis. Vitamin E and atherosclerosis: Beyond prevention of LDL oxidation. *J. Nutr.* 131:366S–368S.

MEYER, H. H. D., A. ROWELL, W. J. STREICH, B. STOFFEL, and R. R. HOFMANN. 1998. Accumulation of polyunsaturated fatty acids by concentrate selecting ruminants. *Comp. Biochem. Physiol.* Part A 120:263–268.

MILLER, G. D., and C. M. WEAVER. 1994. Required versus optimal intakes: A look at calcium. *J. Nutr.* 124:1404S–1405S.

MICICH, T. J., T. A. FOGLIA, and V. H. HOLSINGER. 1992. Polymer-supported saponins: An approach to cholesterol removal from butteroil. *J. Agric. Food Chem.* 40:1321–1325.

MINCHIN, R. F., F. F. KADLUBAR, and K. F. LLETT. 1993. Role of acetylation in colorectal cancer. *Mutat. Res.* 290:35–42.

MOLONEY, A. P. 1995. Immunomodulation of fat deposition. *Livestock Prod. Sci.* 42:239–245.

MOLONEY, A. P., P. ALLEN, and W. J. ENRIGHT. 1998. Passive immunization of sheep against adipose tissue: effects on metabolism, growth and body composition. *Livestock Prod. Sci.* 56:233–244.

MUKHTAR, H. and N. AHMAD. 2000. Tea polyphenols: prevention of cancer and optimizing health. *Am. J. Clin. Nutr.* 71:1698S–1702S.

MULDOON, M. F., S. B. MANUCK, and K. A. MATTHEWS. 1990. Lowering cholesterol concentrations and mortality: a quantitative review of primary prevention trials. *Brit. Med. J.* 301:309–314.

NAIR, S. S. D., J. W. LEITCH, J. FALCONER, and M. L. GARG. 1997. Prevention of cardiac arrhythmia by dietary (ω-3) polyunsaturated fatty acids and their mechanism of action. *J. Nutr.* 127:383–393.

NAKAUE, H. S., R. R. LOWRY, P. R. CHEEKE, and G. H. ARSCOTT. 1980. The effect of dietary alfalfa of varying saponin content on egg cholesterol level and layer performance. *Poult. Sci.* 59:2744–2748.

NASSAR, A. H., and C. Y. HU. 1991. Growth and carcass characteristics of lambs passively immunized with antibodies developed against ovine adipocyte plasma membranes. *J. Anim. Sci.* 69:578–586.

NATIONAL RESEARCH COUNCIL. 1989. *Diet and health. Implications for reducing chronic disease risk.* National Academy Press, Washington, D.C.

NATIONAL RESEARCH COUNCIL. 1991. *Improving America's diet and health. From recommendations to action.* National Academy Press, Washington, D.C.

NATIONAL RESEARCH COUNCIL. 1992. *Eat for life.* Woteki, C.E., and P.R. Thomas (Eds.). National Academy Press, Washington, D.C.

NESTEL, P., M. NOAKES, B. BELLING, R. MCARTHUR, P. CLIFTON, E. JANUS, and M. ABBEY. 1992. Plasma lipoprotein lipid and Lp[a] changes with substitution of elaidic acid for oleic acid in the diet. *J. Lipid Res.* 33:1029–1036.

ODA, Y., M. TANAKA, and I. NAKANISHI. 1994. Relation between the occurrence of K-*ras* gene point mutations and genotypes of polymorphic N-acetyltransferase in human colorectal carcinomas. *Carcinogenesis* 15:1365–1369.

O'DEA, K. 1991. Traditional diet and food preferences of Australian aboriginal hunter-gatherers. *Phil. Trans. R. Soc. Lond.* B 334:233–241.

PAJARI, A. -M., P. HAKKANEN, R. -D. DUAN, and M. MUTANEN. 1998. Role of red meat and arachidonic acid in protein kinase C activation in rat colonic mucosa. *Nutrition and Cancer* 32:86–94.

PARODI, P. W. 1997. Cows' milk fat components as potential anticarcinogenic agents. *J. Nutr.* 127:1055–1060.

PARODI, P. W. 1999. Symposium: A bold new look at milk fat. Conjugated linoleic acid and other anticarcinogenic agents of bovine milk fat. *J. Dairy Sci.* 82:1339–1349.

PARTHASARATHY, S., T. C. KHOO, E. MILLER, J. BARNETT, J. L. WITZUM, and D. STEINBERG. 1990. Low density lipoprotein rich in oleic acid is protected against oxidative modification: Implications for dietary prevention of atherosclerosis. *Proc. Natl. Acad. Sci.* 87:3894–3898.

POPOVICH, D. G., D. J. A. JENKINS, C. W. C. KENDALL, E. S. DIERENFELD, R. W. CARROLL, N. TARIQ, and E. VIDGEN. 1997. The western lowland gorilla diet has implications for the health of humans and other hominoids. *J. Nutr.* 127:2000–2005.

PRENTICE, A. 1997. Is nutrition important in osteoporosis? *Proc. Nutr. Soc.* 56:357–367.

PRIOR, I. A., F. DAVIDSON, C. E. SALMOND, and Z. CZOCHANSKA. 1981. Cholesterol, coconuts, and diet on Polynesian atolls: A natural experiment: The Pukapuka and Tokelau Island studies. *Amer. J. Clin. Nutr.* 34:1552–1561.

REISER, R. 1975. Fat has less cholesterol than lean. *J. Nutr.* 105:15–16.

REISER, R., and F. B. SHORLAND. 1990. Meat fats and fatty acids. pp. 21–62. In: Pearson, A.M., and T.R. Dutson (Eds.). *Meat and health, Advances in meat research,* Vol. 6. Elsevier Applied Science, New York.

REISER, R., J. L. PROBSTFIELD, A. SILVERS, L. W. SCOTT, M. L. SHORNEY, R. D. WOOD, B. C. O'BRIEN, A. M. GOTTO, JR., and W. INSULL, Jr. 1985. Plasma lipid and lipoprotein response of humans to beef fat, coconut oil and safflower oil. *Am J. Clin. Nutr.* 42:190–197.

RENNER, E. 1994. Dairy calcium, bone metabolism, and prevention of osteroporosis. *J. Dairy Sci.* 77:3498–3505.

RHEEDER, J. P., W. F. O. MARASAS, P. G. THIEL, E. W. SYDENHAM, G. S. SHEPHARD, and D. J. VAN SCHALKWYK. 1992. *Fusarium monoliforme* and fumonisins in corn in relation to human esophageal cancer in Transkei. *Phytopath.* 82:353–357.

RIDKER, P. M., N. RIFAI, L. ROSE, J. E. BURING and N. R. COOK. 2002. Comparison of C-reactive protein and low-density lipoprotein cholesterol levels in the prediction of first vascular events. *NEJ Med.* 347:1557–1565.

RIFKIND, B. M., and L. D. GROUSE. 1990. Cholesterol redux. *JAMA* 264:3060–3061.

ROBERTS-THOMSON, I. C., P. RYAN, K. K. KHOO, W. J. HART, A. J. MCMICHAEL, and R. N BUTLER. 1996. Diet, acetylator phenotype, and risk of colorectal neoplasia. *Lancet* 347:1372–1374.

ROTHBLAT, G. H., F. H. MAHLBERG, W. J. JOHNSON, and M. C. PHILLIPS. 1992. Apolipoproteins, membrane cholesterol domains, and the regulation of cholesterol efflux. *J. Lipid Res.* 33:1091–1097.

RUDEL, L. L. 1999. Invited commentary. Atherosclerosis and conjugated linoleic acid. *Brit. J. Nutr.* 81:177–179.

SALONEN, J. T., K. NYYSSONEN, H. KORPELA, J. TUOMILEHTO, R. SEPPANEN, and R. SALONEN. 1992. High stored iron levels are associated with excess risk of myocardial infarction in eastern Finnish men. *Circulation* 86:803–811.

SANDERS, T. A. B. 1991. The Mediterranean diet: Fish and olives, oil on troubled waters. Symposium on "Mediterranean food and health," *Proc. Nutr. Soc.* 50:513–517.

SCANGA, J. A., K. E. BELK, J. D. TATUM, and G. C. SMITH. 2001. Supranutritional oral supplementation with vitamin $D_3$ and calcium and the effects on beef tenderness. *J. Anim. Sci.* 79:912–918.

SCHANBACHER, F. L., R. S. TALHOUK and F. A. MURRAY. 1997. Biology and origin of bioactive peptides in milk. *Livestock Prod. Sci.* 50:105–123.

SCHOKNECHT, P. A., S. EBNER, W. G. POND, S. ZHANG, V. MCWINNEY, W. W. WONG, P. D. KLEIN, M. DUDLEY, J. GODDARD-FINEGOLD, and H. J. MERSMANN. 1994. Dietary cholesterol supplementation improves growth and behavioral response of pigs selected for genetically high and low serum cholesterol. *J. Nutr.* 124:305–314.

SCHROEDER, J. W., D. A. CRAMER, R. A. BOWLING, and C. W. COOK. 1980. Palatability, shelflife and chemical differences between forage- and grain-finished beef. *J. Anim. Sci.* 50:852–859.

SCOLLAN, N. D., N. -J. CHOI, E. KURT, A. V. FISHER, M. ENSER, and J. D. WOOD. 2001. Manipulating the fatty acid composition of muscle and adipose tissue in beef cattle. *Brit. J. Nutr.* 85:115–124.

SCOTT, L. W., J. K. DUNN, H. J. POWNALL, D. J. BRAUCHI, M. C. MCMANN, J. A. HERD, K. B. HARRIS, J. W. SAVELL, H. R. CROSS, and A. M. GOTTO, JR. 1994. Effects of beef and chicken consumption on plasma lipid levels in hypercholesterolemic men. *Arch. Intern. Med.* 154:1261–1267.

SETCHELL, K. D. R. 1998. Phytoestrogens: The biochemistry, physiology, and implications for human health of soy isoflavones. *Am. J. Clin. Nutr.* 68:1333S–1346S.

SHANTHA, N. C., A. D. CRUM, and E. A. DECKER. 1994. Evaluation of conjugated linoleic acid concentrations in cooked beef. *J. Agric. Food Chem.* 42:1757–1760.

SHI, B., and J. E. SPALLHOLZ. 1994. Selenium from beef is highly bioavailable as assessed by liver glutathione peroxidase (EC1.11.1.9) activity and tissue selenium. *Brit. J. Nutr.* 72:873–881.

SHORLAND, F. B. 1988. Is our knowledge of human nutrition soundly based? *World Rev. Nutr. and Diet.* 57:126–213.

SIMOPOULOS, A. P. 1998. Overview of evolutionary aspects of ω-3 fatty acids in the diet. In: Simopoulos, A.P. (Ed.). *The return of ω-3 fatty acids into the food supply. I. Land-based animal food products and their health effects. World Rev. Nutr. and Diet.* 83:1–11.

SIMOPOULOS, A. P. 2000. Symposium: Role of poultry products in enriching the human diet with ω-3 PUFA. *Poult. Sci.* 79:961–970.

SKOG, K. I., M. A. E. JOHANSSON, and M. I. JAGERSTAD. 1998. Carcinogenic heterocyclic amines in model systems and cooked foods: A review on formation, occurrence and intake. *Fd. Chem. Toxic.* 36:879–896.

SPONHEIMER, M., and J. A. LEE-THORP. 1999. Isotopic evidence for the diet of an early hominid, *Australopithecus africanus. Science* 283:368–370.

STADELMAN, W. J. 1999. World's Poultry Science Association Invited Lecture. The incredibly functional egg. *Poult. Sci.* 78:807–811.

STANFORD, C. B., and H. T. BUNN. (Eds.) 2001. *Meat eating and human evolution.* Oxford University Press, New York.

STAVRIC, B. 1994. Biological significance of trace levels of mutagenic heterocyclic aromatic amines in human diet: A critical review. *Fd. Chem. Toxic.* 32:977–994.

STEINBERG, D. 1997. Minireview. Low density lipoprotein oxidation and its pathobiological significance. *J. Biol. Chem.* 272:20963–20966.

STEINBERG, D., and J. L. WITZTUM. 1990. Lipoproteins and atherogenesis. Current concepts. *JAMA* 264:3047–3052.

ST. JOHN, L. C., D. K. LUNT, and S. B. SMITH. 1991. Fatty acid elongation and desaturation enzyme activities of bovine liver and subcutaneous adipose tissue microsomes. *J. Anim. Sci.* 69:1064–1073.

SULLIVAN, J. L. 1989. The iron paradigm of ischemic heart disease. *Am. Heart J.* 117:1177–1188.

SULLIVAN, J. L. 1992. Stored iron and ischemic heart disease. Empirical support for a new paradigm. *Circulation* 86:1036–1037.

SWANEK, S. S., J. B. MORGAN, F. N. OWENS, D. R. GILL, C. A. STRASIA, H. G. DOLEZAL, and F. K. RAY. 1999. Vitamin $D_3$ supplementation of beef steers increases Longissimus tenderness. *J. Anim. Sci.* 77:874–881.

TAUBES, G. 2001. The soft science of dietary fat. *Science* 291:2536–2545.

THOROGOOD, M., J. MANN, P. APPLEBY, and K. MCPHERSON. 1994. Risk of death from cancer and ischaemic heart disease in meat and non-meat eaters. *Brit. Med. J.* 308:1667–1671.

TOPPING, D. L. 1991. Soluble fiber polysaccharides: Effects on plasma cholesterol and colonic fermentation. *Nutr. Rev.* 49:195–203.

TRICHOPOULOU, A., and P. LAGIOU. 1997. Healthy traditional Mediterranean diet: An expression of culture, history, and lifestyle. *Nutr. Rev.* 55:383–389.

TRICHOPOULOU, A., and E. VASILOPOULOU. 2000. Mediterranean diet and longevity. *Brit. J. Nutr.* 84:S205–S209.

VAN DER MEER, R., J. A. LAPRE, M. J. A. P. GOVERS, and J. H. KLEIBEUKER. 1997. Mechanisms of the intestinal effects of dietary fats and milk products on colon carcinogenesis. *Cancer Letters* 114:75–83.

VOSKUIL, D. W., E. KAMPMAN, M. J. A. L. GRUBBEN, R. A. GOLDBOHM, H. A. M. BRANTS, H. F. A. VASEN, F. M. NAGENGAST, and P. VAN'T VEER. 1997. Meat consumption and preparation, and genetic susceptibility in relation to colorectal adenomas. *Cancer Letters* 114:309–311.

WAINWRIGHT, P. 2000. Invited commentary. Nutrition and behaviour: The role of ω-3 fatty acids in cognitive function. *British J. Nutr.* 83:337–339.

WEAVER, B. J., and B. J. HOLUB. 1988. Health effects and metabolism of dietary eicosapentaenoic acid. *Progress in Food Nutr. Sci.* 12:111–150.

WEAVER, L. T. 1997. Significance of bioactive substances in milk to the human neonate. *Livestock Prod. Sci.* 50:139–146.

WEIR, D. G., and J. M. SCOTT. 1998. Homocysteine as a risk factor for cardiovascular and related disease: Nutritional implications. *Nutr. Res. Rev.* 11:311–338.

WEISBURGER, J. H., A. RIVENSON, K. GARR, and C. ALIAGA. 1997. Tea, or tea and milk, inhibit mammary gland and colon carcinogenesis in rats. *Cancer Letters* 114:323–327.

WHEELER, T. L., S. D. SHACKELFORD, E. CASAS, L. V. CUNDIFF, and M. KOOHMARAIE. 2001. The effects of Piedmontese inheritance and myostatin genotype on the palatability of longissimus thoracis, gluteus medius, semimembranosus, and biceps femoris. *J. Anim. Sci.* 79:3069–3074.

WHITE, R., and S. HAVALA. 1995. Commentary. Why America cannot sustain its diet. *Preventive Med.* 24:656–657.

WIDDOWSON, E. M., and A. WHITEN (Eds.). 1991. Foraging strategies and natural diet of monkeys, apes and humans. *Phil. Trans. R. Soc. Lond.* B 334:159–295.

WILLETT, W. C., M. J. STAMPFER, G. A. COLDITZ, B. A. ROSNER, and F. E. SPEIZER. 1990. Relation of meat, fat, and fiber intake to the risk of colon cancer in a prospective study among women. *New Engl. J. Med.* 323:1664–1672.

WINAWER, S. J., and M. SHIKE. 1992. Dietary factors in colorectal cancer and their possible effects on earlier stages of hyperproliferation and adenoma formation. *J. Natl. Cancer Inst.* 84:74–75.

WU, D., M. G. HAYEK, and S. N. MEYDANI. 2001. Vitamin E and macrophage cyclooxygenase regulation in the aged. *J. Nutr.* 131:382S–388S.

YANG, C., J. CHUNG, G. YANG, S. CHHABRA and M. LEE. 2000. Tea and tea polyphenols in cancer prevention. *J. Nutr.* 130:472S–478S.

ZHANG, Y., P. TALALAY, C. G. CHO, and G. H. POSNER. 1992. A major inducer of anticarcinogenic protective enzymes from broccoli: isolation and elucidation of structure. *Proc. Nat. Acad. Sci.* USA 89:2399–2403.

ZHENG, W., D. R. GUSTAFSON, R. SINHA, J. R. CERHAN, D. MOORE, C. -P. HONG, K. E. ANDERSON, L. H. KUSHI, T. A. SELLERS, and A. R. FOLSOM. 1998. Well-done meat intake and the risk of breast cancer. *J. Natl. Cancer Inst.* 90:1724–1729.

ZINN, S. A. 1997. Bioactive components in milk: introduction. *Livestock Prod. Sci.* 50:101–103.

# CHAPTER 3

# Feed vs. Food: Do Livestock Compete with Humans for Food Resources?

**CHAPTER OBJECTIVES**

1. To discuss how livestock production can be integrated with crop production to increase the efficiency of human food production

2. To discuss how livestock production can be based on crop residues and by-products to optimize the utilization of resources

The answer to the question posed by the chapter title is "sometimes yes, sometimes no" (the type of answer expected from a professor!). The question could be rephrased several ways, and the answer would remain the same. Is it justifiable that livestock compete with humans. . . , Must livestock compete with humans. . . , Is it undesirable if. . . , and so on. Much of the argument is theoretical and deals with efficiency of resource utilization, ecological concerns, and idealistic concepts of how livestock should integrate with crop production. In the real world, it is unlikely that

any food shortages, famines, and starvation that occur in human populations are a direct or even fairly remotely indirect consequence of livestock production. Recent episodes of this nature, such as starvation in Ethiopia, Somalia, Rwanda and Yugoslavia, are a result of political struggles. People in areas of dispute among rival tribal groups are used as political pawns. They are starving because they have been driven from their land, herded into refugee camps, and denied food and medicine. The reason they are starving is not related in any way to any competition with livestock for feed. In fact, livestock (and the seed for next year's crop) are the final bulwarks against famine; when times get tough, the animals are butchered and eaten and, in this way, are a **walking larder** (Clutton-Brock, 1989).

In this chapter, two general situations will be discussed: situations where livestock are potentially competitive with humans for food, and situations where livestock and humans are non-competitive.

# DIRECT COMPETITION FOR FOOD RESOURCES BETWEEN LIVESTOCK AND HUMANS

Certain feedstuffs such as grains and pulses (pulses are edible legume seeds such as peas, beans, and lentils) fed to livestock can be directly consumed by humans. Anything that can be used effectively as a food for humans will be used with lower efficiency if used as an animal feed to produce meat, milk, or eggs. This relates to the **food chain** or pyramid. Each step in the pyramid introduces an inefficiency. Strictly on a scientific basis, there can be no dispute that corn and soybean meal are used with more efficiency and can provide food for more people when they are eaten directly by people rather than being fed to swine or poultry to be converted to pork, chicken meat, or eggs for human consumption. However, there is more to life in most human societies than the grim struggle to keep from starving. The people of North America or western Europe are not likely to adopt a diet of corn and soybeans. On the other hand, many people in China, India, or Russia might be pleased to have such a diet. When and if their economic status improves, they would in all likelihood consider the inclusion of more animal products in their diet one of their first priorities.

Thus, in my opinion, a discussion of competition between humans and livestock for grain and other high quality foods is largely of philosophical interest, without much rele-

vance to the real world (a cliche, but it sums it up!). One would be very hard-pressed to identify a human society where the welfare of people has been adversely affected in any major way by competition between humans and livestock for grain. The idea that such would occur is an elitist position, implying that impoverished people in developing countries have so little intellectual capacity that they would starve themselves by feeding their grains and high quality foods to chickens instead of feeding themselves. In fact, these people have developed, over hundreds or thousands of years, systems of raising pigs or poultry so they produce meat and eggs for human consumption without being competitive. Village chickens and pigs have traditionally been raised on table scraps. In China, pigs are fed large amounts of succulent forage harvested from ditch banks and aquatic vegetation from ponds and lagoons. It was largely in the twentieth century that the influx of "experts" led to situations of countries importing grains to feed to livestock, which may or may not be appropriate, depending upon specific local circumstances.

## Meat or Wheat for the Next Millennium

A symposium entitled "Meat or Wheat for the Next Millennium" was published in 1999 (Millward, 1999). An **optimal diet** is one that maximizes health and longevity, prevents nutrient deficiencies, reduces risks for chronic diseases, and is composed of foods that are available, safe, and palatable. An apparent irony is that human nutritionists and the general public in developed countries are emphasizing diets based largely on foods from plant sources, while in developing countries (where most humans live), plant-based diets are associated with extreme poverty and poor health. When economic conditions improve, these populations increase their consumption of meat and display improved health (Nestle,

1999). Meat consumption in many cultures is associated with celebration and ceremony. In developing countries, there is often resentment of what is seen as a somewhat arrogant and proselytizing attitude of vegetarians from developed countries, who have purchasing power and choices enabling the selection of varied and appealing meatless diets that are clearly out of reach of people in developing countries (Millward, 1999). Millward concludes, "For a number of highly complex social, political and economic reasons, meat occupies a pivotal position in the global food chain, which is unlikely to change much in the forseeable future" (p. 9210). Reduction in meat consumption in developed countries (where it is higher than optimal) will do little to improve the health of people in developing countries (through release of grain); significant dietary improvement in developing countries mainly requires economic growth that will allow more people to add animal products to their diets (Rosegrant et al., 1999).

## Cereal Grains in the Human Diet

The major dietary energy sources for humans are the cereal grains (cereal grains are the edible seeds of grasses). The world's top 30 food crops are shown in Table 3-1. Three cereal grains (wheat, maize [corn] and rice) make up over 75 percent of world grain production, providing over half of the food energy consumed by humans. Cordain (1999) has called cereal grains our double-edged sword. He claims that except for the last 10,000 years, for the vast majority of human existence we rarely if ever consumed cereal grains. Humans have existed as non-cereal-eating hunter-gatherers since the emergence of *Homo erectus* 1.7 million years ago. Thus, he claims, "We have had little time (<500 generations) to adapt to a food type which now represents humanity's major source of both calories and protein" (p. 21). Archeological evidence of when cereal grains entered the human diet is the appearance of stone processing

tools (mortar and pestle) needed for grinding and cooking seeds. These artifacts are irrefutable evidence of when and where cultures began to include cereal grains in their diets. According to Cordain (1999), when cereal grains replaced the animal-based diets of hunter-gatherers, there were characteristic negative effects such as reduction in stature, increased infant mortality, reduced lifespan, increase in infectious diseases, increase in iron-deficiency anemia, increased osteomalacia, and increased dental caries and enamel defects. Cereal grains are a "double-edged sword" because without them, civilization as we know it, stemming from the invention of agriculture, would not have developed. On the positive side, the enormous increase in human knowledge would not have taken place without the agricultural revolution. On the other hand, agricultural development has led to societal ills including whole-scale warfare, starvation, tyranny, epidemic diseases, and class divisions. While we are no longer hunter-gatherers, our genetic makeup is still that of a Paleolithic hunter-gatherer who is optimally adapted to a diet of meat, fruits, and vegetables, not to cereal grains.

Cordain (1999) discusses in considerable detail the nutritional shortcomings of cereal grains. For example, cereal grains are deficient in vitamins (vitamin A activity, and several B-complex vitamins such as thiamin, niacin, and vitamin $B_{12}$). Two of the major deficiency diseases that have plagued agricultural humans are pellagra and beriberi, exclusively associated with consumption of cereal grains. Several B vitamins in grains have a low bioavailability to humans (niacin, pyridoxine, biotin). Cereal grains are extremely deficient in calcium, and they contain phytic acid, which greatly reduces bioavailability of phosphorus, zinc, iron, and other trace elements. In populations where cereal grains are a major source of calories, osteomalacia, rickets, osteoporosis, and iron-deficient anemia are common. Zinc deficiency is common in some countries like Iran, where bread contributes over 50 percent of calories. Grains have a high

**TABLE 3-1**  The world's major food crops.
(Adapted from Cordain, 1999.)

| Rank | Crop | Million Metric Tons | Rank | Crop | Million Metric Tons |
|------|------|---------------------|------|------|---------------------|
| 1 | Wheat | 468 | 16 | Beans | 14 |
| 2 | Maize | 429 | 17 | Peanuts | 13 |
| 3 | Rice | 330 | 18 | Peas | 12 |
| 4 | Barley | 160 | 19 | Banana | 11 |
| 5 | Soybean | 88 | 20 | Grape | 11 |
| 6 | Cane sugar | 67 | 21 | Sunflower | 9.7 |
| 7 | Sorghum | 60 | 22 | Yams | 6.3 |
| 8 | Potato | 54 | 23 | Apple | 5.5 |
| 9 | Oats | 43 | 24 | Coconut | 5.3 |
| 10 | Cassava | 41 | 25 | Cottonseed (oil) | 4.8 |
| 11 | Sweet potato | 35 | 26 | Orange | 4.4 |
| 12 | Beet sugar | 34 | 27 | Tomato | 3.3 |
| 13 | Rye | 29 | 28 | Cabbage | 3.0 |
| 14 | Millet | 26 | 29 | Onion | 2.6 |
| 15 | Rapeseed | 19 | 30 | Mango | 1.8 |

$\omega$-6/$\omega$-3 fatty acid ratio, which may lead to essential fatty acid deficiencies. Grains are deficient in essential amino acids. Many people are allergic to proteins in grains. Thus an important role of meat and other animal products in the human diet is to provide a dietary source for which we are evolutionarily adapted in order to make up for problems associated with a food source (grains) for which we are not well adapted.

## Which Livestock Are Most or Least Competitive?

The natural diets of different kinds of livestock are closely intertwined with their feeding strategies, taste and other food perception responses, and digestive tract physiology. Although swine and chickens are both non-ruminant omnivores, they differ in the characteristics just mentioned. Chickens are **gramnivores,** meaning that they are evolu-

tionarily designed for seed eating. Their beaks, crop, gizzard, lack of much microbial digestion and rapid rate of digesta passage all facilitate the eating and digestion of seeds. Plant seeds, especially those large enough for a chicken to eat, are an excellent source of nutrients, containing an abundance of protein and energy. Although chickens are omnivorous and will eat some grass, insects, worms, and even small animals like mice, as well as dead animals (carrion), their primary natural diet is seeds. Swine are quite different. They have a fairly large hindgut, with significant microbial digestion. The natural diet of wild pigs is not based on seeds; they lack the mouth anatomy that would make seed eating feasible. They consume vegetation, roots, and insects and small animals obtained via their rooting behavior, as well as eggs of ground-nesting birds, defenseless newborn animals, and so on. In some countries, such as China, pigs have traditionally been raised on green feeds such as aquatic weeds and waste vegeta-

**FIGURE 3-1** Livestock can be complementary, rather than competitive, in the production of human food. Grain crops produce a large amount of biomass besides grain. In the case of maize, the leaves and stalks (stover) can be utilized by ruminant animals. Large numbers of beef cows in the United States are wintered on corn stover, as illustrated here. (Courtesy of J. L. Johnson.)

bles, as well as table scraps and kitchen wastes. Similarly, chickens are traditionally fed table scraps, kitchen wastes and cull vegetables, but without the coarser forage fed to pigs. Thus the natural diets of swine and poultry are somewhat different, and neither are competitive with human needs. Rather, they increase food production by making use of resources that would otherwise be wasted.

Ruminants, of course, can utilize forages and crop residues (Fig. 3-1) because of microbial digestive processes in the rumen. Theoretically, in countries with high human population density and the need to use food resources efficiently, it might seem that ruminants would be less competitive with humans than swine and poultry. While this is true to some degree, it is interesting to note that swine and poultry are usually more important in these countries. Except in countries with religious rules forbidding consumption of pork, pigs are and have been very important food producing animals in densely populated Asian countries such as China, Vietnam, Thailand, and so on. Ruminants are used mainly for work and milk production in these countries. Their performance on low quality roughages is too low to justify raising them

primarily for meat production, and higher quality feed ingredients are more efficiently used for swine and poultry production. Cropland in most developing countries is too valuable to be used to grow forage or other livestock feed. Livestock are fed mainly on **crop residues.** Low quality residues such as rice straw are fed to mature ruminants, particularly draft animals such as oxen and buffalo, while higher quality residues such as rice bran are fed to swine and poultry (Warren and Farrell, 1990). Cuba is a country where the ruminant vs. nonruminant story has played out. With the collapse of the Soviet Union, Cuba lost its major economic sponsor and was on its own. The Castro government was forced to develop policies to feed the human population as cheaply as possible. Beef cattle production was slashed, while pig and poultry production was expanded ("Cuba Places Limits on Beef Production," *Feedstuffs* Oct. 2, 2000, p. 5).

Intensive production of swine and poultry in both industrialized and developing countries is based on feeding grains and other concentrates. Intensive production systems allow for improvement in efficiency of animal management, disease control, and consistency in feeding. In the developed countries, the feedstuffs used are grains such as corn (maize), sorghum (milo), and to a lesser extent, wheat, barley, and oats. These grains are not currently in high demand for human consumption. Although maize is a desirable food for humans, the majority of the U.S. corn crop consists of varieties grown for animal feed. In Europe, extensive amounts of maize are grown for livestock feed, while very little is consumed by humans. Many Europeans regard the American custom of eating corn (e.g., corn on the cob) as almost barbaric. Feed wheat usually does not meet quality control, baking quality, protein content, or other standards necessary for food wheat. Protein supplements used for livestock include soybean meal, and to a lesser extent, cottonseed and canola meals, meat and bone meal, and fish meal. These products not currently in high demand for human

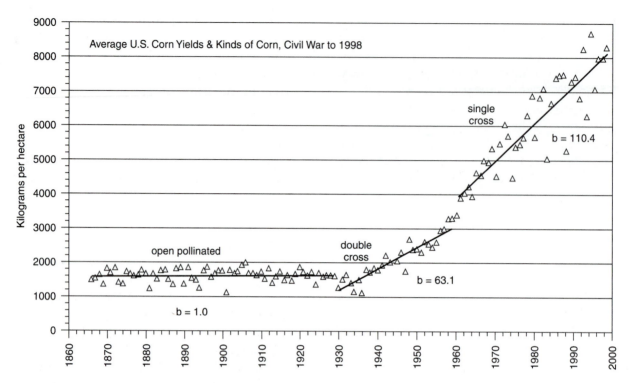

**FIGURE 3-2** Average U.S. corn yields since the Civil War. Note that yields have steadily and dramatically increased, largely because of the development of hybrid corn. (From Troyer, 1999.)

consumption. Cottonseed and canola meals contain inhibitors or toxins (e.g., gossypol and glucosinolates) that limit their direct use in the human diet. Meat and bone meal and fish meal are prepared from renderers' wastes (meat and bone meal) and by-products of fish processing, as well as fish species not generally consumed directly by humans, including so-called "trash fish." Thus in North America and Europe, intensive swine and poultry production is based largely on feed ingredients not currently in demand for human food.

Corn (maize) is grown mainly for livestock and poultry feed and for industrial purposes (corn oil, fructose sweeteners, ethanol, etc.). Yields of corn have increased greatly over the past 150 years (Fig. 3-2). Genetic engineering using biotechnology is likely to result in further improvements in corn yield. For example, corn with an extended photosynthetic capacity during kernel ripening, apportioning a higher percentage of caloric energy to grain production, may be a viable strategy for increasing

grain yield (Hedin et al., 1998). This will allow expansion of livestock and poultry production without direct competition with human food needs. We can eat only so much corn!

For ruminants, various non-competitive sources of nitrogen that can be converted to microbial protein in the rumen include dried poultry waste (DPW) and sewage sludge. DPW is the manure and litter from poultry houses. Birds excrete uric acid, a concentrated source of nitrogen. DPW has been used quite extensively as a nitrogen source for ruminants. Potential drawbacks to the feeding of DPW are the possibilities of drug residues in the excreta, high copper levels (if copper sulfate was used as a growth promotant for the poultry), and consumer rejection of the practice. In some countries, feeding DPW is banned because the practice is perceived as being unwholesome or unclean.

In developing countries, intensive swine and poultry production is often based on imported corn and soybean meal. It could and has been reasonably argued that the use of scarce foreign exchange for purchase of feed ingredients for poultry and livestock production is not wise. Usually there are a number of factors involved in the decision to establish these enterprises, including the availability of grains subsidized by the exporting countries and the desire of governments to meet consumer demand. Citizens of China, the former Soviet Union, and most developing countries have had all the vegetarian diets they want and exhibit strong consumer demand for a higher quality diet with meat, eggs, and other animal products. Thus while China has a very high human population of over 1 billion, which is still increasing, the Chinese government is vigorously developing a feed manufacturing industry and intensive swine and poultry production. If there is any part of the world where swine and poultry production would theoretically be in competition with humans for grains, it might be expected to be China. However, improving the human diet and satisfying consumer demand are more important than the academic issue of whether or not the grain could be more efficiently used when eaten directly by people. The people don't want to eat all of it directly!

It should also be appreciated that intensive swine and poultry production in developing countries can be based on feedstuffs that are reasonably concentrated sources of protein and energy but for various reasons are not, and likely will not be, major items of the human diet. For example, **cassava** (also known as manioc, tapioca) is a tropical tuber that is grown extensively for food and feed throughout the tropics (Fig. 3-3). It is regarded as a poor person's food, and is consumed mainly by very low income people. If their economic status improves, they have no desire to eat cassava. Cassava meal is rich in starch and is an excellent replacement for grain in diets for nonruminants. Cassava will grow in poor, acid soils, and the foliage can also be used as animal feed. In Thailand, extensive cassava plantations have been established to produce cassava meal for export to Europe as feed. This

**FIGURE 3-3** Cassava is a tropical plant that produces tubers rich in starch *(top)*. The tops *(bottom)* may be used as forage. Although cassava can be used directly as human food, it is often socially and economically more promising to use cassava meal as an animal feed.

market is basically an artificial one; cassava is not a grain so is not subject to tariffs imposed by the European Union on feed grains. This results in the apparent absurdity of Thailand exporting cassava to Europe to be used in animal feeds as a replacement for feed grains of which

Europe has a huge surplus. However, within the entangled structure of tariffs, economic assistance to developing countries, generation of hard currency by developing countries, and soothing the guilty consciences of former colonial powers, it all makes sense, or it wouldn't be happening.

There is considerable potential for huge increases in cassava meal production in developing countries. A project in Colombia illustrates the potential. Peasant farmers (campesinos) grow small plots of cassava and sell the roots to a cooperative of which they are members. The cooperative has a chipping machine; the chipped cassava is sun-dried on a large concrete slab. After drying, it is bagged up and sold to feed mills, which use it in their formulation of animal feeds. The manufactured feed is sold to intensive swine and poultry units, which efficiently produce low cost poultry products and pork. As the campesinos improve their economic status by selling cassava, they can begin to buy consumer goods, including poultry products and pork. As the general economic status of the country improves, the middle class expands, increasing the demand for animal-origin foods. Thus despite the fact that humans could use cassava more efficiently by eating it directly rather than feeding it to livestock, they prefer not to, particularly as economic status rises. The competition between humans and livestock for cassava would become important only if people were facing starvation. They would first eat the livestock and then subsist on a cassava diet. This is a **doomsday scenario,** which is unlikely to occur.

Besides cassava, there are many other reasonable quality feedstuffs that can be used for production of nonruminant animals in tropical developing countries. These include energy sources such as rice bran and palm oil by-products, and protein supplements like palm kernel meal, copra (coconut meal), cottonseed meal, and other oilseed meals. Many plant foods contain toxic factors (e.g., cassava contains cyanide). Using them as animal feed puts a "buffer" between the plant toxin and humans.

**FIGURE 3-4** Sugarcane produces huge quantities of biomass. New research in countries such as Colombia, Australia, and Cuba is leading the way to more efficient use of this productive crop as a "tropical grain-substitute" for livestock feeding.

A major potential feedstuff for tropical livestock production is **sugarcane** (Fig. 3-4). Cane can produce more energy per acre than any other crop, in the form of sucrose (table sugar). Various processing methods for sugarcane are being developed to maximize its efficiency of utilization as a feed (McMeniman et al., 1990). The juice can be squeezed out and used as a liquid feed for swine and poultry. The fibrous residue can be used as a ruminant feed. Sugarcane can be regarded as the tropical equivalent of grain. Ruminants can be fed processed sugarcane, with sucrose as the major dietary energy source. Considerable research in Cuba and Colombia has shown the feasibility of sugar-based feeding systems. A sugar-urea feed has been developed for ruminants, to be fed with a small supplement of by-pass protein (such as tropical tree legume forage). Again, on a strictly theoretical basis, sugar could be used with maximum biological efficiency if consumed directly by humans rather than being fed to livestock. It is obvious that in the real world, this will not happen.

An example of effective use of livestock to convert grain to human food is the use of rye grain for poultry feed in Poland. Poland had a surplus of two million tons of rye in 1992, for which there was no buyer (McGinnis, 1992). Rye can be used for human consumption in the form of rye bread. However, most people find rye bread unappetizing; the people of Poland and Russia, who have traditionally consumed rye bread, switch to wheat bread as soon as it is available. Large areas of Poland and Russia are not suitable for growing wheat because of climatic conditions, but rye can be grown. McGinnis (1992) speculates that the efficient use of the unwanted rye grain for poultry production could improve the nutritional status of the Polish people. Various commercial enzyme products with pentosanase activity can improve the feeding value of rye for poultry (Campbell et al., 1991; Friesen et al., 1991). This is an example of how grain can be effectively used by livestock and poultry in a non-competitive way to produce human food. Barley is similarly not well accepted by people as a major portion of the human diet; like rye, it is one of the **coarse grains** traditionally used as animal feed. The use of commercial sources of β-glucanase to improve barley utilization by swine and poultry (see Chapter 5) increases its value as an animal feed.

In summary, nonruminant animals such as swine and poultry are theoretically in direct competition with humans for food resources. In practice, however, they seldom are. In those countries with high human population densities, such as China, the path to improving the human diet is to expand poultry and/or swine production. Although ruminants are theoretically non-competitive, their performance on roughage-based diets is too poor to justify their production, except for use as draft animals and sources of dung for fuel. Their low

efficiency of use of high energy feedstuffs such as grains and cassava results in it being much more profitable and effective to use these feedstuffs for nonruminants. Thus while conventional wisdom might suggest the opposite, in the real world, intensive poultry and swine production has much more potential than ruminant production in meeting food needs in densely populated developing countries. An additional factor is that these are **short cycle animals,** which have a high reproductive rate and reach slaughter size rapidly, in contrast to ruminants, which have a low reproductive rate and take up to several years (for cattle) in the tropics to reach slaughter weight. Among other things, this means that a very high percentage of the lifetime feed intake of ruminants has been used for maintenance, rather than for production.

The industrialized countries of North America and Europe, and other areas such as Argentina and Australia, have huge surpluses of grains and the capacity to greatly increase grain production if market conditions warrant it. The issue of competition between humans and livestock for grains is totally academic under these conditions. The surplus of grain is so huge that it is even practical and economical to use it as a major feed for ruminants. The United States has a very large beef cattle feedlot industry based on this fact. The cost of energy (e.g., $/kcal DE) is much lower in the United States for corn and other grains than for roughages like alfalfa hay and crop residues. The fact that humans can use corn more efficiently by eating corn rather than corn-fed beef is irrelevant. If they choose to do so, they can eat corn. Obviously, most people choose not to and eat a varied diet. The issue of competition between humans and livestock for grain is a nonissue in the developed countries; there isn't any competition. There is so much grain available and potentially available that millions more people and millions more livestock could be fed on grain, and there still would be no competition. Instead, there would be a lot of happy grain farmers who could sell all they could produce, at a profit! Again, with the declining birth rate in developed countries,

eve
cou

suc
not
etai
tha
Live
the:
the
pala
forr

that
fed
com
prov
land
duce
non
proc
arid
topc
coul
curre
tion
land
crop:
plan
dew
be us
low
unlik

T
in a
cient
resul
of str
corn)
crop
grain
woul
prese
devel
rumir
most
conce
integr

what happened many years ago in the United States. At one time, over 90 percent of the U.S. population consisted of farmers. Now it is less than 1 percent, and most people live in cities. They may occasionally express a yearning for the fresh air and quiet of the countryside, but in reality most people prefer city life and its amenities. Cities the world over are growing at a tremendous rate, and the rural-urban ratio is changing rapidly. To support these huge urban centers, some of which (e.g., Mexico City) will soon exceed 30 million people, intensive agriculture is required. In the animal sector, this will involve intensive poultry and dairy production systems (and swine in non-Moslem societies). To be sustainable, these will have to be based on the efficient use of local resources, rather than on imported ingredients. Feeding systems based on low quality roughages cannot support adequate productivity for meat production, even though they may be completely adequate for mature oxen, buffaloes, or donkeys used as work animals.

Efficient meat and dairy production cannot be based on low quality diets. Animals must first meet their maintenance energy requirements before they can grow or produce milk. In the tropics, forages, by-products, and other low quality roughages cannot even meet an animal's maintenance requirements for much of the year. Beef cattle in the Latin American tropics may take five years or more to reach slaughter weight (Fig. 3-5). Most of what they have eaten has simply been used for maintenance. A fast-growing meat animal fed a high energy diet uses a much smaller proportion of its lifetime feed intake for maintenance than a slow-growing animal that takes years to reach slaughter weight. The low quality roughage that sustained the latter animal, while "non-competitive," would have been much better utilized as a component of a feeding system involving supplementation with "competitive" feeds to give a reasonable level of production. The final product, the meat, will be of much greater consumer acceptance if produced by animals fed adequately than by animals several years of age at slaughter.

**FIGURE 3-5** Cattle in the dry season in Brazil. Often they lose more weight in the dry season than they have gained during the previous wet season. Proper nutritional supplementation using high quality ingredients optimizes resource utilization. (Courtesy of Lee R. McDowell.)

# INTEGRATION OF LIVESTOCK INTO SUSTAINABLE CROPPING SYSTEMS

Livestock and crop production have traditionally been integrated, with crop rotations involving forages and use of crop residues (straw, stover) as feed and bedding. **Crop rotations** employing forage legumes improve soil fertility and provide nitrogen to succeeding grain crops. Animal power was used to accomplish the planting, cultivation, and harvesting of crops. This traditional **mixed farm** of Europe and North America was sustainable over the centuries, and effectively integrated crop and animal production. Only in the twentieth century did this change, as agriculture industrialized with the availability of cheap fossil fuels. Chemical fertilizers replaced the use of crop rotations, green manure crops, and animal manure for maintaining soil fertility, and increased grain yields markedly. Now livestock and crop production have become less integrated, with farms specializing in one or the other. In the Corn Belt of the United States, farmers traditionally marketed part of their grain by feeding it to hogs and feeder cattle. Cattle feeding has moved west and southwest, and swine production is rapidly moving to large scale, intensive confinement units, similar to the intensification of the poultry industry that has already occurred. The advantages of this uncoupling of animal and crop production are more efficient production of both crops and livestock, with the cost efficiencies of large-scale enterprises. With intensive grain production techniques, involving high rates of fertilizer and agricultural chemical application, very high yields of grain and soybeans are achieved. Costs of production per unit of product, both for grain and livestock, have been reduced, accounting for the continued movement of agriculture in this direction. The disadvantages of not integrating crop and livestock production are less clear and more based on non-economic factors. These include the relative merits of **factory farms** vs. **family farms** in terms of animal welfare and social values. It is claimed by some that modern agriculture is not sustainable because it is based on the use of high inputs of fossil fuels, which are a depletable resource. Monocultures of grain have increased soil erosion, but this is not inevitable with the use of non-till systems and other techniques. Another disadvantage of large-scale intensive livestock production compared with traditional systems is that the concentration of large numbers of animals in a small area may lead to air and water pollution problems.

The pattern of industrialized agriculture has been that as new technologies that increase production efficiently are introduced, they are rapidly adopted by many farmers. This may result in a temporary increase in income by farmers who adopt the new techniques first, but this is quickly followed by a decrease in the market price for the products. Thus other farmers are either forced to adopt the new technology to remain competitive, or they are gradually squeezed out by the more efficient producers. This has been the pattern of American agriculture throughout most of the twentieth century and during the early

years of the twenty-first century. The predictable results are that there are fewer farmers, larger farms, and steadily decreasing food costs for consumers. For example, since 1946, the cost per dozen eggs in the grocery store, or the cost per lb of chicken meat, has increased very little in absolute dollar amounts, whereas the value of wages and most non-agricultural goods has increased several fold since the 1940s. Left strictly to economic forces, American agriculture will continue to intensify and consolidate into fewer and larger units, for both crop and livestock production. Non-economic factors are likely to play a role in the future, however, particularly with livestock production. There is widespread public opinion, rightly or wrongly, that intensive "factory farming" of animals is inhumane and intrinsically wrong. Several U.S. states now have laws restricting **corporate farming,** in part to prevent the development of large livestock enterprises. Restrictions on raising animals in confinement may remove the present economic advantages of confinement systems. Air and water pollution standards may increase the cost of waste disposal of large units to uneconomically high levels. A 25,000 head cattle feedlot generates more waste than a 100,000 person city. If the same standards for waste disposal had to be met, a return to more dispersed animal feeding systems might be needed. These are intangibles based on factors other than production efficiency. Some people argue that a true measure of production efficiency should include all **societal costs,** including accelerated depletion of petroleum, effects on global warming, pollution and erosion costs, depletion of water supplies, and the costs of relocating displaced farmers and farmworkers. These aspects are considered further in Chapter 8.

My opinion is that livestock production in the industrialized countries will continue to become more intensive and sophisticated, with fewer but larger units. This approach has the highest economic efficiency, leading to lowest food costs. In the long run, low cost of food to the consumer is the bottom line, and gets the most votes.

Particularly with beef cattle, animal and crop production is and will remain integrated to a considerable degree. Millions of acres of wheat pasture are grazed each year in the United States. Corn stover is extensively used for wintering cattle. Wheat stubble is widely used in the western United States for fall and winter feeding of beef cattle. These practices will continue.

In developing countries, **small scale subsistence farming** will likely continue indefinitely. The high human population densities may make it unfeasible to develop large, capital-intensive low labor farming systems typical of the industrial countries. Under these conditions, livestock can be well integrated with crops, utilizing crop residues and by-products. There is increasing interest in agroforestry, incorporating trees into agriculture. One widely used system is alley-cropping (see Chapter 6). Nitrogen-fixing trees are planted in rows, with crops grown in between. The trees provide shade and nitrogen, as well as being used for forage and wood production. The shade is important in the tropics, as many crops (e.g., coffee) are severely damaged when grown in full sunlight.

Thus there is a dichotomy—in the developed countries, with abundant land and capital resources, but expensive labor costs, livestock and crop production will likely become more intensive and specialized and less integrated (in terms of not being integrated on the same farm). In developing countries, increased food production could likely be achieved by increased integration of crops and livestock to obtain maximum conversion of the plant biomass to human food. Thus in one case economic efficiency is the major factor, while in the other, biological efficiency may be more important.

## WHO WILL FEED CHINA (AND INDIA)?

China and India are rapidly industrializing, with fast rates of economic growth and rapidly rising per capita incomes. Characteristically, these changing economic circumstances lead to dietary changes and an increase in consumption of animal products. This is the scenario happening in China. China and India each have over 1 billion people, representing about 50 percent of the world's population. Lester Brown (1995) in his book *Who Will Feed China?* suggested that the rising demand for meat in China, in conjunction with a strong economic status, could cause a world grain crisis. High rates of importation of grain by China to support poultry and swine production could lead to reduced world grain supplies and higher grain prices, which could seriously impact impoverished developing countries. It is worth noting that Brown is director of the Worldwatch Institute (WI), which publishes an annual gloomy assessment of the "State of the World." The WI has a vested interest in painting a bleak picture of the global environment. Brown's book stimulated a lot of discussion, both within China and elsewhere. A major part of Brown's argument is his projected 20 percent reduction in grain production in China over the next three decades. Paarlberg (1997) countered this argument with evidence that Chinese grain production will increase. Alexandratos (1997) believes that even if Brown's projections are correct, the major grain-exporting countries would have no difficulty meeting the anticipated demand. The major grain exporters are North America (United States and Canada), Western Europe, Australia, Argentina, and Thailand. Farmers in these countries would be delighted with a high Chinese demand for grain. The United States has a very large negative trade balance with China; grain exports could help achieve more equity in United States-China relations.

In contrast to Brown's assessment, Simpson et al. (1994) believe that improved efficiencies in China's agriculture can readily meet its future food needs. It may be surprising to many that China ranks number one in the world in swine, poultry, and horse inventories, second in sheep and goats, and fourth in cattle (Simpson et al., 1994). China is the leading producer of major crops, including wheat, rice, cotton, and rapeseed. Thus it has immense agricultural potential. According to Simpson et al. (1994), modest improvements in efficiency of livestock production could meet future needs with even a reduction in animal numbers. This is a well-known phenomenon: as intensification of animal production occurs, there is greater production with fewer numbers. A classic example is the dairy industry in the United States. Dairy cow numbers have been dropping for over half a century, while total milk production has risen dramatically. The same thing can happen in China. Livestock production is rapidly changing from backyard, small-scale production to intensive production. Particularly in the swine and poultry sectors, China is leap-frogging from backyard production to cutting-edge high-tech production techniques, with joint ventures with American and European poultry and swine companies.

Geissler (1999) in a paper entitled "China: The Soyabean-Pork Dilemma" discussed the issue of what food production systems would best meet the needs of China. Regardless of theoretical considerations, she concluded that pork had won out over soybeans[1]: "Consumer demand would appear to strongly favour meat but not beans. It is unlikely in the current economic climate that government policy can effectively counter popular demand. . . . Now that China has become a market economy and government control has been reduced, the soyabean-pork dilemma appears to be disappearing, with the population opting for pork and other animal products and abandoning soyabeans. The opportunity to pre-empt the evolution of the negative effects of the nutrition transition appears to have already been lost" (p. 351). In other words, when given a choice, the Chinese (like us) prefer meat. Campbell et al. (1992) believe that this dietary choice will lead China "from diseases of poverty to diseases of afflu-

ence." Diseases of poverty include infectious diseases (e.g., pneumonia, tuberculosis), parasitic diseases (intestinal parasites, malaria, sleeping sickness), and so on, while diseases of affluence are mainly cancer, diabetes, and heart disease. This choice is a classic example of "out of the frying pan and into the fire"! Campbell et al. (1992) suggest that China has a choice between either Japanese or European-American patterns of consumption. They recommend the Japanese version, characterized by lower intakes of energy and animal products and a higher intake of fish products, "because Japan has the highest average life span in the world" (p. 142).

There are over 800 million peasant farmers in China (Fig. 3-6). What will these people do for a living when China's agriculture industrializes? Concern for their future is sometimes used as a reason for opposing the replacement of water buffaloes with tractors. Perhaps the story about an American businessman visiting China is relevant. The American came upon a team of 100 workers with shovels building a dam. He commented to a local official that an earth-moving machine could build the dam in an afternoon. The official replied, "Yes, but think of the unemployment that would create." The American said, "Oh, I thought you were building a dam. If you want to create employment, take away their shovels and give them spoons!"

The answer to "Who will feed China?" is that Chinese agriculture will feed China (Prosterman et al., 1996). Intensive swine and poultry production will make a major contribution to Chinese food needs. China also has immense rangeland areas. Range livestock production, with feedlot finishing to enhance growth rates, will be important (Simpson et al., 1994).

Like China, **India** is rapidly changing to a consumer-driven market economy. Poultry production will make a particularly important contribution to the improvement of the Indian diet. Religious factors limit beef and pork consumption. Despite the fact that cows are sacred in India, cattle make a significant contribution to human welfare in India. In fact, the evolution of the sacred status of cows is likely due to

---

[1] Soybean is U.S. spelling; soyabean is British spelling.

**FIGURE 3-6** The rice harvest in China *(top)*. Much of the grain in China is still planted, cultivated, and harvested by hand. Primitive but effective detoxification of soybeans for poultry and swine feeding is accomplished by heating the beans in a revolving barrel *(bottom)*. (Bottom photo courtesy of Robert A. Swick.)

their being much more valuable alive than dead. Cattle in India are important sources of milk, animal traction, and fuel (see Chapter 6).

Besides the question "Who will feed China?," perhaps it's appropriate to ask, "What will feed China?" Chinese culinary and medicinal habits are causing the decimation of the world's wildlife. Rhinoceros horn, bear paws and gall bladders, tiger paws, and so on are valuable commodities in China. Vietnam used to have a diversity of turtle species. Chinese entrepreneurs have set up a program in which Vietnamese villagers scour the forests for turtles, which they bring to central collection points, where the turtles are picked up and sent to China as gourmet food items. One turtle can bring more than a year's wages to a villager. As a result, the turtle population has

been decimated (N. E. Forsberg, personal communication). Poaching of wildlife to meet the growing Chinese demand for wildlife parts is a worldwide scourge. The potential extinction of the black rhinoceros and several other rhino species is a direct result of the Chinese demand for rhino horn. In Canada and the United States, bear parts (paws, gall bladders) are especially sought by poachers to sell in China. The Chinese government has experimented with bear and tiger farming to meet the demand for these animals. Caged bears with bile duct catheters are kept for collection of bile. A substantial deer industry in New Zealand supplies the Asian market with deer antlers. While farming these animals may be preferable to poaching them, the ideal solution would be to eliminate the demand. Unfortunately, the rising economic status of the Chinese people suggests that demand for these high-priced "delicacies" will increase. It is disheartening to think that an ancient animal like the rhinoceros is likely to become extinct in the wild because one-quarter of the world's human population thinks that consuming rhino horn enhances sexual function. Obviously, with over 1 billion people, China is not lacking people who function normally! Maybe the only hope is that the global environmental movement will eventually take root in China.

India has about the same human population as China. Contrary to common belief, Indians are not vegetarians. Although Hindus do not eat beef, India has the world's largest cattle population. Many of these are dairy animals; consumption of dairy products is increasing rapidly. Sheep and goat meat are significant (e.g., McDonald's serves mutton burgers instead of beef burgers in India), but the major potential for increased consumption of animal products is poultry and eggs. Intensive poultry production is expanding rapidly in India. The poultry industry will be a key factor in the improvement of the Indian diet. India has about 40 percent of the world's most desperately poor people, and about 60 percent of the Indian population is malnourished.

# STUDY QUESTIONS

1. Claims are made in books like *Diet for a Small Planet* that livestock are taking food out of people's mouths. Discuss the validity of this point of view.

2. Discuss how cassava and sugarcane could be used in a livestock feeding program in tropical countries.

3. Sugarcane can be regarded as the tropical equivalent of grain. Why?

4. What are short cycle animals?

5. Which compete with humans for food to a greater extent: pigs, chickens, or cattle? In densely populated developing countries, which type of livestock are being emphasized?

6. Which is better: feeding dairy cows on high concentrate diets to maximize milk production, or feeding dairy cattle primarily on pasture? Explain your answer.

7. What are some examples of crop residues used in animal feeding? Why aren't beef cattle fattened on crop residues in the United States?

8. Give an example of how animal production systems can be matched with available local resources.

9. In developing countries, people from the countryside are moving to live in cities, in a worldwide trend of urbanization. Why don't they remain on the land and produce their own food, using sustainable agriculture techniques?

10. Discuss the future of intensive, petroleum-based agriculture in North America for the next 100 years. Will the family farm survive?

11. The following letter appeared in a newspaper (*The Oregonian*, March 22, 1997). Critique this letter. Is it a valid point of view?

> *Thursday was Meat-Out Day, a day devoted to refraining from eating meat and all other animal products. The cost of meat-eating to the planet is tremendous.*
>
> *Six billion animals a year are raised on factory farms in the United States. Most animals are subject to inhumane conditions before their brutal slaughter. Each year, the average American family of four consumes 100 chickens, a whole pig, half a steer, 560 eggs, 280 gallons of milk products and 50 pounds of fish.*
>
> *Seventy percent of U.S. deaths are related to diet, particularly the consumption of meats and saturated animal fats. Meat is directly linked to heart disease, strokes and many cancers.*
>
> *More than one-third of all grain grown in the world is fed to livestock. If people were to cut back on their meat consumption by just 10 percent, that would provide enough food to feed the starving worldwide. An acre of land can yield 40,000 pounds of potatoes, but only 250 pounds of beef.*

# REFERENCES

ALEXANDRATOS, N. 1997. China's consumption of cereals and the capacity of the rest of the world to increase exports. *Food Policy* 22:253–267.

BROWN, L. A. 1995. *Who will feed China?: Wake-up call for a small planet.* W. W. Norton and Co., New York.

CAMPBELL, G. L., D. A. TEITGE, and H. L. CLASSEN. 1991. Genotypic and environmental differences in rye fed to broiler chicks with dietary pentosanase supplementation. *Can. J. Anim. Sci.* 71:1241–1247.

CAMPBELL, T. C., C. JUNSHI, T. BRUN, B. PARPIA, Q. YINSHENG, C. CHUMMING, and C. GEISSLER. 1992. China: From diseases of poverty to diseases of affluence. Policy implications of the epidemiological transition. *Ecol. Food Nutr.* 27:133–144.

CLUTTON-BROCK, J. (ED.). 1989. *The walking larder. Patterns of domestication, pastoralism, and predation.* Unwin Hyman, Ltd., London.

CORDAIN, L. 1999. Cereal grains: Humanity's double-edged sword. In: Simopoulos, A. P. (Ed.). *Evolutionary aspects of nutrition and health. Diet, exercise, genetics and chronic disease. World Rev. Nutr. and Diet.* 84:19–73.

FRIESEN, O. D., W. GUENTER, B. A. ROTTERAND, and R. R. MARQUARDT. 1991. The effects of enzyme supplementation on the nutritive value of rye grain (*Secale cereale*) for the young broiler chick. *Poult. Sci.* 70:2501–2508.

GEISSLER, C. 1999. China: The soyabean-pork dilemma. *Proc. Nutr. Soc.* 58:345–353.

HATFIELD, P. G., S. L. BLODGETT, G. D. JOHNSON, P. M. DENKE, and M. W. CARROLL. 1999. Sheep grazing to control wheat stem sawfly. *Proc. West. Sect. Am. Soc. Anim. Sci.* 50:127–129.

HEDIN, P. A., W. P. WILLIAMS, and P. M. BUCKLEY. 1998. Caloric analyses of the distribution of energy in corn plants. *Zea mays L. J. Agric. Food Chem.* 46:4754–4758.

MCGINNIS, J. 1992. Poland's poultry industry ripe for U.S. industry assistance. *Feedstuffs* Vol. 64, No. 11, March 16, 1992.

MCMENIMAN, N. P., R. ELLIOT, and M. O'SULLIVAN. 1990. The use of dried sugar cane fractions as the principal energy source in sheep rations. *Anim. Feed Sci. Tech.* 28:155–168.

MILLWARD, D. J. 1999. Meat or wheat for the next millennium? *Proc. Nutr. Soc.* 58:209–210.

NESTLE, M. 1999. 'Meat or wheat for the next millennium?' Plenary Lecture. Animal *v.* plant foods in human diets and health: Is the historical record unequivocal? *Proc. Nutr. Soc.* 58:211–218.

PAARLBERG, R. L. 1997. Feeding China: A confident view. *Food Policy* 22:269–279.

PRESTON, T. R., and R. A. LENG. 1987. *Matching ruminant production systems with available resources in the tropics and subtropics.* Penambul Books, Armidale, Australia.

PROSTERMAN, R. L., T. HANSTAD, and L. PING. 1996. Can China feed itself? *Sci. Amer.* 275(Nov.): 90–96.

REDMON, L. A., G. W. HORN, E. G. KRENZER, JR., and D. J. BERNARDO. 1995. A review of livestock grazing and wheat grain yield: Boom or bust? *Agron. J.* 87:137–147.

ROSEGRANT, M. W., N. LEACH, and R. V. GERPACIO. 1999. 'Meat or wheat for the next millennium?' Plenary Lecture. Alternative futures for world cereal and meat consumption. *Proc. Nutr. Soc.* 58:219–234.

SIMPSON, J. R., X. CHENG, and A. MIYAZAKI. 1994. *China's livestock and related agriculture.* CAB International, The University of Arizona Press, Tucson.

TROYER, A. F. 1999. Background of U.S. hybrid corn. *Crop Sci.* 39: 601–626.

VAN SOEST, P. J. 1994. *Nutritional ecology of the ruminant.* Cornell University Press, Ithaca, New York.

WARREN, B. E., and D. J. FARRELL. 1990. The nutritive value of full-fat and defatted Australian rice bran. I. Chemical composition. II. Growth studies with chickens, rats and pigs. III. The apparent digestible energy content of defatted rice bran in rats and pigs and the metabolisability of energy and nutrients in defatted and full-fat bran in chickens and adult cockerels. IV. Egg production of hens on diets with defatted rice bran. *Animal Feed Sci. Tech.* 27:219–228; 229–246; 247–257; 259–268.

# CHAPTER 4

# Principles of Animal Nutrition and the Scientific Feeding of Livestock

**CHAPTER OBJECTIVES**

1. To provide sufficient background in the principles of animal nutrition for the discussion of feed additives in Chapter 5 (a major objective)

2. To discuss differences in digestive tract physiology and feeding strategy of herbivores in order to provide a background for the discussion of ecological and environmental effects of livestock and their interactions with wildlife (in Chapters 6 and 7)

3. To attempt to dispel the common misconception that "a ruminant is a ruminant is a ruminant" (explaining why feeding a deer as if it were a cow will kill it)

The nutritional requirements of livestock and their ability to utilize feedstuffs depend to a great extent on what type of animal they are. This in turn is largely determined by the nature of the digestive tract and the role of microbes in digestion. Swine and poultry (**simple nonruminants,** monogastrics) have relatively simple digestive tracts and secrete enzymes to accomplish most of the work of digestion. **Ruminants** (cud-chewing animals with a multi-compartmented stomach) and **nonruminant herbivores,** such as horses and rabbits, have enlarged segments of the gut in which microbes, mainly bacteria and protozoa, accomplish much of the work of digestion. Dependence on microbial digestion is a less efficient way of utilizing feed resources, because it introduces one or more additional steps in the food chain, but it provides animals with the means to utilize plant fiber.

A brief review of general principles of animal nutrition will be given, including digestive tract physiology and function, to provide a background for further discussion of feed additives and resource utilization in later chapters.

# COMPARATIVE FEEDING STRATEGIES AND DIGESTIVE TRACT PHYSIOLOGY

The nutritional requirements of animals are greatly influenced by the nature of their digestive tracts. This influence is at least two-fold: digestive physiology is closely linked to food selection and dietary strategies; it is also linked to the ability of the animal to derive nutritional benefit from particular types of feedstuffs.

Animals have evolved to occupy virtually all ecological niches, and in many cases have specialized feeding strategies. Broad categories of **feeding strategy** include carnivores (meat eaters), herbivores (plant eaters) and omnivores (eat both plant and animal foods). Domestic animal types include carnivores (cats), omnivores (swine, chickens), and herbivores (cattle, sheep, horses, rabbits). Feeding strategies can influence nutrient metabolism and requirements. For example, members of the cat family have an almost exclusively meat-based diet and, as a result, have a substantially different protein and amino acid metabolism than that of other animals. Frugivores (fruit eaters) generally have a dietary requirement for vitamin C (e.g., fruit-eating bat). Some animals are **specialist feeders** and have coevolved with particular plant species. Koalas and several other Australian marsupials have evolved dietary preferences for eucalyptus foliage. It is virtually impossible to raise koalas on any other diet except eucalyptus leaves, making their exhibition in zoos a challenge. Other animals, such as swine, dogs, and numerous other omnivorous species, are very cosmopolitan in their dietary habits and will eat almost anything edible.

Digestive tract physiology has a very important impact on nutritional requirements of animals and on the types of feedstuffs they can utilize. Feeding strategy and gut type are closely linked, for obvious reasons. This can be appreciated by considering the absurdities possible if they were not closely linked, such as a meat-eating cow and a hay-eating cat!

Feeding behavior and gut structure have coevolved, so animals are attracted to feedstuffs that they are capable of utilizing and are not attracted to or are repelled by feedstuffs that they cannot digest efficiently. This relationship applies also to feeds containing toxins. Sheep and goats will readily consume tansy ragwort (*Senecio jacobaea*), a poisonous plant containing liver-damaging alkaloids, while cattle and horses are repelled by the plant. Sheep and goats can detoxify the alkaloids in the liver, while cattle and horses cannot. Vultures and other carrion-eating animals are highly resistant to botulism. There are many other examples of animals that have evolved enzymatic detoxification mechanisms allowing them to utilize and be attracted to certain feeds (e.g., poisonous plants) that are toxic and unpalatable to other animals (see discussion of **coevolution** in Chapter 9, pp. 000). Domestic animals can be classified into three main groups, according to the nature of their digestive tract. These groups are simple nonruminants (monogastrics), ruminants, and nonruminant herbivores.

The **simple nonruminant animals** have a pouch-like, non-compartmentalized stomach and an intestinal tract. The term "nonruminant" is preferred over "monogastric," which means one stomach. All animals have only one stomach, but in some, such as ruminants, it is divided into compartments. The general features of the nonruminant digestive tract are shown in Fig. 4-1. Examples of simple nonruminants are swine, chickens, dogs, cats, and humans. These animals do most of their own work of digestion, by the secretion of digestive enzymes. The stomach functions mainly in the storage of ingested (consumed) feed, releasing it slowly to the small intestine for digestion. Hydrochloric acid (HCl) is secreted into the stomach, producing a low pH (high acidity). The acidity kills most ingested bacteria and also causes some hydrolysis of protein. Some proteolytic (protein-digesting) enzymes such as pepsin are secreted into the stomach. Pepsin is secreted in an inactive form, pepsinogen, which is activated in the stomach

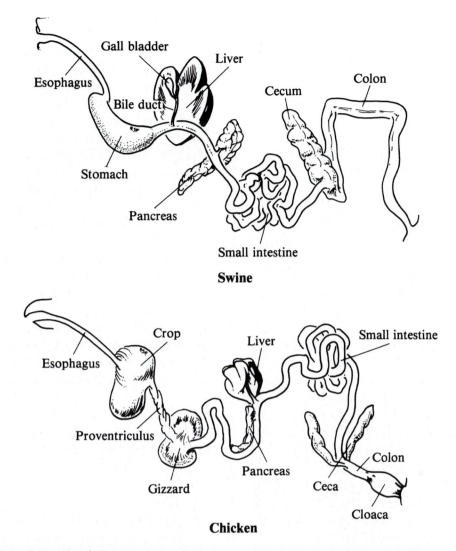

**Swine**

**Chicken**

**FIGURE 4-1** Examples of digestive tracts of simple nonruminant (monogastric) animals. The stomach is a simple pouch, not divided into compartments. The stomach of the chicken is called the proventriculus.

to pepsin by the action of HCl. Thus some protein digestion occurs in the stomach, but most of the digestive processes occur in the small intestine.

In the small intestine, numerous enzymes are secreted that digest the major components of the diet that require digestion. Proteins are digested to amino acids, carbohydrates to simple sugars, and fats to fatty acids and monoglycerides. Minerals and vitamins are absorbed as such and do not require digestion. The source of most of the digestive enzymes is the pancreas gland; it produces the

proteolytic enzymes trypsin and chymotrypsin, carbohydrate-digesting enzymes such as amylase, sucrase, and lactase, and the fat-digesting enzyme lipase. Additional enzymes are secreted by the intestinal mucosa. The small intestine is the major site of digestion and nutrient absorption. The function of the large intestine (hindgut) is the removal of water from the gut contents and the formation of the feces from indigestible residues. Although the hindgut is also a site of microbial growth, this is of relatively minor digestive significance in simple nonruminants.

**Ruminants** are animals that have a complex, compartmentalized stomach, characterized by one large compartment, the rumen, in which microbial fermentation of ingested feed occurs. Ruminants are classified in the order Artiodactyla, suborder Ruminantia. They are even-toed, hooved animals. The term ruminant is derived from *ruminare*, a Latin word meaning to chew again. Thus ruminants are animals that ruminate or "chew their cud" by regurgitation of ingested material. These animals swallow large amounts of poorly chewed vegetation while feeding and then regurgitate and masticate it while resting. Rumination physically breaks down fibrous material to increase its rate and extent of digestion in the rumen.

According to Church (1988) there are 86 living and 333 extinct genera of the order Artiodactyla, with 68 living and 180 extinct genera of ruminants. Only a few of the primitive species remain. Ruminants range in size from the tiny (1 kg) lesser mousedeer to the 1000 kg giraffe.

General features of the ruminant stomach are shown in Fig. 4-2. The four compartments are the rumen, reticulum, omasum, and abomasum. In the post-weaning animal, the **rumen** is the largest compartment. It functions as a fermentation vat. **Rumen microbes,**

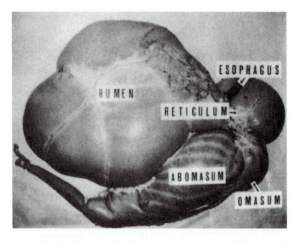

**FIGURE 4-2** The four compartments of the ruminant stomach are illustrated in this dried rumen preparation.

primarily many species of anaerobic bacteria, and to a lesser extent protozoa, secrete enzymes that digest the consumed feed. Because the rumen is an anaerobic environment, microbial fermentation cannot result in complete oxidation of carbohydrates to carbon dioxide and water. Anaerobic fermentation is basically the glycolysis pathway, by which glucose is metabolized to pyruvic acid. In the rumen, the microbes convert pyruvic acid to several short-chain organic acids called **volatile fatty acids (VFA).** These are the main end-product of rumen fermentation and are the primary absorbed energy sources of ruminants. Of major significance to rumen fermentation is the production of the enzyme **cellulase** by rumen microbes, permitting the digestion of fibrous feeds containing cellulose. Other benefits of rumen fermentation include the synthesis of amino acids and water-soluble vitamins by rumen bacteria. As a result, ruminants are largely independent of dietary sources of amino acids and most vitamins. The rumen is lined with projections called **papillae.** The VFA are absorbed into the papillae, analogous to the absorption into the villi of the small intestine.

The rumen is continuous with the **reticulum;** they are often considered together as the **reticulorumen.** The reticulum is lined with honeycomb-shaped projections. One of the functions of the reticulum is the trapping of foreign material such as stones, nails, wire, and so on to prevent puncturing of the digestive tract. This is especially important in cattle, which by their indiscriminate grazing behavior frequently consume foreign objects. Material leaves the reticulorumen via the **omasum.** The omasum is a small compartment containing membranous divisions called omasal leaves. The omasal leaves function as a sieve, retaining material in the rumen until it has been degraded into small particle sizes. Fluids and small particles flow through the omasum to the **abomasum,** or true gastric stomach. The abomasum contains large spiral folds in the fundus gland region. Gastric secretions, such as pepsinogen and HCl, are secreted in the abomasum. The acidity kills rumen microbes in the digesta; they are then subject to digestion in the small intestine, providing a major source of protein to the host animal.

There has been a tendency to regard all ruminant animals as being very similar in digestive physiology and rumen metabolism. This is a fallacy, as well pointed out by Hofmann (1973) in a classic treatise on the stomach structure and feeding habits of East African wild ruminants. Hofmann considered both feeding strategy and stomach anatomy as important determinants of nutritional requirements and digestive capabilities and classified ruminants into three groups based on feeding strategy: concentrate selectors, bulk and roughage feeders, and intermediate feeders. In general, concentrate selectors are small animals, while roughage feeders are large. Small animals have a high metabolic rate and thus must consume a readily digested diet or be capable of a high feed intake and fast rate of passage. Large animals, with a lower metabolic rate per unit of body size, can survive on bulky feeds that are retained for a prolonged period to allow for microbial digestion of fiber.

**Concentrate selectors** consume the low fiber parts of herbage, including tree and shrub leaves, herbs, fruit and other soft, succulent plant parts. This herbage, with a high concentration of cell contents and a low cell wall content, is readily digested. A considerable portion of the material, such as starches and sugars, does not require fermentation but can be digested by the animal itself. Although many concentrate selectors are small animals, such as the dik dik and duiker; others such as deer, moose, and giraffes are large. They tend to nibble on a wide variety of plants and plant parts, rather than feeding intensively on one type of vegetation. The larger concentrate selectors, such as deer and giraffes, consume the leaves and succulent foliage of trees, utilizing an ecological niche unavailable to grazing animals. Concentrate selectors have a stomach anatomy adapted to the use of low fiber forage. They have a relatively small rumen and reticulum so cannot consume a large quantity of feed at one time. The architecture of the rumen is simple, lacking the highly developed sacculations and muscular pillars seen in the roughage eaters. The rumen papillae are elongated and highly developed, facilitating absorption of the high concentration of fermentation end-products produced from their high quality diet. There is a rapid "wash through" of soluble sugars into the abomasum. Unsaturated fatty acids may escape rumen fermentation; thus concentrate selectors have body fat with a higher content of polyunsaturated fatty acids than is typical for grazing ruminants (Meyer et al., 1998). Rumination is less important than in the roughage eaters. The omasum is small and its leaves are less well developed than in the grazers. The omasum can readily become impacted with coarse fibrous material. Thus, concentrate selectors are intolerant of a high fiber diet. Winter feeding of deer with alfalfa hay often results in mass mortalities from rumen and omasal impaction. The feeding of captive ruminants in zoos requires knowledge of their feeding strategy; feeding a high fiber diet to concentrate selectors usually results in mortality from impaction. Thus, the concentrate selectors have a digestive system attuned to their feeding behavior, maximizing the efficiency of utilization of low fiber plant material.

The **ventricular groove** (also referred to as reticular or esophageal groove) is a muscular tube that functions to transport liquids from the base of the esophagus to the omasum, thus bypassing rumen fermentation. This occurs in suckling animals so that the digestion of milk occurs post-ruminally. In concentrate selectors, the ventricular groove and the copious salivary secretion in these species may function in the washing of the soluble cell contents of succulent plant material directly to the omasum, avoiding rumen fermentation of soluble sugars, galactolipids, and so on, which the animal can digest more efficiently itself. This process would allow concentrate selectors to use plant material high in cell contents with a high efficiency. Hofmann (1973, 1988, 1989) and van Hoven and Boomker

(1985) summarized a number of other adaptations of concentrate selectors that aid in efficient use of forages. They have a three–four fold greater salivary gland mass than roughage eaters, facilitating the functioning of the ventricular groove, buffering the high concentration of rumen VFA, and, by a "wash through" effect, reducing ruminal retention times. The glandular tissue of the small intestine is up to 100 percent greater than in roughage eaters, and the liver weight is larger. The hindgut is relatively larger in concentrate selectors. These structural differences in post-ruminal anatomy suggest that more unfermented food escapes from the rumen and its post-ruminal digestion and absorption may be more important for many ruminant species than indicated from data with cattle and sheep. To parody Shakespeare, "a ruminant is a ruminant is a ruminant" is simplistic and incorrect.

**Roughage eaters** or **grazers** have a feeding and digestive strategy based on the utilization of high fiber diets. Their stomach facilitates maximum digestibility of fiber. The volume capacity of the reticulorumen is high, and their grazing behavior permits a rapid intake of a large quantity of fibrous feed. Rumination is pronounced, resulting in physical maceration of ingested forage. The omasum is highly developed with many omasal leaves, thus retaining fibrous feed in the rumen until microbial enzymes have degraded it. The papillae are less prominent than in concentrate selectors because there is a slower production of fermentation end-products requiring absorption. Examples of roughage eaters include cattle, buffalo, camels, and a variety of African antelope species (e.g., wildebeest, topi, water buck, oryx antelope).

**Intermediate feeders** have feeding and digestive strategies that share properties of both concentrate selectors and grazers. Thus, these animals tend to be highly adaptable to varying environments and changing habitats. Sheep and goats are domestic animals of this type; they graze on grass but also feed extensively on shrubs and forbs. Wild intermediate feeders include reindeer, musk oxen, elk, pronghorns, impala, gazelles, and eland.

The camels and other camelids (llama, alpaca, guanaco, and vicuna) are considered to be ruminants but have a substantially different stomach anatomy. The stomach has three compartments, roughly comparable to the rumen, reticulum, and abomasum.

Classification of ruminants into the three feeding-type categories of concentrate feeders, intermediate feeders, and grazers is useful as a generality. However, numerous nutritional and physiological interpretations of anatomical differences of animals by Hoffman (1973, 1988, 1989) are not supportable by the results of rigorous testing of his hypothesis (Robbins et al., 1995). The simple classification of ruminants into three types is an oversimplification. Gordon and Illius (1994, 1996), in studies of grazing and browsing African ruminants, concluded that there is little difference in digestive strategy among African ruminants with different anatomical features of the gut.

In **nonruminant herbivores,** the hindgut or large intestine (cecum plus colon) is enlarged and has a microbial population performing many of the same digestive functions that take place in the rumen. Compared to the rumen, there are several nutritional disadvantages to the hindgut as a fermentation site. Soluble nutrients, such as sugars, amino acids, vitamins and minerals, are absorbed in the small intestine. Thus the composition of material entering the hindgut is less favorable for maximum microbial growth than is the case in the rumen, where the microbes have all the nutrients in the ingested feed as available substrate. The hindgut is a less efficient area for nutrient absorption. Products of rumen fermentation, including the microbes, are digested and/or absorbed in the rumen or small intestine. Microbes in the hindgut are not subject to digestion, unless the feces are consumed, a process known as **coprophagy.** The passage rate through the hindgut is more rapid than through the rumen, leading to a lower efficiency of fiber digestion. There are, however, a number of significant advantages of hindgut fermentation, which will be discussed in the section Evolutionary Aspects of Digestive Physiology.

The rabbit is probably the best known example of a cecal fermentor. The hindgut of the rabbit functions to selectively excrete fiber, and retain the non-fiber components of forage for fermentation in the cecum. This separation is accomplished by muscular activity of the proximal colon. Fiber particles, being larger and less dense than non-fiber components, tend to be concentrated in the lumen of the colon, while fluids and material of small particle size tend to concentrate at the periphery. Peristaltic action propels the fiber particles through the colon rapidly, whereas antiperistaltic contractions of the muscular bands of the proximal colon move the soluble nutrients and fluids "backward" into the cecum. The digestive strategy of the rabbit and other small herbivores is to minimize the digestion of fiber and to concentrate digestive action on the more nutritionally valuable non-fiber constituents. Thus the digestibility of fiber in the rabbit is very low. An additional refinement of this digestive strategy is that at intervals the cecal contents are ingested by the animal. After the colon is emptied of hard fecal pellets, consisting primarily of fiber, the cecum contracts and the cecal contents are squeezed into the proximal colon. Mucin is secreted by goblet cells, producing cecal material covered with a mucilaginous membrane. This material, known as **cecotropes** or "soft feces," is consumed by the animal directly from the anus. Consumption of cecotropes (**cecotrophy**) provides the animal with a means of more efficiently digesting the products of cecal fermentation, digesting the microbial protein, and obtaining microbially synthesized B vitamins. Thus the process of cecal fermentation and cecotrophy allows the animal to utilize forage without the encumbrance of carrying around internally a large mass of slowly digested fiber.

Because cecal fermentation is an adaptation for utilizing fibrous diets without the encumbrance of an overly large gut, most cecal fermentors are small animals. The largest include the koala (about 10 kg) and the capybara (about 50 kg). Small animals have much greater relative metabolic rates and energy re-

quirements than do large animals and cannot meet their energy requirements from the microbial digestion of fiber. Thus cecal fermentors separate out the fiber and excrete it, retaining the more digestible non-fiber components for fermentation in the cecum. Some avian species, such as the ostrich, have cecal fermentation. While retrograde movement of fluid from the colon to the ceca occurs in the goose, the primary reason why this avian species can consume a high forage diet is simply a very high rate of passage and a high forage intake, allowing the bird to meet its energy requirements from the digestion of plant cell contents. The fibrous cell wall material is excreted largely undigested. Another strategy for separating fiber from more digestible non-fiber components of leaves is employed by fruit bats, which chew leaves into a bolus, swallow the liquid fraction, and expel the pellet of fibrous residue from the mouth (Lowry, 1989). Thus small herbivores use a variety of behavioral and digestive strategies to cope with the challenge of surviving on high fiber, low energy feeds.

The capybara is the largest known cecal fermentor, and it uses coprophagy as part of its digestive physiology (Borges et al., 1996). Its digestive efficiency is comparable to that of ruminants on similar diets.

In all large (over 50 kg) hindgut fermentors, the enlarged proximal colon is the primary site of fermentation. Examples include the horse and other equids (zebra, donkey, etc.), the elephant and rhinoceros (Fig. 4-3). The cecum is often enlarged as well, but acts as an extension of the colon as a fermentation site, rather than functioning in selective retention of small particles as in the cecal fermentors. Ceco-colonic fermentors also include a number of New World monkeys, lemurs, and rodents such as the porcupine and beaver. The digestive strategy of colon fermentors is similar to that of ruminants; hindgut fermentation functions in digestion of plant cell wall constituents. The percent digestibility of fiber fractions is generally lower in colon fermentors than in ruminants, due primarily to the greater rate of passage and less optimal environment for microbial growth.

**FIGURE 4-3** The discipline of animal nutrition is increasingly concerned with the nutrition and feeding of wild animals, both in their natural but often shrinking habitats and in zoological gardens and game farms. Critical to understanding their nutritional needs is a knowledge of digestive physiology and feeding strategy. Is an elephant like a cow (a ruminant) or a horse (hindgut fermentor)? (Courtesy of Leith Pemberton.)

Not all nonruminant herbivores are hindgut fermentors. Some nonruminant animals have a chambered stomach with microbial fermentation. Examples include a number of Australian marsupials (kangaroo, wallaby), the hippopotamus, peccary, colobus monkey, and tree sloth. The hoatzin is the only bird known to have foregut fermentation (Grajal et al., 1989; Dominguez-Bello et al., 1993a, b). This bird, native to South America, feeds on young leaves plucked from the canopy of tropical forests. The leaves are fermented in the crop by a variety of bacteria and protozoa (Dominguez-Bello et al., 1993a). Another unusual feature of the hoatzin is that it has functional claws on the wings, which aid in feeding in tree canopies.

# EVOLUTIONARY ASPECTS OF DIGESTIVE PHYSIOLOGY

Current scientific evidence suggests the age of the universe to be about 15 billion years and the age of the Earth about 4.6 billion years. The Earth was born as a result of im-

mense explosions, with the solar system formed by the condensation of clouds of galactic dust and gas, collapsing under intense gravitational forces. Under the reducing conditions of the Earth's early atmosphere, rich in water, $CO_2$ and $N_2$ and lacking free oxygen, chemical reactions could occur to produce simple organic molecules such as methane, ammonia, amino acids, simple carbohydrates and nucleotides. These reactions have been experimentally produced under conditions believed to closely simulate those early in the Earth's history. Over a period of about 300 million years, these simple organic compounds accumulated in warm, shallow seas, forming a **primordial soup.** Self-replicating molecules, probably formed on templates of clay minerals, developed and gave rise to simple prokaryote cells (single cells with no membrane-bound organelles; all bacteria are **prokaryotes**). The oldest known fossils, dated at about 3.5 billion years, are of prokaryote cells in tidal mudflats. Over the next 2 billion years, **eukaryotes** (cells with membrane-bound organelles) evolved, such as red and green algae and fungi. By 700 million years ago, multicellular green algae appear in the fossil record, while 600 million years ago, the first multicellular animals appeared. At the dawn of the Cambrian epoch, about 570 million years ago, animals with mineralized skeletons were abundant. The Mesozoic era (240–195 million years ago) was the age of the dinosaurs. The first mammals appeared at this time, and after the abrupt extinction of the dinosaurs about 65 million years ago, mammals became the dominant fauna of Earth.

Digestion began in a primitive manner with the first eukaryotes. With the evolution of vertebrates, microbial habitats were abundant in all regions of the gut. Microbial involvement in digestion became critical to animals consuming plant foods containing appreciable fiber. The cellulase enzyme, necessary for cellulose (and fiber) digestion, is produced only by microbes. Surprisingly, and perhaps because of the evolution of symbiotic relationships between microbes and herbivores, no vertebrates evolved the ability to produce cellulase.

The earliest mammals were small nocturnal insectivores. The digestive system of early mammals would no doubt contain the same components as found in their reptilian ancestors: stomach, small intestine, cecum, and colon. Hume and Warner (1980) have discussed the evolution of the digestive tracts of modern mammals. Because the fossil record provides no information on the morphology, physiology, biochemistry, or microbiology of the gut, their deductions are unquestionably speculative. An intriguing question concerns the development of microbial digestion. Which came first, foregut or hindgut digestion? Hume and Warner (1980) suggest it is likely that early insectivorous mammals began to ingest seeds, fruits, and succulent vegetation, perhaps during periods of low prey availability. This would cause an increased intake of indigestible bulk, stimulating greater secretions of saliva and mucous and greater sloughing of epithelial cells from the mechanical effects of fiber. This could lead to development of a mechanism to reabsorb fluid, electrolytes, and metabolites from the lower tract. This would provide a selective advantage for prolonged transit time to permit reabsorption to occur. The hindgut would enlarge, because of the presence of indigestible fiber. (Hindgut enlargement is readily demonstrated in pigs, rats, etc., when a diet high in indigestible fiber is fed.) The presence of fiber, fluids, and nutrients in the hindgut would provide an environment favorable to microbial growth. Thus, Hume and Warner (1980) believe that microbial digestion probably began in the hindgut, although the initial impetus for its development was the selective nutritional advantage of conservation of electrolytes and endogenous nutrients. Continued intake of a high fiber diet would place selection pressure on high salivary secretion, mucus secretion, and hindgut enlargement. Coprophagy might develop from the habit of many omnivores of eating virtually every accessible edible item (as readily appreciated by observing the eating habits of dogs). The habit could be reinforced if a nutritional benefit accrued.

Foregut fermentation could arise from enlargement of the stomach or formation of diverticula to provide pouches where mixing of acid with the ingesta would be impeded. With the effects of stomach acid minimized, the foregut is an ideal site for microbial growth. Enlargement, elongation, or sacculation of the stomach is very common in mammals, and appears to have independently evolved in diverse lines, including the marsupials (e.g., kangaroo), ruminants, rodents (e.g., hamster), and primates (e.g., colobus monkey). A primitive rodent-like animal, the rock hyrax of Africa, has a stomach with three fermentation chambers.

The large modern herbivores, the ruminants and equids, evolved in the Miocene and Pliocene epochs in association with climatic changes that led to extensive development of grasslands. Ruminants and equids represent different evolutionary strategies for dealing with low energy, high fiber roughage. The ruminant is well adapted for achieving maximum digestive efficiency of high fiber diets. The reticulo-omasal orifice retains fibrous feed in the fermentation chamber until it has been digested. This maximum efficiency of fiber digestion would be of adaptive significance if the quantity of feed were limiting and if the food were of low digestibility. Thus ruminant digestive strategy probably evolved in regions where quality or quantity of forage was at least periodically limiting, such as deserts or deciduous forests (Hume and Warner, 1980). Under conditions of a high availability of poor quality forage, hindgut digestion may be a superior adaptation (Janis, 1976), because intake is not limited by rate of digestion. Thus on the Serengeti plains of Africa, the zebra can survive on low quality roughages (Fig. 4-4), selecting the most fibrous plant parts (Bell, 1971), on which wildebeest, antelope and other ruminants cannot survive. The equids are able to meet their protein requirements on high fiber, low protein forage because feed intake is not dependent on

**FIGURE 4-4** Hindgut fermentors such as the equids (horses, donkeys, and zebras) can survive on poorer quality roughage than can the ruminant animals. On the plains of East Africa, such as the Serengeti, the most mature, fibrous forage is consumed by zebras. (Courtesy of W. D. Hohenboken.)

the rate of fiber digestion, so they can have a very high intake of very low quality forage. Equids will graze almost continuously during daylight when on range with low quality, mature forage. Further information on the evolution of herbivory and microbial digestion is presented by Janis (1976, 1989), Hume and Warner (1980), Illius and Gordon (1992), and Langer (1984).

# NUTRIENTS REQUIRED BY ANIMALS

All nutrients required in the diet, in addition to water, can be put into one of the following major nutrient categories: proteins, carbohydrates, lipids, minerals, and vitamins.

## Proteins and Amino Acids

**Proteins** are composed of individual units called amino acids. All **amino acids** have at least one amino group, which contains nitrogen. There are 20 amino acids of major importance in the structure of animal proteins, while there are several hundred amino acids found in plants. Many of these function as plant chemical defenses against insects and other herbivores. Not surprisingly, many of them (e.g., mimosine in Leucaena) are toxic to animals. Dietary proteins consumed by animals are digested to release amino acids in the gut, which are absorbed into the blood. They are taken up by the animal's cells, and re-assembled into the characteristic sequence of amino acids for each type of tissue protein. The sequence and arrangement of amino acids in a protein are controlled genetically by the animal's DNA.

Certain amino acids (about 10 in most animals) cannot be synthesized by animal tissues and so must be obtained by way of the diet. These are known as the **essential amino acids,** meaning that they are essential in the diet. The essential amino acids for most species are the following:

| | |
|---|---|
| Arginine | Methionine |
| Histidine | Phenylalanine |
| Isoleucine | Threonine |
| Leucine | Tryptophan |
| Lysine | Valine |

The so-called **non-essential amino acids** are not needed per se in the diet, because they can be synthesized in the animal by interconversion reactions from one amino acid to another.

There is a fundamental difference between ruminant and nonruminant animals in the manner in which their amino acid requirements are met. Nonruminants, such as swine, poultry and humans, require the essential amino acids in their diets. They need to be present in the required amount of each one for the proteins that the animal is synthesizing (e.g., for growth, lactation, etc.). Since they are not stored in the body, the feed must contain adequate amounts of the essential amino acids at all times. In ruminants, however, the dietary protein is digested by bacteria in the rumen, with the nitrogen of the amino group released as ammonia ($NH_3$). The microbes then use the $NH_3$ as the starting point for synthesis of their amino acids. Thus the protein in the diet is converted to microbial protein of a different amino acid composition. The microbial amino acids are absorbed by the animal following digestion of the microbes in the small intestine. The advantage of this for the ruminant animal is that it can utilize dietary proteins that are deficient in essential amino acids, because the microbes synthesize microbial protein that contains them. The disadvantage, in terms of efficiency, is that the microbes introduce a further step in the food chain. The cow eats dietary protein, rumen bacteria digest the protein and convert it to bacterial protein, some of the bacteria are digested by rumen protozoa and converted to protozoal protein, and finally the microbial protein (bacteria and protozoa) is digested by the cow. With each step, there is a loss of protein and energy. This represents a food chain or pyramid in the gut, with each step introducing inefficiency in nutrient utilization.

Thus the rumen microbes have both positive and negative effects on protein utilization. They permit the animal to utilize poor quality proteins and non-protein nitrogen, but they are wasteful of high quality proteins, which could be better utilized directly by the animal. (This is the concept behind bypass proteins, to be discussed later.)

## Improvements in Protein Efficiency in Animal Feeding

Traditionally, poultry and swine diets have been "balanced" by using a mixture of protein supplements so that all of the essential amino acid requirements are met. This inevitably results in some amino acids being in excess and thus being wasted as far as protein synthesis is concerned. Several of the essential amino acids, such as lysine, methionine, and tryptophan, are available as **synthetic amino acids.** They are produced commercially by microbes in a fermentation process, rather than being chemically synthesized. New developments in

biotechnology, involving genetic modifications of bacteria to cause them to produce large amounts of specific amino acids, are increasing the numbers of amino acids available commercially and lowering their cost. Increasingly, swine and poultry diets will be formulated using synthetic amino acids rather than protein supplements. This will reduce the dependence of these livestock on high quality protein sources. Feeding synthetic amino acids to livestock is probably a more acceptable means of using them than adding them directly to human food.

Replacement of protein supplements with synthetic amino acids will alter the supplementation needs of several other nutrients. For example, replacement of soybean meal in swine diets with synthetic lysine (lysine-HCl) reduces potassium and increases chloride, causing changes in electrolyte balance (Patience, 1990). Protein supplements provide many nutrients in addition to essential amino acids, so these will need to be considered when synthetic amino acids are used in place of intact protein.

It is widely recognized that total replacement of dietary protein with an apparently adequate mixture of amino acids reduces growth of poultry and swine. The concept that dietary protein must be completely digested to free amino acids is an oversimplification. Significant amounts of peptides (di, tri, and oligopeptides) are absorbed. **Peptides** are polymers of amino acids; for example, a tripeptide consists of three amino acids. Peptides are absorbed more rapidly than free amino acids (Webb, 1990), which may facilitate their metabolism. Peptides are a more important form of amino acid absorption in ruminants than free amino acids, and the rumen and omasum may be major sites of peptide absorption (Webb et al., 1992). Animal tissues can utilize peptides as a source of amino acids for protein synthesis (Webb, 1990). In the future, specific peptides or peptide mixtures may be developed as feed additives to improve animal productivity (Webb, 1990).

As discussed in Chapter 5, future advances in biotechnology may include implanting amino acid-secreting bacteria in the gut, reducing or eliminating the dietary requirement.

A new development in swine nutrition is the concept of **ideal protein,** in which both essential amino acids and amino acid digestibility (pre-hindgut, or ileal digestibility) are used in diet formulation. This allows more precise balancing of diets, which is important in reducing costs and also in minimizing nitrogen excretion, thus reducing pollution.

Although not a new concept, the partitioning of dietary nitrogen for ruminants into **fermentable and non-fermentable nitrogen categories** is being applied much more frequently than in the past. This partitioning offers much potential for improving the efficiency of protein utilization by ruminant animals, particularly in developing countries where fermentable nitrogen (urea) and highly fermentable carbohydrate (sugarcane juice, molasses) can be very effectively combined in a feeding program. The rumen bacteria need two major types of nutrients for their growth, nitrogen (N) and a carbon source. The N is used for the synthesis of bacterial amino acids, which are then used to synthesize the bacterial proteins. The carbon is used as a source of energy and for biosynthesis reactions. Bacteria ferment sugars, for example, to produce ATP. They also use parts of the carbon structure of sugars (the carbon skeleton or backbone) to synthesize amino acids and other components of their structure.

The least expensive source of the ammonia nitrogen used by the rumen bacteria to synthesize amino acids is usually a source of **non-protein nitrogen** (NPN) such as urea. Proteins that are digested by bacteria in the rumen, releasing ammonia, are referred to as rumen-degradable proteins, degradable proteins, or fermentable proteins. The ammonia nitrogen thus released is known as **fermentable N.** The fermentable N is used by the bacteria in synthesizing the bacterial amino acids.

By knowing what fraction of the dietary protein is fermentable, the nutritionist can calculate whether or not additional fermentable nitrogen in the rumen is needed. If it is, it should be provided using NPN because NPN sources are cheaper than proteins. If the dietary protein meets or exceeds the fermentable nitrogen needs of the rumen, then supplementation with NPN would be of no value. NPN is useful only when there is a need for additional ammonia N in the rumen.

Ideally, the rumen requirements for ammonia N would be met with dietary NPN, and all of the dietary protein would be digested directly in the small intestine by the animal. Protein that is not digested in the rumen is called **non-degradable, escape** or **bypass protein,** meaning that it escapes from or bypasses rumen digestion. Also, under ideal conditions, the bypass protein would have an amino acid composition that would complement the microbial protein. Microbial protein is first-limiting in sulfur amino acids, and is also limiting in lysine. Bypass protein should provide the lysine and methionine necessary to optimize the composition of the mixture of absorbed amino acids.

Currently, adequate information on the bypass protein content of protein sources is not available. The degradability is not a constant for a protein, but depends on processing conditions and other factors. Heat treatment of proteins increases their bypass value, by altering the physical structure of the protein to make it more resistant to microbial enzymes in the rumen. Some of the proteins with the highest bypass potential are of animal origin, such as fish meal and meat meal. These proteins also usually have a good composition of essential amino acids.

A goal for ruminant nutritionists is to formulate diets that provide the bypass amino acids in the amounts necessary to supplement the microbial protein so that the absorbed amino acids meet the animal's requirements. For example, for a high-producing dairy cow, to do this would require knowing the amounts of microbial essential amino acids liberated in the small intestine each day and the animal's metabolic requirements for essential amino acids. The difference between these two values is the apparent deficiency of each amino acid. Then the dietary protein program would be calculated so that the bypass protein provided the amount of each amino acids required to make up the apparent deficiency. New systems of diet formulation (NRC, 1989) are directed towards attaining this goal but are not there yet.

## Carbohydrates and Their Utilization

**Carbohydrates** make up the major part of animal feeds (except for diets of carnivores). They are the products of photosynthesis in plants. The photosynthetic reactions produce some organic acids from which glucose and other simple sugars are synthesized in the plant chloroplasts. In crop and forage plants, the major carbohydrates are starch and cellulose. These are both polymers of glucose, differing only in the chemical nature of the bond joining the glucose molecules together. **Starch** is the main carbohydrate in swine and poultry diets. It is digested by enzymes (e.g., pancreatic amylase) to liberate glucose in the gut. The glucose is absorbed and is the principal absorbed energy source supporting metabolism. The primary source of starch in swine and poultry diets is cereal grain— mainly corn (maize) and sorghum. In some areas, wheat and barley are also important energy sources for these animals. It is difficult to identify potential feedstuffs that could replace cereal grains as energy sources for nonruminants, without markedly reducing animal performance.

The other major plant carbohydrate is **cellulose.** Cellulose comprises over 50 percent of all of the organic compounds on Earth, as the major component of vegetation. Cellulose, along with a phenolic compound called **lignin,** makes up the cell wall of plant cells. The cellulose-lignin complex, lignocellulose, provides the structural rigidity of plant stems. Wood is basically lignocellulose. Forage plants contain cellulose as their primary car-

bohydrate. The lignocellulose complex is known as **plant fiber.** Although fiber is of interest in human nutrition for its physical properties, only animals with microbial fermentation in the gut can effectively utilize fiber as an energy source. Because of the special importance of ruminants in the conversion of plant fiber to human food, carbohydrate digestion in the rumen will be discussed in more detail.

## Carbohydrate Digestion in Ruminants

The rumen is a huge fermentation vat, uniquely well suited for its job. New substrate is continually being added, the fluid portion undergoes continual turnover, and the microbial wastes are rapidly and continually removed from the system by being absorbed from the rumen into the blood. The vat is kept at a more or less constant temperature and kept well mixed. There is an abundance of nutrients needed for microbial growth, continually being replenished as the animal eats.

**Rumen fermentation** is carried out by anaerobic microbes, with rumen bacteria the main fermentors. **Rumen protozoa,** which are small one-celled animals, consume feed particles and bacteria. Many species of **rumen bacteria** are involved in the fermentation process. They can be grouped into two main categories: the starch digestors (amylolytic bacteria) and cellulose digestors (cellulolytic bacteria). On grain-based diets high in starch, amylolytic organisms predominate, while with forage-based diets, the rumen flora is cellulolytic. The bacteria secrete amylase, which breaks down starch to release glucose, and cellulase, which converts cellulose to its glucose units. The released glucose is then metabolized (fermented) by the bacteria to produce ATP, which they use as their energy source (as do virtually all other organisms). Because the rumen is an anaerobic environment, the bacteria do not carry out the aerobic phase of metabolism (Kreb's cycle, citric acid cycle). Aerobic metabolism, such as that of higher animals, converts glucose to carbon dioxide and water, with energy liberated as ATP and heat. Rumen bacteria cannot completely oxidize glucose to $CO_2$ and water (fortunately for the host animal!). They metabolize glucose by the anaerobic **glycolysis pathway,** which breaks glucose down to two pyruvic acid molecules. This 3-carbon compound is then converted to other short-chain organic acids, which are excreted by the bacteria into the rumen fluid. These short-chain acids are called **volatile fatty acids (VFA),** because they are volatile in steam and can be steam-distilled out of a sample of rumen fluid. The three major VFA, acetic, propionic, and butyric acids, are the major absorbed energy sources of ruminant animals.

The amylolytic bacteria produce mainly propionic acid as their waste product, while the cellulolytic organisms produce acetic acid. Only small amounts of butyric acid are produced by either type. Propionic acid is used more efficiently in animal metabolism than acetic acid, partially explaining the superior performance of grain-fed vs. forage-fed ruminants.

The secretion of **cellulase** by rumen bacteria is of tremendous significance. Ruminant animal production represents the only feasible method of converting the glucose in plant fiber (i.e., the cellulose) into human food. Perhaps some day there will be artificial fermentation vats to accomplish this, and we can drink the vat liquor instead of milk! For the foreseeable future, conversion of cellulose to human food will continue to depend on ruminants.

Because of the complexity of rumen fermentation, it has been difficult for animal nutritionists to accurately predict the effects of dietary conditions and changes on animal performance. Cornell University scientists have developed a Net Carbohydrate and Protein System for quantitatively estimating the fermentation end-products (VFA, microbial protein, $NH_3$) and constituents that escape fermentation (carbohydrates, protein, undegraded peptides), to more accurately predict the effects of diet composition on metabolizable nutrients absorbed by the animal (Rus-

sell et al., 1992). The refinement and use of such prediction equations in ration formulation should improve the competitive position of ruminant animal production as compared with swine and poultry feeding systems, which are highly sophisticated.

## Lipids

The principal **lipids** of importance in animal feeding are fats and oils. They are triglycerides, meaning that they are composed of glycerol combined with three fatty acids. Saturated and unsaturated fatty acids are discussed in Chapter 2.

Fats and oils are important in animal nutrition because they are concentrated sources of energy. High energy diets for poultry and swine contain added fat to boost their energy content. On a weight basis, fats have about 2.25 times as much energy as carbohydrates.

## Minerals

Numerous mineral elements are dietary essentials for animals. Some, such as calcium and phosphorus, are needed in relatively large quantities. These are the **macro-elements.** The micro- or **trace elements** are needed in very small quantities. The macro-elements function as part of the tissue structure, such as calcium and phosphorus in bone. The trace elements have a catalytic role as "cofactors" or activators of enzymes. The macro- and micro-elements are as follows:

| Macro-elements | Micro-elements |
| --- | --- |
| Calcium | Manganese |
| Phosphorus | Zinc |
| Sodium | Iron |
| Potassium | Copper |
| Chlorine | Molybdenum |
| Magnesium | Selenium |
| Sulfur | Iodine |
|  | Cobalt |
|  | Chromium |

### Vitamins

**Vitamins** are organic compounds that are essential for normal metabolism. Like the trace elements, they function as cofactors or activators of enzymes. Vitamin deficiencies result in a loss of the activity of the particular enzymes involved, resulting in a specific deficiency symptom.

Two groups of vitamins are recognized, the fat-soluble and water-soluble groups.

| Fat-soluble | Water-soluble |
|---|---|
| Vitamin A | Thiamin ($B_1$) |
| Vitamin D | Riboflavin ($B_2$) |
| Vitamin E | Pyridoxine ($B_6$) |
| Vitamin K | Cyanocobalamin ($B_{12}$) |
| | Pantothenic acid |
| | Nicotinic acid (niacin) |
| | Folic acid (folacin) |
| | Biotin |
| | Choline |
| | Vitamin C (ascorbic acid) |

The water-soluble vitamins, with the exception of vitamin C, are members of the B-complex vitamins. This terminology stems from their original discovery. The first vitamin discovered was called vitamin A, and the second one vitamin B. It was soon realized that vitamin B was actually a mixture of several vitamins, now collectively called the B-complex group. It was also discovered that the original vitamin A preparation isolated from butter was a mixture of two substances, now called vitamin A and vitamin D.

# THE ENERGY CONCEPT IN ANIMAL NUTRITION

The concept of **energy** in nutrition is one of the most important and most difficult to appreciate. It is important to any discussion of the role of livestock in agriculture and human welfare because most of what we and animals eat is used to meet our energy requirements.

Energy is the nutritional component that is usually limiting the productivity of livestock. Most people probably would say "protein" when asked what is the most important nutritional component of feeds. However, sources of energy make up about 80 percent of the total diet of most animals. This varies with the level and type of production; the protein requirements are highest for young, rapidly growing animals and lactating animals. For most livestock, except for young poultry and for fish, the percent protein needed in the diet is less than 20 percent of the total diet.

Nutritional energy is measured and expressed in the United States as calories. A **calorie** is the amount of heat required to raise the temperature of 1 g of water by 1° C. Calories are measured in an instrument called a **bomb calorimeter.** In animal nutrition, kilocalories (kcal) and megacalories (mcal) are often used. These are equal to 1,000 and 1,000,000 calories, respectively. In human nutrition, the Calorie (large calorie) is the same as a kcal. When people talk about how many calories there are in a serving of cake or mashed potatoes, they are talking about Calories rather than calories. Animal nutritionists avoid this confusion by not using large and small calories.

In most countries other than the United States, nutritionists use the **joule** rather than the calorie as the unit of energy expression. A calorie is equal to 4.184 joules; one kcal is 4.184 kjoules (kj). This change has occurred as countries have adopted the metric system. Ironically, the joule is not a true metric unit; it is a unit of electrical energy. The only way that the joule content of feeds can be measured is by burning the feed in a bomb calorimeter and then converting the calories measured to joules by using a conversion factor. There is no direct way of measuring chemical (or food) energy in joules (joules = power in watts × time in seconds). Thus U.S. scientists have resisted the use of joules in nutrition, because it is fundamentally illogical; we don't eat electricity!

The **gross energy** (GE) content of a feed is measured in a bomb calorimeter. The total heat released from the complete burning of a known amount of feed is measured and expressed as kcal GE per g or kg of feed. What does burning a feed have to do with an animal or person? We burn feed (burn = oxidation by oxygen) but fortunately not all of the chemical energy in the feed is converted to heat, although most of it is. All animals give off heat as a waste product of energy metabolism. Only about 20 percent of the chemical energy metabolized in the body is converted to useful chemical energy (as ATP); the remainder is given off as heat.

## Energy Metabolism of Living Organisms

All life requires **chemical energy.** Life by definition involves growth and replication (reproduction), which involve formation of new tissue. Tissues such as proteins do not form spontaneously. Proteins are synthesized by the joining together of amino acids by peptide bonds. The formation of peptide bonds requires the input of chemical energy as **adenosine triphosphate (ATP).** The original source of the energy in the ATP is the sun. To be technically correct, the original source was the **big bang** when the universe was formed. The sun and the planets were formed by the condensation of cosmic particles. The sun is giving off heat energy as it cools; eventually, the sun will have lost all of its heat energy and will "die" and become what astrophysicists call a "white dwarf." Our planet will lose its source of energy and will approach absolute zero in temperature. Fortunately, this event is about 5 billion years in the future! All life on earth (with the exception of a few bacteria and deep sea primitive organisms that obtain energy from the oxidation of iron or sulfur) ultimately derives its energy from the sun. Plants trap solar energy (electromagnetic radiation) in the process of photosynthesis and use it to synthesize carbohydrates (sugars) from carbon dioxide and water. The carbohydrates or

**photosynthates** are used as the starting materials for the synthesis of complex carbohydrates, proteins, lipids, vitamins, and other plant compounds. When plants are eaten by animals, the nutrients synthesized by plants are digested and utilized directly in animal tissue synthesis (e.g., amino acids), converted to other compounds, or burned as a source of energy. The main sources of energy are carbohydrates and lipids. Protein can be used as an energy source, but it is generally less expensive to use carbohydrates than proteins as sources of energy in animal diets.

## Animal Combustion

The French scientist Lavoisier in 1777 recognized that "la vie est une fonction chemique" or "life is a chemical function." (In spite of this being known for over 200 years, it is still difficult to convince many animal science students of the importance of chemistry!) Lavoisier demonstrated that animals convert chemical energy in food to biologically useful chemical energy and waste heat energy.

When a feed sample is burned in a bomb calorimeter, all of the organic material that can be oxidized is burned, giving off carbon dioxide, water, and heat. Animals burn feeds also. **Animal combustion** involves biochemical pathways of metabolism, in which carbohydrate (mainly in the form of glucose) is burned (oxidized) to carbon dioxide and water, in a series of steps in which energy is released in small quantities, rather than all at once as in a flame. The major metabolic pathways in the oxidation of glucose are called **glycolysis** and the **citric acid cycle** (tricarboxylic acid cycle [TCA cycle], Kreb's cycle). Each step in these pathways involves an enzyme. At various steps, energy is transferred from compounds in the pathway to be used for the synthesis of ATP. ATP is then used in other biochemical reactions that require the input of energy (e.g., protein synthesis). These reactions form the basis of modern biochemistry. A person cannot truly understand nutrition without a good understanding of biochemistry; nutrition is applied biochemistry.

## Energy Categories of Feed

The useable energy in a feedstuff can be estimated by measuring energy losses in a metabolism trial. Animals are fed a known amount of feed of known GE content. The energy actually available for ATP production is the amount of GE consumed, corrected for all of the losses that occur. These include fecal energy (indigestible material), urinary energy (absorbed but not metabolized), the energy in rumen gases (mainly methane), and the heat energy lost as body heat. The heat is produced in all the cells of the body by the glycolysis and TCA cycle reactions that are exothermic (giving off heat). In ruminants, there is also the heat of rumen fermentation. When these losses are subtracted, the **net energy** (NE) is obtained.

This is the amount of energy in the feed actually used for productive functions. Because of the difficulties involved in determining all of these losses, especially heat and the rumen gases, the categories of **digestible energy (DE)** (GE-fecal energy) and **metabolizable energy (ME)** (DE-urinary energy, rumen gas energy) are often used in preference to NE.

## SUMMARY

The importance of livestock and other animals in the food chain can be appreciated through a knowledge of animal nutrition. Herbivorous animals such as the domestic ruminants have a role in the conversion of plant biomass to human food. Livestock production does not need to compete with humans for food; rather, it should enhance the efficiency of food production through utilization of crop residues, by-products, and forages.

## STUDY QUESTIONS

1. Are any livestock species specialist feeders? What types of feeding strategy are represented with common livestock species?

2. Discuss the role of salivary secretion in rumen function.

3. Domesticated ruminants are all roughage feeders and intermediate feeders, according to the classification scheme of Hofmann. Is it a coincidence that there are no concentrate-selector livestock species? Discuss.

4. Explain the differences between coprophagy and cecotrophy.

5. What is the largest rodent known? Discuss its digestive strategy.

6. What are the behavioral and anatomical characteristics of equids that give them an advantage over ruminants in some conditions? What are those conditions?

7. Discuss the concepts of fermentable and non-fermentable nitrogen in ruminant nutrition.

8. Under what dietary conditions would you expect no ruminant response from supplementation with urea? Under what dietary conditions is urea likely to be useful?

9. Discuss the nutritional significance of dietary peptides.

10. Is the elephant a ruminant or a hindgut fermentor? What are the advantages of the type of digestive tract of an elephant for very large animals?

# REFERENCES

BELL, R. H. V. 1971. A grazing system in the Serengeti. *Sci. Am.* 225:86–93.

BORGES, P. A., M. G. DOMINGUEZ-BELLO, and E. A. HERRERA. 1996. Digestive physiology of wild capybara. *J. Comp. Physiol.* B 166:55–60.

CHURCH, D. C. (Ed). 1988. *The ruminant animal. Digestive physiology and nutrition.* Prentice Hall, Englewood Cliffs, New Jersey.

DOMINGUEZ-BELLO, M. G., M. LOVERA, P. SUAREZ, and F. MICHELANGELI. 1993a. Microbial digestive symbionts of the crop of the hoatzin (*Opisthocomus hoazin*): An avian foregut fermenter. *Phys. Zool.* 66:374–383.

DOMINGUEZ-BELLO, M. G., M. C. RUIZ, and F. MICHELANGELI. 1993b. Evolutionary significance of foregut fermentation in the hoatzin (*Opisthocomus hoazin*; Aves: Opisthocomidae). *J. Comp. Physiol.* B 163:594–601.

GORDON, I. J., and A. W. ILLIUS. 1994. The functional significance of the browser-grazer dichotomy in African ruminants. *Oecologia* 98:167–175.

GORDON, I. J., and A. W. ILLIUS. 1996. The nutritional ecology of African ruminants: A reinterpretation. *J. Anim. Ecol.* 65:18–28.

GRAJAL, A., S. D. STRAHL, R. PARRA, M. G. DOMINGUEZ, and A. NEHER. 1989. Foregut fermentation in the hoatzin, a neotropical leaf-eating bird. *Science* 245:1236–1238.

HOFMANN, R. R. 1973. *The ruminant stomach.* East African Literature Bureau, Nairobi, Kenya.

HOFMANN, R. R. 1988. Anatomy of the gastro-intestinal tract. pp. 14–43 In: Church, D. C. (Ed.). *The ruminant animal.* Prentice Hall, Englewood Cliffs, New Jersey.

HOFMANN, R. R. 1989. Evolutionary steps of ecophysiological adaptation and diversification of ruminants: A comparative view of their digestive system. *Oecologia* 78:443–457.

HUME, I. D., and A. C. I. WARNER. 1980. Evolution of microbial digestion in mammals. pp. 665–684 In: Ruckebusch, Y., and P. Thivend. (Eds.). *Digestive physiology and metabolism in ruminants.* AVI Publ. Co., Westport, Conn.

ILLIUS, A. W., and I. J. GORDON. 1992. Modelling the nutritional ecology of ungulate herbivores: Evolution of body size and competitive interactions. *Oecologia* 89:428–434.

JANIS, C. 1976. The evolutionary strategy of the Equidae and the origins of rumen and cecal digestion. *Evolution* 30:757–774.

JANIS, C. M. 1989. A climatic explanation for patterns of evolutionary diversity in ungulate mammals. *Palaeontology* 32:463–481.

LANGER, P. 1984. Comparative anatomy of the stomach in mammalian herbivores. *Quarterly J. Exp. Physiology* 69:615–625.

LOWRY, J. B. 1989. Green-leaf fractionation by fruit bats: Is this feeding behaviour a unique nutritional strategy for herbivores? *Aust. Wildl. Res.* 16:203–206.

MEYER, H. H. D., A. ROWELL, W. J. STREICH, B. STOFFEL, and R. R. HOFMANN. 1998. Accumulation of polyunsaturated fatty acids by concentrate selecting ruminants. *Comp. Biochem. Physiol.* 120:263–268.

NATIONAL RESEARCH COUNCIL. 1989. *Nutrient requirements of dairy cattle.* National Academy Press, Washington, D.C.

PATIENCE, J. F. 1990. A review of the role of acid-base balance in amino acid nutrition. *J. Anim. Sci.* 68:398–408.

ROBBINS, C. T., D. E. SPALINGER, and W. VAN HOVEN. 1995. Adaptation of ruminants to browse and grass diets: Are anatomical-based browser-grazer interpretations valid? *Oecologia* 103:208–213.

RUSSELL, J. B., J. D. O'CONNOR, D. G. FOX, P. J. VAN SOEST, and C. J. SNIFFEN. 1992. A net carbohydrate and protein system for evaluating cattle diets. I. Ruminal fermentation. *J. Anim. Sci.* 70:3551–3561.

VAN HOVEN, W., and E. A. BOOMKER. 1985. Digestion. pp. 103–120. In: Hudson, R. J., and R. G. White (Eds.). *Bioenergetics of wild herbivores.* CRC Press, Boca Raton, Florida.

WEBB, K. E., Jr. 1990. Intestinal absorption of protein hydrolysis products: A review. *J. Anim. Sci.* 68:3011–3022.

WEBB, K. E., Jr., J. C. MATHEWS, and D. B. DIRIENZO. 1992. Peptide absorption: A review of current concepts and future perspectives. *J. Anim. Sci.* 70:3248–3257.

# CHAPTER 5

# Feed Additives and Growth Promotants in Animal Production

**CHAPTER OBJECTIVES**

1. To describe some of the major feed additives used in livestock production and to indicate why they are effective in improving some aspect of animal performance

2. To discuss the risks versus benefits of major feed additives

3. To present background information on feed additives prior to discussion (Chapter 9) of societal concerns regarding use of feed additives

**Feed additives** are non-nutritive substances added to feeds to improve the efficiency of feed utilization and feed acceptance or to be beneficial to the health or metabolism of the animal in some way (Cheeke, 1999). Thus they are substances other than nutrients or sources of nutrients. By this definition, synthetic amino acids, vitamin preparations, and mineral supplements are not considered feed additives.

A comprehensive, but not necessarily complete, list of various types of feed additives (Cheeke, 1999) is as follows:

1. Additives that influence feed stability, feed manufacturing, and non-nutritive properties of feeds

   A. Mold inhibitors (antifungals)
   B. Antioxidants (preservatives)
   C. Pellet binders

2. Additives that modify feed intake, digestion, growth, feed efficiency, metabolism, and performance

   A. Feed flavors
   B. Digestion modifiers

C.   Metabolic modifiers

D.   Growth promotants

3.   Additives that modify animal health

A.   Drugs

B.   Environmentally active substances

4.   Additives that modify consumer acceptance of animal products

A.   Pigmenting agents

The use of feed additives in livestock feeding is a controversial subject. Critics of modern agricultural production techniques often claim that feed additives are hazardous to human health, may be inhumane to animals by forcing them to achieve unnatural levels of production, and are a symptom of undesirable reliance on chemicals and technology. Some additives are more or less controversial than others. Antibiotics, hormones, preservatives and growth promotant chemicals are suspect to many people (Fig. 5-1). Enzymes, flavoring agents, herbs, and probiotics are probably not perceived as threatening by nearly as many people.

**FIGURE 5-1** No hormones, preservatives or antibiotics! This type of "natural" advertising capitalizes on the public's concern about feed additives and "chemicals" in our food supply. What are the hazards, if any, from the use of feed additives? Chapter 5 discusses the roles of feed additives in animal production, while the risks and benefits are weighed in further detail in Chapter 9.

The objective of this section will be to discuss some of the major feed additives, emphasizing the advantages and disadvantages of their use. Inevitably, personal bias is involved. I hope my interpretations will provide a framework for further discussion by students and producers.

# ADDITIVES THAT INFLUENCE STABILITY AND NON-NUTRIENT PROPERTIES OF FEEDS

Consider corn grain. The kernel has a tough, waxy outer shell. If whole corn is fed to cattle, many of the kernels will pass undigested through the digestive tract and be excreted as whole corn kernels in the feces. Thus corn is usually processed by being rolled or ground before it is used in preparing animal feeds. This, however, introduces a further complication. In the whole seed, the corn oil is in a different part of the seed than an enzyme called lipoxidase. When the seed is rolled or ground, the corn oil and lipoxidase are mixed together. The oil in the intact seed is also protected from exposure to oxygen, but when it is processed, the oil is exposed to air. The lipoxidase causes the double bonds of the oil to become oxidized or take up oxygen. This process is known as **rancidity** (autooxidation), and the oil becomes rancid. As rancidity develops, the unsaturated fatty acids take up oxygen, **peroxides** are formed, and these undergo further reaction to produce toxic and foul-smelling aldehydes and ketones. These reduce palatability of the feed, destroy vitamin A and other easily oxidized vitamins, and may be toxic. Animals fed rancid feeds "back off" the feed, and growth, milk or egg production, and so on is reduced. To prevent this from occurring, an antioxidant feed additive can be mixed with the feed. **Antioxidants** inactivate peroxides and inhibit the chain reactions that cause rancidity to develop rapidly. Which is more undesirable: to prepare feed that will quickly become rancid or to add a preservative (antioxidant) to prevent rancidity? It is well established that rancid oils are undesirable and can be toxic. What are the hazards of antioxidants?

Antioxidants occur naturally; vitamin E and vitamin C function as antioxidants. We could add vitamin E to prevent the corn from becoming rancid, but this would be quite expensive. We could also use **synthetic antioxidants,** which include substances such as ethoxyquin and butylated hydroxytoluene (BHT). To the average person, butylated hydroxytoluene sounds ominous. If you've had organic chemistry, you recognize the simple nature of butyl and hydroxy groups and recognize toluene as simply a 6-carbon ring (benzene ring) with a methyl group. It may sound less threatening if you have some knowledge of its chemical structure. Synthetic antioxidants prevent peroxide formation and are very effective at preventing rancidity. At the levels approved for use by the U.S. Food and Drug Administration (FDA), they are nontoxic. Extensive animal testing was required before they were approved for use. However, the public perception of preservatives is bad. A common advertising claim is "no preservatives." The alternative to their use is a short shelf life for feed and food (e.g., potato chips), off-flavors in feed and food, and potential toxicity of rancid fats. Which is better (or worse): eating or feeding rancid feed or using an antioxidant? Also, bear in mind that by retarding autooxidation, antioxidants are believed to reduce the risk of CHD and cancer and retard aging to increase longevity, as discussed in Chapter 2.

Another group of preservatives is the mold inhibitors, which inhibit the growth of undesirable molds in feeds. Many molds (fungi) produce extremely toxic substances called **mycotoxins** (myco = mold or fungus). Examples of mycotoxins that may occur in moldy feed are aflatoxin, the most potent cancer-causing substance known, zearalenone, a potent estrogen causing reproductive impairment, and vomitoxin, which causes feed refusal and vomiting in swine. The most common **mold inhibitors** are products containing propionic acid. Propionic acid at an added level of 1 percent of the feed or grain will prevent mold growth. The lack of hazard of propionate can be appreciated by the fact that it is produced as one of the major VFAs in rumen fermentation and is a major absorbed energy source for ruminants. It readily enters the TCA (Kreb's cycle) metabolic reactions of animal tissue.

Ammonia is another very effective mold inhibitor. **Ammoniation** of stored grains prevents mold growth and in addition causes the destruction (due to the alkaline conditions) of aflatoxin. Ammoniation is used to detoxify aflatoxin-contaminated grains. Ammonia can be used as a source of non-protein nitrogen (NPN) in the rumen.

In the author's opinion, the benefits of using preservatives such as antioxidants and mold inhibitors are readily apparent. The hazards seem to be slight. The benefits from their use seem to far exceed any potential hazards. Perhaps the best way (in the long run) of convincing the general public of this is through education; the more people who take chemistry courses, the less the scientific illiteracy.

Another type of feed additive that modifies properties of feeds is **pellet binders.** Pelleted feeds are widely used for livestock and poultry. Pelleting increases the density of feeds, reducing storage and transportation costs. It reduces feed wastage and sorting and often improves animal performance and feed conversion efficiency. It reduces dustiness of feed and facilitates automation of feed handling and feeding.

Pelleting is accomplished by forcing the mixed feed through a rotating die that extrudes the feed. As the extruded feed leaves the die, it is cut off at a predetermined length by stationary knives. The diameter of the pellets depends on the size of the holes in the die, while the length depends on the knife setting. Some ingredients make good quality pellets, while others tend to crumble and fall apart. Wheat and wheat milling by-products (e.g., bran) improve **pellet quality,** because the wheat gluten has a gooey consistency when moist, helping to bind the other ingredients together. Steam is injected into the feed prior to pelleting to cause the starch to gelatinize, helping to bind the feed together. In many cases, however, the ingredients may not form a good tight pellet, and the pellets crumble and produce a lot of "fines." Animals like pelleted feed, but they do not like poor quality pellets with fines and dust. For this reason, good pellet quality is very important to both feed manufacturers and livestock producers. A number of pellet binders are available as feed additives. One of the most commonly used is bentonite, which is a clay mineral (montmorillonite, hydrated aluminum silicate). Bentonite is mined in various parts of the United States and marketed as a pellet binder. Hemicellulose extract from hardboard manufacture from wood is also used as a binder. These additives have no nutritional value but improve the quality of pelleted feed.

Summarizing the properties of these feed additives that are used to improve stability and quality of livestock feeds, it is my conclusion that pellet binders and preservatives such as antioxidants and mold inhibitors are useful and important and free of discernible hazards or undesirable properties.

## ADDITIVES THAT MODIFY FEED INTAKE OR DIGESTION

A number of **feed flavors** are used in the feed industry. Their intended use is to increase feed intake, particularly of diets lacking good palatability. Their beneficial effects are generally not great, but there is no hazard from their use.

Various **enzymes** are used as feed additives. The most useful are sources of β-glucanase. β-glucans are gummy carbohydrates in certain grains, mainly barley, rye, and triticale. Animals do not produce β-glucanases, the enzymes which digest these compounds. Thus swine and poultry are not able to efficiently utilize the glucans in these grains. This is particularly important in poultry fed barley-based diets. Some barley varieties contain fairly high levels of β-glucans, which are not

digested. The glucans are viscous, hydroscopic compounds, which when excreted cause the excreta to be wet. This causes wet, soupy litter in poultry houses, which normally should be dry. The wet litter causes elevated humidity and ammonia in the house, causing respiratory disease and problems with manure removal. The viscous glucans in the gut interfere with the digestion and absorption of nutrients. The addition of a feed additive containing β-glucanases improves the litter and air quality of the poultry house and improves bird performance (Pettersson et al., 1991).

Besides β-glucans, there are other **nonstarch polysaccharides** (NSP) in grains that may adversely affect animal performance (Annison and Choct, 1991). Arabinoxylans (pentosans) are viscous, poorly digested polysaccharides found in the cell walls of wheat and rye grain. Commercial enzyme preparations with pentosanase activity are available to the feed industry. The use of these enzymes increases the performance of poultry fed rye, wheat, and wheat by-products (Annison and Choct, 1991; Choct and Annison, 1992). The mode of action appears to be to hydrolyze the NSP to smaller polymers that do not have the viscosity to inhibit nutrient absorption (Annison and Choct, 1991). The pentose sugars of which the NSP are composed, such as arabinose, are poorly absorbed and metabolized. The use of enzyme products to improve utilization of grains by swine and poultry has been reviewed by Campbell and Bedford (1992).

There are numerous other enzyme products, such as **cellulases,** which are used as feed additives. Cellulases digest cellulose, increasing the ability of nonruminants to utilize fibrous feeds. This has obvious and far-reaching implications in livestock and food production. The use of cellulases could eventually reduce the role of ruminants in converting fiber to human food.

**Phytases** are important feed additives in some circumstances. Phytates are organic molecules containing bound phosphorus, which is of low bioavailability to nonrumi-

**FIGURE 5-2** Pollution of waterways with animal wastes may cause oxygen depletion and massive fish kills. (Courtesy of C. E. Bond.)

nant animals. In many countries, phosphorus pollution of ground water and waterways is becoming a major environmental problem. Phosphorus and nitrogen as water pollutants are very important, because they stimulate algae growth, which can have serious impacts on water quality and fisheries. Algae blooms in rivers and lakes use up dissolved oxygen, causing massive fish kills (Fig. 5-2). Although algae produce oxygen via photosynthesis during the day, at night their respiration depletes dissolved oxygen. When the algae die and sink to the bottom, their decomposition uses up oxygen and can also cause fish losses. The drinking quality and aesthetic and recreational values of the water are diminished. Since the banning of phosphates in detergents, the major sources of phosphorus pollution are agricultural sources such as fertilizers and animal manure. In the Netherlands, government regulations have been implemented to reduce the number of pigs produced in order to reduce nitrogen and phosphorus pollution. Singapore has banned swine production, and Taiwan is implementing programs to reduce pig numbers to reduce pollution. Phytase feed additives are of potential value in dealing with this pollution problem, as a means of increasing the availability of the phosphorus in the feed, so less of it is excreted in the feces (Simons and Versteegh, 1990).

There are no apparent hazards or disadvantages of using enzymes as feed additives, and there are numerous benefits. They are also increasingly used directly by humans. For example, lactase is available for people with lactose intolerance, and commercial enzyme preparations are available (e.g., Beano) to improve the digestion of oligosaccharides in food products such as beans.

There are a number of feed additives, such as buffers, specifically intended for use with ruminants. **Buffers** are compounds that resist changes in pH. When ruminants are fed high concentrate diets, there is vigorous microbial growth in the rumen. The microbes produce large amounts of organic acids as waste products (VFA), as discussed earlier. The rate of production of VFA may exceed their rate of absorption, causing the acidity of the rumen to increase. This may cause acidosis, characterized by corrosion of the rumen wall (rumenitis or parakeratosis) followed by liver abscesses due to the invasion of bacteria through lesions of the rumen wall. The use of dietary buffers can help to buffer the VFA and keep the rumen from becoming highly acidic. Sodium bicarbonate is the most commonly used. Others include sodium sesquicarbonate, magnesium oxide, and calcium carbonate (limestone, cement kiln dust). Buffers are particularly useful for feedlot cattle fed high concentrate diets and for high-producing dairy cows fed very high levels of grain. The use of buffers as feed additives provides obvious benefits, while no hazards are evident.

Antibloat agents are compounds used as feed additives to prevent bloat in ruminants. **Bloat** is caused by the formation of a stable foam in the rumen, causing rumen gases to accumulate. The presence of foam at the base of the esophagus inhibits the eructation (belching) of gases from the rumen. This is a protective reflex action, because eructated foam would enter the lungs and cause suffocation. As the gases accumulate in the rumen, a tremendous pressure can develop, leading to the point where the animal can no longer breathe because of pressure on the diaphragm.

The rumen foam is caused by the cytoplasmic or soluble proteins in plant cells, which are good foaming agents. Lush, succulent high-protein forages such as legumes (clover, alfalfa) and heavily fertilized grasses (e.g., wheat pasture) cause pasture bloat. This is because of the rapid release of the proteins into the rumen from the succulent plant material, causing formation of foam. The rapid fermentation of the soluble carbohydrates in the forage causes vigorous gas production, further contributing to the foam formation. The combination of profuse gas bubbles and abundant foaming agents results in a stable rumen foam.

The main antibloat additive is poloxalene, marketed as Bloat Guard, usually in the form of blocks. Poloxalene is a surface-active agent (detergent), which reduces the surface tension of the foam bubbles in the rumen, causing the bubbles to break down and release the trapped gas. Bloat Guard blocks should be distributed in pastures likely to cause bloat (Fig. 5-3). Because of the salt and molasses in the blocks, livestock will lick them frequently, thus keeping an adequate amount of poloxalene in the rumen to prevent bloat. There are no hazards associated with the use of this additive, and the benefits, allowing the grazing of pastures with high nutritional value (e.g., alfalfa) without the danger of bloat and distress to the animal, are substantial.

**Defaunating agents** are substances that will kill off the protozoa in the rumen. Examples include copper sulfate, nonionic and anionic detergents, and saponins. Saponins are naturally occurring detergent-like compounds that occur in many plants. Saponins from the desert plant *Yucca schidigera* are commonly used as feed additives (Fig. 5-4). They are very toxic to rumen protozoa (Wallace et al., 1994.) On high-energy, low protein diets such as those based on molasses or sugarcane, defaunation improves ruminant animal performance (Bird and Leng, 1984). With conventional diets, there seems to be little effect (Viera, 1986). The rationale of defaunation is that the rumen protozoa feed on rumen bacteria and thus introduce another

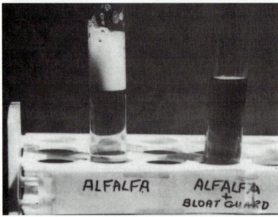

**FIGURE 5-3** A steer licking a Bloat Guard block (*top*). When a water extract of alfalfa is shaken, a stable foam forms. When Bloat Guard is added, the alfalfa extract does not foam (*bottom*).

trophic level (another step in the food chain). Therefore eliminating protozoa should theoretically improve the efficiency of protein utilization. Defaunation is still an experimental procedure, and its benefits have not been unequivocally demonstrated.

**Ionophores** are a class of feed additives used extensively in cattle production. They are antibiotics that inhibit the growth of certain bacteria by disrupting the normal transport of mineral ions across the cell membrane of gram-positive bacteria. Gram-positive bacteria lack the protective outer membrane that is characteristic of gram-negative bacteria. Some common ionophores are rumensin (monensin), lasalocid, and salinomycin.

**FIGURE 5-4** *Yucca schidigera* is harvested in the deserts of Baja California, Mexico, for the production of saponin-containing yucca extract used in the feed and beverage industries.

Ionophores consistently improve the feed conversion efficiency of ruminants and often improve growth rate as well. Their effect is due to a modification of the rumen bacterial population, resulting in a greater proportion of propionate and lower acetate in the fermentation end-products. Ionophores inhibit gram-positive organisms. The gram-negative organisms, which then proliferate, produce propionate as their major waste product. When a given amount of carbohydrate is fermented to propionate, there is a greater amount of carbon remaining to be absorbed and metabolized by the animal than when acetate is produced:

The six carbons in the propionate will yield more ATP when metabolized by the animal than the four carbons from acetic acid. The $CO_2$ and $CH_4$ (methane) cannot be metabolized and represent an energy loss. Thus from the same quantity of ingested carbohydrate, the animal gets more net energy when ionophores are fed, improving the feed to gain ratio.

Ionophores reduce methane production, as shown in the preceding reaction. This has implications in global warming (see Chapter 6). Ionophores have a number of other beneficial effects besides improving feed conversion. They reduce lactic acidosis, are toxic to fly larvae in the manure, and aid in the control of coccidiosis, feedlot bloat, and acute bovine pulmonary emphysema (Goodrich et al., 1984; Carlson and Breeze, 1984; Herald et al., 1982). Ionophores are coccidiostats and are used to control coccidiosis in cattle and poultry. **Coccidiosis** is a disease caused by protozoa (one-celled animals) that damage the gut lining. Rumensin and monensin are the same compound; rumensin is the name used for the ruminant additive and monensin is the poultry coccidiostat.

Ionophores such as rumensin are toxic to horses and may cause poisoning at the levels used for cattle. Horses should not have access to feed or blocks containing ionophores. Signs of toxicity in horses include weakness, staggering, profuse sweating, and head pressing (pressing the head into a wall or other solid object), with liver and kidney damage. The minimum lethal intake is 1 mg rumensin per kg body weight. There is no treatment.

Ionophores improve feed conversion efficiency in cattle by about 10 percent, providing a means of significantly improving the utilization of natural resources. Reduction in rumen methane production may have beneficial ecological effects. Other than their toxicity to horses, which can be avoided by feed

$$2\ CH_3CH_2COOH \longleftarrow C_6H_{12}O_6 \longrightarrow 2\ CH_3COOH + CO_2 + CH_4$$
Propionic acid                      Glucose                          Acetic acid

and animal management, there seem to be no hazards to their use. The use of ionophores and their modes of action have been reviewed by Bergen and Bates (1984) and Russell (1987).

# ADDITIVES THAT MODIFY METABOLISM

Based on stories presented in the visual (TV) and print media, popular opinion seems to be that livestock are raised on "factory farms" and fed diets "laced with chemicals and hormones." In reality, **hormones** are neither permitted nor used to any extent as feed additives in the United States. The only hormone permitted as a feed additive for beef cattle is **melengestrol acetate (MGA),** a synthetic progesterone, which suppresses estrus in heifers. It is used to some extent with heifers in feedlots to maintain normal behavior and feed intake by eliminating the presence of heifers in heat.[1]

At one time, **diethylstilbestrol (DES),** a synthetic estrogen, was used as a feed additive for cattle, but its FDA approval was withdrawn in the late 1970s. DES was implicated as a carcinogen in the daughters of women who had been administered high levels of DES during pregnancy to prevent miscarriage. No hazards to human health as a result of its use as a feed additive were demonstrated. Estrogens stimulate the growth rate of ruminants, particularly steers. They can be administered as slow-release **implants** in the ear. One of the most

---

[1] Another incident in a long list of contaminated animal products occurred in Europe in 2002. Thousands of swine farms in Europe were closed and tainted animals and products were destroyed. Feed was contaminated with MGA, traced back to a mislabelled waste sugar solution from a pharmaceutical factory in Ireland. The contaminated syrup was also sold to some soft drink manufacturers (*Feedstuffs,* July 22, 2002, p. 3. "Tainted Feed Contamination Closes European Hog Farms.").

commonly used implants is zearalenone (trade name Ralgro), which is an estrogenic compound produced by molds growing on corn. While it may be skirting the issue, these substances are not used as feed additives, even though they are administered to livestock as implants. It is true that livestock are administered hormones but not as feed additives. **Growth hormone** (bST, pST; bovine and porcine (pig) somatotropin, respectively) is a controversial substance. When administered to animals by injection, growth hormone increases growth rate and milk production. Growth hormone is *not* a feed additive; as a protein, it would simply be digested in the gut if fed and would have no biological activity. Growth hormone as an injectable growth promotant is discussed later in this chapter. The potential use of bST in the dairy industry is quite controversial. The ethical, social, and political implications of use of bST are discussed in Chapter 9. These aspects pertaining to DES are also covered in Chapter 9.

In summary, hormones are used in animal production as implants (estrogens) and injectables (bST). However, they are *not* feed additives.

# ANTIBIOTICS AS FEED ADDITIVES

Commercial production of antibiotics began in the 1940s after the discovery of penicillin. The term antibiotic means "against life." Specifically, it refers to the ability to kill or suppress bacteria, many of which cause disease. **Antibiotics** are defined as substances produced by living organisms (generally fungi) that inhibit the growth or survival of micro-organisms (generally bacteria). There are literally thousands of different antibiotics produced by fungi growing in soil and other environments. Antibiotics are produced by fungi as chemical defenses against bacteria. Almost all fungi produce protective chemicals. Those that we have put to beneficial use we refer to as antibiotics; those that are mainly of concern because of their toxicity we call

**mycotoxins.** Commercial production of antibiotics involves isolating fungi from soil or other materials, determining if they have desirable antibacterial effects, and then growing the fungi in fermentation vats to produce large quantities of the antibiotic. Usually a grain like corn is used as the fermentation medium. When commercial production of antibiotics began in the 1940s, the grain fermentation residues were tested as animal feeds. It was observed that animal performance, including growth rate and feed conversion efficiency, was improved by feeding fermentation residues. This was a result of the residual antibiotics in the fermentation substrate. The era of antibiotics as feed additives was born.

The mechanism of action of antibiotics in their growth promotant activity is not fully known. Some of the possibilities are the following:

1. Antibiotics may suppress mild but unrecognized subclinical infections.

2. Antibiotics may inhibit the growth of toxin-producing gut microbes.

3. Antibiotics may reduce microbial destruction of essential nutrients in the gut or improve the gut synthesis of vitamins.

4. Antibiotics may improve nutrient absorption by causing a thinning of the intestinal mucosa.

5. Two or more combinations of the preceding.

There is evidence that all of these factors may be involved. It is well-documented that the intestinal mucosa of antibiotic-fed animals is thinner than for controls. This may be due to inhibition of pathogenic microbes that colonize the gut lining. A thinner gut lining could improve nutrient absorption. Also, the cells of the intestinal mucosa are the most rapidly growing tissue of the body. The cells (enterocytes) lining the intestinal villi are sloughed off and constantly replaced by new cells. This cell turnover is increased by exposure to bacterial toxins and metabolites, including ammonia

(Visek, 1978). Dietary antibiotics reduce the turnover rate of intestinal mucosal cells. Therefore, antibiotics could stimulate growth by reducing the nutritional demand for synthesizing intestinal mucosal tissue (reducing endogenous protein losses), thus diverting protein and energy to the synthesis of other tissues.

Klasing (1988) and Roura et al. (1992) have proposed another mechanism of action of antibiotics. Dietary antibiotics may reduce the number of pathogenic gut microbes. These microbes produce **immunogens,** or substances that induce an immune response. The immune response involves the increased production by the animal of **cytokines** (interleukins, interferons) and antibodies. By reducing gut pathogens, antibiotics reduce the quantity of absorbed immunogens, reducing the animal's need for an immune response. Thus nutrients are diverted from the immune response to be available for synthesis of new body tissue.

Bacteria can rapidly develop resistance to antibiotics by mutation. It is interesting that while the antibiotic-resistant gut microbes increase when antibiotics are routinely fed in the diet, this does not change their growth-promoting activity. The magnitude of growth promotant activity of antibiotics has not diminished since their use as feed additives began in the 1950s.

Use of antibiotics as feed additives is controversial. The major concerns have been with the potential implications in human health. The scenario presented is that routine inclusion of antibiotics in animal feeds may lead to development of **antibiotic-resistant bacteria.** Either these bacteria or other microbes carrying their resistance factors might infect humans, which could then reduce the effectiveness of antibiotics in treating human disease. The ability of bacteria to develop resistance to antibiotics is carried in nonchromosomal pieces of genetic material called **plasmids.** Plasmids may be transferred between bacteria. The plasmids carrying antibiotic resistance are called R-factors. The main concern with antibiotics as feed additives is that R-factors conferring resistance to certain antibiotics could be transferred to bacteria that are human pathogens (e.g., *Salmonella* spp.), making them resistant to treatment. Holmberg et al. (1984) of the Centers for Disease Control (U.S.) reviewed the etiology of outbreaks of antibiotic-resistant *Salmonella* infections in the United States in the period 1971–1983. They identified several documented cases of transmission of antibiotic-resistant *Salmonella* from livestock and poultry to humans. The human fatality rate was 21 times higher with antibiotic-resistant *Salmonella* infection than when antimicrobial-sensitive bacteria were involved. They concluded that antimicrobial-resistant enteric bacteria frequently arise in food animals and can cause serious infections in humans. The feed industry is moving in the direction of livestock-specific antibiotics as feed additives, using antibiotics that are not used in human medicine. This would minimize the potential for adverse effects on human health. There is also an increasing use of alternatives to antibiotics as feed additives, such as direct-fed microbials.

Growth promotant antibiotics are used primarily for poultry and swine. Other than the previously discussed ionophores, ruminants are not routinely fed antibiotics, because of the severe disruptions in rumen microflora which could occur.

The major groups affected by the use of antibiotics as feed additives are livestock producers, feed manufacturers, drug companies, and consumers. The primary beneficiaries are the two latter groups: drug companies and consumers. Drug companies obviously profit by making and selling the products. Consumers benefit from lower-cost meat and eggs. Neither livestock producers nor feed manufacturers derive major financial benefits. As the production of animals becomes more efficient, the savings in costs of production are passed onto the consumer. Similarly, the feed manufacturer doesn't make a greater profit by adding antibiotics to feed. In fact, with the stringent regulations and accountability involved with the use of drugs in animal feeds, life would be a lot simpler for feed manufac-

turers if they didn't produce medicated feeds. In some cases, they decide to opt out of the use of antibiotics and other regulated additives to avoid the paperwork necessary with their use.

In summary, antibiotics as feed additives are economically beneficial because they improve animal performance and feed utilization, thus reducing the cost of animal products to the consumer. They have not lost their effectiveness with time and continue to provide the same degree of benefit as when first introduced. For what it's worth, antibiotics are "natural," being produced by fungi inhabiting soil. Antibiotics have existed for eons. The hazards associated with their use are primarily with the development of antibiotic-resistant bacteria. These can have adverse effects on human health. The risk vs. benefit equation of feed additive use of antibiotics continues to be debated by agricultural and medical authorities (Witte, 1998). The hazards to human health appear to be slight if proper sanitation procedures are used in food preparation. Many of the *Salmonella* outbreaks discussed by Holmberg et al. (1984) involved consumption of raw milk so are basically self-inflicted situations. Although there is no medically sound reason to object to pasteurization of milk, some people are opposed to it. Food safety aspects of antibiotic-resistant bacteria are further discussed in Chapter 9.

Rollin (2001) argues that subtherapeutic use of antibioitcs in animal feeds should be curtailed to force agriculture "to resume a decent standard of animal care." He maintains that the cheap cost of animal products in the United States has come at the expense of animal welfare. His conclusion follows:

> The issue of antibiotic resistance and agriculture may well be a blessing in disguise. It may well be a last call for us to back off from chasing high-technology fixes in agriculture and cheap food, prevent its ingression into developing countries, and to begin to pursue a more thoughtful, sus-

tainable, humane, community-preserving agriculture, one that stresses harmony between humans, animals, health, and nature, not perpetual overcoming of nature and chasing of high-tech solutions, generating high-tech problems. The specter of an eternal, spiraling, escalating race between pathogens and drug companies is a chilling and endlessly costly project. (p. 37)

Thus Rollin (2001) claims that the decision as to whether or not to use antibiotics in feeds is an ethical one, not a scientific one: "We argue that society has all the data it needs to make a reasonable ethical decision, which would be curtailing such use" (p. 29).

On September 11, 2001, the United States was bombed by terrorists, followed by anthrax bioterrorism. The specter of bioterrorism has altered the antibiotics-as-feed-additives debate. The possibility that large numbers of people may need to receive antibiotic treatment for bacterial diseases such as anthrax raises the likelihood of increased development of antibiotic-resistant bacteria. As a matter of national security, widespread use of antibiotics in livestock production seems irresponsible. A further blow to the pro-feed additive use of antibiotics came from the publication of three papers (White et al., 2001; McDonald et al., 2001; Sorensen et al., 2001) in the highly respected *New England Journal of Medicine* that document the occurrence of antibiotic-resistant bacteria in meat and in the digestive tracts of humans consuming meat from antibiotic-fed animals. This prompted an editorial in the *New England Journal of Medicine* (Gorbach, 2001) titled "Antimicrobial Use in Animal Feed—Time to Stop." Gorbach called these three papers the "smoking gun" and concluded: "The subtherapeutic use of these agents to promote growth and feeding efficiency should be banned—a move that would decrease the burden of antimicrobial resistance in the environment and provide health-related benefits to both humans and animals" (p. 1203).

# ALTERNATIVES TO ANTIBIOTICS

With the controversies and negative perceptions regarding antibiotic use as feed additives, there is much interest in development of alternatives, such as probiotics. The term **probiotics** was coined to contrast these products with antibiotics. Probiotic means "pro life." They are live microbial supplements that may beneficially affect the host animal by improving its gastrointestinal microbial balance (Fuller, 1989). One of the major microbes used in probiotics is *Lactobacillus acidophilus*, which is also found in fermented products such as yogurt. Other organisms such as *Bacillus subtilis, Streptococcus faecium*, and yeasts are used in commercial probiotic products. The preferred term in the feed industry for probiotics is **direct-fed microbials.** A marketing tool extensively employed for probiotics is that they are "natural" alternatives to antibiotics and drugs. However, evidence for beneficial effects of these products has tended to be more of the testimonial type than hard scientific data. The preponderance of the positive results attributed to probiotics has been presented in testimonials and popular press articles, neither of which are necessarily authoritative or definitive sources of information. There is considerable skepticism in the scientific community regarding the usefulness of probiotics, because of the lack of data showing statistically significant responses in papers published in peer-reviewed scientific journals.

A postulated mode of action of probiotic preparations is suppression of gut pathogens (e.g., *E. coli*) by **competitive inhibition.** Most enteric pathogens attach to the intestinal mucosa and cause enteric disease by severely damaging the mucosa. Live microbial cultures contain microbes that attach to the mucosa, and by so doing, may prevent attachment of pathogenic bacteria. Competitive inhibition is being used to prevent *Salmonella* infection of poultry. Baby chicks are inoculated orally with preparations of normal gut bacteria, which colonize the gut effectively, thus denying colonization sites to pathogens. In essence, the technique swamps *Salmonella* in the gut with massive quantities of beneficial microbes.

While the efficacy of currently available probiotics is still controversial, the concept would seem to have a lot of merit. Using biotechnology, it should be possible to "design" microbes that have beneficial effects in the gut. For example, Newman et al. (1988) developed genetically modified strains of *Lactobacillus* bacteria that secreted lysine-rich polymers. It may be possible in the future to feed probiotics containing microbes that secrete lysine, methionine, and other essential amino acids, reducing the need for providing these amino acids in the diet. Because of the importance of rumen microbes in ruminant digestion, there is potential for modification of rumen microbes to alter their enzymes and digestive efficiency, their composition or digestive end-products (Hespell, 1987; Russell and Wilson, 1988). Some microbes degrade plant toxins. An interesting example of this involves a tropical legume forage, *Leucaena leucocephala*, which contains a toxic amino acid, mimosine. Livestock in some parts of the world (e.g., Australia) develop severe toxicity signs when grazing leucaena, whereas in other areas (e.g., Hawaii, Indonesia), animals can feed on leucaena without any signs of poisoning. Australian researchers demonstrated that ruminants in Hawaii and Indonesia had a rumen microbe that degraded mimosine. When rumen fluid from these animals was inoculated into Australian cattle, they became resistant to mimosine (Jones and Megarrity, 1986; Quirk et al., 1988). This has also been demonstrated in the United States (Hammond et al., 1989). It is hoped that it will be possible to genetically modify rumen microbes to detoxify other toxins in poisonous plants and feedstuffs.

A variety of substances that are not antibiotics have growth-promotion activity. **Copper sulfate** has been used quite extensively as a feed additive for swine and poultry, at dietary levels of 125–250 ppm. It apparently acts in the same way as antibiotics, through modification of gut flora. It improves growth and feed conversion and reduces enteric disease. There are several valid concerns with the use of copper sulfate for this purpose. The excreta of copper-fed animals contains high copper levels, and may be a pollution problem. Sheep are especially sensitive to copper toxicity; pastures fertilized with high copper manure could lead to copper toxicity in grazing animals. **Dried poultry waste (DPW)** is often used as a feedstuff for ruminants; manure from copper-fed birds has caused copper toxicity in cattle (Banton et al., 1987). Tissue levels of copper in animals fed feed additive levels of copper sulfate are not elevated, except for liver. There appear to be no human health implications of tissue residues of copper.

Other chemicals with growth promoting properties are **arsenicals, nitrofurans,** and **sulfonamides** (sulfa drugs). Withdrawal times have been established by the FDA to prevent tissue residues. Residues of sulfa products (e.g., sulfamethazine) in pork have been a problem in the swine industry. At high dietary levels, sulfa drugs are thyroid inhibitors and can cause cancer in laboratory animals. This finding invokes the Delaney Clause (see Other Growth Promotants That Are Not Feed Additives). The continued approval of sulfa drugs in animal feeds is under review by the FDA (1993). Arsenicals and nitrofurans are used in poultry and swine diets for growth promotion and prevention of enteric disease. Arsenicals are organic compounds containing arsenic. They function in the gut and are not absorbed, but the mere hint of arsenic makes them sound ominous to many people.

# REPARTITIONING AGENTS

**Repartitioning agents** are substances that cause a change in the nutrient profile of body tissues, or a "repartitioning" of nutrients from fat to protein. They stimulate protein formation (either by increasing protein synthesis or

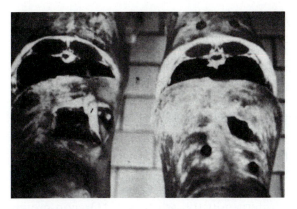

**FIGURE 5-5** The effect of feeding a repartitioning agent on protein and fat deposition in lambs. The carcass on the left is from a lamb fed a repartitioning agent; a control carcass is on the right. Note the larger loin eye area (muscle tissue) and thinner backfat deposition in the carcass on the left. (Courtesy of American Cyanamid Co.)

decreasing its degradation) and decrease body fat (either by decreasing fat synthesis or increasing its mobilization). The net result is to increase muscle tissue (protein) and reduce body fat (Fig. 5-5). Examples of repartitioning agents that have been under development include clenbuterol, cimaterol, and ractopamine. Only ractopamine has been approved by the FDA for use in the United States. Unfortunately, illicit use of these compounds has occurred with animals raised for the show ring to improve body composition and carcass traits (Kuiper et al., 1998; Mitchell and Dunnavan, 1998). Illicit use in humans has also occurred to improve athletic skills or physique. Mechanisms of action of these substances, also known as β-**agonists,** have been reviewed by Mersmann (1998) and Smith (1998).

It is estimated that at least 5 billion pounds of unwanted fat are trimmed from carcasses each year in the United States, most of which is rendered to be recycled in animal feeds. This is obviously a wasteful and inefficient process. The use of repartitioning agents, if no toxic or adverse effects are found, could have a major impact on the efficiency of animal production.

# MISCELLANEOUS FEED ADDITIVES

Some feed additives are used because of their favorable effects on the environment or because they modify consumer acceptance of animal products. Extracts of the yucca plant (**sarsaponins, yucca extract**) are being used as feed additives to control ammonia levels in the air of livestock confinement facilities. Yucca extract contains compounds that bind ammonia and prevent its release into the air. When included in the diet, the yucca extract is excreted with the feces and binds ammonia released by bacterial action in the manure. The active ingredient in yucca extract was originally thought to be sarsaponins but now is believed to be a glycoprotein fraction. **Zeolites,** which are clay minerals, have a similar effect and are used as feed additives. Reduction in ammonia is very important, because levels above 20 ppm in the air cause damage to the eyes (conjunctivitis) and respiratory system, leading to respiratory disease and pneumonia. This is harmful to the livestock (usually swine or poultry in confinement buildings) and the farmworkers.

Phytases, discussed earlier under Enzymes, also have favorable effects on the environment by decreasing excretion of phosphorus, which is a major pollutant of groundwater and streams.

There is interest in developing feed additives that have **hypocholesterol effects,** such as the lowering of the cholesterol content of meat and eggs. To date, no effective and nontoxic additive for this purpose has been developed. A seaweed product (kelp) with a high iodine content has been developed as a feed additive for layers to produce eggs that reputedly have cholesterol-lowering properties, but FDA approval has not been given because of the potential problem of iodine toxicity in humans consuming the eggs. Apparently the seaweed contains an iodinated compound that lowers serum cholesterol. The use of **modified eggs** containing more vitamin E, iodine, and unsaturated fat than "generic eggs" in cholesterol-lowering diets has been described by Garwin et al. (1992). Consumption of 12 modified eggs per week did not affect the cholesterol-lowering property of a low fat, low cholesterol diet over a six week test period in a group of healthy human subjects. "Designer eggs" enriched in ω-3 fatty acids have been described in Chapter 2.

Elkin and Rogler (1990) fed a hypocholesterolemic drug, Lovastatin, to laying hens and demonstrated a reduction in egg cholesterol levels. Lovastatin inhibits the rate-limiting enzyme in the cholesterol biosynthetic pathway. The authors suggest that it might be possible to develop "feed grade" analogs of this drug, which is used in human medicine, to use as egg cholesterol-lowering agents. Beyer and Jensen (1992) attempted to block cholesterol synthesis in layers by feeding α-ketoisocaproic acid, an inhibitor of a key enzyme in cholesterol biosynthesis. However, this approach was unsuccessful. Plant sterols, such as β-sitosterol and campesterol, are structurally similar to cholesterol and may compete with it for membrane transfer. Numerous studies have been conducted on the effects of dietary plant sterols to layers on egg cholesterol levels, with negative results (Hargis, 1988).

Potential exists for removal of cholesterol from liquid eggs used industrially in the food industry, in a manner similar to the removal of cholesterol from milk. A compound under development for this purpose is β-cyclodextrin, which binds cholesterol (Awad et al., 1997).

A very important additive in the poultry industry and potentially in aquaculture production is xanthophyll. **Xanthophyll pigments** cause the yellow color of egg yolks and the skin and shanks of broilers. In many areas, there is consumer preference for highly pigmented poultry products. Xanthophylls are carotenoid pigments very similar in structure to β-carotene, the plant precursor of vitamin A. Xanthophylls have no vitamin A activity and no nutritional value. They are used entirely to enhance consumer appeal. A yellow egg yolk is not "richer" than a pale yolk but try convincing consumers of that! Xanthophylls are found in green plants (e.g., alfalfa meal, grass meal) and yellow plant material (yellow corn). The richest natural source of xanthophyll is marigold petals. Marigold petal meal is used as a pigmenting agent in poultry diets. There are now a number of synthetic carotenoids (e.g., canthaxanthin) that are used as feed additives. In some areas, such as the U.S. Pacific Northwest, the consumer preference is for nonpigmented poultry products, so low xanthophyll diets are used. To contrast their product with broilers from Arkansas, the Oregon broiler industry advertises "Oregon-grown broilers have no ugly yellow fat"! The history of use of **pigmenting agents** and their current applications in marketing are reviewed by Williams (1992), while a review of their chemistry and physiological effects is provided by Hencken (1992).

Pigmenting agents are also important in salmon and trout aquaculture, to provide the pink coloration of the flesh of salmon and the skin colors of rainbow trout. Natural sources include crustacean products (krill, crab waste, shrimp waste)—which contain astaxanthin pigments, or synthetic pigments such as canthaxanthin.

# OTHER GROWTH PROMOTANTS THAT ARE NOT FEED ADDITIVES

**Growth hormone** is produced in the anterior pituitary gland of all mammals. It is a protein hormone, or more correctly, a polypeptide hormone consisting of about 190 amino acids joined together in a single polypeptide molecule. The exact amino acid composition varies slightly among species. Growth hormone (GH) is also called **somatotropin** (ST) and is identified according to species. The main growth hormones of current interest in livestock production are bovine somatotropin (bST, bGH) and porcine (pig) somatotropin (pST, pGH). GH has numerous physiological functions influencing metabolic responses involved in growth and lactation. It stimulates protein synthesis, in part by increasing amino acid transport into muscle cells. It influences carbohydrate metabolism, in general having opposite effects of insulin. Insulin promotes carbohydrate storage as glycogen; GH stimulates liver glucogenesis from amino acids. It promotes mobilization of body fat and increases the oxidation rate of fatty acids. GH also influences mineral metabolism, increasing the retention of Ca, Mg, P, Na, K, and Cl. It promotes the growth of the cartilage in long bones of growing animals. A GH deficiency results in dwarfism, while an excess causes gigantism. Both of these abnormalities occur in humans. Another effect of GH is to mimic the effect of prolactin, a structurally related polypeptide hormone, by binding to prolactin receptors in the mammary gland. Thus it has a lactogenic, or milk-stimulating, effect. Many of the effects of GH are mediated by a protein hormone called insulin-like growth factor (IGF-I) (McGuire et al., 1992).

Although it has been known for many years that administration of GH to animals would stimulate growth and milk production, this procedure was not practical because of the extremely high cost of the hormone. Since it is quite species-specific, it had to be extracted from the pituitaries of the target animal. Thus bST could only be obtained in very small quantities by extracting a large number of cattle pituitary glands. This situation changed abruptly when recombinant DNA biotechnology was developed, allowing the insertion of GH genes into bacteria, which then secrete the hormone. This has led to the capability of producing large amounts of bST and pST commercially at a comparatively low cost. The major responses to the administration of these products to livestock are an increase in milk yield of dairy cattle and increased growth rate in pigs, beef cattle, and sheep. Poultry have shown much less response. In experimental work, GH administration has increased milk production of dairy cows by 10–40 percent and growth rate of pigs by 10–25 percent. Under farm conditions, responses are likely to be somewhat less. Carcass quality of pigs is improved, with increased

lean and less fat. The commercial application of GH in livestock production is likely to alter nutrient requirements (Collier et al., 1992), because of the increased protein and amino acid demands of accelerated protein synthesis. In ruminants, the response to bST is often limited by insufficient ruminal bypass protein (Houseknecht et al., 1992).

Several companies in the United States are marketing bST. This development has been accompanied by unprecedented controversy (Fig. 5-6). The possible use of bST in dairy cattle has led to legislation in some U.S. states to prohibit the use of bST and has led to widespread public controversy. These concerns are unique, because they deal with issues other than efficacy and safety of the product. There is little doubt about efficacy, since nearly all studies show a marked response in milk production to bST administration (Collier et al., 1992). There is little legitimate concern about safety, as discussed shortly. The concerns are about social and ethical issues: will bST benefit large farms at the expense of small farms, will family farms be driven out of business, is bST treatment cruel or inhumane (at present, animals are injected daily with bST), is it cruel or inhumane to force cows to produce more milk by injecting a hormone, will it cause cows to "burn out" and have reduced longevity, should we introduce this technology to produce more milk when the United States already has huge surpluses of dairy products, will it make cows more susceptible to mastitis and other diseases, and on and on. These social and ethical issues will be discussed in Chapter 9. Even the name growth hormone scares many people, because of widespread concern about "hormones and chemicals" used in animal production. Interestingly, the public concern over bST seems to have been short-lived, and at the time of this writing seems to have largely disappeared. The bST issue has been replaced by protests over **GMO (genetically modified organisms) crops.**

Growth hormone cannot be given orally since it would simply be digested to amino

# Dairy cows can't say no.

Drug companies are pushing a new product — genetically engineered Bovine Growth Hormones (BGH).

BGH forces cows to produce milk in quantities unnatural and detrimental to their bodies. The result: hormone-injected cows suffer from increased susceptibility to infection and disease.

As infection and disease increase in BGH injected cows, so does the need for additional antibiotics and other drugs which can be passed on to consumers.

Would you drink hormone-laced milk?

Unfortunately, you may already have.

Test milk from cows injected with genetically engineered BGH has been placed on the market and sold to unsuspecting consumers.

It's too soon to know all the effects of drinking hormone-laced milk. Like BGH, the hormone

DES was originally promoted as "safe." It took years for its cancer-causing side effects to be acknowledged. For many, that news came too late.

The fact is, the dairy industry is already flooded with excess milk. As taxpayers, we pay *billions* of dollars in price supports due to this wasteful overproduction.

Now agri-drug companies are pushing hormones that will increase overproduction even further — while harming animals and endangering the safety of our food supply.

Dairy cows can't say no to genetically engineered hormones. But we can. **Please join us.**

THE HUMANE FARMING ASSOCIATION

*• Campaign Against Factory Farming •*

**FIGURE 5-6**   The development of bovine somatotropin (bST) has resulted in an unprecedented public backlash against intensive animal production, as exemplified by this advertisement in *U.S. News and World Report.*

acids like any other dietary protein. Experimental methods to improve delivery have been developed, such as sustained-release implants that can be given at monthly intervals. Of all the technological developments in animal production, GH is probably among the safest. Since it is a protein, it is digested in the gut. Proteins are also denatured by heat. Thus if bST is administered to dairy cows, any increase in the GH level normally in milk would be very unlikely to affect consumers. It is de-

natured when milk is pasteurized; if raw milk were consumed, the GH would be digested in the human gut. In pigs given pST, the hormone would be inactivated in cooking. If in the unlikely event that a person ate raw pork, pST in the pork would be digested in the intestine. It would appear to be virtually impossible for GH given to livestock to enter the human body in a physiologically active form. Thus there would be no health risk to consumers from the use of GH.

The somatropins are species-specific, with minor amino acid differences. Neither swine nor bovine GH are active in humans; for example, the amino acid sequence of bST is about 35 percent different from that of human GH. Also, humans are currently exposed to endogenous bST in milk and beef, and to pST in pork. Any small increases due to exogenous GH are not likely to be significant.

An alternative to administering GH to animals is to produce **transgenic animals** with the insertion of a GH gene into their genome to increase GH production. Pursel et al. (1989) inserted the bGH gene into pig embryos and produced two generations of transgenic pigs (*transgenic* meaning that genes from one species are introduced into another). The transgenic pigs exhibited improved growth and feed conversion. However, adverse pathological effects were noted, including enlarged livers, reduced lifespan, gastric ulcers, skin lesions, kidney damage, reproductive problems, and degenerative joint disease. Clearly, the production of transgenic animals is in the early stages of development. The rapid developments in this area pose many philosophical and ethical questions. Biotechnology is moving faster than its ramifications can be agreed upon by society.

The adverse public perception of hormone administration to farm animals has led to the search for more consumer-friendly methods of improving animal production. A potential method of changing an animal's GH status is to modify the rate of secretion of the substances that control GH secretion. These are GH releasing factor (GRF), which stimulates GH secretion, and somatostatin, which has an inhibitory effect. Several techniques for altering the activity of GRF and somatostatin show potential, including immunization against somatostatin. These techniques have been reviewed by Holder et al. (1991).

Steroid hormones (**anabolic steroids**) such as **androgens** (e.g., testosterone) and **estrogens** (e.g., estradiol) influence animal growth and development. Estrogenic hormones such as estradiol and diethylstilbestrol (DES) stimulate

the growth rate of ruminants. At one time, DES was widely used as a feed additive and an implant for beef cattle. After it was shown to be carcinogenic, its use was banned by the FDA. This does not mean that DES was a health hazard when administered to cattle. U.S. law (the **Delaney Clause**[2]) mandates that any substance shown to cause cancer in one or more higher animals cannot be administered to livestock. Animal scientists made valiant calculations to the effect that a person would have to eat a ton of beef liver every day for 300 years to get the same dose of estrogen as secreted by an adult woman, but all for naught. DES has been permanently banned.

Other estrogens, including estradiol and zearalenol, are administered as implants. **Implants** are small pellets inserted beneath the skin of the ear, allowing for a slow release (over several weeks or months) of the hormone into the animal's circulatory system. **Zearalenol** is a mycotoxin produced by corn infected with *Fusarium* fungi. Swine are particularly sensitive to its estrogenic effects and develop reproductive problems when fed corn containing low amounts of zearalenol. However, since ruminants respond favorably to estrogen administration, zearalenol has been commercially developed as an implant (Ralgro). A number of other estrogenic implant compounds are available commercially in the United States. Androgens are anabolic steroids, increasing muscle development in ruminants. Trenbolone acetate is a synthetic androgen that is effective as a growth promotant (DeHaan et al., 1990).

Implants have substantial effects as growth promotants and in the improvement of feed

---

[2] The Delaney Clause was attached to the Food Additive Amendment of 1958 to the Food, Drug and Cosmetic Act. It states that "no additive shall be deemed to be safe if it is found to induce cancer when ingested by man or animal, or if it is found, after tests which are appropriate for the evaluation of food additives, to induce cancer in man or animal." It has been interpreted in terms of "zero tolerance," meaning that if a substance has been shown to be carcinogenic, even if only at high doses, it cannot be used at any level.

conversion efficiency. The responses are usually greater in steers than in heifers. Steers have very little estrogen and have low androgen due to castration. Heifers have fluctuating estrogen and low androgen levels. Thus steers respond best to estrogen administration while heifers respond most to androgens.

Anabolic steroids increase the efficiency of ruminant animal production. Their use by the livestock industry is controversial and probably contributes negatively to the image of beef. Although their use in livestock does not seem to be a hazard to human health (in contrast to their direct application in humans, both for legitimate medical reasons and for illicit purposes), their negative effects on public perceptions of animal products may be of greater significance than their positive effects on animal performance. The consumer is the primary beneficiary of the use of these substances, in the form of lower meat costs. The net income of livestock producers is not likely to be influenced either way if anabolic hormone and growth hormones are banned from use, although consumers will pay more for meat and milk. If collectively that is what consumers prefer, so be it. However, consumer opinion on such matters is heavily influenced by publicity generated by activists opposed to increased use of technology in agriculture. This publicity can be counterbalanced by animal industry promotional activity, although this is often viewed as self-serving. Achieving a successful strategy for responding to challenges to the use of safe technological developments in livestock production is an as yet unrealized goal of the animal and poultry industries.

# STUDY QUESTIONS

1. Why are preservatives added to feeds? How do they function? Are there any hazards from their use?

2. Whole soybeans can be stored for long periods without becoming rancid, whereas ground raw soybeans become rancid quite quickly. Explain.

3. Explain how enzymes added to feeds may increase nutrient bioavailability. What impact could inexpensive cellulase have on the importance of livestock in food production?

4. Algae blooms in waterways can cause fish kills due to oxygen depletion. Explain how this can occur, in view of the fact that algae produce abundant oxygen during photosynthesis. What time of day would you expect algae-induced fish kills to occur? Why?

5. Explain how ionophores such as rumensin can improve feed conversion efficiency in cattle.

6. What hormones are fed to cattle in feedlots?

7. What are antibiotics? Why are they used as feed additives? Are there any hazards associated with their use?

8. Define the term "probiotic." How are probiotics used in animal nutrition?

9. Are bright yellow egg yolks more nutritious than pale yolks? What causes the egg pigmentation?

10. Discuss the health hazards to consumers arising from the use of bovine somatotropin in the dairy industry.

11. Do you believe that the FDA should ban the use of antibiotics for use as feed additives (growth promotants)? Explain your choice.

# REFERENCES

ANNISON, G., and M. CHOCT. 1991. Anti-nutritive activities of cereal non-starch polysaccharides in broiler diets and strategies minimizing their effects. *World's Poult. Sci. J.* 47:232–242.

AWAD, A. C., M. R. BENNICK, and D. M. SMITH. 1997. Composition and functional properties of cholesterol reduced egg yolk. *Poult. Sci.* 76:649–653.

BANTON, M. I., S. S. NICHOLSON, P. L. H. JOWETT, M. B. BRANTLEY, and C. L. BOUDREAUX. 1987. Copper toxicosis in cattle fed chicken litter. *J. Am. Vet. Med. Assoc.* 191:827–828.

BIRD, S. H., and R. A. LENG. 1984. Further studies on the effects of the presence or absence of protozoa in the rumen on live-weight gain and wool growth of sheep. *Br. J. Nutr.* 52:607–611.

BERGEN, W. G., and D. B. BATES. 1984. Ionophores: Their effect on production efficiency and mode of action. *J. Anim. Sci.* 58:1465–1483.

BEYER, R. S., and L. S. JENSEN. 1992. Cholesterol concentration of egg yolk and blood plasma and performance of laying hens as influenced by dietary $\alpha$-ketoisocaproic acid. *Poult. Sci.* 71:120–127.

CAMPBELL, G. L., and M. R. BEDFORD. 1992. Enzyme applications for monogastric feeds: A review. *Can. J. Anim. Sci.* 72:449–466.

CARLSON, J. R., and R. G. BREEZE. 1984. Ruminal metabolism of plant toxins with emphasis on indolic compounds. *J. Anim. Sci.* 58:1040–1049.

CHEEKE, P. R. 1999. *Applied animal nutrition: Feeds and feeding.* Prentice Hall, Upper Saddle River, New Jersey.

CHOCT, M. and G. ANNISON. 1992. The inhibition of nutrient digestion by wheat pentosans. *Brit. J. Nutr.* 67:123–132.

COLLIER, R. J., J. L. VICINI, C. D. KNIGHT, C. L. MCLAUGHLIN, and C. A. BAILE. 1992. Impact of somatotropins on nutrient requirements in domestic animals. *J. Nutr.* 122:855–860.

DEHAAN, K. C., L. L. BERGER, D. J. KESLER, F. K. MCKEITH, and D. L. THOMAS. 1990. Effect of prenatal trenbolone acetate treatment on lamb performance and carcass characteristics. *J. Anim. Sci.* 68:3041–3045.

ELKIN, R. G., and J. C. ROGLER. 1990. Reduction of the cholesterol content of eggs by the oral administration of lovastatin to laying hens. *J. Agric. Food Chem.* 38:1635–1641.

FULLER, R. 1989. Probiotics in man and animals: A review. *J. Appl. Bacteriol.* 66:365–378.

GARWIN, J. L., J. M. MORGAN, R. L. STOWELL, M. P. RICHARDSON, M. C. WALKER, and D. M. CAPUZZI. 1992. Modified eggs are compatible with a diet that reduces serum cholesterol concentrations in humans. *J. Nutr.* 122:2153–2160.

GOODRICH, R. D., J. E. GARRETT, D. R. GAST, M. A. KIRICK, D. A. LARSON, and J. C. MEISKE. 1984. Influence of monensin on the performance of cattle. *J. Anim. Sci.* 58:1483–1498.

GORBACH, S. L. 2001. Antimicrobial use in animal feed—Time to stop. *N. Engl. J. Med.* 345:1202–1203.

HAMMOND, A. C., M. J. ALLISON, M. J. WILLIAMS, G. M. PRINE, and D. B. BATES. 1989. Prevention of leucaena toxicosis of cattle in Florida by ruminal inoculation with 3-hydroxy-4-(IH)- pyridone-degrading bacteria. *Am. J. Vet. Res.* 50:2176–2180.

HARGIS, P. S. 1988. Modifying egg yolk cholesterol in the domestic fowl—A review. *World's Poult. Sci. J.* 44:17–29.

HENCKEN, H. 1992. Chemical and physiological behavior of feed carotenoids and their effects on pigmentation. *Poult. Sci.* 71:711–717.

HERALD, F., F. W. KNAPP, S. BROWN, and N. W. BRADLEY. 1982. Efficacy of monensin as a cattle feed additive against the face fly and horn fly. *J. Anim. Sci.* 54:1128–1131.

HESPELL, R. B. 1987. Biotechnology and modifications of the rumen microbial ecosystem. *Proc. Nutr. Soc.* 46:407–413.

HOLDER, A. T., R. ASTON, and D. J. FLINT. 1991. Potential of immunization for increasing animal production. *J. Agric. Sci.* 116:175–181.

HOLMBERG, S. D., J. G. WELLS, and M. L. COHEN. 1984. Animal-to-man transmission of antimicrobial-resistant *Salmonella*: Investigations of U.S. outbreaks, 1971–1983. *Science* 225:833–835.

HOUSEKNECHT, K. L., D. E. BAUMAN, D. G. FOX, and D. F. SMITH. 1992. Abomasal infusion of casein enhances nitrogen retention in somatotropin-treated steers. *J. Nutr.* 122:1717–1725.

JONES, R. J., and R. G. MEGARRITY. 1986. Successful transfer of DHP-degrading bacteria from Hawaiian goats to Australian ruminants to overcome the toxicity of leucaena. *Aust. Vet. J.* 63:259–262.

KLASING, K. C. 1988. Nutritional aspects of leukocytic cytokines. *J. Nutr.* 118:1436–1446.

KUIPER, H. A., M. Y. NOORDAM, M. M. H. VAN DOOREN-FLIPSEN, R. SCHILT, and A. H. ROOS. 1998. Illegal use of β-adrenergic agonists: European Community. *J. Anim. Sci.* 76:195–207.

MCDONALD, L. C., S. ROSSITER, C. MACKINSON, Y. Y. WANG, S. JOHNSON, M. SULLIVAN, R. SOKOLOW, E. DEBESS, L. GILBERT, J. A. BENSON, B. HILL, and F. J. ANGULO. 2001. Quinupristin-dalfopristin-resistant *Enterococcus faecium* on chicken and in human stool specimens. *N. Engl. J. Med.* 345:1155–1160.

MCGUIRE, M. A., J. L. VICINI, D. E. BAUMAN, and J. J. VEENHUIZEN. 1992. Insulin-like growth factors and binding proteins in ruminants and their nutritional regulation. *J. Anim. Sci.* 70:2901–2910.

MERSMANN, H. J. 1998. Overview of the effects of β-adrenergic receptor agonists on animal growth including mechanisms of action. *J. Anim. Sci.* 76:160–172.

MITCHELL, G. A., and G. DUNNAVAN. 1998. Illegal use of β-agonists in the United States. *J. Anim Sci.* 76:208–211.

NEWMAN, C. W., D. C. SANDS, M. E. MEGEED, and R. K. NEWMAN. 1988. Replacement of soybean meal in swine diets with L-lysine and *Lactobacillus fermentum*. *Nutr. Rep. Int.* 37:347–355.

PETTERSSON, D., H. GRAHAM, and P. AMAN. 1991. The nutritive value for broiler chickens of pelleting and enzyme supplementation of a diet containing barley, wheat and rye. *Anim. Feed Sci. Tech.* 33:1–14.

PURSEL, V. G., C. A. PINKERT, K. R. MILLER, D. J. BOLT, R. G. CAMPBELL, R. D. PALMITER, R. L. BRINSTER, and R. E. HAMMER. 1989. Genetic engineering of livestock. *Science* 244:1281–1288.

QUIRK, M. F., J. J. BUSHELL, R. J. JONES, R. G. MEGARRITY, and K. L. BUTLER. 1988. Live-weight gains on leucaena and native grass after dosing cattle with rumen bacteria capable of degrading DHP, a ruminal metabolite from leucaena. *J. Agri. Sci.* 111:165–170.

ROLLIN, B. 2001. Ethics, science, and antimicrobial resistance. *J. Agric. Environ. Ethics* 14:29–37.

ROURA, E., J. HOMEDES, and K. C. KLASING. 1992. Prevention of immunologic stress contributes to the growth-permitting ability of dietary antibiotics in chicks. *J. Nutr.* 122:2383–2390.

RUSSELL, J. B. 1987. A proposed mechanism of monensin action in inhibiting ruminal bacterial growth: Effects on ion flux and protonmotive force. *J. Anim. Sci.* 64:1519–1525.

RUSSELL, J. B., and D. B. WILSON. 1988. Potential opportunities and problems for genetically altered rumen microorganisms. *J. Nutr.* 118:271–279.

SIMONS, P. C. M., and H. A. J. VERSTEEGH. 1990. Improvement of phosphorus availability by microbial phytase in broilers and pigs. *Brit. J. Nutr.* 64:525–540.

SMITH, D. J. 1998. The pharmacokinetics, metabolism and tissue residues of β-adrenergic agonists in livestock. *J. Anim. Sci.* 76:173–194.

SORENSEN, T. L., M. BLOM, D. L. MONNET, N. FRIMODT-MOLLER, R. L. POULSEN, and F. ESPERSEN. 2001. Transient intestinal carriage after ingestion of antibiotic-resistant *Enterococcus faecium* from chicken and pork. *N. Engl. J. Med.* 345:1161–1166.

VIERA, D. M. 1986. The role of ciliate protozoa in nutrition of the ruminant. *J. Anim. Sci.* 63:1547–1560.

VISEK, W. J. 1978. The mode of growth promotion by antibiotics. *J. Anim. Sci.* 46:1447–1469.

WALLACE, R. J., L. ARTHAUD, and C. J. NEWBOLD. 1994. Influence of *Yucca shidigera* extract on ruminal ammonia concentrations and ruminal microorganisms. *Appl. Environ. Microbiol.* 60:1762–1767.

WHITE, D. G., S. ZHAO, R. SUDLER, S. AYERS, S. FRIEDMAN, S. CHEN, P. F. MCDERMOTT, S. MCDERMOTT, D. D. WAGNER, and J. MENG. 2001. The isolation of antibiotic-resistant salmonella from retail ground meats. *N. Engl. J. Med.* 345:1147–1154.

WILLIAMS, W. D. 1992. Origin and impact of color on consumer preference for food. *Poult. Sci.* 71:744–746.

WITTE, W. 1998. Medical consequences of antibiotic use in agriculture. *Science* 279:996–997.

# CHAPTER 6

# Environmental Concerns Involving Livestock Production

**CHAPTER OBJECTIVES**

To discuss some of the major environmental issues that involve livestock production:
- global warming and methane from ruminants
- tropical deforestation and "the hamburger connection"
- water and air pollution arising from livestock production

Cattle production, and especially the consumption of beef, provokes many environmental activists to a virtual frenzy. This is exemplified by the publication in 1992 of *Beyond Beef: The Rise and Fall of the Cattle Culture* by the infamous activist, Jeremy Rifkin. This book begins with what I consider an absurd statement that cattle "take up nearly 24 percent of the land mass of the planet" and proceeds to explain how cattle production is responsible for most of the economic, ecological, and social ills of the world, including, besides all the common ones, gender hierarchies and dominance by human males of European origin, class struggles, colonialism, nationalism, and "cold evil," perpetuated by (mainly) males of European, American, and Japanese origin. After graphically describing various evils such as murder, robbery, violent crimes in general, and other forms of personal evil, Rifkin states:

> A new dimension of evil, however, has been incorporated into the modern cattle complex—a cold evil. . . . Cold evil is evil inflicted from a distance; evil concealed by layer upon layer of technological and institutional

garb [*sic*]. Today, Europeans, Americans and Japanese eat atop an artificial food chain, laced with marbled beef, gorging themselves on animal fat. . . . Their bodies clogged with cholesterol, their arteries and organs choked with animal fat, they fall victim to "diseases of affluence," often dying in excruciating pain from heart disease, colon and breast cancer, and diabetes. . . . Teenagers gobbling down cheeseburgers at a fast-food restaurant will likely be unaware that a wide swath of tropical rain forest had to be felled and burned to bring them their meal. (p. 284)

Rifkin (1992) quotes from *Newsweek*, "The water that goes into a 1000 lb steer would float a destroyer" (p. 219). (A toy destroyer, presumably!) This ignores the fact that almost all of the water that went into a 1000 lb steer has come back out, and much of it is reclaimed or irrigates grass via urination.

The beef cattle industry should breathe a collective sigh of relief about Rifkin's book. Rather than being a thought-provoking analysis of the issues, it is simply sensationalism. The citations consist of one activist quoting another. It is hard to imagine Rifkin's book being taken seriously, because of what I believe are distortions of fact, lack of objectivity, and the shrill, almost desperate attempts to portray cattle as the most evil, destructive animals ever to soil the planet. (After a brief period of pre-publication publicity, the book has largely been ignored.) However, many of the issues raised by Rifkin (1992) and others critical of livestock production are legitimate and are worthy of discussion and debate. This chapter provides a framework for discussing some of these issues. Ecological issues involving livestock grazing on rangelands are discussed in Chapter 7.

## LIVESTOCK PRODUCTION AND GLOBAL WARMING

Global warming, due to the so-called greenhouse effect, is of widespread concern. While it is not absolutely certain that global warming is in progress (only a few years ago there was concern that global cooling and a new Ice Age were developing!), there is general agreement in the scientific community that it is a real phenomenon. Briefly, the scenario involved is this: some of the solar energy that reaches the surface of the earth is radiated back into the atmosphere where it is absorbed by various gases such as water vapor, carbon dioxide ($CO_2$), methane ($CH_4$), nitrous oxide ($N_2O$), chlorofluorocarbons (CFCs), and many other gases. The common feature of these gases is that they absorb infrared radiation. The absorbed radiation is eventually emitted back into space, a process that may take several years. The atmosphere bears a superficial resemblance to a greenhouse. Solar energy penetrates the glass of a greenhouse and is absorbed by the benches, plants, pots, and so on. Energy is emitted as infrared radiation from these objects in the greenhouse. The greenhouse glass does not absorb the incoming radiation from the sun but does absorb the infrared radiation emitted when the solar radiation reaches the solid objects inside the greenhouse. The result is that heat energy (infrared) is trapped inside the greenhouse, heating it up. The **greenhouse effect,** by analogy, involves the trapping of radiation by infrared-absorbing gases in the lower atmosphere (troposphere). Over eons of time, an equilibrium was established, keeping the earth's atmospheric temperature more or less constant. (Although over millions of years there are cycles of temperature changes, producing either ice ages or worldwide tropical climates.) The present concern over global warming has come about because of well-documented increases in **greenhouse gases,** primarily carbon dioxide, methane, nitrous oxide, and CFCs. An increase in the level of these gases in the atmosphere will trap more heat, causing a rise in the temperature of the atmosphere and global climate changes. These changes will involve not only temperature but also precipitation patterns. While it is difficult to predict exactly what these changes might be, they could have devastating effects on the climates of some countries, while some arid countries might have increased rainfall and would benefit (at least from an agricultural perspective) from global warming. These changes in global temperature and precipitation patterns are known collectively as **global climate change.**

The increases in greenhouse gases are occurring mainly as a result of human activities. The internal combustion engine is a major contributor to this increase. Since the Industrial Revolution began, huge quantities of carbon, previously sequestered as coal and petroleum, have been extracted from the earth and burned, releasing $CO_2$ into the atmosphere. During the same period of time, global deforestation has occurred and is continuing with the destruction of the tropical rain forests. In many cases, the trees are simply burned as the land is cleared for farming or ranching. The destruction of large areas of forestland has contributed to a marked net increase in atmospheric $CO_2$ when the forests are burned. During the past 100 years, the atmospheric $CO_2$ level has increased from about 270 ppm to 350 pm (Post et al., 1990).

Carbon dioxide is the major greenhouse gas, while methane ($CH_4$) is the next in importance. The **CFCs** (chlorofluorocarbons) are synthetic compounds, used in refrigeration systems, aerosol propellants, and other industrial applications. Besides being greenhouse gases, they also destroy the ozone in the stratosphere, which can also affect global climate (the hole in the **ozone layer**). The CFCs are not relevant to livestock production. However, livestock production can influence $CO_2$ and $CH_4$ emission into the atmosphere. The main effect on $CO_2$ is an indirect one. One of the major reasons for the destruction of tropical rain forests, particularly in Central America and the Amazon region, is for the conversion of jungle to cattle pasture. Thus cattle ranching may be an incentive for tropical deforestation, resulting in a net increase in $CO_2$ when the forests are burned. However, deep rooted perennial grasses such as *Andropogon gayanus* and *Brachiaria humidicola* that are planted for cattle pasture on cleared tropical forestland may sequester large

amounts of carbon in their roots, counteracting the effects of $CO_2$ released when the forests are burned (M. J. Fisher et al., 1994). A contrary assertion is made by Garcia-Oliva et al. (1994), who measured soil organic carbon in tropical deciduous forest and in forestland converted to pasture. They observed high rates of net loss of carbon from the grassland systems. Kauffman et al. (1998) claim that typically the Brazilian Amazon pastures are burned about every two years, causing extensive loss of carbon. Realistically, conversion of forests with deep-rooted trees to grasslands is likely to result in a net flux of carbon into the atmosphere. Nevertheless, grasslands are a significant sink for carbon dioxide (van Ginkel et al., 1999). Perennial forage crops are especially effective in sequestering carbon in the soil (Bruce et al., 1999). Thus livestock production can have a favorable effect on reducing global warming, because of the linkage of forages with animal production. Lal et al. (1999) suggest that tax credits could be used as incentives to encourage farmers to adopt farming practices that promote soil retention of carbon. A carbon tax could be used to encourage reductions in carbon emissions.

**Methane** is a by-product of anaerobic fermentation in wetlands (marshes, swamps, rice paddies), land-fills, and in various animals, notably ruminants and termites. In the rumen, methane represents a "hydrogen sink." The biochemistry of methane production in the rumen and the hydrogen sink concept are discussed by Hungate (1966), Van Soest (1994), and Fahey and Berger (1988). In the period 1989–1991, the role of beef cattle as sources of methane contributing to global warming received a lot of media attention, much of it of a derogatory sort. This occurred when researchers at Washington State University received a "government grant" to study methane emissions from cattle and their influence on global warming. This for some reason attracted media attention, and it was incorrectly referred to as "cow flatulence" or sometimes more correctly as "cow belching." Newspaper accounts were accompanied by

## ANCIENT CARBON DIOXIDE AND AN ANCIENT TREE

*(Review of G. J. Retallack. 2001. "A 300-Million-Year Record of Atmospheric Carbon Dioxide from Fossil Plant Cuticles." Nature 411:287–290.)*

There is currently much interest in the role of atmospheric carbon dioxide in global climate change, and particularly global warming. Evidence for a linkage of $CO_2$ and climate can be deduced by examining their relationships in the past. In the fairly short term (hundreds of years), climate can be assessed from width of tree rings (wider in wet years, narrow in periods of drought). Ice cores of ancient ice in Arctic and Antarctic regions and glaciers contain air trapped when the ice was formed, which can be analyzed for $CO_2$ content. Ice cores of over 400,000 years have been obtained. In this paper, the author was able to estimate atmospheric $CO_2$ by examination of fossilized leaves of the *Ginkgo* tree. The *Ginkgo* is an ancient species, extending back to over 300 million years ago. The author determined that the number of stomata (leaf pores) on the leaves of modern *Gingkos* is influenced by atmospheric $CO_2$. With higher $CO_2$, plants respond by decreasing the density of stomata on their leaves. Examination of the stomata of fossilized *Ginkgo* leaves, dated isotopically to different ages, provides evidence of atmospheric $CO_2$ over the past 300 million years and supports evidence obtained using other techniques.

headlines such as "Your tax dollars at work." The public and media perception was that the whole idea of cattle, methane, and global warming was ridiculous, and studying it was a typical example of how the government wastes taxpayers' money by pouring it into stupid projects. As far as the livestock industry is concerned, this was probably a more desirable outcome than the alternative, which would be widespread alarm that cattle production is contributing in a major way to global destruction, by promoting global warming. (From an agricultural scientist's point of view, it is distressing to have the media continually reinforce the common misconception (e.g., Senator Proxmire's Golden Fleece Awards) that government grants are given out willy-nilly for ridiculous projects. Anyone who has written a research proposal knows how far from reality this perception is.)

In terms of global warming, some of the major sources of methane emissions to the atmosphere are rice paddies, wetlands, biomass burning, and the oil and natural gas industries (Table 6-1). In many oil fields, the methane is simply flared (burned) off. Other major sources include landfills, tundra (arctic accu-

**TABLE 6-1**  Estimates of the principal global methane sources (Adapted from Johnson and Johnson, 1995).

| All figures in million metric tons per year | | | | | |
|---|---|---|---|---|---|
| **Natural** | | **Energy/Refuse** | | **Agricultural** | |
| Wetlands | 115 | Gas and oil | 50 | Rice | 60 |
| Oceans | 15 | Coal | 40 | Livestock | 80 |
| Termites | 20 | Charcoal | 10 | Manure | 10 |
| Burning | 10 | Landfills | 30 | Burning | 5 |
| | | Waste water | 25 | | |
| Total | 160 | | 155 | | 155 |

mulations of vegetation) decomposition, coal mines, and leaking Russian natural gas pipelines. Ruminant animals and termites (Fig. 6-1) are significant but relatively minor sources. Byers (1990) estimated that on a worldwide basis, cattle production accounts for about 7 percent of total annual methane emissions. Johnson and Johnson (1995) estimate that about 17 percent of total methane emissions arise from livestock (Table 6-1), in agreement with the 15–20 percent estimate of

**FIGURE 6-1** A termite mound in South Africa. Termites have anaerobic fermentation in their digestive tracts; bacteria and protozoa secrete cellulase to digest wood consumed by termites. Methane is produced as a by-product, as it is with all anaerobic fermentation.

Leng (1992) and the 15–25 percent value of Crutzen et al. (1986).

Methane production by ruminants is nutritionally significant. Methane, also known as natural gas, represents a major loss of energy in rumen metabolism. From 4–10 percent of the gross energy intake is lost to the animal as a result of rumen methane production (McAllister et al., 1996). Development of ways to reduce ruminal methane would increase the efficiency of ruminant production. The amount of methane produced in ruminants is determined in large part by diet. There are two major types of rumen fermentation: cellulolytic and amylolytic. Cellulolytic bacteria produce mainly acetic acid as their end-product of fermentation, while amylolytic (starch-digesting) bacteria produce propionic acid as the major VFA end-product. The stoichiometry of acetate and propionate production is indicated by the following equations:

Equation 1 (amylolytic bacteria):

glucose $\longrightarrow$ 2 propionic acid ($CH_3CH_2COOH$) ($C_6H_{12}O$)

Equation 2 (cellulolytic bacteria):

glucose $\longrightarrow$ 2 acetic acid ($CH_3COOH$)+ $CO_2$ + $CH_4$ ($C_6H_{12}O_6$)

Thus with fiber digestion, methane is produced whereas with starch digestion it is not. When propionate is produced, all of the carbon and hydrogen present in the glucose are still present in the VFA (see Equation 1 and add up the C and H). In contrast, when acetate is produced, both carbon and hydrogen are lost as $CO_2$ and $CH_4$ (see Equation 2). The rumen contains a vast array of microbes, so the actual overall reactions are more involved than these shown above. However, with high fiber diets, more acetate and methane are produced; with high grain diets, more propionate and less acetate and methane are produced. For example, suppose a high grain diet is fed, giving an acetate:propionate molar ratio of 1:1:

Equation 3:

3 glucose $\longrightarrow$ 2 acetate + 2 propionate + butyrate + 3 $CO_2$ + $CH_4$ + 2 $H_2O$

In contrast, a high forage diet might give an acetate:propionate ratio of 3:1:

Equation 4:
5 glucose $\longrightarrow$ 6 acetate + 2 propionate + butyrate + 5 $CO_2$ + 3 $CH_4$ + 6 $H_2O$

Thus per mole of glucose, Equation 3 would yield 1/3 mole of methane, whereas with Equation 4, the methane yield would be 3/5 mole methane per mole of glucose. For each mole of glucose, the high forage diet yielded almost twice as much methane as the high concentrate diet (Equations 3 and 4 are from Fahey and Berger, 1988).

The **ionophore feed additives** (Chapter 4) improve the efficiency of rumen fermentation by stimulating propionate production and reducing methane. Thus the use of ionophores can help to reduce methane emissions from cattle and have a small beneficial effect on alleviating global warming. Intensive cattle production, using high concentrate diets and ionophore feed additives, can reduce the already slight contribution of cattle to global warming by the following means:

1. Methane production per unit of feed is reduced on high concentrate diets.

2. Ionophores reduce methane production in the rumen by altering rumen fermentation.

3. Ionophores reduce the amount of feed required to produce a given amount of weight gain, further reducing methane production.

4. Both high concentrate diets and ionophores (and other feed additives and implants) increase growth rate, reducing the age at slaughter. Thus with a shorter lifespan, the animal has a lower lifetime methane production.

For these reasons, a five-year-old steer in the tropics that has finally reached slaughter weight will have eructated much more methane in its life than a U.S. feedlot steer in its shorter lifetime. Ruminants in the tropics are generally fed low quality roughages and by-products or graze on poor-quality tropical pastures. These dietary situations result in inefficient rumen fermentation and maximization of methane production. Relatively inexpensive and simple supplementation programs, with bypass proteins for example, have the potential of greatly increasing efficiency of tropical cattle production and reducing methane emissions (Leng, 1993). Additionally, tropical cattle are particularly important in methane production, because over 50 percent of the world's cattle are in tropical developing countries (Kurihara et al., 1999). A number of review papers on methane production by cattle are published in the *Australian Journal of Agricultural Research* Vol. 50, Number 8, 1999.

Rumen protozoa are a significant source of methane (Hegarty, 1999). Methanogens (methane-producing bacteria) live within the protozoa. Certain medium chain fatty acids are toxic to rumen protozoa, and when used

as feed ingredients, may reduce methane emission. For example, Machmuller and Kreuzer (1999) found that feeding coconut oil at 3.5 and 7 percent of the diet reduced methane production by sheep.

According to the calculations of Byers (1990), about 40 g of methane are generated in the production of the beef for a quarter-pound hamburger. A five mile drive to and from a hamburger shop to buy the hamburger, in a car getting good gas mileage, will generate about 100 times the amount of greenhouse gases (mainly $CO_2$) as involved in the production of the beef. This is a further example of where our society gets worked up about a minor matter (methane from beef production) while ignoring a major problem for which individuals actually could do something meaningful (i.e., the pollution and resource-depleting effects of excessive and unnecessary use of automobiles). This is like getting all worked up about pesticide residues on vegetables (little or no health hazard) while puffing away on a cigarette (extreme health hazard over which an individual has total control). It is easier to be upset about something "they" are responsible for than making changes in one's own unhealthy behaviors that might demand some will power (e.g., eliminating tobacco, alcohol and illicit drug use, reducing use of automobiles, reducing obesity and engaging in exercise, and so on).

It should also be mentioned that wild ruminants and other herbivores produce methane. The number of wild herbivores in many areas of the world was much higher historically than after European settlement. The vast herds of American bison and other North American ungulates probably produced about as much methane as currently produced by cattle in North America. On a geologic time basis, numbers of herbivores have been high prior to ice ages, suggesting that they weren't a major factor contributing to global warming. In fact, grasslands may help to reduce global warming by absorbing methane and reducing nitrous oxide emissions, through the action of ammonium-oxidizing bacteria in the soil

(Mosier et al., 1991). These authors suggest that conversion of grasslands and pastures to crop production has decreased methane uptake and increased nitrous oxide release to the atmosphere. Thus the methane production of ruminants is offset by removal of atmospheric methane by grassland soils.

Rice production is a significant source of atmospheric methane (Mitra et al., 1999). Methane is produced by anaerobic bacteria growing on straw and other crop residue in flooded rice paddies. Root exudates of rice plants are significant substrates for methane production. The methane is released when the paddies are drained and cultivated. Methane can also travel from the soil to the atmosphere through the vascular system of rice plants (Mitra et al., 1999).

There is one aspect of methane production that doesn't seem to have received much consideration. Carbon dioxide has about 100 times as much "greenhouse effect" as an equal amount of methane, although methane molecules persist longer in the atmosphere. The carbon in methane is from $CO_2$ removed from the air by plants in photosynthesis. Calculations of the amount of methane produced by ruminants seem to have ignored this fact and treat the methane production as if it were a continuous net release of methane from ruminants. In fact, each molecule of methane eructated by a ruminant had its origin as a molecule of $CO_2$ removed from the air by a plant. Thus methane generation by ruminants cannot cause a net increase in atmospheric carbon; it is exactly balanced by photosynthesis. Automobile emissions, on the other hand, cause a net increase in $CO_2$ because the carbon is derived from petroleum, which represents a sequestered form of carbon that has not been in the atmosphere for millions of years. Thus the effect of rumen fermentation on global warming is much less than commonly reported.

The upper atmosphere has an **ozone layer** that absorbs some of the ultraviolet (UV) radiation of the sun. For several years, the ozone layer above Antarctica has become thin (the hole in the ozone layer), presumably because of the decomposition of ozone by CFCs. A similar thinning of the ozone layer over the northern hemisphere appeared in 1991. The reduction in the ozone layer, which allows more UV to reach the earth and may increase skin cancer rates, is not caused by $CO_2$ or methane. As yet, the thinning of the ozone layer has not been blamed on beef cattle! The CFCs are almost entirely of industrial origin, although there are natural sources of simple halocarbons such as $CH_3Cl$ and $CH_3Br$ (Butler et al., 1999). Thus the thinning of the ozone layers over the North and South Poles is due almost entirely to human industrial activities.

Modern intensive ruminant production systems with high energy diets and feed additives minimize methane generation and effects of cattle on global warming. In contrast, forage-based systems maximize methane production. While there may be many valid reasons for encouraging forage-based feeding systems for ruminants, reducing global warming is not one of them.

Another agricultural factor influencing global warming is **nitrogen fertilizer.** The German chemist Fritz Haber received the Nobel prize in 1919 for his development of the Haber-Bosch method of synthesizing ammonia from nitrogen and hydrogen, allowing the production of inexpensive chemical fertilizers containing nitrogen. Lack of nitrogen was a bottleneck in attempts to increase food production. Since the development of the Haber-Bosch method of ammonia synthesis, the human population of the world has quadrupled, with most of this population dependent on chemically synthesized nitrogen (as dietary protein of plant or animal origin). The widespread use of nitrogen-containing fertilizers has led to extensive pollution of soil and water. Of relevance to global warming is the production of **nitrous oxide** gas by soil microbes. Nitrous oxide is a "greenhouse gas," trapping outgoing radiation in the troposphere and also reacting with reactive oxygen to cause destruction of ozone in the stratosphere. The atmospheric lifetime of nitrous

## OCEAN FERTILIZATION TO MANIPULATE GLOBAL CLIMATE

*(Review of A. J. Watson, D. C. E. Bakker, A. J. Ridgwell, P. W. Boyd, and C. S. Law. 2000. "Effect of Iron Supply on Southern Ocean $CO_2$ Uptake and Implications for Glacial Atmospheric $CO_2$." Nature 407:730–733.)*

Almost half of the photosynthesis on the earth is carried out by phytoplankton in the sea. Phytoplankton play an important role in the global carbon cycle by transferring the greenhouse gas $CO_2$ from the atmosphere to the deep ocean. When plankton die, the carbon they have incorporated into their bodies is transported to the bottom of the sea via sinking dead cells. Plankton growth is stimulated by nutrients (nitrogen, phosphorus, silicon) upwelling from the deep sea to sunlit surface waters where photosynthesis occurs. In the equatorial and southern Pacific ocean, nutrient supply exceeds that needed for current phytoplankton growth rates. It has been identified that iron deficiency is the factor limiting phytoplankton growth. A major source of iron for phytoplankton growth in the southern oceans is dust storms from continental land masses such as Africa. Oceanographers hypothesized that fertilization of oceans with iron might be a feasible way of stimulating massive growth of phytoplankton, thus removing $CO_2$ from the atmosphere to the deep seas and reducing global climate change caused by elevated $CO_2$. Phytoplankton carbon remains on the ocean floor for about 1,000 years. In 1999, an experiment involving fertilization of a patch of ocean with iron was conducted. The result was a dramatic increase in phytoplankton growth. Experimental findings were reported in *Nature* 407:695 and 407:730, 2000. In an accompanying editorial in *Nature* 407:685, the point is made that biogeochemical cycles involving land, air, and oceans have evolved over millions of years. Using ocean fertilization to manipulate climate could have huge unintended and unforeseen consequences, and by the time these were recognized, it would probably be too late to reverse them. This is an example of a simplistic technological fix to a complex human-induced ecological problem (global climate change).

oxide is more than a century, and every molecule of it absorbs about 200 times more outgoing radiation than does a single carbon dioxide molecule (Smil, 1997). Nitrous oxide also reacts with other pollutants to produce **photochemical smog.** An interesting effect of nitrous oxide is that it acts as a trigger to induce seed germination in plant species adapted to periodic fires, such as the chaparral vegetation of California. Nitrous oxide as an air pollutant can cause premature germination of seeds under non-fire conditions, so that when a fire does occur, the seedbank has been destroyed (Malakoff, 1997; Keeley and Fotheringham, 1997). This issue is particularly important in southern California, which is home to many smoke-germinated species and also to a very high level of nitrous oxide pollution.

President Ronald Reagen is said to have claimed that trees pollute the air, so there is no point in enacting and enforcing environmental laws. Trees do in fact emit huge quantities of hydrocarbons into the atmosphere. Their role, if any, in global climate regulation has not been established. One hydrocarbon emitted in large amounts by many trees is isoprene, a simple five-carbon structure. The amount of isoprene emitted by plants equals or exceeds all of the methane entering the atmosphere (Sharkey, 1996).

Humans are conducting a huge **geochemical experiment,** by pumping into the atmosphere vast quantities of carbon dioxide from fossil fuels and nitrous oxide from chemically synthesized ammonia. The prudent course of action would be to reduce these emissions as much as feasible.

# LIVESTOCK PRODUCTION AND FORESTRY RESOURCES

## *The "Hamburger Connection" and Tropical Deforestation*

One of the major global ecological crises is **deforestation,** particularly of tropical rain forests. Loss of these forests is important for a number of aesthetic, economic, and ecological reasons (Lal, 1986). Tropical forests have a high diversity of plant and animal species (biodiversity) compared with temperate regions (Wilson, 1988). In addition to overall biodiversity, many of these plant and animal species are unique to a small geographical area, making them susceptible to extinction when the forest is cleared. The greatest reason for clearing tropical forests is for small-scale farms, including slash and burn farming, by the rapidly increasing human populations of developing countries (Myers, 1991a, b). The next two largest causes of deforestation are commercial timber harvesting and cattle ranching. The clearing of tropical forests for conversion to cattle pastures (Fig. 6-2) has been termed **the hamburger connection** by environmentalists. Consumption of hamburgers by Americans is blamed for the loss of tropical rain forest because of cattle ranching to produce beef for export (Uhl and Parker, 1986). According to them, much of Central America has been deforested for conversion to cattle pastures. Cattle ranching is also expanding into the Amazon rain forests in Brazil. However, the largest single cause of deforestation is small-scale farming. It accounts for more deforestation (60%) than commercial logging and cattle ranching combined (Myers, 1991a, b). The reasons for conversion of jungles to cattle pastures have little or nothing to do with hamburger consumption in the United States. The U.S. beef industry can produce enough beef, using U.S. resources, for every U.S. resident to eat hamburgers three

**FIGURE 6-2** (*Top*): Tropical rain forest in Costa Rica. These tropical forests are rapidly diminishing because of logging, slash and burn farming, and cattle ranching. (*Bottom*): Conversion of rain forest in Costa Rica to cattle pasture. This has been called "your steak in the jungle" and "the hamburger connection." However, the main cause of tropical deforestation is slash and burn farming by land-less small farmers. Conversion of these cleared areas to pasture often occurs when the land loses its fertility in 3–5 years, and the farmers move on to burn another patch of forest. (Bottom photo courtesy of Anne C. Ayers Fanatico.)

times per day, every day. In fact, the beef industry would be delighted to have this opportunity. The development of cattle ranching in the tropics has been largely the result of social, political, and economic pressures in the countries themselves. Most Third World countries have high rates of inflation and a desperate need to earn foreign exchange to provide their people with the amenities that people all over the world expect and seek. Central American countries have few resources other than their forests and agriculture. Large areas of land have been cleared to produce crops for export, such as coffee, cacao, bananas, and beef. With few exceptions, the governments of these Central American countries, Brazil, and most other developing countries have actively promoted the expansion of agriculture to produce cash crops for export. Tax incentives are provided to encourage these developments. Frequently, political corruption is involved, with government officials and their friends and relatives making fortunes as deforestation proceeds. Until recently, development agencies such as the World Bank have actively promoted development schemes of this type, including a highway system in the Amazon region that hastened rapid environmental degradation.

Thus cattle ranching has been promoted in the tropics to produce beef for export, not as a result of an insatiable American appetite for beef, but because the developing countries themselves and their foreign advisors have perceived it as a means of generating income. My impression, gained in travels to several Latin American countries, including Brazil, Chile, Colombia, Ecuador, and Costa Rica, is that entrepreneurs in these countries regard the United States as a bottomless pit for exports. They generate all sorts of schemes to produce commodities, with the explanation that it's for export to the United States. This mind-set is found in developed countries as well. I experienced in Australia this same type of entrepreneurship; whatever we produce, the Americans or the Japanese will buy it!

A major cause of tropical deforestation is **slash and burn agriculture** by small-scale farmers. This involves cutting down the trees in a small area, burning them, and planting subsistence crops for 2–5 years, and then abandoning the site and moving on. A high proportion of the minerals in many tropical soils are in the vegetation itself and are quickly recycled back into vegetation by decay when leaves and other plant parts fall to the forest floor. Thus when the forest is burned, the minerals are released and the fertility is high for a season or two. Then as the minerals are leached away, crops fail and the subsistence farmer moves on. In traditional slash and burn farming, these small patches eventually return to forest, and the system is sustainable (Kleinman et al., 1995). In some areas, slash and burn farming has been practiced for millennia. However, with rapidly increasing populations, the accelerated cycle of slash and burn farming results in permanent loss of the forest. In many cases, cattle ranching follows on the abandoned land. Conversion of this land to pasture delays its regeneration into forest and subsequent suitability for another slash and burn cycle (Loker, 1993). The pace of Amazon forest clearing, in spite of concerns about its negative effects, is actually increasing (Laurance et al., 2000).

The "hamburger connection" will not be ended by an American boycott of fast-food restaurants, most of which use only American-raised beef anyway (Bonti-Ankomah and Fox, 1998). Clearing of tropical forests for cattle production is unnecessary and is an artifact of poorly thought-out development programs. On a recent visit to Costa Rica, I saw extensive and often illegal clearing of tropical rain forest, primarily for growing coffee, and to a lesser extent, for other plantation crops, as well as for the sale of the illegally cut logs. As Buschbacher (1986) has described for the Amazon region, the economic returns from clearing forest for cattle ranching are so low that it can be profitable to do so only with tax incentives and subsidies. Prevention of the "hamburger connection" can best be achieved by working on the elimination of these incentives where they exist, rather than boycotting fast-food restaurants that don't get their beef from these sources anyway (although this action did draw attention to the deforestation problem so was useful in that respect).

The major **exporters of beef** in world trade are Argentina, Australia, Brazil, the European Economic Community, New Zealand, and Uruguay. These countries produce grass-fed beef on natural grasslands. Even in Brazil,

**TABLE 6-2** Total beef imports into the United States, 1985–1996. Source: USDA Economic Research Service and U.S. Department of Commerce.

| Source of Imported Beef | % of Total Beef Imports | | |
|---|---|---|---|
| | 1985 | 1990 | 1996 |
| Argentina | 8.6 | 9.0 | 7.4 |
| Australia | 38.0 | 46.9 | 26.3 |
| Brazil | 6.6 | 0* | 4.2 |
| Canada | 11.5 | 9.6 | 28.3 |
| Mexico | 0.1 | 0.4 | 0* |
| New Zealand | 24.7 | 25.0 | 24.3 |
| Other Countries | 10.4 | 9.1 | 9.5 |
| Total Beef Imports (billion lbs) | 2.091 | 2.312 | 2.070 |

*Statistically insignificant amount.

the bulk of the beef comes from natural southern grasslands rather than cleared Amazon rain forest. "Hamburger connection" beef is a very minor component of world beef trade. **U.S. beef imports** are mainly from Canada, Australia, and New Zealand, with very little from tropical countries (Table 6-2).

Hecht (1989) has made a detailed analysis of the role of cattle ranching in the deforestation of the Brazilian Amazon region. Her conclusions are the following:

> Thousands of hectares of forest are cleared each year in Amazonia to make way for cattle. Although misguided subsidies have played a role in the expansion of cattle ranching, they are not the root cause of the problem. Nor, in the case of the Amazon, is the "hamburger connection." Rather, the attraction of cattle lies in their ancillary benefits: it is the things cattle do besides producing meat that make them so profitable. (p. 229)

This paper should be read by anyone particularly interested in the "hamburger connection." Briefly, for both large-scale cattle

ranchers and small-scale peasant farmers, cattle raising has a number of desirable attributes compared to crop production. For ranchers, the attractions include the low cost of pasture development compared to agricultural crops like cocoa, coffee or rice, the low labor demands, and the rapidity with which cattle occupy large areas of land, providing the advantages of land ownership. For peasant farmers, cattle provide economic benefits such as a ready market, a hedge against inflation, the highest per kilo value of any food commodity, and the ability to accumulate wealth until it is needed. Cattle provide these benefits with less labor and expertise than growing cash crops of rice, beans, maize, cassava, or fruit. The timing of animal harvest is determined by the owner's need or market opportunity, whereas crops are perishable and for a given crop the farmers are marketing their crops simultaneously and driving down the price. Cattle production extends the life of a cleared area, providing a return from land no longer able to produce crops. For both large- and small-scale farmers, pasture is the cheapest and easiest way to claim land rights. Finally, Hecht (1989, 1993) dismisses the "hamburger connection" in the Amazon by pointing out that the Amazon is a net beef importing rather than exporting region, and Amazon beef is denied entry into the U.S. market because of enzootic diseases. The entire subject of cattle ranching, deforestation, and development has been comprehensively reviewed in a book by Faminow (1998).

As the preceding discussion suggests, it has been fashionable to demonize cattle production in the tropics. Loker (1996) somewhat facetiously suggested that two essential components of the issue of tropical deforestation as expressed in "the prevailing wisdom of many academics regarding tropical deforestation" are (1) the displacement, ethnocide, and/or assimilation (under unfavorable terms) of indigenous people; and (2) the contribution of tropical deforestation to the emission of greenhouse gases. "With the inclusion of indigenous people and the negative environmental effects of greenhouse gases, the 'demonization' of tropical deforestation becomes complete" (Loker, 1996, p. 52). In fact, it is a common perception among animal scientists, and a correct one in my opinion, that the prevailing mood of many development agencies is "pro-crops" and "anti-livestock." This may in part explain the failure of many aid programs, because at the "ground level," the farmers see the overwhelming advantages to them of animal production as opposed to crops, even though this might not be the optimal land usage in ecological terms (Pichon, 1996). Deforestation can be slowed by development of infrastructure and management systems that increase the economic viability of already cleared land (Smith et al., 1997). Thus there is a need for more, not less, research on cattle production and management in the tropics.

Although Amazonian cattle ranching has been bashed by academics (political scientists, economists, ecologists) for many years, a reversal in perspective seems to be taking place. Christopher Uhl, an earlier critic of Amazon ranching (see Uhl and Parker, 1986), now seems to view it in a more favorable light. Ranching is becoming more sophisticated (Mattos and Uhl, 1994), with some ranches becoming more specialized, focusing exclusively on calf production, range fattening, or dairy production. New, better adapted forages are now commonly planted and pasture lifetime has been extended. Old, degraded pastures are being rehabilitated and are becoming more economically viable. The productivity of ranching in the Brazilian Amazon region can be markedly enhanced by more intensive methods (fertilizing, pasture improvement, cattle management, etc.); the ranching sector can continue to grow simply by intensifying ranching practices on land that has already been cleared (Arima and Uhl, 1997). They conclude that "cattle could play an important role in the diversification of small-scale agriculture and in the development of sustainable family farms" (p. 445).

Ranching in the Amazon involves both dry-land (terra ferme) and seasonally flooded conditions. Arima and Uhl (1997) found that traditional floodplain ranching could be integrated with terra ferme ranching by using the seasonally flooded areas for calf production and by moving the weaned calves to terra ferme pastures for fattening. They also observed successful use of water buffalo in seasonally flooded ecosystems. Under these conditions, water buffalo have higher birth rates and lower death rates than those of cattle. For seasonally flooded areas of the Amazon, and the llanos (seasonally flooded savannahs of the Orinoco River in Venezuela), cattle production can be integrated with harvesting of capybara and caiman (small crocodiles). Hoogesteijn and Chapman (1997) make the case that the mass extinctions of megafauna in the last Ice Age left South America with no large herbivores. The only native grazer that remained was the capybara. The lack of native grazing herbivores has left large areas of grasslands (the llanos of Venezuela, pampas of Argentina) available for cattle production. Management of the llanos for cattle production has improved the habitat for both capybara and caiman. Sustained harvesting of capybara and caiman for meat and leather on cattle ranches is both sustainable and profitable (Hoogesteijn and Chapman, 1997).

Ironically, cattle ranching in Costa Rica may play an important role in **regenerating tropical forests.** Many cleared areas are now vegetated with vigorous tropical grasses, which prevent forest trees from reseeding. Grazing with cattle and horses, in moderation, is an effective way of controlling the grass, allowing regeneration by trees. In Guanacaste National Park in Costa Rica, controlled grazing by cattle and horses is an integral part of a reforestation program (Allen, 1988). In fact, these large herbivores may be playing a role in returning the forest to its original state of 10,000 years ago. In the Pleistocene, before the last great Ice Age of 10,000 years ago, there were many **megaherbivores** in North, Central and South America, including mammoths, mastodons, giant rhinoceros, giant ground sloths, glyptodonts, horses, and numerous other large herbivores that became extinct during or just after the last Ice Age. (Horses became extinct in the Americas at that time and were reintroduced by the Spanish thousands of years later.) Since that time, until the arrival of Columbus, the forests of Central America have lacked large herbivores. According to Janzen and Martin (1982), the Central American forests contain many tree species that appear to have coevolved with the large herbivores and to have depended upon them for seed dispersal. Typically, these trees have large fruits with large hard seeds, which formerly were dispersed in the feces of large herbivores that ate the fruit. With the extinction of large herbivores, the fruit simply rotted on the ground under the trees, and most seedlings died from shading. Some of these trees became uncommon in Central America. The arrival of cattle and horses with the Spanish provided a means of seed dispersal, as these herbivores ate the fruit and distributed seeds. According to Janzen and Martin (1982), "Plant distributions in neotropical and grassland mixes that are moderately and patchily browsed by free-ranging livestock may be more like those before megafaunal extinction than were those present at the time of the Spanish conquest" (p. 19). As an interesting aside, the giant flightless **dodo** bird was hunted to extinction by sailors a few hundred years ago, on the island of Mauritius. Some Mauritian trees coevolved with the dodo and apparently depended on it for seed germination by the cracking of the shell in the dodo's gizzard. No new seedlings of the tree *Calvaria major* have been seen in the wild since the extinction of the dodo, and the population was down to 13 surviving trees. Temple (1977) demonstrated that the seeds would germinate after the fruits were fed to turkeys and has been able to establish new trees by using this procedure.

Several approaches have been suggested as alternatives to beef cattle production in the tropical forests. In Costa Rica, programs to encourage farmers to convert from beef to dairy cattle production have increased farmer

income with fewer animals, with a better integration of the animals into cropping systems (Gradwohl and Greenberg, 1988; Simpson and Conrad, 1993). In Central America, the iguana (an herbivorous lizard) is a common item of the diet. **Iguana farming** is a means of producing meat without destroying the forest. Young iguanas are started in captivity and then released on farms. When feeding stations are provided, they remain near the area of release, and can be harvested as needed. Further information on iguana farming is provided by Gradwohl and Greenberg (1988) and NRC (1991).

As an alternative to deforestation to create pastures, Preston (1990) and Murgueitio (1990) recommend **intensive livestock production in confinement systems,** using local feed resources such as sugarcane and tree foliage. Nitrogen-fixing forage trees such as *Gliricidia sepium* can be used in alley cropping systems to provide shade, fertilizer, and forage. **Alley cropping** (Fig. 6-3) is a technique of planting hedgerows of fast-growing, repeatedly pruned, nitrogen-fixing trees, with crops cultivated in the spaces between the hedges (Kang et al., 1990). The manure produced by livestock in confinement can be effectively

**FIGURE 6-3** Alley cropping offers a means of integrating livestock and crop production. Crops are planted between rows of nitrogen-fixing trees. The trees provide nitrogen to the crops, and the high-protein tree foliage can be harvested for feeding to animals. Note the rows of corn seedlings between the hedges of leucaena. (Courtesy of R. C. Rosecrance.)

by grazing livestock (Chong et al., 1997). In Western Australia, alley cropping with trees and shrubs is being investigated as a means of reducing soil salinity. An important ecological function of eucalyptus trees in Western Australia has been the transpiration of water to lower the water table. Land that has been cleared for agriculture is increasingly becoming highly saline, because the water table has risen to the surface, where water evaporates and leaves minerals as saline deposits in the surface soil. Alley cropping with trees offers a means of lowering the water table, while permitting continued crop production. The shrub tagasaste or tree lucerne (*Chamaecytisus proliferus*) is used in Western Australia alley cropping systems, providing among other benefits a palatable and nutritious forage (Lefroy et al., 2001).

Logging is becoming an increasingly controversial issue globally. In the U.S. Pacific Northwest, clear cutting of old-growth forests is a major political, ecological, and economic issue (Norse, 1990). An alternative to deforestation of remaining virgin forests around the world is to intensify wood production on land that has already been deforested or on marginal agricultural land. For example, fast-growing hybrid poplars can be grown for pulp and paper production (Fig. 6-5). Other **fast-growing trees** such as the black locust also have potential. In the tropics and semitropics, eucalyptus species can be used. The production of plantation trees can involve livestock in several ways. Grazing in tree plantations is possible. When the trees are harvested, the branches, foliage, and bark could be used as feedstuffs. Poplars and black locust produce leaves with a high protein content; they could be used to prepare a dehydrated leaf meal. Both of these species contain tannins that adversely affect protein utilization (Bas et al., 1985; Ayers et al., 1996), but these effects can be overcome by utilizing tannin-binding feed additives such as propylene glycol. Selection of plantation species for low tannin content would also be useful. Many trees and shrubs contain toxins for chemical defense against herbivory. This needs to be considered in agroforestry systems. For example, several

pine species including Ponderosa pine (*Pinus ponderosa*) contain substances which cause abortion in cattle (Panter et al., 1992), so grazing of pine plantations should be managed carefully. In the arid and semiarid tropics, *Acacia* spp. are useful for fuel production, livestock forage, and stabilizing soil and sand dunes (Tiedeman and Johnson, 1992). *Acacia* spp. are shrubby leguminous trees native to Australia and Africa. Many species contain tannins, which reduce palatability and protein digestibility but also help protect the plant from overbrowsing. With any tree grown in an agroforestry system, good management is required to prevent overharvesting or excessive grazing, which will kill the plants.

Besides trees, other crops can be grown to produce cellulose for pulp and paper production. One promising plant is **kenaf** (*Hibiscus cannabinus*), a high-yielding semitropical annual plant that grows 12 feet or more in height in a single season. The woody stalks can be mechanically harvested and used for pulp and paper production. By-products including leaves of kenaf or other biomass producing plants might be potential livestock feeds.

## Animal Dung and Deforestation

Fuel for cooking is in short supply in many Third World[1] countries. This has several implications involving livestock production, deforestation, and the **social status and roles of women.** The search for fuel leads to extensive deforestation and failure of reforestation efforts. When the main concern of people is their day-to-day survival, they are not likely to give much consideration to the long-term ecological implications of their actions. Cooking and collection of fuel are generally carried out by women in these cultures. When they must spend an ever-increasing proportion of their time searching for cooking fuel, they are less

---

[1] The terms "Third World country" and "developing country" are not intended to be derogatory but are used in this book to be consistent with common terminology.

**FIGURE 6-7** Cattle and buffalo in India are important sources of fuel for cooking. These animals feed on rice straw and other crop residues. Their manure is collected and made into dung patties for fuel. Lack of cattle for fuel production results in increased deforestation as the people are forced to cut down trees for fuel. (Courtesy of R. A. Leng.)

able to pursue education, employment, or other means of self-improvement. The birth rate all over the world negatively correlates with these opportunities. Thus with increasing deforestation, women often must spend an inordinate amount of time seeking fuel, reducing their opportunity for improvement in economic or social status, with the result that the population growth rate is high, the lives of women are impoverished, and the destruction of the environment intensifies.

In countries such as India and Pakistan, cattle and buffalo are an important source of fuel (Fig. 6-7). The dung is collected and used as fuel for cooking (Aggarwal, 1989). The livestock are fed primarily on rice straw and other crop residues, converting these products to cooking fuel while providing milk and work. This process can be made more efficient by using the animal excreta in biodigestors to produce methane for fuel. The development of "appropriate technology" biodigestors and methane-fueled cooking stoves will aid the adoption of this process. In the grasslands of China and Mongolia, animal dung is the primary—and often the only—cooking and heating fuel available to pastoralists (D. P. Sheehy, personal communication).

Thus the use of dung as a fuel, either directly or in **biogas generation,** can help prevent de-

forestation. Also, the planting of N-fixing fodder trees such as gliricidia and leucaena provides both fuel and livestock feed.

# LIVESTOCK PRODUCTION AND WATER RESOURCES

Water utilization and pollution by livestock will be discussed in this chapter. Issues involving water, livestock and rangelands, especially the effects of grazing on riparian zones, will be discussed in Chapter 7. Waterborne diseases in humans linked to pollution of water by livestock, such as giardiasis and *Cryptosporidia* infection, are discussed in Chapter 9.

## Water Utilization by Livestock

Animals differ in the efficiency with which they utilize water. Animals adapted to arid or desert areas have physiological or behavioral adaptations that minimize water requirements. Water is increasingly a limiting resource in many parts of the world and is another resource, besides food and land, for which humans and livestock may be in competition. For example, in Arizona and California, the rapidly increasing human population is in direct conflict with agriculture for water. Beef feedlots and dairies are large consumers of water and are under increasing pressure to vacate the Sun Belt states.

Pimentel et al. (1997) claim that producing 1 kg of animal protein requires approximately 100 times more water than producing 1 kg of vegetable protein. For example, they state that producing 1 kg of beef requires approximately 100 kg of hay-forage and 4 kg of grain. Producing this much forage and grain requires approximately 100,000 liters of water to produce 100 kg of plant biomass plus 5,400 liters to produce 4 kg of grain. Pimentel et al. (1997) claim that on rangeland more than 200,000 liters of water are required to produce 1 kg of beef. In contrast to beef, 1 kg of broiler chickens can be produced with approximately 2.5

kg of grain requiring about 3,500 liters of water for its production. A problem with this type of calculation is that much of the forage used to produce beef is from rain falling on grasslands. For example, the water used to grow plants on rangelands is from rainfall and would occur whether or not the range was grazed by cattle. Most of the water consumed by cattle returns to the rangeland via urination. Including this water in the cost of producing beef does not seem a fair means of calculating efficiency of water use. The calculations would be legitimate if all of the grain and forage produced for livestock were grown entirely with irrigation. Irrigation is the major cost of water in U.S. beef production, and irrigated pasture is a major component of the total water cost (Beckett and Oltjen, 1993). These authors calculated a U.S. cost of 3,682 liters of water per kg of boneless beef, a much lower figure than quoted by anti-beef enthusiasts (Rifkin, 1992).

Maloiy et al. (1979) classified domestic and wild large herbivores into three main physiological types with respect to water requirements:

1. Herbivores with high water requirements and poor urine concentrating ability, adapted to wet climates. Examples are cattle, buffalo, pig, horse, eland, elephant.

2. Herbivores with intermediate water requirements, and medium urine concentrating ability. Examples are sheep, donkey and wildebeest. They are adapted to warm, dry, semiarid savanna.

3. Herbivores with low water requirements and high urine concentrating ability, adapted to arid climates. Examples are camel, goat, oryx, ostrich, and gazelle.

Although cattle are relatively inefficient in water utilization, local conditions and adaptations may modify and reduce their requirements. Cattle raised in an area, as opposed to newly introduced animals, are more familiar with the grazing conditions and terrain and drink less frequently than introduced cattle. Sheep are more efficient in water utilization

**FIGURE 6-8** The oryx is a ruminant well adapted to arid areas. It has very low water requirements. There is potential for domestication of the oryx as a new livestock species to make more efficient utilization of arid rangelands than currently used species such as cattle.

than are cattle, and under temperate conditions, can sometimes meet their water needs from the forage consumed. Adaptations of sheep include production of a concentrated urine, excretion of dry fecal pellets, and insulatory properties of the wool that reduce evaporative cooling. Goats are even more efficient in water conservation than are sheep and can tolerate considerable tissue dehydration. The same is true of the camel.

In some arid areas of the world, perhaps there are animals with lower water requirements than those of domestic livestock that might have potential for domestication. The oryx (Fig. 6-8) is a grazing ruminant native to East Africa that is adapted to extremely arid conditions because of its low water requirements. A program to domesticate the oryx as a meat animal is in progress (Stanley-Price, 1985). Oryx can be quickly tamed, move as a herd, and have a very low rate of evaporative water loss. Because they ruminate primarily at night, the heat of rumen fermentation has minimal effects on evaporative losses. With the widespread problems of desertification and diminishing water resources, consideration of efficiency of water utilization could lead to changes in animal production, with newly domesticated species such as the oryx playing a greater role. Domestication and/or game ranching of species such as the oryx, whose survival in the wild is unlikely, could save them from extinction (Ostrowski et al., 1998; Spalton et al., 1999). Utilization of water by ruminants in hot, arid environments has been reviewed by Silanikove (1992).

# ENVIRONMENTAL EFFECTS OF ANIMAL WASTES

Animal waste (manure, excreta) is increasingly viewed as a major environmental pollutant, mainly because of the trend for livestock production to become highly concentrated with large numbers of animals confined to a small area of land. Modern intensive production systems for swine, poultry and dairy cattle, and beef feedlots generate large quantities of animal wastes. Environmental concerns with animal waste include soil, water, and air pollution with nutrients and microbes.

## Livestock and Water Pollution

Another means (besides water consumption) by which livestock interact with water resources is from pollution of waterways (Fig. 6-9) and groundwater with animal excreta. Pollution

**FIGURE 6-9** An issue of current concern is water pollution by livestock. Putting livestock handling facilities over a stream, as shown in this picture, is inappropriate. Such practices should be eliminated.

concerns are with excess nutrients and with pathogenic microbes causing diseases in humans. Waterborne diseases linked to contamination of water with animal wastes will be discussed in Chapter 9. In many countries, regulations are becoming increasingly strict relative to confinement systems and waste disposal. In countries such as the Netherlands, the number of pigs that can be raised is regulated and is being progressively decreased because of pollution of the waterways and groundwater. Nitrogen and phosphorus are of particular concern because they stimulate growth and subsequent die-off of algae in waterways (**eutrophication**). Increased precision in diet formulation, including use of highly digestible proteins, synthetic amino acids and bioavailable phosphorus sources, will help to minimize excretion of these elements. Use of phytases as feed additives increases the availability of phytate phosphorus in grains and plant protein supplements, reducing the amount of phosphorus excreted (see Chapter 5). New corn varieties with reduced levels of phytic acid and higher bioavailability of phosphorus have been developed. Formulating diets to minimize nutrient excretion has been termed **ecological nutrition** or **precision nutrition**. In the past, it was common to formulate swine and poultry diets with mineral and vitamin concentrations above the NRC suggested levels, on the basis that stressful conditions on farms increased requirements above what was measured under university conditions. In swine production, finishing pigs (the feeding period prior to market) consume the greatest amount of feed and have the lowest feed conversion efficiency. McGlone (2000) reported that supplemental minerals and vitamins could be deleted from swine finisher diets without adversely affecting pig performance. Adoption of this practice would reduce mineral excretion and thus help mitigate negative environmental effects of pig production.

A novel means of reducing phosphorus excretion by pigs was reported by Golovan et al. (2001). They developed a line of **transgenic pigs** that secrete salivary phytase. A bacterial phytase gene was introduced by pronuclear embryo microinjection. The transgenic animals were capable of complete digestion of dietary phytate phosphorus.

Another type of water pollution linked to livestock is increased sediment loads due to soil erosion. In an Ohio study, Owens et al. (1996) found that fencing livestock out of a stream reduced annual sediment concentration by over 50%.

Tamminga (1992, 1996) has reviewed the pollution problem in Holland with particular relevance to dairy production. In the past, dairy nutritionists in Holland, as elsewhere, have used diet formulation to maximize output of milk. Now they must include undesirable outputs and ways to minimize them, including nitrogen, phosphorus, and methane excretion, in computer diet formulation. Complete, on-farm estimates of incoming and outgoing nutrients are estimated, a process called **nutrient management** (Van Dyke et al., 1999). Thus diets are optimized with respect to maximizing production and minimizing pollution. Nitrogen contributes to environmental pollution both as atmospheric ammonia and as nitrate in groundwater. **Atmospheric ammonia** creates health hazards, undesirable odor, and contributes to damage to vegetation from air pollution. **Nitrates** in groundwater are a health hazard in drinking water and cause eutrophication of waterways. In dairy cattle nutrition, use of new systems of evaluating protein requirements, such as the degradable:nondegradable nitrogen as proposed by NRC (1989) will help to maximize efficiency of protein utilization by the animal and minimize the nitrogen excretion.

Legislation in Holland limits the amount of **phosphorus** that can be applied to soils, which by 1999 was reduced to 175 kg of P as $P_2O_5$ per hectare. This is equivalent to a stocking rate of four dairy cows per hectare. One option being considered in Holland is to export the manure to countries that can use it for fertilizer. The position of the government is that if farmers are going to raise livestock to produce pork and dairy products for export, then they'll also have to export the wastes generated by this activity. Similar legislative action to reduce livestock numbers has taken place in several countries, including the United States. In parts of Florida, dairying is being phased out to reduce pollution of the Everglades. Another option for manure disposal is the construction of waste treatment plants, similar to those used for sewage treatment. Obviously, this option is likely to markedly increase the costs of animal production and of the products produced. Farmers cannot permanently produce commodities at a loss, so if there is a major, industrywide increase in cost of production, this is inevitably passed on to the consumer, or less likely, the industry would cease to exist. Increasingly, whole-farm nutrient balance will be monitored to ensure that livestock production units are not causing environmental degradation (Van Horn et al., 1996). A new concern with nitrogen and phosphorus pollution from animal production has been the emergence of a highly toxic algae-like dinoflagellate, *Pfiesteria piscicida,* in contaminated estuaries on the U.S. Atlantic coast (Burkholder and Glasgow, 1997, 2001). Health effects of *Pfiesteria* are also discussed in Chapter 8. Human health problems in people exposed to *Pfiesteria* toxins include headaches, skin lesions, and learning and memory difficulties (Grattan et al., 1998, 2001). Eutrophication of estuaries and oceans is becoming a global problem, affecting the North American Atlantic coastal waters, the Caribbean, Baltic and North Seas, the western Mediterranean, and the Yellow Sea (Paerl, 1997; Paerl and Whitall, 1999). New harmful algae are being found (Bourdelais et al., 2002), supporting the notion of global malfunction of marine ecosystems, because of pollution. Chesapeake Bay in Maryland is highly polluted with nutrients, especially phosphorus, from intensive poultry production on the Maryland eastern shore (Staver and Brinsfield, 2001). Nitrogen from anthropogenic (human-caused) sources, such as intensive animal production, fertilizers, and so on, and industrial emissions of nitrogen oxides cause algae "blooms," which can deplete water oxygen, alter food webs, and cause other ecological problems. Nitrogen, phosphorus, and iron in animal wastes contribute significantly to eutrophication. For example, there is evidence of a "zone of anoxia" in the Gulf of Mexico, attributed to nitrogen from animal waste and fertilizers from the Midwest, transported down the Mississippi River (Paerl, 1997). Atmospheric ammonia deposition is also a significant source of nitrogen in coastal waters. As discussed under Livestock and Air Pollution, the major source of atmospheric ammonia is animal wastes. The situation is further influenced by another human activity, the building of dams. Normal phytoplankton growth requires soluble silicates, in addition to nitrogen and phosphorus. Silicates (silicon) gets trapped behind dams so that polluted river water reaching the coastal seas is enriched in nitrogen and phosphorus but is lacking sufficient silicate. This change in nutrient status promotes the development of toxic flagellate algae blooms (Ittekkot et al., 2000).

The estuaries on the Atlantic coast of North Carolina have been particularly hard-hit by pollution from the rapidly expanding swine industry of that state (Burkholder et al., 1997; Mallin et al., 1997), primarily because of the siting of hog farms in floodplains. Rupture of lagoons has been a problem and reached crisis proportions in 1999 when a so-called 500-year event, Hurricane Floyd, slammed into North Carolina.

Although much of the concern with animal waste has involved nitrogen, phosphorus is also of concern. Swine and poultry diets frequently contain excess phosphorus, added to diets to compensate for the low bioavailability of phytate phosphorus in grains and soybean meal. Many soils in the southeastern United States are accumulating excess phosphorus from the widespread land application of poultry litter (Sharpley, 1999). Phosphorus enters waterways as sediment-bound phosphorus associated with eroding soil particles and as dissolved phosphates. In conjunction with dissolved nitrogen, phosphates stimulate algal growth and eutrophication of water courses and estuaries (Bennett et al., 2001).

**Copper sulfate** is used as a feed additive in swine and poultry production for its antibiotic-like action (see Chapter 5). Concerns about its use include the copper in the excreta as an environmental contaminant. The European Economic Community has restricted the level of added copper sulfate in feeds to 125 ppm copper because of soil and water pollution concerns. High dietary levels (2000–4000 ppm) of zinc as **zinc oxide** are being used as growth promotants and antibacterial agents in pig diets (Hill et al., 2001). With the low bioavailability of zinc oxide, much of the zinc is excreted in the feces, with resultant environmental concerns.

**Selenium supplements** for livestock have also been identified as being of potential concern as environmental pollutants. This concern arose from an incident of selenium toxicity in aquatic birds at the Kesterson National Wildlife Refuge in California. In the early 1980s, severe pathology in adult birds and embryonic deformities and death were noted in aquatic birds (coots and grebes) nesting at the Kesterson Refuge. The reservoir was an artificial one, created by forming evaporation ponds to trap the drainage water from irrigated agricultural land in the San Joaquin Valley (Fig. 6-10). The birds and other organisms in the reservoir were found to have very high tissue levels of selenium. The source of the selenium was from dissolved selenium salts in irrigation water, which were concentrated in the evaporation ponds. Outcroppings of Cretaceous marine shale formations contain high levels of selenium. The irrigation water arising from hills surrounding the San Joaquin Valley dissolved soluble selenium salts out of the shale, to eventually arrive at the evaporation ponds. The selenium toxicity problems induced in wildlife at Kesterson have been described in detail by Ohlendorf et al. (1986, 1988).

The Kesterson incident received national attention. A reporter for the newspaper *Sacramento Bee*, Tom Harris, published a series of newspaper articles claiming that the U.S. Bureau of Reclamation and the Fish and Wildlife Service were involved in a cover-up of the selenium toxicity to protect corporate

**FIGURE 6-10** (*Top*): The termination of the San Luis Drainage Canal at the former Kesterson Reservoir. The canal was originally intended to take San Joaquin Valley drainage water to evaporation ponds. Now it simply dead-ends as shown in the photo. (*Bottom*): The former Kesterson Reservoir, which has been drained because of selenium contamination. Ironically, the land was originally a cattle ranch, and the land has now been returned to cattle grazing. The cattle may even have a bioremediation effect, because selenium is metabolized in the rumen to volatile metabolites. The road on the left is on the dike that formerly formed the reservoir, which occupied the land to the right (i.e., the fence is on land that was under water when the reservoir was functioning as an evaporation pond).

farmers and the mammoth California water irrigation projects. A full account of the newspaper investigation is provided by Harris (1991) whose motivations were probably summed up in the first two sentences of the Preface:

> Every reporter dreams of being at the very edge of some riveting new issue, there

at the beginning to discover, interpret, and record a cascade of events so fresh and startling, so important and unique, that they become a watershed to an emerging national crisis and, hopefully, correction. We fantasize not just about writing the first, exclusive accounts but of riding the crest of the subsequent wave of events to their conclusion (p. xi).

(Subsequent events confirmed the fantasy aspect of Harris's expose!). The notoriety given to the Kesterson incident inevitably called attention to the use of selenium in animals feeds. Harris (1991) stated, ominously, in the preface to his book: "Selenium supplements are being added, without government supervision, to most of the nation's beef and dairy cattle, swine and poultry" (p. xiv). This association of selenium as a livestock supplement with the Kesterson problem has created public anxiety about the safety of meat and the use of feed additives. Of course, the word "laced" is always used: selenium supplements were being used "even in areas where the native feed was already laced with potentially dangerous levels" (p. 82). In fact, the use of selenium for livestock is closely regulated and monitored by the FDA. Selenium supplements to feeds are prepared under veterinary supervision. Selenium is used in those parts of the world, including much of the United States, where soils are selenium-deficient and livestock develop selenium deficiency diseases such as white muscle disease. It is amazing that environmental and consumer activists can convince themselves and others that livestock producers would "lace" animal feeds with substances that cause severe tissue damage, infertility, fetal deformities and death. Rifkin (1992) states "Food and Drug Administration (FDA) officials say that it's not uncommon for some feedlot operators to mix industrial sewage and oils into the feed to reduce costs and fatten animals more quickly" (p. 13). How can people believe that feeding industrial sewage and oils, and toxic levels of selenium, could in any way increase productivity and profitability of livestock?

Part of the current concern about selenium is due to geological differences between the eastern and western United States. Much of the west has alkaline soils and deposits of seleniferous marine sediments. These alkaline soils are effective at oxidizing and mobilizing selenium as water-soluble selenate salts. The process is accelerated by irrigation, which is widespread throughout the western United States. Drainage water from the irrigated land may concentrate selenium salts, with adverse environmental impacts. In the eastern United States, the soils tend to be acid to neutral, and the climate is humid, favoring the formation of less soluble, reduced forms of selenium, such as ferric selenite. Norman et al. (1992) studied the effect of selenium supplementation of cattle on ranches in northern California on the concentrations of the element in streams and aquatic plants and animals. There was no evidence that selenium supplementation at the maximum FDA-permitted level caused selenium accumulation in the aquatic ecosystems. According to Ullrey (1992), if all of the livestock and poultry in the United States were supplemented with the legal limit of 0.3 ppm selenium in their diets, the annual selenium contribution to the environment would be less than 0.5% of that originating from natural sources and other human uses of the element.

Cattle grazing on selenium-contaminated land such as the Kesterson reservoir (which has now been drained) might even have potential as a restorative technique (bioremediation). Dietary selenium is metabolized in the rumen, with the production of volatile metabolites. Thus cattle grazing might be a way of harmlessly discharging some of the excess soil selenium into the air. Another possible means of decontamination might be growing selenium-accumulating crops such as rapeseed or other cruciferous plants. These plants have a high sulfur content, some of which can be substituted for by selenium. Tall fescue, a common pasture species, is particularly tolerant of high-selenium soils (Wu et al., 1988).

Another nutrient in manure, which may be a pollutant, is **potassium.** Levels of potassium in forages and corn silage produced on soils heavily fertilized with dairy cattle manure may contain potassium levels as high as 5 percent of the dry matter (L. J. Fisher et al., 1994). Alfalfa is high in potassium and is widely used in dairy cattle feeding in the United States. Over a period of time, as the manure from dairies is dispersed on a limited land area, the soil potassium level may build up, leading to high forage concentrations of the element. Excess dietary potassium can lead to metabolic disorders in dairy cattle.

**Chromic picolinate** has been approved (1996) by the FDA as a feed additive. Arsenic-containing organic compounds (arsenicals) such as **arsanilic acid** are used as growth promotants in poultry. Excretion of **arsenic** and heavy metals such as **chromium** may be of some potential concern when large amounts of animal waste are disposed of on croplands.

Chromium exists in two elemental forms: trivalent and hexavalent chromium. The trivalent form is used as a feed additive and food supplement. **Trivalent chromium** has very low toxicity. **Hexavalent chromium** is the valence state of the element as it is used industrially for chrome plating, stainless steel, leather tanning, and so on. Hexavalent chromium is toxic, a carcinogen, and an environmental pollutant. The use of chromium as a feed additive suffers from the perception that chromium is toxic, but the form used nutritionally (trivalent chromium) is of little concern. Chromium is a component of an organic molecule called the **glucose tolerance factor,** which is essential for normal insulin function (Mertz, 1993). In livestock, chromium supplementation may increase resistance to stress and stimulate immune responses.

In areas of intensive swine and poultry production, such as Arkansas and North Carolina, disposal of animal waste is a major problem. The principal means of waste disposal has been application to fields as a fertilizer and soil amendment. This has been very useful in promoting crop and forage production. For example, in Arkansas, the beef industry has expanded mainly because of the availability of inexpensive poultry waste for fertilizing tall fescue pastures (Sims and Wolf, 1994). However, in many areas, waste applications must be reduced or stopped because of contamination of surface and groundwater with nitrogen and phosphorus. **Grass filter strips** are often kept along stream banks in pastures and croplands fertilized with manure. The filter strips help prevent runoff of sediment and nutrients into waterways. Nichols et al. (1998) found that grass filter strips reduced the runoff of estrogen (17β-estradiol) from poultry litter. Animal waste contains natural estrogens excreted in urine. Excreted estrogens can contribute to the pool of "**environmental estrogens**" (see Chapter 9), which may have adverse effects on wildlife and fish. Waste management is and will continue to be one of the major challenges for intensive livestock production. Diet formulation to minimize nutrient excretion is one of the major means of dealing with this problem. In essence, waste disposal problems are a consequence of the intensification of animal production.

Historically, impacts of agriculture on water quality were considered to be natural and uncontrollable. The development of intensive animal production in the 1960s led to numerous ecological problems caused by discharges of animal wastes. These incidents led to increased regulatory scrutiny, and regulations such as the Clean Water Act (Martin, 1997). Many U.S. states have Departments of Environmental Quality or other such agencies, which regulate animal waste discharges. This is an inevitable result of the changes in the structure of agriculture; where once animals were dispersed among many farms, they are now concentrated in high density on relatively few farms.

### Aquatic Vegetation as a Feedstuff

In the southern United States and in much of the tropics, **aquatic weeds** are a major problem. The worst offender is the water hy-

acinth. This rapidly growing plant can quickly cover a waterway, lake, or reservoir with a dense mat of vegetation, interfering with navigation, fishing, hydroelectric turbines, and so on. Decomposing plant material depletes oxygen from the water, causing fish kills. Water hyacinth and other aquatic weeds are mechanically removed at considerable expense. One means of recovering value from them is to utilize them as animal feed. Numerous studies have been conducted with use of aquatic weeds as feedstuffs (e.g., Lizama et al., 1988). Plants such as water hyacinth have also been used in the treatment of sewage and animal wastes by growing the plants on waste water lagoons. They use the nitrogen and phosphorus to support an explosive growth of vegetation, thus purifying the water. They are then harvested at regular intervals and can be used for compost or animal feed. Haustein et al. (1990) reported that dietary levels as high as 40 percent sewage-grown duckweed (*Lemna* spp.) could be used in poultry diets without affecting the performance of layers.

## LIVESTOCK AND AIR POLLUTION

Odors and gases from confinement livestock operations can cause environmental problems. Odors from swine manure are particularly obnoxious and are the source of much controversy, particularly with the emergence of swine mega-farms (see Chapter 8).

The major gas pollutant emitted from livestock operations is **ammonia**. Ammonia is released from the microbial metabolism of nitrogenous products (urea, uric acid, undigested protein) in animal excreta. Ammonia can be reduced by lowering dietary protein concentrations, precise balancing of dietary amino acid levels with amino acid requirements, and use of synthetic amino acids in place of dietary protein. Feed additives such as zeolites and yucca extract can bind ammonia in animal wastes and reduce ammonia emissions (see Chapter 5). Under alkaline or basic conditions (above pH 7.0), ammonia exists as

$NH_3$, a volatile gas. Under acidic conditions, it exists as the non-volatile ammonium ion, $NH_4^+$. If animal waste is acidic, either because of the nature of the diet or from the application of acids to the waste, ammonia remains as the non-volatile ammonium ion. One dietary modification that facilitates this in pigs is feeding a source of hindgut-fermentable carbohydrate, such as soybean hulls. The volatile fatty acids (VFA) produced by the microbial fermentation of the carbohydrate increases the VFA content of the feces, reducing the pH and suppressing the evolution of ammonia (Cahn et al., 1998). Sutton et al. (1999) reviewed the potentials for reducing swine waste odor by diet modification. Methods include more precision in diet formulation to reduce nutrient excretion, enhancement of microbial metabolism in the lower tract to reduce excretion of odor-causing compounds, and changing physical characteristics of the feces and urine to reduce odor emissions.

Miner (1999) reviewed methods of lagoon management to reduce swine odors. Lagoons function as areas for anaerobic bacteria to decompose residual solids, converting them to liquids and gases. Lagoon covers or use of enclosed digesters aid in reducing odiferous emissions.

Although ammonia is alkaline in reaction, it can contribute to **acid rain**. This is because ammonia in the air can react with atmospheric sulfuric acid to form ammonium sulfate. Ammonium sulfate can react with oxygen in soil to produce nitric and sulfuric acids:

$$(NH_4)_2SO_4 + 4O_2 \longrightarrow 2HNO_3 + H_2SO_4 + 2H_2O$$

**Ammonia emission** from animal waste is the major factor limiting livestock production in Holland (DeBoer et al., 1997; Jongbloed and Lenis, 1998). The Dutch government has mandated reductions in pig numbers in that country to reduce ammonia emissions. Ammonia is considered a serious air pollutant, with the major source of atmospheric ammonia being animal agriculture (Buijsman et al., 1987). Control of ammonia emissions will likely emerge as one of the major regulatory issues in animal production in many countries, including the United States.

In some animal production systems, especially for swine and dairy cattle, waste is handled as a liquid or slurry and stored in tanks or lagoons. Loss of ammonia from storage can be reduced by the use of surface covers (Sommer et al., 1993), such as straw, peat, or oil. Both ammonia and nitrous oxide are volatilized to a large extent when liquid animal waste is pumped and sprayed onto pastures or croplands (Sharpe and Harper, 1997).

Airborne particles (dust) are a significant source of complaints about intensive livestock facilities. Dust can act as a carrier of pathogenic microbes.

## Animal Wastes and Microbial Contaminants

In addition to soil, water, and air pollution with nutrients caused by animal wastes, there are concerns with pathogenic microbes. An excellent review on manure and microbes is provided by Pell (1997) with particular emphasis on dairy cattle.

Pathogenic organisms of concern in animal wastes include bacteria, protozoa, and viruses. Some of the major bacteria include *Salmonella* spp., *E. coli*, *Mycobacterium* spp., *Bacillus anthracis* (anthrax), *Brucella abortis*, *Leptospira* spp., *Listeria monocytogenes*, *Campylobacter jejuni* and *Yersinia enterocolitica*. Food safety and human health implications of these organisms are discussed in Chapter 9, as are waterborne protozoa associated with animal wastes (*Giarida, Cryptosporidia.*)

A number of viruses pathogenic to humans may be excreted in animal waste. Some of these include bovine adenovirus, enterovirus, parvovirus, coronavirus, and papillomavirus, bovine virus diarrhea (BVD), foot and mouth, cowpox, and parainfluenza.

Composting of animal waste is an effective means of destroying pathogens. **Compost** should be turned so that all parts of the pile reach a temperature of 60° C. Another method of reducing environmental effects of animal waste is to burn it for power generation. Poultry litter, being dry, is especially amenable to use in this manner. Several power plants in Europe use poultry litter as their fuel source. The ash from such "bioenergy" plants retains phosphorus and other plant nutrients and can be used as a fertilizer. Similar poultry litter-fired power plants are planned for the United States. Dairy cattle wastes can be fermented in biogas digesters, producing methane that is burned to generate electricity.

### Odors and Livestock Wastes

An important aspect of animal waste disposal and animal confinement systems is air pollution with unpleasant and foul-smelling odors. The odor from swine manure is particularly offensive. Swine odor has emerged as one of the main contentious issues in the swine industry. Virtually nobody wants a confined swine facility in his or her neighborhood. In contrast to pollutants in animal waste such as nitrogen and phosphorus compounds, odors are very difficult to measure and regulate. Different people vary in their perceptions of unpleasant odors. Various attempts to develop odor monitoring systems have been made; one of the most successful is an "electronic nose," which measures certain key volatile compounds (Mackie et al., 1998). Some of the main odiferous compounds in swine waste are ammonia, amines, sulfur-containing compounds, volatile fatty acids, indoles, skatole, phenols, alcohols, and carbonyls (Mackie et al., 1998). These compounds are produced by anaerobic microbes during manure storage in pits and lagoons (Zhu and Jacobson, 1999). Mackie et al. (1998) offer an excellent review of the production of these compounds. Many of the odiferous compounds are produced by the bacterial fermentation of amino acids. Sulfur-containing compounds such as hydrogen sulfide ($H_2S$) are produced from methionine and other sulfur amino acids. Phenols and cresol are derived from the amino acid tyrosine. In-

doles such as skatole (3-methyl indole) are produced from tryptophan. Several of the swine odor volatiles adversely affect human health and well-being (Schiffman, 1998). They cause eye, nose, and throat irritation, headache, drowsiness, depression and mood disorders.

# ECOLOGICAL BENEFITS FROM INTENSIVE LIVESTOCK PRODUCTION (FACTORY FARMING)

Contrary to the impression that many doomsday scenarios give, the world food production capacity is adequate to meet present needs and foreseeable needs, even with rapid human population growth until stabilization eventually occurs. In most of the developed and semideveloped (eastern Europe, former Soviet Union) countries, human population growth has stabilized, and in numerous countries there is even a negative population growth rate. In most of these, such as the United States, Canada, Australia, and the European Economic Community (EEC), agriculture is plagued by surpluses. The same probably will be true of eastern Europe when a free-market economy is established. The problem is not one of competition between livestock and humans for grain and other foodstuffs but rather surpluses of grains, protein concentrates, oil crops, meat, and dairy products. This results in such absurdities as the extensive use of dried milk powder in dairy cattle diets in the 1980s in England. This was an attempt to reduce huge surpluses of dairy products, so dried milk was fed to dairy cattle to produce more milk!

An alternative to continual surpluses would be economic policies that promote a reduction in agricultural output so that supply more evenly matches demand. Second, by encouraging **intensive agriculture** (factory farming), including livestock production, output per unit of land could be maximized. This would allow retirement of surplus agricultural land into forests or native grasslands, increasing wildlife habitat, watersheds, and sequestering carbon dioxide in plant biomass to have a favorable effect on global warming. Such proposals have received serious (in some quarters at least) consideration. For example, a well-publicized proposal has been to retire surplus grain-producing land from the northern great plains of the United States and establish grasslands and a large bison preserve (Matthews, 1992). Such proposals are not well received by the agricultural community, needless to say, but with the trend to fewer and larger farms in the United States, the agricultural community has less and less political power (one person, one vote) to influence such decisions.

Thus a case can be made that intensive crop and livestock production, making optimal use of herbicides, pesticides, feed additives, growth promotants, bST, and other products of biotechnology, offers the best solution to minimizing adverse ecological and environmental impacts of agriculture. This proposal has been made by Avery (1995) in his book *Saving the Planet with Pesticides and Plastic*. The agricultural products could be produced using a much smaller land base, allowing much of presently farmed land to revert to forests or grasslands. A similar case could be made for intensive forestry using plantation trees and kenaf. Social and ethical aspects of intensive agriculture and factory farming are further considered in Chapters 8 and 10.

# STUDY QUESTIONS

1. What is the greenhouse effect? What are the major "greenhouse gases"?

2. Describe how livestock production could influence global warming and the "hole in the ozone layer."

3. How does tropical deforestation contribute to global warming?

4. How could methane emissions by cattle be reduced? Why does dietary fiber level affect methane production?

5. What is the "hamburger connection"? Is boycotting hamburgers a useful way of protesting tropical deforestation?

6. What countries are the main sources of beef imported into the United States? Is the United States a net importer of beef?

7. David Janzen is a well-known environmentalist in Costa Rica. He urges people to buy Costa Rican beef. Why?

8. What is the "cut and carry" system of forage utilization? Under what conditions is it used?

9. What is the role of cattle in India in helping to prevent deforestation?

10. It has been claimed that producing 1 kg of animal protein requires about 100 times as much water as producing 1 kg of plant protein. Discuss the merits of this claim.

11. How do sheep and cattle differ in their water requirements and efficiency of water utilization?

12. Algae in water are photosynthetic; they release oxygen. If you examine algae-enriched water on a sunny day, you can observe oxygen bubbles in the water, coming from the algae. How do you explain the observation that there are often massive fish kills due to oxygen starvation in algae-infested water? (Hint: The fish kills usually occur on cloudy mornings after a period of sunny weather.)

13. What relationship, if any, was there between selenium poisoning of waterfowl in the Kesterson reservoir in California and the use of selenium as a feed additive?

14. How can animal diets be formulated to minimize nitrogen and phosphorus excretion?

15. Define "ecological nutrition."

16. How can livestock production contribute to acid rain?

17. What are the main nutrients excreted by livestock that are of potential concern as soil, water, or air pollutants?

18. What is your opinion of the merits of intensive agriculture as a means of "Saving the Planet with Pesticides and Plastic"?

# REFERENCES

AGGARWAL, G. C. 1989. Judicious use of dung in the Third World. *Energy* 14:349–352.

ALLEN, W. H. 1988. Biocultural restoration of a tropical forest. *BioScience* 38:156–161.

ARIMA, E. Y., and C. UHL. 1997. Ranching in the Brazilian Amazon in a national context: Economics, policy, and practice. *Society and Natural Resources* 10:433–451.

AVERY, D. T. 1995. *Saving the planet with pesticides and plastic.* Hudson Institute, Indianapolis.

AYERS, A. C., R. P. BARRETT, and P. R. CHEEKE. 1996. Feeding value of tree leaves (hybrid poplar and black locust) evaluated with sheep, goats and rabbits. *Anim. Feed Sci. Tech.* 57:51–62.

BAS, F. D., F. R. EHLE, and R. D. GOODRICH. 1985. Evaluation of pelleted aspen foliage as a ruminant feedstuff. *J. Anim. Sci.* 61:1030–1036.

BECKETT, J. L., and J. W. OLTJEN. 1993. Estimation of the water requirement for beef production in the United States. *J. Anim. Sci.* 71:818–826.

BENNETT, E. M., S. R. CARPENTER, and N. F. CARACO. 2001. Human impact on erodable phosphorus and eutrophication: A global perspective. *BioScience* 51:227–234.

BONTI-ANKOMAH, S., and G. FOX. 1998. Hamburgers and the rainforest—A review of issues and evidence. *J. Agric. Environ. Ethics* 10:153–182.

BOURDELAIS, A. J., C. R. TOMAS, J. NAAR, J. KUBANEK, and D. J. BADEN. 2002. New fish-killing alga in coastal Delaware produces neurotoxins. *Environ. Health Perspectives* 110:465–470.

BRUCE, J. P., M. FROME, E. HAITES, H. JANZEN, R. LAL, and K. PAUSTIAN. 1999. Carbon sequestration in soils. *J. Soil and Water Conservation* 54:382–389.

BUIJSMAN, E., H. F. M. MAAS, and W. A. H. ASMAN. 1987. Anthropogenic $NH_3$ emissions in Europe. *Atmosph. Environ.* 21:1009–1022.

BURKHOLDER, J. M. 1999. The lurking perils of *Pfiesteria. Sci. Amer.* August:42–49.

BURKHOLDER, J. M., and H. B. GLASGOW. 2001. History of toxic *Pfiesteria* in North Carolina estuaries from 1991 to the present. *BioScience* 51:827–841.

BURKHOLDER, J. M., and H. B. GLASGOW, JR. 1997. *Pfiesteria piscicida* and other *Pfiesteria*-like dinoflagellates: Behavior, impacts, and environmental controls. *Limnol. Oceanogr.* 42:1052–1075.

BURKHOLDER, J. M., M. A. MALLIN, H. B. GLASGOW, JR., L. M. LARSEN, M. R. MCIVER, G. C. SHANK, N. DEAMER-MELIA, D. S. BRILEY, J. SPRINGER, B. W. TOUCHETTE, and E. K. HANNON. 1997. Impacts to a coastal river and estuary from rupture of a large swine waste holding lagoon. *J. Environ. Qual.* 26:1451–1466.

BUSCHBACHER, R. J. 1986. Tropical deforestation and pasture development. *BioScience* 36:22–28.

BUTLER, J. H., M. BATTLE, M. L. BENDER, S. A. MONTZKA, A. D. CLARKE, E. S. SALTZMAN, C. M. SUCHER, J. P. SEVERINGHAUS, and J. W. ELKINS. 1999. A record of atmospheric halocarbons during the twentieth century from polar firn air. *Nature* 399:749–755.

BYERS, F. M. 1990. Beef production and the greenhouse effect—The role of methane from beef production in global warming. *Proc. West. Sect. Am. Soc. Anim. Sci.* 41:144–147.

CAHN, T. T., A. L. SUTTON, A. J. A. AARNINK, M. W. A. VERSTEGEN, J. W. SCHRAMA, and G. C. M. BAKKER. 1998. Dietary carbohydrates alter the fecal composition and pH and the ammonia emission from slurry of growing pigs. *J. Anim. Sci.* 76:1887–1895.

CHONG, D. T., I. TAJUDDIN, ABD. M. S. SAMAT, W. W. STUR, and H. M. SHELTON. 1997. Stocking rate effects on sheep and forage productivity under rubber in Malaysia. *J. Agric. Sci.* 128:339–346.

CRUTZEN, P. J., I. ASELMANN, and W. SEILER. 1986. Methane production by domestic animals, wild ruminants, other herbivorous fauna, and humans. *Tellus* 38B:271–284.

DE BOER, I. J. M., H. T. A. PETERS, M. GROSSMAN and W. J. KOOPS. 1997. Nutrient flows in agriculture in The Netherlands with special emphasis on pig production. *J. Anim. Sci.* 75:2054–2063.

FAHEY, G. C., JR., and L. L. BERGER. 1988. Carbohydrate nutrition of ruminants. pp. 269–297 In: Church, D. C. (Ed.). *The ruminant animal.* Prentice Hall, Englewood Cliffs, New Jersey.

FAMINOW, M. D. 1998. *Cattle, deforestation and development in the Amazon. An economic, agronomic and environmental perspective.* CAB INTERNATIONAL, Wallingford, United Kingdom.

FISHER, L. J., N. DINN, R. M. TAIT, and J. A. SHELFORD. 1994. Effect of level of dietary potassium on the absorption and excretion of calcium and magnesium by lactating cows. *Can. J. Anim. Sci.* 74:503–509.

FISHER, M. J., I. M. RAO, M. A. AYARZA, C. E. LASCANO, J. I. SANZ, R. J. THOMAS, and R. R. VERA. 1994. Carbon storage by introduced deep-rooted grasses in the South American savannas. *Nature* 371:236–238.

GARCIA-OLIVA, F., I. CASAR, P. MORALES, and J. M. MAASS. 1994. Forest-to-pasture conversion influences on soil organic carbon dynamics in a tropical deciduous forest. *Oecologia* 99:392–396.

GOLOVAN, S. P., R. G. MEIDINGER, A. AJAKAIYE, M. COTTRILL, M. Z. WIEDERKEHR, D. J. BARNEY, C. PLANTE, J. W. POLLARD, M. Z. FAN, M. A. HAYES, J. LAURSEN, J. P. HJORTH, R. R. HACKER, J. P. PHILLIPS, and C. W. FORSBERG. 2001. Pigs expressing salivary phytase produce low-phosphorus manure. *Nature Biotech.* 19:741–745.

GRADWOHL, J., and R. GREENBERG. 1988. *Saving the tropical forests.* Island Press, Washington, D.C.

GRATTAN, L. M., D. OLDACH, T. M. PERL, M. H. LOWITT, D. L. MATUSZAK, C. DICKSON, C. PARROTT, R. C. SHOEMAKER, C. L. KAUFFMAN, M. P. WASSERMAN, J. R. HEBEL, P. CHARACHE, and J. G. MORRIS, JR.

1998. Learning and memory difficulties after environmental exposure to waterways containing toxin-producing *Pfiesteria* or *Pfiesteria*-like dinoflagellates. *The Lancet* 352:532–539.

GRATTAN, L. M., D. OLDACH, and J. G. MORRIS. 2001. Human health risks of exposure to *Pfiesteria piscicida. BioScience* 51:853–857.

HARRIS, T. 1991. *Death in the marsh.* Island Press, Washington, D.C.

HAUSTEIN, A. T., R. H. GILMAN, P. W. SKILLICORN, V. VERGARA, V. GUEVARA, and A. GASTANADUY. 1990. Duckweed, a useful strategy for feeding chickens: Performance of layers fed with sewage-grown lemnacea species. *Poul. Sci.* 69:1835–1844.

HECHT, S. B. 1989. The sacred cow in the green hell: Livestock and forest conversion in the Brazilian Amazon. *The Ecologist* 19:229–234.

HECHT, S. B. 1993. The logic of livestock and deforestation in Amazonia. *BioScience* 43:687–695.

HEGARTY, R. S. 1999. Reducing rumen methane emissions through elimination of rumen protozoa. *Aust. J. Agric. Res.* 50:1321–1327.

HILL, G. M., D. C. MAHAN, S. D. CARTER, G. L. CROMWELL, R. C. EWAN, R. L. HARROLD, A. J. LEWIS, P. S. MILLER, G. C. SHURSON, and T. L. VEUM. 2001. Effect of pharmacological concentrations of zinc oxide with or without the inclusion of an antibacterial agent on nursery pig performance. *J. Anim. Sci.* 79:934–941.

HOOGESTEIJN, R., and C. A. CHAPMAN. 1997. Large ranches as conservation tools in the Venezuelan llanos. *Oryx* 31:274–284.

HUNGATE, R. E. 1966. *The rumen and its microbes.* Academic Press, New York.

ITTEKKOT, V., C. HUMBORG, and P. SCHAFER. 2000. Hydrological alterations and marine biogeochemistry: A silicate issue? *BioScience* 50:776–782.

JANZEN, D. H. 1973. Tropical agroecosystems. *Science* 182:1212–1219.

JANZEN, D. H. 1988a. Tropical ecological and biocultural restoration. *Science* 239:243–244.

JANZEN, D. H. 1988b. Dan Janzen's thoughts from the tropics. 10. Buy Costa Rican beef. *Oikos* 51:257–258.

JANZEN, D. H., and P. S. MARTIN. 1982. Neotropical anachronisms: The fruits the Gomphotheres ate. *Science* 215:19–27.

JOHNSON, K. A., and D. E. JOHNSON. 1995. Methane emissions from cattle. *J. Anim. Sci.* 73:2483–2492.

JONGBLOED, A. W., and N. P. LENIS. 1998. Environmental concerns about animal manure. *J. Anim. Sci.* 76:2641–2648.

KANG, B. T., L. REYNOLDS, and A. N. ATTA-KRAH. 1990. Alley farming. *Advances in Agronomy* 43:315–359.

KAUFFMAN, J. B., D. L. CUMMINGS, and D. E. WARD. 1998. Fire in the Brazilian Amazon 2. Biomass, nutrient pools and losses in cattle pastures. *Oecologia* 113:415–427.

KEELEY, J. E., and C. J. FOTHERINGHAM. 1997. Trace gas emissions and smoke-induced seed germination. *Science* 276:1248–1250.

KLEINMAN, P. J. A., D. PIMENTEL, and R. B. BRYANT. 1995. The ecological sustainability of slash-and-burn agriculture. *Agric. Ecosyst. Environ.* 52:235–249.

KURIHARA, M., T. MAGNER, R. A. HUNTER, and G. J. MCCRABB. 1999. Methane production and energy partition of cattle in the tropics. *Brit. J. Nutr.* 81:227–234.

LAL, R. 1986. Conversion of tropical rainforest: Agronomic potential and ecological consequences. *Advances in Agronomy* 39:173–264.

LAL, R., R. F. FOLLETT, J. KIMBLE, and C. V. COLE. 1999. Managing U.S. cropland to sequester carbon in soil. *J. Soil and Water Conservation* 54:374–381.

LAURANCE, W. F., H. L. VASCONCELOS, and T. E. LOVEJOY. 2000. Forest loss and fragmentation in the Amazon: Implications for wildlife conservation. *Oryx* 34:39–45.

LEFROY, E. C., R. J. STIRZAKER, and J. S. PATE. 2001. The influence of tagasaste (*Chamaecytisus proliferus* Link). trees on the water balance of an alley cropping system on deep sand in south-western Australia. *Aust. J. Agric. Res.* 52:235–246.

LENG, R. A. 1992. Ruminant production and greenhouse gas emissions. *Proc. N.Z. Anim. Prod.* 52:Supplement 15–23.

LENG, R. A. 1993. Quantitative ruminant nutrition—A green science. *Aust. J. Agric. Res.* 44:363–380.

LIZAMA, L. C., J. E. MARION, and L. R. MCDOWELL. 1988. Utilization of aquatic plants *Elodea canadensis* and *Hydrilla verticillata* in broiler chick diets. *Anim. Feed Sci. Tech.* 20:155–161.

LOKER, W. M. 1993. The human ecology of cattle raising in the Peruvian Amazon: The view from the farm. *Human Organization* 52:14–24.

LOKER, W. M. 1994. Where's the beef?: Incorporating cattle into sustainable agroforestry systems in the Amazon basin. *Agroforestry Systems* 25:227–241.

LOKER, W. M. 1996. Cowboys, Indians and deforestation: Ethical and environmental issues associated with pastures research in Amazonia. *Agriculture and Human Values* 13:52–58.

MACHMULLER, A., and M. KREUZER. 1999. Methane suppression by coconut oil and associated effects on nutrient and energy balance in sheep. *Can. J. Anim. Sci.* 79:65–72.

MACKIE, R. I., P. G. STROOT, and V. H. VAREL. 1998. Biochemical identification and biological origin of key odor components in livestock waste. *J. Anim Sci.* 76:1331–1342.

MALAKOFF, D. A. 1997. Nitrogen oxide pollution may spark seeds' growth. *Science* 276:1199.

MALLIN, M. A., J. M. BURKHOLDER, M. R. MCIVER, G. C. SHANK, H. B. GLASGOW, JR., B. W. TOUCHETTE, and J. SPRINGER. 1997. Comparative effects of poultry and swine waste lagoon spills on the quality of receiving streamwaters. *J. Environ. Qual.* 26:1622–1631.

MALOIY, G. M. O., W. V. MACFARLANE, and A. SHKOLNIK. 1979. Mammalian herbivores. pp. 185–209 In: Maloiy, G. M. O. (Ed.). *Comparative physiology of osmoregulation in animals,* Vol. 11. London, Academic Press.

MARTIN, J. H., JR. 1997. The Clean Water Act and animal agriculture. *J. Environ. Qual.* 26:1198–1203.

MATTHEWS, A. 1992. *Where the buffalo roam.* Grove Press, New York.

MATTOS, M. M., and C. UHL. 1994. Economic and ecological perspectives on ranching in the eastern Amazon. *World Development* 22:145–158.

MCALLISTER, T. A., E. K. OKINE, G. W. MATHISON, and K.-J. CHENG. 1996. Dietary, environmental and microbiological aspects of methane production in ruminants. *Can. J. Anim. Sci.* 76:231–243.

MCGLONE, J. J. 2000. Deletion of supplemental minerals and vitamins during the late finishing period does not affect pig weight gain and feed intake. *J. Anim. Sci.* 78:2797–2800.

MERTZ, W. 1993. Chromium in human nutrition: A review. *J. Nutr.* 123:626–633.

MINER, J. R. 1999. Alternatives to minimize the environmental impact of large swine production units. *J. Anim. Sci.* 77:440–444.

MITRA, S., M. C. JAIN, S. KUMAR, S. K. BANDYOPADHYAY, and N. KALRA. 1999. Effect of rice cultivars on methane emission. *Agric. Ecosyst. Environ.* 73:177–183.

MOSIER, A., D. SCHIMEL, D. VALENTINE, K. BRONSON, and W. PARTON. 1991. Methane and nitrous oxide fluxes in native, fertilized and cultivated grasslands. *Nature* 350:330–332.

MURGUEITIO, E. 1990. Intensive sustainable livestock production: An alternative to tropical deforestation. *Ambio.* 19:397–400.

MYERS, N. 1991a. Tropical deforestation: The latest situation. *BioScience* 41:282.

MYERS, N. 1991b. Tropical forests: Present status and future outlook. *Climactic Change* 19:3–32.

NATIONAL RESEARCH COUNCIL. 1989. *Nutrient requirements of dairy cattle.* National Academy Press, Washington, D.C.

NATIONAL RESEARCH COUNCIL. 1991. *Microlivestock. Little-known small animals with a promising economic future.* National Academy Press, Washington, D.C.

NEPSTAD, D. C., C. UHL and E. A. S. SERRAO. 1991. Recuperation of a degraded Amazonian landscape: Forest recovery and agricultural restoration. *Ambio* 20:248–255.

NICHOLS, D. J., T. C. DANIEL, D. R. EDWARDS, P. A. MOORE, JR., and D. H. POTE. 1998. Use of grass filter strips to reduce 17β-estradiol in runoff from fescue-applied poultry litter. *J. Soil and Water Cons.* 53:74–77.

NORMAN, B., G. NADER, M. OLIVER, R. DELMAS, D. DRAKE and H. GEORGE. 1992. Effects of selenium supplementation in cattle on aquatic ecosystems in northern California. *J. Am. Vet. Med. Assn.* 201:869–872.

NORSE, E. A. 1990. *Ancient forests of the Pacific Northwest.* Island Press, Washington, D.C.

OHLENDORF, H. M., D. J. HOFFMAN, M. K. SAIKI, and J. W. ALDRICH. 1986. Embryonic mortality and abnormalities of aquatic birds: Apparent impacts of selenium from irrigation drainwater. *Sci. Total Environ.* 52:49–63.

OHLENDORF, H. M., A. W. KILNESS, R. K. STROUD, D. J. HOFFMAN, and J. F. MOORE. 1988. Selenium toxicosis in wild aquatic birds. *J. Toxicol. Environ. Health* 24:67–92.

OSTROWSKI, S., E. BEDIN, D. M. LENAIN, and A. H. ABUZINADA. 1998. Ten years of Arabian oryx conservation breeding in Saudi Arabia—Achievements and regional perspectives. *Oryx* 32:209–222.

OWENS, L. B., W. M. EDWARDS, and R. W. VAN KEUREN. 1996. Sediment losses from a pastured watershed before and after stream fencing. *J. Soil and Water Cons.* 51:90–94.

PAERL, H. W. 1997. Coastal eutrophication and harmful algal blooms: Importance of atmospheric deposition and groundwater as "new" nitrogen and other nutrient sources. *Limnol. Oceanogr.* 42:1154–1165.

PAERL, H. W., and D. R. WHITALL. 1999. Anthropogenically-derived atmospheric nitrogen deposition, marine eutrophication and harmful algal bloom expansion: Is there a link? *Ambio* 28:307–311.

PANTER, K. E., L. F. JAMES, and R. J. MOLYNEUX. 1992. Ponderosa pine needle-induced parturition in cattle. *J. Anim. Sci.* 70:1604–1608.

PELL, A. N. 1997. Manure and microbes: Public and animal health problem? *J. Dairy Sci.* 80:2673–2681.

PICHON, F. J. 1996. The forest conversion process: A discussion of the sustainability of predominant land uses associated with frontier expansion in the Amazon. *Agriculture and Human Values* 13:32–51.

PIMENTEL, D., J. HOUSER, E. PREISS, O. WHITE, H. FANG, L. MESNICK, T. BARSKY, S. TARICHE, J. SCHRECK, and S. ALPERT. 1997. Water resources: Agriculture, the environment, and society. *BioScience* 47:97–106.

POST, W. M., T.-H. PENG, W. R. EMANUEL, A. W. KING, V. H. DALE, and D. L. DEANGELIS. 1990. The global carbon cycle. *Amer. Scientist* 78:310–326.

PRESTON, T. R. 1990. Future strategies for livestock production in tropical third world countries. *Ambio* 19:390–393.

RETALLACK, G. J. 2001. A 300-million-year record of atmospheric carbon dioxide from fossil plant cuticles. *Nature* 411:287–290.

RIFKIN, J. 1992. *Beyond beef: The rise and fall of the cattle culture.* Dutton, New York.

SCHIFFMAN, S. S. 1998. Livestock odors: Implications for human health and well-being. *J. Anim. Sci.* 76:1343–1355.

SHARKEY, T. D. 1996. Isoprene synthesis by plants and animals. Endeavour 20:74–78.

**SHARPE, R. R., and L. A. HARPER.** 1997. Ammonia and nitrous oxide emissions from sprinkler irrigation applications of swine effluent. *J. Environ. Qual.* 26:1703–1706.

**SHARPLEY, A.** 1999. Agricultural phosphorus, water quality, and poultry production: Are they compatible? *Poult. Sci.* 78:660–673.

**SHARROW, S. H., D. H. CARLSON, W. H. EMMINGHAM, and D. P. LAVENDER.** 1992. Direct impacts of sheep upon Douglas-fir trees in two agrosilvopastoral systems. *Agroforestry Systems* 19:223–232.

**SHARROW, S. H., W. C. LEININGER, and B. RHODES.** 1989. Sheep grazing as a silvicultural tool to suppress brush. *J. Range Management* 42:2–4.

**SILANIKOVE, N.** 1992. Effects of water scarcity and hot environment on appetite and digestion in ruminants: A review. *Livestock Production Science* 30:175–194.

**SIMPSON, J. R., and J. H. CONRAD.** 1993. Intensification of cattle production systems in Central America: Why and when. *J. Dairy Sci.* 76:1744–1752.

**SIMS, J. T., and D. C. WOLF.** 1994. Poultry waste management: Agricultural and environmental issues. *Advances in Agronomy* 52:1–83.

**SMIL V.** 1997. Global population and the nitrogen cycle. *Sci. Amer.* 277:76–81.

**SMITH, J., J. V. CADAVID, A. RINCON, and R. VERA.** 1997. Land speculation and intensification at the frontier: A seeming paradox in the Colombian savanna. *Agric. Systems* 54:501–520.

**SOMMER, S. G., B. T. CHRISTENSEN, N. E. NIELSEN, and J. K. SCHJORRING.** 1993. Ammonia volatilization during storage of cattle and pig slurry: Effect of surface cover. *J. Agric. Sci.* 121:63–71.

**SPALTON, J. A., M. W. LAWRENCE, and S. A. BREND.** 1999. Arabian oryx reintroduction in Oman: successes and setbacks. *Oryx* 33:168–175.

**STANLEY-PRICE, M. R.** 1985. Game domestication for animal production in Kenya: The nutritional ecology of oryx, zebu cattle and sheep under free-range conditions. *J. Agric. Sci.* 104:375–382.

**STAVER, K. W., and R. B. BRINSFIELD.** 2001. Agriculture and water quality on the Maryland Eastern Shore: Where do we go from here? *BioScience* 51:859–868.

**SUTTON, A. L., K. B. KEPHART, M. W. A. VERSTEGEN, T. T. CANH, and P. J. HOBBS.** 1999. Potential for reduction of odorous compounds in swine manure through diet modification. *J. Anim. Sci.* 77:430–439.

**TAMMINGA, S.** 1992. Nutrition management of dairy cows as a contribution to pollution control. *J. Dairy Sci.* 75:345–357.

**TAMMINGA, S.** 1996. A review of environmental impacts of nutritional strategies in ruminants. *J. Anim. Sci.* 74:3112–3124.

**TEMPLE, S. A.** 1977. Plant-animal mutualism: Coevolution with dodo leads to near extinction of plant. *Science* 197:885–886.

**TIEDEMAN, J. A., and D. E. JOHNSON.** 1992. *Acacia cyanophylla* for forage and fuelwood in North Africa. *Agroforestry Systems* 17:169–180.

**UHL, C., and G. PARKER.** 1986. Our steak in the jungle. *BioScience* 36:642.

**ULLREY, D. E.** 1992. Basis for regulation of selenium supplements in animal diets. *J. Anim. Sci.* 70:3922–3927.

**VAN DYKE, L. S., J. W. PEASE, D. J. BOSCH, and J. C. BAKER.** 1999. Nutrient management planning on four Virginia livestock farms: Impacts on net income and nutrient losses. *J. Soil and Water Cons.* 54:499–505.

**VAN GINKEL, J. H., A. P. WHITMORE, and A. GORISSEN.** 1999. *Lolium perenne* grasslands may function as a sink for atmospheric carbon dioxide. *J. Environ. Qual.* 28:1580–1584.

**VAN HORN, H. H., G. L. NEWTON, and W. E. KUNKLE.** 1996. Ruminant nutrition from an environmental perspective: Factors affecting whole-farm nutrient balance. *J. Anim. Sci.* 74:3082–3102.

VAN SOEST, P. J. 1994. *Nutritional ecology of the ruminant.* Cornell University Press, Ithaca, New York.

WATSON, A. J., D. C. E. BAKKER, A. J. RIDGWELL, P. W. BOYD, and C. S. LAW. 2000. Effect of iron supply on southern ocean $CO_2$ uptake and implications for glacial atmospheric $CO_2$. Nature 407:730–733.

WILKINSON, J. M., J. HILL, and C. T. LIVESEY. 2001. Accumulation of potentially toxic elements in the body tissues of sheep grazed on grassland given repeated applications of sewage sludge. *Animal Science* 72:179–190.

WILSON, E. O. (Ed.). 1988. *Biodiversity.* National Academy Press, Washington, D.C.

WILSON, J. R. 1996. Shade-stimulated growth and nitrogen uptake by pasture grasses in a subtropical environment. *Aust. J. Agric. Res.* 47:1075–1093.

WU, L., Z.-Z. HUANG, and R. G. BURAU. 1988. Selenium accumulation and selenium-salt cotolerance in five grass species. *Crop Sci.* 28:517–522.

ZHU, J. 2000. A review of microbiology in swine manure odor control. *Agric. Ecosyst. Environ.* 78:93–106.

ZHU, J., and L. D. JACOBSON. 1999. Correlating microbes to major odorous compounds in wine manure. *J. Environ. Quality* 28:737–744.

# CHAPTER 7

# Livestock Grazing and Rangeland Issues

**CHAPTER OBJECTIVE**

To discuss contentious issues involving livestock grazing on arid and semiarid rangelands:
—desertification
—watersheds, wetlands, and riparian areas
—public lands grazing in the western United States
—Yellowstone National Park as an example of grazing management
—African grasslands
—Asian grasslands
—Australian grasslands
—wildlife habitat and livestock
—preservation of endangered species
—predator and pest control programs
—wild horse issues

## OVERVIEW—LIVESTOCK GRAZING AND RANGELAND HEALTH

Livestock grazing on rangelands in western North America is controversial, with environmentalists and ranchers occupying opposite ends of the opinion spectrum. We should delight in this controversy for the sake of future generations. We still have time in North America to alter our patterns of resource utilization to avoid the devastation that has occurred in some other parts of the world.

The Middle East—The Mediterranean area—is the cradle of civilization and the site of the mythical **Garden of Eden.** Five thousand years ago, it was a land of milk and honey. Today, much of the area is desert, with severe erosion and total loss of soil with nothing but exposed bedrock. A major contributing factor to the destruction first of forests, then of shrubs and brush, and then perennial grasses has been livestock grazing. Migratory herders and their flocks of goats, sheep, cattle, camels, and equines have been responsible for much environmental degradation. A grazier's wealth is traditionally determined by the size of his herds and flocks and not by their productivity. In the last hundred years, the great increase in human population and consequently in numbers of domestic animals has resulted in an almost total destruction of the remaining forests and rangelands in the Mediterranean region (Thirgood, 1981). He describes the Jordan as follows:

> Grazing means an intense concentration of sheep, goats, cattle, camels, and some equines. There is remarkably little vegetation that can escape such an onslaught. The lean and hungry goats, after years of practice, have developed arboreal habits, whilst the camels already have a high reach. . . . The effects on soil are equally disastrous. The surface is maintained as a slippery mass of rubble and soil exposed to the ravages of accelerated erosion . . . (p. 76)

In the mountains, the remaining forests are hacked down to provide goat fodder. Such has been the fate of the Garden of Eden.

Perevolotsky and Seligman (1998) have a somewhat different perspective. They suggest that traditional grazing practices in the Mediterranean region are an efficient form of land use and are ecologically sound. They claim that heavy grazing of woody vegetation in that region is necessary for fire prevention and maintenance of habitat diversity. Heavy grazing of tree species such as oaks converts them to a topiary-like shrub form, creating a large surface area that provides forage during the long, dry summers. When grazing is eliminated, these areas convert to fire-susceptible impenetrable thickets.

Aldo Leopold, the renowned American conservationist, forester, and founder of The Land Ethic, said this about parts of the U.S. Southwest:

> This region, when grazed by livestock, reverted through a series of more and more worthless grasses, shrubs and weeds to a condition of unstable equilibrium.

Each recession of plant types bred erosion; each increment to erosion bred a further recession of plants. The result today is a progressive and mutual deterioration, not only of plants and soils, but of the animal community subsisting thereon. . . . So subtle has been its progress that few residents of the region are aware of it. It is quite invisible to the tourist who finds this wrecked landscape colorful and charming. (Leopold, 1949) (p. 242)

Milton et al. (1994) have described this process of stepwise rangeland degradation in more scientific terms, but the end result remains: **human-made deserts.**

In this chapter, controversial rangeland-livestock issues will be discussed, with major emphasis on the rangelands of western North America. Brief consideration will be given to rangeland issues in Africa, Asia, and Australia as well.

# RANGELANDS, LIVESTOCK, AND DESERTIFICATION

Rangelands are defined as those areas where the native vegetation is predominantly grasses, shrubs, or open woodlands (Kauffman and Pyke, 2001). Rangelands account for 61 percent of the land surface of the United States and 70 percent of all the land surface of the world (Fuhlendorf and Engle, 2001). They include grasslands, shrublands, savannas, open woodlands, and most desert, tundra (arctic and alpine), meadow, wetland, and riparian ecosystems. They tend to be lands unsuited for purposes other than grazing—they are too wet, too dry, too cold, or too high for cultivation. Livestock grazing is the most ubiquitous human activity on Earth, occurring on more area than any other land use (Kauffman and Pyke, 2001). Hence livestock grazing is one of the main activities affecting global biodiversity. An estimated 84 percent of the mammal species and 74 percent of bird species in the

United States are associated with rangeland ecosystems (Kauffman and Pyke, 2001). Public lands in the United States are home to more than 3,000 wildlife species, including about a third of the nation's threatened or endangered species. Therefore it is not surprising that livestock production, and especially cattle ranching, is at the forefront of environmental debates and battles.

In addition to direct effects of livestock grazing on vegetation and soil factors, there are other effects that have profound environmental repercussions. Livestock production has facilitated the invasion of rangeland by exotic plant species and pathogens. In much of the U.S. range communities, areas formerly dominated by perennial bunch grasses are now dominated by exotic annual grasses (e.g., cheatgrass, medussa head). Grazing has altered fire regimes by removing fuels (dried grasses) that carry fire in rangelands and savannas, allowing invasion of trees (e.g., juniper) and shrubs (e.g., mesquite). A major consequence of livestock grazing has been the elimination of predators and other animal "pests" such as prairie dogs by ranchers and "pest control" specialists. The wolf and grizzly bear have been almost completely eliminated from the continental United States by stock raisers. Ironically, this has led to a proliferation of the wily coyote across the entire country (see Predator and Pest Control Programs, this chapter).

Grasslands and large grazing herbivores coevolved. Evidence for this, mainly from the fossil record, is discussed in detail by Stebbins (1981). The fossil remains of both grasses and mammals allow a detailed interpretation of their changes throughout geologic time. The fossilized seed parts, pollen, and leaf epidermis provide information on the grass species and structure, while the teeth and hooves of fossilized herbivores are highly diagnostic of their feeding behavior and mode of life. High crowned (hypsodont) teeth with complex patterns of enamel evolved in response to eating the hard, silica-containing leaves of grasses, while long, slender legs terminating in one or two-toed hooves are characteristic of grazing animals that inhabit grasslands or lightly forested areas (savanna) (Stebbins, 1981; MacFadden, 1997).

The concept of **coevolution** of grasses and herbivores implies that grasslands and large grazing herbivores are mutually dependent, and each benefits the other. This is obvious in terms of the herbivores: grasslands provide grazing herbivores with food and habitat. What is not so easily recognized is that grasslands may benefit from the periodic removal of the vegetative material by grazing (or fire). By periodically removing the vegetative material (grazing) and breaking up the mature plant growth and soil crust (trampling), grazing animals invigorate and renew grasslands and have done so for the past 45 million years of grassland-herbivore coevolution. Herbivores help to create and maintain the environment in which they live (Jacobs et al., 1999). The diverse array of herbivores on the plains of Africa, the vast bison herds of North America, and the mobs of kangaroos and other grazing marsupials in Australia have maintained these grasslands over eons of time. Thus the calls for elimination of grazing livestock on rangelands are not based on biological principles. Grazing per se does not destroy grasslands and rangelands. Both grazing and periodic fires have maintained grasslands and prevented invasion by trees and shrubs. The major threat to grassland ecosystems today is their potential for conversion to farmland; for example, the native grasslands of North America have largely been converted to cropland. Most rangelands remaining in the United States are unsuitable for crop production (Fig. 7-1). The grasslands of Africa, with their vast herds of many species of wild herbivores, are under increasing pressure for conversion to cropland to feed Africa's rapidly increasing human population.

Grazing by domestic livestock has without question caused tremendous degradation of the world's grasslands in the past 200 years (Fleischner, 1994). On the other hand, in North America, the principal degradation of

**FIGURE 7-1** Much of the world's land area is too arid or otherwise unsuitable for growing crops. Livestock production is the primary means of utilizing these areas for the production of human food. However, use of rangelands by livestock is an increasingly controversial issue.

grasslands of the prairies and great plains was accomplished not by livestock but by other human activity, mainly farming and urbanization. The homesteading of the American West, facilitated by John Deere's invention of the sod-busting moldboard plow, resulted in the reduction of native tall- and short-grass prairie grasses to a tiny remnant. The vast grasslands and their coevolved herds of bison (and Native American peoples) were largely destroyed and replaced by homesteads. In the 1930s, the rains didn't come, the wind blew incessantly, and the topsoil formed over thousands of years was blown east as far as the Atlantic Ocean. The **dustbowl** of the American West was an ecological disaster, caused not by grazing but the destruction of grasslands by plowing and cultivation to grow wheat and other crops (Fig. 7-2).

Although soil erosion today is not so dramatic as in the dust bowl days, it continues to be a problem. It remains to be seen whether conservation tillage and other farming practices can sustain crop production on these lands indefinitely. It is known that grazing animals sustained the grasslands, with little apparent environmental damage, for thousands of years. Before the European settlement of the North American grasslands, it is estimated that there were about 60 million bison, 40

million white tailed deer, 40 million pronghorn, 10 million elk, and innumerable small herbivores such as prairie dogs, jackrabbits, and cottontail rabbits (Heath and Kaiser, 1985).[1] The Serengeti plains of East Africa have supported millions of various antelope species, wildebeest, giraffe, elephant, and zebra. If nature could manage these grasslands with intensive grazing programs, it is not inconceivable that humans could eventually learn to do so also! (Actually, according to Frank et al. [1998], there are large functional similarities between the Serengeti and Yellowstone ecosystems. In both cases, indigenous peoples played a critical role in shaping these ecosystems [Kay, 1998]. Perhaps the earlier statement should be reworded: "If indigenous peoples could manage these grasslands, it is not inconceivable that people of European origin could eventually learn to do so also!")

These remarks are intended to provide a background for the current controversy between the range livestock industry and environmentalists in the western United States. "Cattle-free by ninety-three" was a rallying cry for environmentalists (countered by "cattle galore in ninety-four" by ranchers). Livestock grazing on public lands has become a major controversial issue. Radical environmentalists have engaged in acts of vandalism and violence, including the burning and destruction of ranch property and livestock auction yards, shooting holes in water tanks and vandalizing windmills, threatening ranchers and their families, and so on. Ferguson and Ferguson (1983), in their book *Sacred Cows at the Public Trough*, declared: "No industry or human activity on

---

[1] Historical estimates of animal numbers should be viewed prudently, recognizing the inherent difficulties of making such estimates, even with modern technology. Shaw (1995) evaluated numerous presettlement estimates of bison numbers. He concluded that the most accurate estimate that could be made was that the bison population numbered in the millions, and probably in the tens of millions. Shaw (1995) states "Any greater accuracy seems unlikely" (p. 150).

**FIGURE 7-2** (*Top*): The 1930s Oklahoma dust-bowl. The soil has taken to the air (note low visibility) and formed drifts around fences and buildings. (Courtesy of USDA Soil Conservation Service.) (*Bottom*): Farm machinery in South Dakota buried by the blowing dust. Conservation practices, including crop rotations with forages for livestock, can prevent such tragedies from happening again. (Courtesy of USDA Soil Conservation Service.)

earth has destroyed or altered more of nature than the livestock industry" (p. 4). Aside from the fact that this statement is clearly incorrect (it can reasonably be argued that the human activity that has destroyed more of nature than anything else is our reproductive activity!), there is no question that improper livestock production has caused extensive environmental damage and rangeland degradation. Why has it done so, if grasslands and herbivores co-evolved? The answers to this question vary in different parts of the world, so a few particular areas will be discussed later in this chapter.

## Livestock, Watersheds, and Riparian Areas

A **watershed** is the area drained by a distinct stream or river system and separated from other similar systems by ridgetop boundaries. The entire land surface from which water ultimately drains into a particular stream or river system is a watershed. Watersheds function in capturing water, storing it in the soil, and slowly releasing it into springs and seeps and ultimately into streams and rivers. Watershed quality is very important to human existence, because it will influence flooding, river levels, water storage for human use, and vegetation patterns. The ability of a watershed to capture, store, and safely release water may be heavily influenced by livestock production, mainly in those areas that are rangelands. This would include most of western North America and much of Australia, non-equatorial Africa (the Sahel and savanna of southern Africa), the grasslands of South America, and the steppes and plains of Europe and Asia.

Livestock on rangelands can influence watershed quality by their effects on vegetation and soil structure (Trimble and Mendel, 1995). The infiltration and retention of moisture by rangelands depend upon their vegetative cover. Plant cover and plant litter on the soil surface improve water infiltration by reducing raindrop impact on the soil surface and reducing soil crusting. The aboveground vegetation and surface litter, as well as roots and other below-ground organic matter, improve water infiltration and soil moisture storage. Poor grazing management can seriously impair these processes, leading to less water infiltration and more runoff and erosion. The desirable vegetation characteristics can be maintained by vigilant grazing management. Adequate plant cover also helps to trap and retain snow. Water may penetrate to lower depths by following plant root channels. The process is aided by burrowing animals and insects associated with the plant cover. Friable, non-compacted soil aids water infiltration. Plant species composition is important. Some plants such as western juniper on western U.S. rangelands are notorious for wasting water by high rates of transpiration and outcompeting other plants that better encourage water infiltration and prevent soil erosion. Generally, many years are required for natural recovery of soil physical conditions after cessation of grazing in drier climates (Greenwood and McKenzie, 2001).

**Grazing management** is an important determinant of watershed and rangeland quality. Overgrazing or improperly timed grazing reduces plant cover, may change the species composition to plants less effective in promoting water capture and retention, exposes bare ground and trails to encourage rapid water runoff, and may cause soil compaction, reducing infiltration and increasing runoff. Overgrazing results in the simplification of stream channels, which may include channel straightening, increased channel width, and decreases in deep pools (Kauffman and Pyke, 2001). Improved grazing management is needed on many rangelands to improve watershed quality. Grazing should be managed on the basis of the watershed and range condition, rather than on the needs of the livestock. This management can be achieved by consideration of livestock type (e.g., sheep vs. cattle, cows and calves vs. yearlings, etc.), stocking density, grazing duration, recovery period, time of year grazing allowed, and so on. The use of rangelands by livestock is coming under increasing societal scrutiny and concern. Livestock producers and range management specialists must have an appreciation of multiuse of rangelands, even when the land is privately owned. The management practices on private land affect watershed quality and have an impact on adjacent properties and downstream effects.

In the western United States, heavy livestock grazing has resulted in considerable changes in the vegetation. In the late 1800s and early 1900s, uncontrolled grazing by sheep and cattle negatively affected grasslands, allowing sagebrush and juniper trees to invade. Juniper trees have an extensive root system and are inefficient in use of water. Rangeland invaded by juniper becomes unproductive for livestock and wildlife, and its watershed value is decreased, because surface water fails to infiltrate the soil and may cause severe erosion (flash floods), permanently reducing soil productivity. Control of the juniper by fire or herbicides often results in dramatic increases in forage production and the renewed flow of water from springs that have been dry for years. The proliferation of sagebrush and juniper on western ranges has been due to suppression of fire as well as to overgrazing in the past (Belsky, 1996).

A specific aspect of watershed management that is also receiving much scrutiny, concern, and criticism is the impact of livestock on riparian areas. A **riparian area** is the zone between a river or stream and the nearby upland areas. On semiarid rangelands, the riparian areas are usually zones of green, lush vegetation compared to the surrounding arid rangeland. Compared to the surrounding upland area, riparian areas are much more productive and are major sources of forage for livestock and wild animals. They are a major habitat for many types of wildlife, fish, and other aquatic organisms. They help to control flooding and increase groundwater recharge by providing a floodplain to accumulate water when streams overflow their banks. They trap sediments and slow flow rates, thus reducing downstream flooding and silt deposition.

Riparian areas are important in livestock production because they are productive, have water for drinking, usually have trees for shade, and the land is usually flat whereas surrounding areas may be steep. Cattle often will heavily overuse riparian areas and avoid the surrounding uplands. Heavy use of riparian areas by cattle can (and usually does) cause severe environmental damage. The streambanks are eroded, with caving in of the banks, widening the channel (Fig. 7-3). Fish habitat is destroyed. Trees and shrubs are destroyed or prevented from regeneration. For example, aspen trees in

**FIGURE 7-3** A riparian area on an Oregon range. Note the instability of the bank, with caving-in and erosion. A controversial issue in rangeland management is whether or not livestock should be excluded completely from such areas, by fencing, to allow growth of stream bank vegetation. The trees in the background are Ponderosa pine; cattle graze on grass growing in the undergrowth of the pine forest.

riparian areas cannot regenerate with heavy cattle grazing pressure, because the palatable saplings are eaten and killed. The loss of streambank vegetation reduces the amount of shade, often causing water temperatures to rise to unacceptable levels for cold water fish such as trout and young salmon (Beschta, 1997). Unmanaged and mismanaged livestock grazing generally results in overuse and degradation of riparian areas (Fitch and Adams, 1998).

The effect of livestock on riparian zones is a major area of controversy in western U.S. range cattle production (Belsky et al., 1999). Environmentalist groups urge or demand the complete removal of livestock from riparian zones, whereas ranchers naturally wish to continue to utilize these productive areas. For example, in 1996, Oregon ranchers and environmentalists battled for the public mind over a legislative ballot initiative (Ballot Measure 38) that would have required that livestock be fenced away from virtually all streams and rivers in the state. Although ranchers won this one (the measure was defeated), environmentalists prepared a "son of Measure 38" for the next election (also defeated). Probably a more productive route than the legislative one would be for ranchers and environmentalists to work together to try to solve these types of issues. The restoration and maintenance of riparian ecosystem functions and values will depend not on general grazing prescriptions, but on developing grazing prescriptions for specific riparian areas (Clark, 1998; Green and Kauffman, 1995). The best prescription for stream recovery is a long period of rest from livestock grazing (Belsky et al., 1999). The lack of landscape-scale riparian systems that have never been grazed by livestock means that there is a lack of clear ecological benchmarks against which the effects of grazing can be measured (Dobkin et al., 1998).

By use of fencing and properly timed grazing, livestock and riparian areas can usually coexist. In the western United States, sheep have less impact on riparian areas than do cattle (Kauffman and Krueger, 1984). They are normally under the control of a herder and can be herded away from riparian areas. Also, sheep have a natural inclination to climb to higher elevations when grazing and so would tend to move up on surrounding hillsides. Cattle can be herded out of riparian zones, and by intensive management the area can be used for grazing without major adverse effects (Butler, 2000).

Development of off-stream stock watering tanks can attract cattle away from riparian areas. Solar-powered pumps can be used to pump water from a stream to a water tank some distance away (Chamberlain and Doverspike, 2001). Nose pumps are also used. Cows bump their noses on a metal flap while drinking; the flap moves, moving a cylinder to pump water. Both solar-powered and nose pumps have the advantage of not requiring an external power source, and the disadvantage of being fairly expensive (compared with letting cattle drink out of a stream). McInnis and McIver (2001) also noted that provision of off-stream water and salt reduced the time that cattle spent in riparian areas, but it was difficult to document an actual improvement in riparian condition.

## Water Wars: Fish, Cows, Farms, and People

In the U.S. Pacific Northwest, the marked decline in salmon populations is of major concern. Salmon runs (spawning migrations) on the Columbia and Snake Rivers and their tributaries are declining precipitously. Some wild populations are virtually extinct; one run on the Snake River system was reduced to a single female salmon in 1991. Various causes for the marked decline in salmon populations have been advanced, including climatic change (El Niño), drift netting on the high seas, commercial fishing off-shore and in the rivers, sports and Indian fisheries, hydroelectric dams and turbines, irrigation, and damage to riparian zones by livestock grazing. The latter two factors are related to agriculture and livestock production. Large numbers of young fish are killed when they are drawn into irrigation canals and pumps. Salmon spawn in gravel beds in small streams. Damage to riparian zones by livestock grazing (and logging activities) can cause damage to the gravel beds, silting of the gravel, and increased water temperature due to reduced stream flow and loss of streambank vegetation. Improvements in grazing management of riparian areas, including exclusion fencing in some cases, can be used to minimize the impacts of livestock on the Pacific salmon (Armour et al., 1994; Elmore and Beschta, 1987; Kauffman et al., 1997).

The reduction in, and threatened extinction of, salmon populations is a major social, political, and economic issue in the Pacific Northwest states, with finger-pointing and assigning blame to commercial and sports fishers, farmers, ranchers, environmentalists, Indian tribes, hydroelectric companies, river transportation (barges) and even city dwellers (whose runoff of lawn fertilizers and chemicals enters storm drains and ultimately rivers). The **Endangered Species Act (ESA)** is a favorite target of resource-based industries. As this is being written (2001), the region is experiencing a severe drought and California is

enduring energy (electricity) shortages and blackouts. Demands for hydroelectric power from the Columbia River are conflicting with the water needs of salmon for migration (upstream for adults; downstream for juveniles) and spawning. In some areas, such as the Klamath Basin of Oregon and Northern California, irrigation water has been turned off to retain water for fish survival. Farmers and ranchers blame the fish and the ESA for their plight. The Water Wars have begun throughout the American west. It is becoming increasingly obvious that demand exceeds supply, particularly during droughts. Rapid urbanization is occurring, further increasing water demand. Farmers and ranchers will increasingly feel that they are "up against the wall." For this, they tend to blame environmentalists and endangered species. Who needs fish? Who cares if they go extinct?

It is being recognized that the ecological importance of salmon is profound. They are a "keystone" species, on which the health of an entire ecosystem depends (Willson et al., 1998). Spawning salmon migrate hundreds and even thousands of miles up Northwest rivers. After spawning, they die. In this process, they transport nutrients from the ocean far inland (Fig. 7-4). Eagles, bears, and other animals that feed on salmon deposit nutrients throughout the forests. Fish carcasses in rivers are a major nutrient source for invertebrates, which are at the base of the food chain (Hilderbrand et al., 1999). These invertebrates serve as food for fish in the streams, including juvenile salmon. Bald eagles (America's national symbol) depend upon salmon, as do grizzly bears (Cederholm et al., 1999; Hilderbrand et al., 1999). Gresh et al. (2000) estimated that just 6–7 percent of the marine-derived nitrogen and phosphorus once delivered to the rivers of the Pacific Northwest is currently reaching those streams. They suggest that this nutrient deficit may be one indication of ecosystem failure that has contributed to the downward spiral of salmon abundance and may diminish the possibility of salmon recovery. The cultures of many Native Americans are closely linked to salmon. Extinction

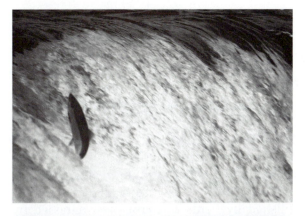

**FIGURE 7-4** Pacific salmon migrate hundreds and even thousands of miles upriver to spawn, surmounting huge obstacles such as waterfalls *(top)*. They are a food source for many species such as the grizzly bear *(middle)*, so that nutrients from the ocean are distributed far inland. Thus the salmon are a key link in a biogeological cycle (see Chapter 11). Construction of hydroelectric dams *(bottom)* on major salmon-producing rivers has resulted in the listing of some species of Pacific salmon as endangered species. (Courtesy of Winifred B. Kessler.)

of salmon would cause a complex food web to collapse. On this, as well as on many other environmental fronts, everyone is going to have to "bite the bullet." Ranchers will be involved in such issues as riparian zone health, other effects of grazing, elevated stream temperature issues, use of water for irrigation, and so on. The Water Wars are just beginning in earnest, and already they're not a pretty sight!

Another current political issue in the United States is that of **wetlands.** The U.S. government has made a commitment to preserve wetlands and to have no net decrease in wetland area. Wetlands are important habitats for wildlife and have watershed value. Livestock grazing on wetlands can have impacts on wildlife habitat, bird nesting, and so on. In the western United States, use of wetlands for livestock production is another rancher-environmentalist battle zone. As with other issues, there are conflicting effects and viewpoints. Livestock on these areas may adversely affect some wildlife species and benefit others. Grazing, for example, often improves the feeding and nesting habitat for ducks and geese. Most of the loss of wetlands in North America has been due to urbanization, creation of industrial sites in estuaries, and draining of prairie potholes to grow wheat for human consumption.

The Netherlands (Holland) is famous for its reclamation of farmland from the sea and low-lying areas. Diking and reclamation of coastal salt marshes began in Holland about 1,000 years ago (Wolff, 1992). A large part of Holland, before human intervention, was wetlands. The two major Dutch reclamation projects in the twentieth century were the Delta and the Zuiderzee projects. Both were opposed by some people because of potential damage to fisheries but there was no significant opposition based on environmental considerations. Since about 1960, several new reclamation projects have been proposed, all of which have been blocked by environmental concerns. According to Wolff (1992), it is unlikely that any further reclamation of wetlands in the Netherlands will occur, and in fact the trend is now to reconvert reclaimed land back into wetlands. This is justifiable on the basis that even though Holland is a densely populated country, its agricultural production greatly exceeds domestic needs. Less than 5 percent of the populace is involved in agriculture, and the remaining 95 percent of the people have an increasing demand for recreational activities. It appears that the 1,000 year Dutch tradition of reclaiming land from the North Sea is over (Wolff, 1992).

## Western North American Rangelands

As previously discussed, the major North American grasslands of the Great Plains were changed mainly by the plow, not by livestock. In the intermountain and far-west rangelands, the land was too dry, too high in elevation with a short growing season, or too rugged in terrain to support farming, although homesteaders tried valiantly to make farms. Some of the degradation of these rangelands was caused by homesteaders and the plow. The other major factor was uncontrolled grazing. In the late 1800s and early 1900s, cattle and sheep were grazed unmercifully on western rangelands, with huge herds of free-ranging cattle and thousands of bands of sheep under the control of itinerant sheepherders. It was a classic case of **the tragedy of the commons** (Hardin, 1968). The land belonged to no one, so it belonged to everyone. Each rancher or sheepherder tried to get to the forage first before his competitors. The land was overgrazed, regrazed too soon, and grazed continuously without any opportunity for the plants to recover. The rangeland vegetation suffered serious damage, from which it has not yet fully recovered. Suppression of fire has also contributed to changes in rangeland vegetation and the proliferation of shrubs (e.g., sagebrush) and trees (e.g., juniper).

Perhaps the obvious solution to reverse the degradation would seem to be to eliminate livestock grazing from these rangelands. Like most obvious solutions, this one probably won't work either. Savory (1988) has discussed in detail the practical relationships between grazing management and maintenance and recovery of rangelands. The Savory concepts are intuitively logical but not accepted in their entirety by many range specialists. In brief, the Savory concepts are built upon grassland:herbivore coevolution. Grasses require periodic grazing to remove the vegetative material, and trampling to break down old mature material and break the soil crust to allow water infiltration and growth of new seedlings. This historically was accomplished by herds of herbivores, harassed by predators. Predators keep grazing animals in a compact herd, so they intensively graze the areas they are in, removing the forage thoroughly and trampling detritus (accumulations of dead plant tissue) into the soil. Periodic attacks by the predators may cause the herd either to stampede or to bunch together closely. This results in a lot of hoof action that breaks up old dead plant material and the soil surface. For example, on the plains of Africa, herds of antelope, wildebeest, and zebra migrate in an established pattern, following the rains. Numerous predators (lions, cheetahs, hyenas, etc.) periodically harass the herds, causing considerable trampling and hoof action. The essential features of this natural system are the periodicity of grazing (Fig. 7-5), allowing the plants to recover, and the physical herd effect of trampling. In North America, huge herds of bison behaved in a similar way. They had seasonal migrations and frequent stampedes when harassed by predators such as wolves or the Plains Indians. Savory (1988) has built upon observations of natural grasslands to develop a methodology for **short-duration planned grazing.** Animals are kept at a high density by creating small paddocks using electric fencing (Fig. 7-6). The high animal density provides for short-duration grazing pressure and the trampling effects. A series of pastures (paddocks or grazing cells) are used, allowing the grazed forage sufficient recovery time before being grazed again. His entire program is called **holistic resource management (HRM).** A brief biography of Allan Savory and a description of his philosophies of range management are given by

**FIGURE 7-5** An example of intensive grazing management on semiarid rangeland in South Africa. The range on the left portion of the hillside is managed by short-duration high intensity grazing. The land on the right and in the foreground, owned by a different rancher, is grazed continuously. The forage and other vegetation has been largely eliminated by overgrazing in this poorly managed pasture.

**FIGURE 7-6** Intensive grazing of dairy cattle in New Zealand, using a self-propelled electric fence. (Courtesy of Leith Pemberton.)

Hadley (2000), in an article, "The Wild Life of Allan Savory." A new edition of Savory's book, now entitled *Holistic Management*, has been published (Savory and Butterfield, 1999).

The HRM concept is somewhat controversial among range specialists and animal scientists, some of whom are highly negative to certain aspects of it (Holechek et al., 2000). Savory, considered a maverick and nonconformist, is not above strongly criticizing

the research of many range scientists. They, in turn, are not above taking delight in proving that HRM doesn't work. One of the major contributions of HRM has been to stimulate ranchers and resource managers to think about grazing management and how they can improve what they are doing. The planning, monitoring, and goal setting involved in HRM have improved the lives of ranchers as well as improving range management.

In my opinion, many aspects of HRM are valid in concept, and proper grazing management is critical in maintaining the vigor and integrity of rangelands. Elimination of grazing, in most cases, is not a means of accomplishing these goals. Savory (1988) and others (e.g., Knapp and Seastedt, 1986) have documented that exclusion of grazing animals from degraded rangeland not only does not enhance its recovery, but results in further degradation. Without periodic grazing and soil disturbance, the grasses become degenerate and smothered by detritus, and regeneration is inhibited. The concept of whether or not grasslands benefit from grazing, especially by livestock, is very controversial. Savory (1988) is a leading proponent of the idea that herbivory and soil disturbance by hoof action are essential for maintaining healthy grasslands. However, studies such as that of Brady et al. (1989) are not supportive. These authors found that exclusion of livestock from a relatively brittle (see Savory for discussion of brittle vs. non-brittle grasslands) grassland for 16 years did not show any evidence of loss of biodiversity or deterioration of the plant community. Belsky and colleagues (1993; Painter and Belsky, 1993) are leading proponents of the concept that grasses do not benefit from being grazed. They claim that ecological research suggesting such benefits has been used to justify what they term "heavy livestock grazing on western North American rangelands" (p. 2). McNaughton (1993), one of the most prominent researchers studying grassland-herbivore interactions, refutes Belsky's claims by pointing out that certain low-growing grasses disappear from grasslands protected from grazing:

We must conclude that plant species and genotypes that disappear when grazers are removed are, defacto, benefitted by grazing. . . . Therefore, grazing benefits many grasses and other plants in grassland ecosystems, including those studied by Belsky. . . . Furthermore, moderate grazing promotes the productivity of many grasslands above the levels that prevail in the absence of grazing. If these statements are not true, then I believe the sun rises in the west, Earth is flat, and water flows uphill unaided. (pp. 19–20)

Clearly, the benefit or lack thereof of grazing on grasses is controversial!

Savory (1988) contends that livestock grazing on arid lands is necessary to break up the surface crust of lichens in order to allow grasses to become established. On the contrary, environmentalists contend that this surface crust, variously called the **cryptogamic soil crust,** microbiotic crust, microphytic crust, and cryptobiotic crust, plays very important roles in rangeland ecology (Evans and Johansen, 1999). These crusts (Fig. 7-7) are composed of nitrogen-fixing cyanobacteria (blue-green algae) associated with lichens, mosses, microfungi, and bacteria, which are members of the family Cryptophyceae. Environmentalists claim that the crusts are a major source of soil nitrogen in arid regions and are important for the growth of higher plants. They claim that cattle trampling destroys the crust, leading to soil erosion and loss of fertility (Beymer and Klopatek, 1992). The crusts bind soil particles together, increasing surface soil stability and resistance to erosion. Hoof impact has been observed to adversely affect cryptogamic crusts in Australia (Hodgins and Rogers, 1997). The postulated importance of the cryptogamic crust and the impacts on it of human activities, including livestock grazing, are reviewed in a symposium on soil crust communities (St. Clair and Johansen, 1993) and in a review article by Evans and Johansen (1999). This is another unresolved contentious issue involving livestock production on rangelands.

**FIGURE 7-7** The cryptogamic soil crust is composed of N-fixing blue-green algae associated with lichens, microfungi, mosses, and bacteria. They are believed to be a major source of soil nitrogen in arid regions. Environmentalists claim that the crusts are destroyed by cattle. In the photo, the crust is the cracked, bare ground-appearing material between the clumps of bunchgrass. (Courtesy of J. A. Belsky.)

One of the keys to improved rangelands is good grazing management, not lack of grazing. This concept is not new. I can't improve on the remarks of a pioneer Oregon rancher on overgrazing (Oliver, 1961):

All kinds of things are called overgrazing, some of which are just poor management. That is, one man can hurt a range with 100 cattle, whereas another man can have the same set-up and his range may improve. A lot of so-called overgrazing is grazing too early. Let's say a range on April 1 has enough feed for 50 cattle. By April 20, it may have enough for 100 cattle, and by May 20 enough for 150. But if 100 are turned in early, then the animals take all the feed there is, trample the soft, muddy ground, damage the grass, and the cows will run foot races for each new spear as it shows up. Two or three years of this may pretty well ruin a range, and people say it is overgrazed. Actually, it has been grazed at the wrong time, because another man might see the range improve with 150 cattle on it, turned in later. (p. 114)

In the western United States, much of the land is not privately owned but is the property of the U.S. government and is administered mainly by the Bureau of Land Management (BLM) and the U.S. Forest Service. In some western states, such as Nevada, over 70 percent of the land is federally owned. Much of the BLM-administered land is semiarid rangeland. In the latter part of the nineteenth century, the western United States was opened up for homesteading. A person could claim a tract of uninhabited land and would receive title after fulfilling some requirements, including living on the farm, clearing some land for farming, and so on. Typically, a homesteader would stake his claim near a spring or water source. Most of the rangeland was unsuitable for farming, being too dry, too erodible, growing season too short, and so on. The homesteaders were doomed to failure, and fail they did. In the meantime, the remaining rangeland that was unsuitable for even attempts at farming was used for grazing by cattle and sheep. Ranchers and homesteaders fought over the springs and available water. After the homesteaders failed, the ranchers bought the abandoned farms to get the water rights for their livestock. The grazing land was not purchased by the ranchers, because of its extremely low carrying capacity (10–100 acres per cow), and there was no competition for its use—no one else wanted it. This publicly owned land eventually came to be administered by the BLM, and ranchers were sold permits for grazing rights to control livestock numbers. Thus many western ranches now operate with a privately owned base, often derived from land once homesteaded, with much of their grazing land leased from the BLM or the Forest Service.

In recent years, **public lands grazing** of livestock has become very controversial (Vavra, 1998). One issue is economic. Critics claim that the BLM and Forest Service fees are too low and are a subsidy to livestock producers. They claim that the BLM spends more in administering the leases than it takes in as grazing fees. Ranchers and many range spe-

cialists claim that the fees are fair, because ranchers incur numerous extra expenses such as fencing and development of water sources that would be paid by the landlord if they leased private land. Thus they claim that it is inevitable that BLM fees are lower than those charged by private land owners, but when all costs are allocated, the fees are similar and fair. This argument will no doubt rage on for years: are the grazing fees a subsidy or not? This issue has been reviewed from the ranchers' perspective by Collins and Obermiller (1992).

Holechek and Hess (1995) advanced the position that subsidies to ranchers by way of the **Emergency Feed Program (EFP)** contribute to overgrazing and rangeland degradation. This subsidy reimburses ranchers on both private and public land for 50 percent of the cost of additional feed needed to sustain their livestock during drought and other disasters. Holechek and Hess (1995) suggest the EFP gives ranchers a financial incentive to stock their ranges for high precipitation years instead of for below average years. It makes non-sustainable grazing profitable, by having the government (taxpayers) pay for extra feed when the range is overstocked, rather than forcing ranchers to reduce animal numbers to what the range can support on a sustainable basis. Holechek and Hess (1995) conclude "We believe the long term welfare of ranchers will best be served by allowing market forces to function freely without government intrusion" (p. 136). Ironically, although the American West has a reputation for independence and a desire to be free of government interference, ranchers have been as willing as anyone else to take advantage of government handouts such as the EFP. In a more recent report, Holechek et al. (1999) concluded that conservative grazing of about 30–35 percent use of forage will give higher livestock productivity and financial returns than stocking at grazing capacity. In favorable precipitation years, the extra income from stocking at capacity is small compared with the losses when drought inevitably occurs. Drought is an inevitable part of normal climate fluctuation and is a recurring, but unpredictable, environmental feature. Uncertainty in determining the beginning of a drought usually results in a lagged response in reducing stocking rates (Thurow and Taylor, 1999), leading to lowered vegetative cover and accelerated erosion following the drought. Hence, for long-term rangeland health, conservative stocking rates are essential.

Another basis for the movement to restrict or eliminate livestock grazing from public lands is ecological. Environmentalists claim that the rangelands are overgrazed, that livestock are causing extensive damage to the natural resources, and that domestic animals should be removed from public lands to allow the native fauna and flora to return to their **pristine state.** They claim that the amount of beef produced on western rangelands is only about 3 percent of total U.S. beef production and that elimination of the western range livestock industry would make little difference to U.S. agricultural productivity but would have a major beneficial impact on the environment (Ferguson and Ferguson, 1983). What constitutes a pristine state is debatable. In North America, the American Indians have had a major impact on many ecosystems for millennia. Their use of fire, for example, was an important factor in maintaining grasslands. One hypothesis to account for the extinction of megaherbivores such as woolly mammoths (Fig. 7-8) in North America after the last Ice Age is that they were hunted to extinction by the ancestors of American Indians, who came to North America across the Siberian land bridge. What many people regard as the pristine state in North America is the pre-European period. A common perception is that Native American management of ecosystems was natural, but European management was not (Hunter, 1996). Hunter (1996) suggests that we do not blindly accept the management practices of other people as sacrosanct just because they were here first. While the overall ecological impact of Native Americans was probably much less than that of European colonizers, it is perhaps relevant that Native American hunters played a major

**FIGURE 7-8** Woolly mammoths were megaherbivores in North America that became extinct after the arrival of the native Americans across the Siberian-Alaska land bridge. The probable cause of megaherbivore extinction was hunting by humans along with climatic change. (Courtesy of Royal British Columbia Museum, Victoria, B.C.)

role in the extinction of over 30 genera of large mammals, such as antelopes, mammoths, horses, and ground sloths, while the only large mammal extinctions attributable to European colonization were the Steller's sea cow and the Caribbean monk seal (Hunter, 1996). Because of continuous climatic changes, the pristine state is a moving target anyway. Attempts to manage or eliminate livestock grazing on western rangelands so we can return to their pristine, pre-European state are illogical and scientifically untenable.

The majority of the U.S. public lands where livestock grazing is a controversial issue are in the intermountain region, comprising parts of the states of Washington, Oregon,

Idaho, Utah, Montana, Wyoming, Nevada, Colorado, Arizona, and New Mexico. The ecological nature of these intermountain grasslands is considerably different from that of the plains, and the differences are at the base of a considerable amount of the controversy surrounding the livestock grazing issue (Jones, 2000). The ecological factors are reviewed in detail by Mack and Thompson (1982); a brief synopsis will be presented here.

The grasslands of the North American plains coevolved with bison (Fig. 7-9). These grasslands are dominated by rhizomatous and stoloniferous grasses, which are resistant to the grazing and trampling of large ungulates (the predictable result of coevolution). The intermountain grasslands, however, did not coevolve with herds of large ungulates such as bison.[2] These grasslands were dominated by bunch grasses such as *Agropyron* spp. (e.g., wheatgrasses). These grasses were very intolerant of heavy grazing pressure by introduced cattle, sheep, and horses. Furthermore, the introduced livestock species were very effective in widely disseminating alien European annual grasses such as cheatgrass (*Bromus tectorum*) and medussa head (*Elymus caput-medusae*). These annual grasses quickly became the dominant grass species throughout vast areas of intermountain rangelands,

**FIGURE 7-9** The grasslands of the North American Great Plains coevolved with millions of migratory bison. These grasslands are resistant to the grazing and trampling of large ungulates, the predictable result of coevolution. (Courtesy of Winifred B. Kessler.)

replacing native bunchgrasses (Gillis, 1991). The disruption of the microbiotic soil crust by large herds of cattle and sheep facilitated the invasion by cheatgrass (Evans and Johansen, 1999). For an interesting popular press description of the cheatgrass problem, see Devine (1993). Cheatgrass and medussa head grow quickly in the spring and mature and dry up early in the season. Except for a short window during early growth, they are unproductive and unpalatable. Thus as a consequence, in large part, of overgrazing in the 1800s by domestic livestock, large areas of western rangelands have been permanently altered. It would be virtually impossible to convert these lands back to their pre-1850 ecological state. However, because improper livestock grazing management may have been involved in the original degradation of the native grasslands does not mean that cessation of livestock production in these areas will be beneficial to grassland restoration. As discussed earlier, proper grazing management can be an important factor in the recovery of these lands. Climatic conditions (wet springs) favorable for survival of perennial grass seedlings occur very rarely (Young et al., 1999). Medusa head and cheatgrass are

---

[2] As with virtually every other issue discussed in this book, there is a contrary view! Burkhardt (1996) reviews evidence suggesting that the intermountain rangelands coevolved over several million years with megafauna (mammoths, mastadons, ground sloths, equines, camelids, rhinoceros, etc.) comparable to the mix of herbivores found in the Serengeti plains of Africa today. These animals became extinct in a massive extinction event at the close of the Pleistocene, about 11,000 years ago. Burkhardt argues that this is a very short period of geological time, and the grasses at the time of European arrival in North America were those that had coevolved with the extinct megaherbivores. He suggests that livestock are a reasonable substitute for the extinct herbivores.

highly competitive for limited soil moisture, making conversion of these sites back into perennial grasslands very problematic.

Finally, some critics of public land grazing believe that since the land belongs to everyone, people should be able to enjoy a wilderness camping or hiking experience without the presence of cows.

On a scientific basis, concerns about livestock grazing on rangelands can be dealt with. It is possible to have **multiple use** programs, with livestock production a part of the use of public lands. However, it is not necessarily a scientific issue. Harvesting of old-growth forests and use of publicly owned rangelands by domestic animals are social and political issues that will not be decided on purely logical, scientific grounds. Forestry, agriculture, and mining are receiving much criticism worldwide. People view these extractive industries as exploitive, a plundering of the planet. What is sometimes not recognized by many people is that these **primary industries** are the ultimate source of most wealth. The world's economy is based on a foundation of primary industries. While we can sell each other automobiles, electronic goods, computers and software, fast food, tourism, and other services, this system is sustained by the food, wood, paper, and metals produced by primary industries. A state or national economy cannot be adequately maintained by generating revenue from taxes on incomes of people selling each other services. Cities and urban centers exist as congregations of labor forces to efficiently process the products of the countryside into manufactured goods and services. The human population of the world has increased exponentially ever since our early ancestors learned to exploit natural resources through agriculture. It is not possible to eliminate the extraction and use of natural resources and return the environment to its pristine condition. Our challenges are to use these resources with as little adverse impact as possible and to reduce the growth rate of the human population to one that is sustainable on a long-term basis. We must ensure that our

extractive activities, which are essential to modern society, are carried out with as little disturbance of the natural environment as possible. This will require good management, not prohibition of resource utilization.

Livestock production plays an important role in the **generation of wealth.** A decision by the American people that livestock production should be banned on public lands could have substantial effects on the economies of several western states and hundreds of towns and communities. Besides the social costs, there would be the costs of absorbing these displaced people into service or manufacturing industries. On the other hand, much of the economic impact might be reduced by the increased use of the public lands for tourism and recreational purposes (Fig. 7-10). It has been argued that continued reliance on extractive and natural resource industries in the west is delaying the transition to a more stable and vibrant economy based on tourism, recreation, small business, and information services industries (Power, 1996).

Use of livestock on public lands has also been criticized because of effects on tree regeneration. In California, it is claimed that the native oak trees, which grow in a savanna-like setting on sidehill grasslands, are being prevented from reseeding by cattle grazing. On ri-

**FIGURE 7-10** Tourism, such as white-water rafting, and other recreational activities, is becoming increasingly important in the western United States. Ranchers are under pressure to accept that other uses, besides livestock grazing, are important priorities of the public.

parian areas in the western United States, the numbers of aspen, willow, and cottonwood trees are declining, because cattle browsing has prevented new trees from becoming established. Improved grazing management in riparian areas can allow their recovery. Trees and shrubs are very important in riparian areas in trapping sediment, lowering water temperature, and providing fish habitat. These concerns are legitimate but can be addressed by improved grazing management.

Livestock grazing can be used to control problem vegetation. Sheep and goats are used to control broadleaved weeds in newly planted reforested areas and to control brush under powerlines. They can be used in **weed control,** especially with noxious plants such as leafy spurge on western rangelands (Bangsund et al., 2001).

Livestock grazing on rangelands and grasslands is important in **fire suppression**. Sheep and goats are often used for brush control for fire prevention. Removal of vegetative growth of grasses by cattle grazing prevents buildup of dried plant material. Large amounts of dead grass litter can have explosive fire potential, leading to calamitous range and forest fires. This has increasing significance with the continued building of homes and cabins in range and forest areas. Under drought conditions, such as the 1987–1992 drought of the western United States, the importance of grazing in the prevention of buildup of highly flammable plant litter became quite apparent to many homeowners. On the other hand, the prolonged suppression of fire on rangelands and the removal of fine fuels by livestock grazing have caused a dramatic shift in the vegetation in many areas (Smith, 1995). An increase in woody vegetation such as sagebrush and juniper as a result of fire exclusion and grazing can increase the potential for uncontrollable, catastrophic wildfires. Thus one view is that grazing has contributed to increased fire potential.

This view is further supported by claims of Belsky and Blumenthal (1997). (Dr. Joy Belsky, ecologist with the Oregon Natural Desert Asso-

ciation, has been a vocal and influential critic of livestock grazing on western rangelands.) Historically, forests of the semiarid western United States have consisted of large, widely spaced fire-tolerant Ponderosa pines and other conifers, underlain by dense grass swards (Fig. 7-11). Belsky and Blumenthal (1997) claim that as a result of livestock grazing of the grass in these forests, the fuel (dry grass) for frequent, low-intensity fires has been removed. This has allowed many tree seedlings to survive, causing the forest structure to change to one dominated by thickets of closely spaced, fire-susceptible small trees. These altered forests are susceptible to disease, insects, and catastrophic fires. These fires kill even the large trees that are resistant to low-intensity ground fires because of their thick bark and clear trunks. According to these authors, the decline in forest health over the past 100 years coincides with the introduction of livestock grazing. They conclude that cattle and sheep grazing contributed to the conversion of the original parklike forests into dense stands of less fire-tolerant tree species and altered the physical environment by compacting soils, reducing water infiltration

**FIGURE 7-11** Forests of the semiarid western United States were dominated by widely spaced trees such as fire-tolerant Ponderosa pines (foreground), with an understory of grasses. Removal of the grass by livestock grazing, thus removing the fuel for frequent, low-intensity fires, has allowed tree seedlings to survive, changing the forest structure to one dominated by thickets of closely spaced, fire-susceptible small trees (background).

rates, and increasing erosion. There appears to be much validity in these assertions. At the same time, it would seem that altered grazing management, including periods of rest to allow fuel accumulation followed by prescribed low-intensity burning, could be used to maintain open forests. Browsing animals such as goats might also be useful in controlling overabundant tree seedlings. The concerns raised by Belsky and Blumenthal (1997) can be satisfactorily addressed with modified grazing management.

Predictably, the president-elect of the Oregon Cattlemen's Association, Sharon Beck, dismissed the ideas of Belsky and Blumenthal (1997) in an article "Cows Cause Forest Fires" (Beck, 1997). Her article begins with "Stay with me here and stop trying to figure out where a cow would carry a book of matches" and ends with "One thing for sure we'll breed cows with pockets before we'll see an end to this nonsense." So long as farmers and ranchers continue to respond to environmentalists in this manner, there is little hope for resolution of these conflicts in favor of the livestock industries.

The population of western North America is increasing rapidly. Most western towns and cities are expanding, with urbanization of the surrounding countryside occurring at a rapid rate. Tourist and vacation-related activities, including ski facilities, white-water rafting, vacation homes in the wilderness, and so on, are steadily encroaching on formerly isolated areas. Cattle ranchers in many areas face constant battles with environmentalists and urbanites, on the one hand, and the temptation of a high price for selling their property for "development," on the other. Livestock production on western rangelands is a great deal more environmentally friendly than most of the alternatives, and especially those related to urbanization and development (Fig. 7-12). Both plant and animal biodiversity is high on rangelands grazed by livestock. Urbanization of rangelands is a terminal step, which is unlikely to ever be reversed. Ranching, whatever its faults or limitations, does not permanently change land use possibilities. If at some future

**FIGURE 7-12** Livestock ranching on western rangelands is more environmentally friendly than urbanization. Urban development has destroyed much of the winter range of large ungulates such as deer and elk in the American west. (Courtesy of M. Vavra.)

time the prevailing public view is that western rangelands should be converted to wildlife habitat, parks, and recreational areas, with resource-extraction industries phased out, land that has been used for livestock grazing will be more suitable for these purposes than areas subjected to urbanization, summer cottages, condominium sites, ski lodges, or other such uses that are currently squeezing out cattle ranches. It is hoped that before the west is completely blacktopped, environmentalists and ranchers will realize that they are on the same team. In some instances, ranchers and environmentalists have been effective in forming coalitions so that ranching is not just tolerated as a means of environmental preservation but is valued and planned for, ecologically, socially, and economically (Huntsinger and Hopkinson, 1996). Of course, not all critics of livestock production agree. Wuerthner (1994), a self-styled critic of cattle ranching and the founder and sole member of the National Wolf Growers Association (!), asserts, " . . . proponents of the livestock industry repeatedly threatened that they would have to subdivide their ranches if grazing fees were raised. This skeleton rattles every time anyone criticizes agriculture. The message is clear: no matter how bad you think cows are, condos are worse" (p. 905). Further, he states, "Agriculture—both livestock production and farming—rather than being 'compatible with environmental protection' has had a far greater impact on the western landscape than all the subdivisions, malls, highways, and urban centers combined" (p. 905). Knight et al. (1995) provide a good discussion of the opposing view. Their conclusions seem sound: "Whether ranches make better neighbors than condominiums adjacent to public lands and other protected areas needs to be studied empirically . . . it is not a question of whether or not agriculture or subdivisions should exist. In any society, achieving the proper balance of urban, pastoral, and wild landscapes is an important task" (p. 461).

As urbanization of rangelands proceeds, a tipping point is reached where ranchers realize that it is the end of the road. When the presence of urban dwellers causes sufficient disruption of ranching activities, ranchers concede the social, ecological, and economic landscape as urban, and the most rational decision is to sell out for "development" of suburbs and ranchettes, and relocate in a more isolated area (Liffmann et al., 2000). For land conservation efforts to be successful, they must be initiated before significant urban development occurs. Conservation organizations such as **The Nature Conservancy**[3] are recognizing that working cattle ranches are compatible with conservation (Jensen, 2001). The Ecological Society of America's annual meeting in 2000 included a symposium entitled "Cattle and Conservation: A Role for Ranching in Protecting Biodiversity." The effects of urbanization are more subtle than simple elimination of habitat. Dogs are 60 percent more common and cats 20 percent more common on ranchettes than in ranching areas (Jensen, 2001). These are essentially "subsidized predators" of birds and other wildlife. The remaining wide open spaces of

the west are either owned or managed by ranching families. For conservation efforts to succeed, these people (ranchers) need to be on board, and not regarded as "the enemy" by conservation groups. In a study of why Colorado ranchers decide to stay or sell out, Rowe et al. (2001) found that changes in public land policy regulations are a major factor. Cutting grazing rights on public land or substantially increasing grazing fees will almost certainly mean additional pressure on ranchers to subdivide and convert working ranches into recreational properties (Rowe et al., 2001). This conversion fragments and degrades wildlife habitat, reduces open space, and in general is not consistent with maintaining ecosystem health.

The field of range management has traditionally been mainly concerned with management to maximize production of domestic livestock. Fuhlendorf and Engle (2001) suggest that range specialists should enlarge their worldview to the ecosystem level, managing rangelands for heterogeneity (biodiversity). Two disturbances—grazing and fire—can be used to produce shifting mosaics of heavily grazed and ungrazed areas, each supporting different types of organisms. Grazing distribution and intensity can be maximized over several years but minimized within individual years, promoting structural and compositional heterogenicity. In contrast to systems promoting uniform distribution of livestock grazing for maximum forage harvest, focal points of intensive herbivory can be rotated across the landscape over several years. Thus appropriate grazing management can be used to enhance biodiversity and enrich wildlife habitat in grasslands. Adoption of such techniques could reduce or eliminate efforts by environmentalists who are opposed to livestock production on rangelands.

Holechek (2001) estimates that the area of western rangelands will shrink by 25–40 percent over the next 100 years, due to urban encroachment. This would still leave a substantial land base of 700 million acres. Sophisticated range management will be needed to ensure

_____
[3] The Nature Conservancy, 4245 North Fairfax Drive, Suite 100, Arlington, VA, 22203 (*www.tnc.org*).

that this diminished rangeland area can continue to support ecosystem functions for livestock, water, wildlife habitat, and other amenities.

## The Greater Yellowstone Ecosystem: Livestock Interactions and a Model for Natural Regulation

The **Greater Yellowstone Ecosystem (GYE)** encompasses an area of about 18 million acres of public and private lands in Wyoming, Montana, and Idaho. It includes two national parks (Yellowstone and Grand Teton), six national forests, and the headwaters of three major river systems (Snake-Columbia, Green-Colorado, and Yellowstone-Missouri). A major component, **Yellowstone National Park (YNP)**, is the flagship and first park in the Na-

tional Parks System and the world's first national park. In the western United States, there are several major implications and impacts of YNP and GYE on the livestock industries of Wyoming, Montana, and Idaho:

1. Public vs. private management of rangelands and herbivores

   If, as proposed by many environmentalists, livestock grazing on public lands in the western United States were eliminated, what are the implications of changing from domestic to wild herbivores? In other words, can the government manage rangelands and grazing better than it can be done by ranchers? The answer, as will be discussed in more detail, appears to be that wildlife management by the National Parks Service in YNP has been a disaster for the animals and for the

---

### THE WESTERN RANGE REVISITED. REMOVING LIVESTOCK FROM PUBLIC LANDS TO CONSERVE BIODIVERSITY

*(Review of D. L. Donahue. 1999. University of Oklahoma Press, Norman, Oklahoma, 388 pp.)*

The "guts" of this book are that the author proposes that livestock (cattle and sheep) should be removed from western U.S. rangelands, in the interests of ecological restoration and enhancement of biodiversity. The author, Debra L. Donahue, is a professor of law, University of Wyoming. Needless to say, her proposal to eliminate livestock grazing generated intense controversy and numerous attempts to get her dismissed from The University of Wyoming. Donahue argues that public (BLM) lands grazing produces relatively little in the way of livestock products, and the grazing programs cost the government more to administer than revenue generated. Second, she argues that livestock grazing adversely affects all other uses of public land, causing potentially irreversible damage to native wildlife and vegetation. She maintains that her aims can be achieved under existing law.

I never finished reading this book. I found it extremely ponderous and sleep inducing. I was amazed to see this assessment by Thad Box, a long time and well-known range scientist: "The book is an easy read, unfolding like a brief prepared by an excellent lawyer" (Box, 2000, p. 27). Based on my reading experience with it, I suspect that many of the politicians and ranchers who blasted the book and the author never read it. Nevertheless, this book, like the publications of Allan Savory, serves a purpose in promoting discussion and dialogue and perhaps opens a few minds. The "siege mentality" of the Western livestock industry must eventually open up to a bit of introspection.

A more balanced view of western ranching is provided by Starrs (1998): Starrs, P. F. 1998. *Let the Cowboy Ride: Cattle Ranching in the American West*. Johns Hopkins University Press, Baltimore; and by Knight et al. (2002). *Ranching West of the 100th Meridian: Culture, Ecology and Economics*. Island Press, Washington, D.C.

rangeland, although this view is not shared by all observers (Schullery, 1999). This has important implications in the livestock grazing controversies of the western United States.

2. Public vs. private management of riparian zones

   Similar to item 1, the state of riparian zones in YNP is hardly a reassuring model of the fate of rivers and streams if public land grazing by livestock were eliminated.

3. Bison-cattle interactions

   The YNP bison herd is infected with the disease brucellosis, which causes abortion in cattle. Movement of bison from YNP on to adjacent livestock grazing areas has precipitated disputes between the state of Montana, ranchers, and the National Parks Service.

4. Wolf reintroduction program

   The reintroduction of wolves into YNP in 1995 has been a major source of concern to livestock producers who believe that their animals will be subject to wolf predation. A major impediment to wolf reintroduction was the vocal objections of the ranching community.

These controversies will be discussed. The management of the GYE and YNP has been the subject of much debate among wildlife biologists, range management specialists, and environmentalists. Many books, conferences, and symposia have dealt with the Yellowstone controversies. Some of these include Alston Chase's book *Playing God in Yellowstone. The Destruction of America's First National Park* (1987) and the Keiter and Boyce (1991) book *The Greater Yellowstone Ecosystem. Redefining America's Wilderness Heritage.* Some of the issues involving Yellowstone that have attracted great public attention are the 1988 wildfires that burned over 750,000 acres, and the starvation or shooting of about 75 percent of the park's bison herd in the winter of 1996–97.

Both events were extensively covered by television and attracted widespread debate and condemnation of park management.

## Grazing and Riparian Issues in YNP

Yellowstone National Park is highly visible, and several agencies are involved in different aspects of its management. Much of the management is based on public opinion and political considerations. Legislators make threats implying "do this or we'll cut your funds." Thus there are great external pressures that impact on management of the park. In the early part of the twentieth century, park policy was to eliminate predators to produce abundant numbers of large ungulates, mainly elk and bison, for the benefit of tourism. Park visitors wanted the experience of readily viewing large numbers of these animals. Elk (Fig. 7-13) numbers increased dramatically. In the 1960s, in response to severe ecological problems caused by overgrazing, a policy of elk reduction by shooting was initiated. Not surprisingly, this caused public outrage. The solution to the elk problem was development of a new policy called **natural regulation,** which was a complete reversal of park policy. Basically, the natural regulation policy implies that herbivores and vegetation reach an ecological equilibrium called ecological carrying capacity. In this model, predators do not have a significant role in regulating herbivore numbers. Thus by a policy change, YNP changed from having an overpopulation of elk to having a desirable number of the animals, in equilibrium with their environment. Unfortunately, the evidence suggests otherwise. The natural regulation policy has been vigorously attacked by professionals in range management, ecology, and conservation. The problems are discussed at length in Alston Chase's book *Playing God in Yellowstone. The Destruction of America's First National Park.*

An outspoken critic of the YNP natural regulation policy is Charles Kay (Kay, 1994b,

**FIGURE 7-13** Large numbers of elk in the Greater Yellowstone Ecosystem require winter feeding (*top*) and cause overgrazing of summer range (*bottom*). (Top photo courtesy of Charles E. Kay; bottom photo courtesy of R. E. Dean.)

**FIGURE 7-14** (*Top*): Company D of the Minnesota National Guard on patrol in Yellowstone National Park in 1893. (In the early years of the park, management was provided by the U.S. Army, mainly to control poachers.) Note that the aspens and willows in riparian areas are abundant and little browsed. (*Bottom*): The same scene in 1988. Excessive populations of elk have eliminated virtually all willows and aspen. (Top photo courtesy of Charles E. Kay and the Montana Historical Society; bottom photo courtesy of Charles E. Kay.)

1996). Some of his major viewpoints are the following:

1. Excessive numbers of elk have caused severe overgrazing. They have severely overbrowsed willows and aspen, especially in riparian areas (Fig. 7-14). Narrowleaf cottonwood will likely be completely eliminated from the YNP northern range unless elk browsing is reduced (Keigley, 1997).

2. The loss of riparian area trees (Fig. 7-15) has decimated **beaver** populations. Beaver are ecologically extinct in much of the park, because they lack a food source (willows and aspens).

3. Beaver are **keystone species.** They dramatically alter the hydrology, energy flow, and nutrient cycling of aquatic and riparian systems. Ecosystems with beaver are completely different from systems without beaver (Fig. 7-16). Among the contributions of beaver are the following:

a) Beaver dams impound water, trap sediments, and extend riparian zones.

b) Beaver create and maintain riparian areas that are critical habitats for many other species.

c) Beaver activity can restore riparian systems that have been dam-

**FIGURE 7-15** (*Top*): Aspen in this 1952 photo were declining due to excessive browsing by elk and other ungulates. (*Bottom*): By 1997, these aspen (same site) have been completely eliminated by overbrowsing by elk. The loss of aspens results in the elimination of beaver as well. (Courtesy of Charles E. Kay.)

**FIGURE 7-16** Aquatic mammals such as beaver and muskrat are keystone species, dramatically altering ecosystems by their presence or absence. (Courtesy of A. G. Hollister.)

aged by normal herbivore browsing and grazing.

    d)    The elimination of beaver can result in lowered water tables, reduced summer stream flows, and a change in vegetation of an entire ecosystem.

4. Elk numbers in prepark times were low. A major reason for this was predation by Native Americans and large carnivores such as wolves and mountain lions.

5. Beaver numbers in YNP were very high in historical times. There are factual fur company records of thousands of beaver trapped per year. It is estimated that there were over 10,000 beaver in YNP in the early 1900s (trapping ended years before).

6. There is photographic evidence of abundant willow and aspen communities throughout YNP in the late 1800s and early 1900s (Fig. 7-14). There is now abundant evidence that these plant communities have been decimated by elk, and there is very little recruitment of new aspens and willows. Young plants have little chance of surviving heavy elk grazing pressure (Fig. 7-17).

Near Yellowstone, the USDA has operated the U.S. Range Sheep Experiment Station in Dubois, Idaho, for many years. Since 1922, USDA range sheep have been grazed in the Centennial Mountains of Montana and Idaho. An independent 1991 inventory revealed thriving beaver populations and abundant willows and aspen in riparian areas (Anonymous, 1995). There were more active beaver colonies along one creek in this sheep summer range than in the entire northern range of YNP (Charles Kay, personal communication). In contrast to YNP, the rangelands of the Centennial Mountains are thriving under well-regulated grazing by domestic sheep (Kay and Walker, 1997).

**FIGURE 7-17** This exclosure was built in 1962 on Yellowstone National Park's northern range, showing the abundance of willows within the exclosure and their complete absence outside due to grazing by elk. (1986 photo courtesy of Charles E. Kay.)

Evidence that there were few elk in the Yellowstone area prior to formation of YNP comes from explorers' records and archaeological data. According to Kay (1994a), hunting and fire use by Native Americans kept elk numbers low. Few elk bones are found in archaeological sites of Native American hunting areas; bones of bison, bighorn sheep, and mule deer predominate. According to Kay (1994a), Native Americans were the ultimate keystone species who structured entire ecosystems. Elk have large hooves and move in herds. They have the same effects on streambanks and riparian areas as cattle. Thus the recent phenomenon of elk becoming the dominant large herbivore in YNP can be expected to be at least as damaging to riparian areas as uncontrolled cattle grazing. Large ungulates can have profound effects on ecosystems, influencing the nitrogen cycle, composition of plant communities, successional processes, fire, and spatial heterogeneity (patchiness of vegetation). Hobbs (1996) reviewed in depth the effects of ungulates on ecosystem patterns and processes.

Although winter starvation has some effect on controlling elk numbers (Singer et al., 1997), the only practical and politically possible solution to the overabundance of elk in YNP is the reintroduction of wolves (Singer and Cates, 1995). Other solutions, such as opening the park to hunting, shooting of elk by park personnel, allowing mass starvation when the forage is exhausted, fencing elk away from riparian areas, and so on, are not feasible. Wolves were introduced in 1995, and their diet since their release has been almost exclusively elk (Fritts et al., 1997).

There has been much public and scientific concern about the decline of trembling aspen (*Populus tremuloides*) in YNP (Romme et al., 1995). **Aspen** stands in the park are deteriorating, with mainly only large, mature trees that are gradually succumbing to disease and old age and are not being replaced by new recruits. An overabundance of elk is often cited as the main cause (White et al., 1998). Besides browsing the leaves and sprouts, elk eat the bark of mature trees, which facilitates invasion by fungi and other pathogens (Romme et al., 1995). The major way that new aspen trees arise is from root sprouts, rather than from seed. Thus most aspen groves are actually clones of a single or a few plants. Repeated browsing prevents the sprouts from surviving. Fires also promote sprouting, while killing the mature stems. Romme et al. (1995) cite evidence that regeneration of aspen trees has always been episodic in YNP. They suggest that a complex of factors are involved, including elk abundance, climatic variation, fire, mammalian predators (e.g., wolves), and interactions with other herbivores (e.g., bison, which are not browsers). The last major episode of aspen tree regeneration occurred in 1870–1890, a historically unique period when numbers of elk, beaver, and moose were low, fires had recently occurred, climatic conditions were moist, and wolves were present (Romme et al., 1995). This combination of events has not occurred since. Restoration of wolves, with their impacts on browsers like elk and moose, may result in improved prospects for aspen regeneration (White et al., 1998; Ripple and Larsen, 2000).

Elk browsing has also contributed to the decline of **willows** (*Salix* spp.) in YNP (Singer et al., 1998). Willow declines in YNP have been dramatic. Declines in **beaver** numbers, by reducing wetlands that are willow habitat, may have also contributed to the willow decline, as well as climatic changes trending to increased aridity of the region. Keigley (2000) disputed the hypothesis that declines in beaver numbers contributed to willow decline and places the blame solely on an overpopulation of elk. A rebuttal was provided by Singer et al. (2000). Kay (1998) maintains that prior to YNP establishment, predation by Native Americans kept elk numbers low, limiting herbivory on willows and other vegetation.

Beaver build dams and create ponds on streams within forested areas. When the beaver abandon an area or are killed, the ponds usually drain and the former pond area becomes a grass-dominated meadow that may last for many decades before being eventually invaded by trees. Beaver pond meadows resist invasion of conifers such as spruce trees even if seed sources are nearby. Spruce and fir species are obligate ectomycorrhizal, meaning that they require the association of fungi (mycorrhizae) in their roots for growth and survival. Prolonged flooding, such as with beaver ponds, kills the soil fungi. Reinvasion of the beaver pond meadows by conifers depends on reintroduction of the fungi. Small mammals such as voles are the major dispersal mechanism for the fungal spores. Small mammalian species consume large amounts of mycorrhizal fungi and disperse the spores in their feces. Thus conifer succession in beaver meadows depends on gradual invasion of the meadows by forest-dwelling small mammals, distributing fungal spores that are necessary for conifer survival (Terwilliger and Pastor, 1999). This is another interesting example of the subtle "interconnectedness of things."

The dominant vegetation type on the YNP Northern Range (main site of elk overpopulation controversies) is **big sagebrush** (*Artemisia tridentata*). It is an important winter forage for elk and is a preferred browse of mule deer and pronghorn. Excessive elk browsing is responsible for sagebrush declines (Wambolt, 1998), negatively impacting pronghorn, mule deer, and small animals and birds dependent on big sagebrush. It is apparent that excessive numbers of elk have far-reaching effects on the health of the YNP ecosystem (Alt and Frisina, 2000). The restoration of wolves will likely have many positive effects on other plant and animal life, by keeping elk numbers in check. The successful reintroduction of wolves to Yellowstone is described in the next section.

The grazing management of native herbivores in YNP has implications for the management of western rangelands. If cattle are excluded from public lands grazing, their niche will be filled by native herbivores, primarily elk, and other species such as wild horses. Any efforts to regulate numbers of these species by hunting, and the like, generates tremendous controversy with lawsuits and animal rights-promoted legislation. Cattle numbers and cattle grazing management are much easier to regulate. The most plausible means of regulating elk and wild horse numbers if they replaced cattle would be with wolves and mountain lions. The tolerance of the public for these species is not boundless. These animals kill pets (e.g., dogs, llamas). In Minnesota, the killing of dogs by wolves has caused considerable public animosity towards wolves (Mech, 1995). Mountain lions in California have killed pets and several people. If the management of YNP is any indication of how western rangelands would be managed if livestock were removed, it seems apparent that ecosystem destruction would be the obvious result. "Natural regulation" is not an option. Cattle grazing on western rangelands may be essential if severe overgrazing, as has happened in Yellowstone, is to be avoided. The health of western rangelands may rest on maintaining a viable cattle ranching industry.

## Yellowstone, Aboriginal Overkill, War Zones, and Game Sinks

There is considerable controversy as to whether YNP is overpopulated with large ungulates such as elk. Part of the controversy is based on historical perceptions—what was the situation before the park was established? Many clues can be obtained from diaries of explorers, trapping records and furs purchased by fur companies, and vegetative characteristics. Kay (1994a, 1998) is a strong proponent of the viewpoint that Native Americans were responsible for keeping ungulate numbers low. He maintains that Native Americans were the ultimate "keystone species," whose activities created the ecosystems that we now consider natural or pristine. Kay (1998) disputes the widely held view that the religious beliefs of Native Americans prevented them from overutilizing their resources. A scarcity of game was not viewed as a biological or ecological phenomenon, but as a spiritual consequence because the gods were displeased. Because Native Americans saw no connection between their hunting and animal numbers, their system of religious beliefs actually fostered the overexploitation of ungulate populations (Kay, 1994a). Kay (1994a, 1998) refers to the excessive hunting by Native Americans as "**Aboriginal Overkill.**" Native Americans also had a large influence on western ecosystems by repeatedly burning the vegetation. The majority of their fires were in the spring, which tended to produce low-intensity fires. In Yellowstone, one of the reasons for burning was to promote favorable conditions for berries, such as service berries (*Amelanchier alnifolia* and choke cherries (*Prunus virginiana*). According to Kay (1998), large ungulate populations and abundant berries are mutually exclusive, because these shrubs are highly preferred forage. The fact that berries were harvested and consumed in large quantities by Native Americans suggests that ungulate populations were low and were kept low by something other than food limitation. High elk numbers, for example, would have decimated berry-producing vegetation. Kay uses the argument of "Aboriginal Overkill" to dispute the YNP policy of "Natural Regulation," or allowing animal numbers to be limited by availability of feed.

It is widely believed that large mammals (mastadons, etc.) native to North America became extinct because of Native American hunting. If the concept of "Aboriginal Overkill" is correct, then a logical follow-up question is why didn't the modern game animals also become extinct? A plausible explanation is that large ungulates survived and were abundant in buffer zones between warring tribes of Native Americans. It is well documented in various parts of North America that game was scarce in areas dominated by an Indian tribe and plentiful in the "no-man's-land" between tribes (Martin and Szuter, 1999). This was documented in the Lewis and Clark expedition from St. Louis to the Pacific Coast. Many areas, with abundant forage, were devoid of game. Animals of all types were abundant only in the "war zones" between tribes. Even the herds of migratory bison on the plains favored intertribal "no-man's-lands" (Martin and Szuter, 1999). Wolves form wolf pack buffer zones; ungulates are often more common in the areas between packs that wolves avoid to minimize the chance of fatal encounters with other wolves (Ripple and Larsen, 2000).

Aboriginal overkill has been implicated in the extinction of large animals on several continents. Alroy (2001) attributes the loss of megafauna in North America to human hunters, based on results with computer modeling. Roberts et al. (2001) came to a similar conclusion for Australia; all Australian land mammals, reptiles and birds weighing more than 100 kg, and six of the seven genera with a body mass of 45–100 kg became extinct following the arrival of humans in Australia. Alroy (attributed to Dayton, 2001) concluded, "The results (of these two studies) show how much havoc our species can cause without anyone at the time having the slightest idea of what is going on, much less any intention of causing harm" (p. 1819). This message has obvious contemporary relevance.

A comprehensive book *Wilderness and Political Ecology Aboriginal Influences and the Original State of Nature* edited by Kay and

Simmons (2002) provides an in-depth review of the roles of indigenous peoples in structuring ecosystems. Kay and Simmons (2002) maintain that prior to the arrival of European colonizers in North America, the continent was not "pristine" but was markedly altered by management practices of Native Americans, particularly the controlled use of fire. Kay and Simmons (2002) believe that these activities have not been generally recognized, because of ingrained racist beliefs that native peoples were simply part of the landscape and were too primitive to have had an impact on the nature of North American ecosystems. They provide compelling evidence that the opposite is true.

## The Top Carnivores: Wolves and Grizzly Bears

In the early days of YNP, a major function of the park was perceived to be having an abundant supply of photogenic large animals (**photogenic megafauna**) near the headquarters, readily viewable by the large number of eastern tourists who arrived by train. Bison and elk were the preferred species. Predators of these species, including hunters, were viewed negatively. In the early 1900s, YNP was placed under the control of the U.S. Army to combat market hunters (Fig. 7-14). In 1916, the National Park Service was formed, and a program was initiated to control predators (wolves and mountain lions) of elk and bison. By 1926, wolves were eliminated from YNP. Numbers of bison and elk increased dramatically. It is now widely believed, as discussed previously, that elk have caused severe ecological damage in the park. As early as 1944, the "Father of Wildlife Management," Aldo Leopold, proposed restoration of wolves in YNP (Leopold, 1944). During the 1970s and 1980s, there were various proposals for the introduction of wolves from Canada into YNP. The proposals were condemned and opposed by livestock interests, who feared wolf predation of sheep and cattle outside the park.

A widely held view of ranchers is that wolf recovery is not about wolves at all, but is a "Trojan horse" to eliminate public lands grazing of livestock. Fritts et al. (1997) reviewed the complex web of events by which Canadian wolves were finally introduced into YNP in 1995, on an experimental basis, after a twenty year struggle by wolf reintroduction proponents. The restoration is considered a resounding success, far exceeding expectations. The wolves are reproducing, feeding, and causing little damage to livestock. The sky has not fallen! Fritts et al. (1997) conclude:

> Wolf restoration is not for the impatient . . . The inherent interest in wolves, the strong and conflicting attitudes about them, and their symbolic nature, when added to the current acrimony about how public-owned lands in the American West should be used, created difficulty every step of the way. Reintroducing wolves was far more than just a biological issue. (p. 23)

The wolf has become an environmentalist's and animal rights advocate's "poster animal"; the vile, vicious wolf has been replaced in the public mind by the unjustly persecuted wolf (Mech, 1995).

Aldo Leopold, the founder of the discipline of wildlife management in North America and the concept of "The Land Ethic" (see Chapter 10), played an important role in extermination of the wolf from the continental United States in the early part of the twentieth century. He began to question the wisdom of predator elimination after seeing the results. In his short essay "Thinking Like a Mountain," Leopold (1949) described the seminal moment when his opinion of wolves dramatically changed, after he had just shot a wolf with pups:

> My own conviction on this score dates from the day I saw a wolf die. . . .
>
> We reached the old wolf in time to watch a fierce green fire dying in her eyes. I realized then, and have known ever since, that there was something new to me

in those eyes—something known only to her and to the mountain. I was young then, and full of trigger-itch; I thought that because fewer wolves meant more deer, that no wolves would mean hunters' paradise. But after seeing the green fire die, I sensed that neither the wolf nor the mountain agreed with such a view. Since then I have lived to see state after state extirpate its wolves. I have watched the face of many a newly wolfless mountain, and seen the south-facing slopes wrinkle with a maze of new deer trails. I have seen every edible bush and seedling browsed, first to anaemic desuetude, and then to death. I have seen every edible tree defoliated to the height of a saddlehorn. Such a mountain looks as if someone had given God a new pruning shears, and forbidden Him all other exercise. In the end the starved bones of the hoped-for deer herd, dead of its own too-much, bleach with the bones of the dead sage, or molder under the high-lined junipers. I now suspect that just as a deer herd lives in mortal fear of its wolves, so does a mountain live in mortal fear of its deer. And perhaps with better cause, for while a buck pulled down by wolves can be replaced in two or three years, a range pulled down by too many deer may fail of replacement in as many decades. So also with cows. The cowman who cleans his range of wolves does not realize that he is taking over the wolf's job of trimming the herd to fit the range. He has not learned to think like a mountain. Hence we have dustbowls, and rivers washing the future into the sea. (pp. 138–140)

Leopold's account of too many deer rings true in eastern North America, and the Midwest, where white-tailed deer have become major pests in some areas (Waller and Alverson, 1997). They cite numerous examples of instances where excessive deer numbers are preventing forest regeneration, almost completely eliminating young seedlings. Various endangered plant species, such as certain rare lilies and orchids, are heavily browsed by deer. Ground-nesting birds and birds nesting at sev-

eral levels in the forest are affected by the loss of cover. Deer compete with birds and small mammals for acorns. Of course, excessive numbers of deer have direct impacts on humans, including automobile accidents and destruction of home gardens. Reintroduction of top predators such as wolves and mountain lions is obviously not an option in most of these areas. The primary means of adjusting deer populations is by hunting or through winter starvation: hunting is heavily constrained by local restrictions and animal rights groups. As is true for virtually all areas of resource management today, game management must consider a multitude of interactions and is faced with almost guaranteed controversy.

The return of wolves to Yellowstone was not merely an issue about wolves. Wilson (1997) discusses in detail the conflict between environmentalists and the "Wise Use" movement. **The "Wise Use" movement** is a coalition of rural resource-extraction organizations opposed to what they perceive to be a powerful, oppressive enemy: radical urban environmentalists working in collusion with the federal government to "lock up" the western landscape with restrictive environmental regulations with the ultimate aim of eliminating private property. The "enemy" is the urbanized, middle class world view of the mainstream environmental agenda. To environmentalists, return of the wolf to Yellowstone is a symbol of ecological reconciliation, the restoration of ecosystem integrity. To the Wise Use movement, the wolf is a Trojan horse, that will bring with it land use restrictions and regulations of the environmentalists and preservationists. The debate is no longer about wolves; it is about whose values win. Yellowstone wolves have united and mobilized rural communities for protection of an endangered western way of life. Nevertheless, the wolves are back, apparently thriving, while the struggle for "control of the West" goes on. Endangered species, such as the spotted owl, the wolf, Pacific salmon, and so on, are usually just symbols of conflict; the real conflicts are opposing world views of nature: environmental preservation vs. resource extraction (wise use).

Assuming that there is validity to the belief that Yellowstone ungulates require "trimming," can wolves do the job? Experience with wolves and moose on Isle Royale in Lake Superior suggests that wolf predation can regulate ungulate populations (Eberhardt and Peterson, 1999). Likewise, ungulates are important agents of change in ecosystems (Hobbs, 1996).

The Yellowstone **grizzly bears** are among the few remnant populations of *Ursus arctos* in the continental United States. In the 1800s, as many as 100,000 grizzlies roamed the American west; less than 1,000 are left. The number of grizzlies in YNP is difficult to measure because of technical difficulties in studying these reclusive animals. Most grizzly bear mortality is from interaction with humans, including road accidents and destruction of "problem bears." A major variable in the long-term survival of the park bears is variation in the cone crop of the whitebark pine, a major food source for grizzlies (Pease and Mattson, 1999). Threats to the whitebark pine, such as white pine blister rust, global warming, and fire suppression, are also threats to grizzly survival. White pine blister rust is caused by a fungus, whose alternate hosts include wild gooseberries. A long-time program to "seek and destroy" wild gooseberries has been abandoned in YNP, as an ineffective control of blister rust. This is another example of the interconnectedness of ecological issues and the complexity of management decisions of park administrators. Pease and Mattson (1999) pessimistically conclude, "The threats posed by diminishing whitebark pine and increasing numbers of people are inconsistent with an optimistic prognosis for the Yellowstone grizzly bear population" (p. 969).

The presence of the top predators (wolves and grizzly bears) has an influence on the behavior of prey animals like large ungulates (Berger, 1999). Predators keep prey "on their toes." Ripple and Larsen (2000) suggest that the absence of wolves in the park for many years altered the behavior of elk. As discussed earlier, heavy browsing by elk is believed to have suppressed regeneration of aspen trees. Elk could loiter undisturbed in groves of young aspen and repeatedly browse them. With wolves back in the ecosystem, elk can't "hang out" in one area very long. Also, wolves may increase aspen recruitment indirectly through effects on grizzly bears. Wolf-killed elk carcasses benefit grizzlies (Ripple and Larsen, 2000), which could increase grizzly populations, increasing grizzly predation on elk. The bears are particularly significant predators of elk calves (Singer et al., 1997). So the reintroduction of the wolf may have diverse, unexpected ecological consequences (probably beneficial).

## Bison, Brucellosis, and Cattle

*Brucella abortus* is a bacterial infection that causes the disease **brucellosis** in livestock and **ungulate fever** in humans. Clinical signs in cattle include abortion in the last trimester of gestation, retained placenta, arthritis, and bursitis. Bison, elk, moose, and white-tailed deer can be carriers. Many U.S. states have brucellosis control programs, requiring by law the vaccination of female calves against *Brucella abortus*. Control has involved the quarantine of infected herds and slaughter of animals testing positive for exposure. Many states require the quarantine of cattle entering from a non-brucellosis free state. The disease is transmitted by exposure of animals to contaminated placentas, aborted fetuses, or amniotic fluid.

The essence of the Yellowstone bison-brucellosis controversy is that the Yellowstone bison herd is infected with brucellosis, and Montana ranchers fear transmission of the disease to cattle in rangelands adjacent to YNP. Montana is a brucellosis-free state, and there would be large negative economic repercussions to the Montana beef cattle industry if the disease were transmitted from bison to cattle. Because (in large part) of mismanagement of bison within YNP, the animals face winter starvation and leave the park in large numbers during the winter, possibly exposing cattle to brucellosis. Although there is controversy as to whether the disease can be transmitted from bison to cattle under field conditions, the likelihood that it could is high (Meyer and Meagher, 1995; McLeod and Tassell, 1996). A relatively new factor is that YNP has been opened for winter recreation, including snowmobiling. The snowmobiles make trails, which, along with snowplowed roads, give the bison direct routes out of the park. In the severe winter of 1996–97, many bison left the park and were shot by National Park Service personnel. Resulting television coverage produced national outrage (one protester even doused the Secretary of the Interior with a bucket of bison guts!). By the end of the winter, about 75 percent of the Yellowstone bison herd had perished (while bad public relations, this event may bring bison numbers down to a sustainable level that the YNP range can adequately support). Although most of the YNP brucellosis controversy has involved bison, the 100,000 head elk herd is also infected (McLeod and Tassell, 1996).

In 1995, the Greater Yellowstone Interagency Brucellosis Committee was formed involving Wyoming, Idaho, Montana, the USDA, and the U.S. Department of the Interior. The group is charged with the responsibility to develop and implement a brucellosis management plan that will protect public interests as well as the economic viability of the livestock industry, while sustaining the bison and elk populations (Keiter, 1997). Elimination of brucellosis from the area by 2010 is proposed. The likely solution will be an ecosystem management approach, involving changes in cattle management (mandatory brucellosis vaccination, altered grazing programs, and intensive cattle management to avoid wildlife contact in susceptible [calving] periods), management of elk and bison populations to improve winter range conditions, and regulation of numbers in accordance with carrying capacity of the range. These management programs would encompass the entire Greater Yellowstone Ecosystem.

## Current Status of the Yellowstone Ecosystem

As the preceding discussion has indicated, many aspects of the management of YNP are controversial. Some of these involve livestock, such as brucellosis in bison and elk, and effects of wolf reintroduction on ranching adjacent to the park. The management practice of natural regulation has received heavy criticism. While many of the criticisms seem valid, the basic premise of natural regulation, which is allowing natural ecological processes to run their course as unhampered by human influences as possible (Boyce, 1998) also seems valid. Boyce (1998) concludes, "As a general rule, it is not necessary to control ungulate populations in our national parks" (p. 392). Reintroduction of the top predator, the wolf, should be a positive development. Natural regulation is often discussed solely in reference to elk management but in fact is a multifaceted attempt to restore ecological equity to management of all plant and animal communities (Huff and Varley, 1999; Singer et al., 1998; Schullery, 1999). Huff and Varley (1999) point out that "In addition to its contributions to scientific understanding, the natural regulation era has witnessed an incredible flourishing of creative, argumentative science" (p. 25). Shades of Allen Savory! Finally, Schullery (1999) offers an insider's view, as a National Park Service employee, of the history of YNP and the difficulties of responding to a very wide spectrum of public opinion. He maintains that YNP administration is in general responding appropriately to very diverse inputs and on-the-ground realities.

Although the Greater Yellowstone Ecosystem sounds very large at 18 million acres, large carnivores such as grizzly bears and wolves may require even larger areas of land to maintain viable populations with adequate genetic diversity. One way of providing habitat is to link non-contiguous areas by **wildlife corridors,** which allow individuals to move readily from one area to another. One interesting proposal is the **Y2Y (Yukon to Yellow-**

stone) Conservation Initiative, which would link national parks, national forests, state and provincial parks, and so on, with corridors, to provide continuous habitat from the Yukon Territory in northern Canada to YNP. Ultimately, it may be possible to provide continuous habitat from Alaska to Mexico. One of the most significant barriers to accomplishing this is the objections raised by livestock industries, which view increased grizzly and wolf populations as threats.

## Return of the Bison: The Buffalo Commons and Other Proposals

The destruction of the vast herds of bison[4] on the Great Plains of North America has been described in detail by Isenberg (2000). At first glance, the bison slaughter appears to be a matter of simple greed. Millions of bison were killed mainly for their hides and secondarily for their tongues, which were a gourmet item "back East." Another usual explanation is that the U.S. government employed the army to destroy bison to aid in the pacification of the Plains Indians. Isenberg (2000) argues that the history of the bison near-extinction was a complex phenomenon involving climate change, acquisition of horses by Plains Indians, European conquest of North America, industrialization, and diseases such as smallpox. Isenberg believes that the fate of the buffalo was sealed when nomadic people of the Plains acquired horses. Native American harvest of buffalo had reached unsustainable proportions before the arrival of Euroamerican market hunters. By 1900, fewer than 1,000 buffalo remained. Many prominent Americans wrote off the bison and believed the species should be slaughtered to extinction. Fortunately, a few ranchers captured bison calves and raised

them, and a few zoos kept bison on exhibit. The bison alive today, now numbering over 200,000, are descendants of these scattered remnants. Even the bison of Yellowstone were slaughtered by market hunters; the present-day animals were established from private sources, including the New York Zoo. In the early years of the twentieth century, the Yellowstone bison were managed intensively in corrals, until they were eventually released and allowed to roam free.

A husband and wife team from Rutgers University in New Jersey, Frank and Deborah Popper, caused a firestorm of controversy when they proposed creation of a huge "Buffalo Commons" on the Great Plains. The development of this idea is chronicled by Matthews (1992) in a book, *Where the Buffalo Roam*, and in an article by the Poppers (Popper and Popper, 1999). In simple terms, the Poppers, who are, respectively, an urban and regional planner (Frank) and a geographer (Deborah), point out that many counties in the Great Plains states are losing population, and have a very low human population density (Fig. 7-18). Farming on the plains is increasingly unprofitable and tenuous. Small towns are dying. Nearly 60 percent of the counties in the Great Plains lost population in the decade of the 1990s. The remaining population is ageing. The current situation is summed up in a popular press article, "The Broken Heartland," *U.S. News and World Report* (May 7, 2001). The Buffalo Commons idea is basically a large-scale ecological-economic restoration project involving the conversion of large areas into prairie ecosystems, with reestablishment of all indigenous components, including buffalo. A non-profit organization, the Great Plains Restoration Council[5], is working to restore 1 million contiguous acres of functioning prairie ecosystem. The Nature Conservancy[6] has established a Tallgrass

---

[4] Bison and buffalo are terms used interchangeably in common usage. Correct terminology uses the term buffalo only for the Asian and African buffalo, while the North American animals are bison.

---

[5] Great Plains Restoration Council, P.O. Box 46216, Denver, CO, 80201. *www.gprc.org*.

[6] The Nature Conservancy, Tallgrass Prairie Preserve, P.O. Box 458, Pawhuska, OK, 74056.

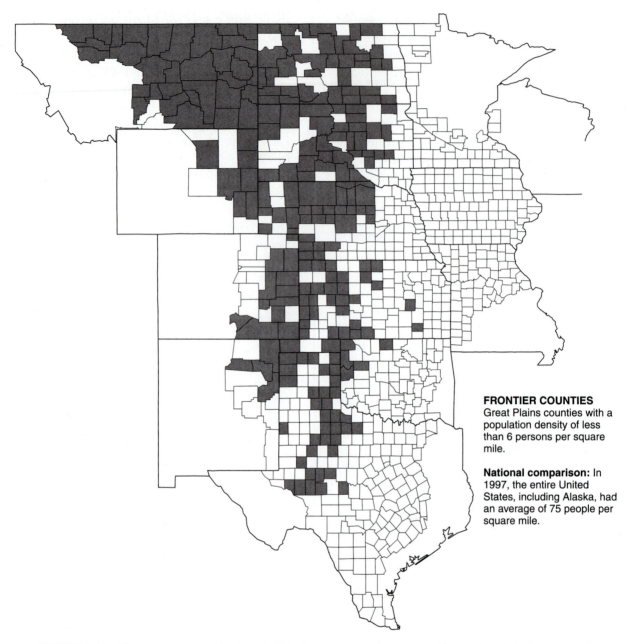

**FRONTIER COUNTIES**
Great Plains counties with a
population density of less
than 6 persons per square
mile.

**National comparison:** In
1997, the entire United
States, including Alaska, had
an average of 75 people per
square mile.

**FIGURE 7-18**  The area proposed for the "Buffalo Commons" is characterized by small and declining population. On this map, the counties in black have a population density of less than 6 persons per square mile; in the entire United States, including Alaska, the average is 75 persons per square mile. (Courtesy of Frank Popper.)

**FIGURE 7-19**   A bison ranch, with 3,000 animals, established in Montana by media mogul Ted Turner, founder of CNN.

**FIGURE 7-20**   Wind-powered turbines for generating electricity are compatible with cattle and buffalo ranching, offering ranchers a source of supplementary income, as well as providing a public service.

Prairie Preserve in Oklahoma. Media magnate Ted Turner has established bison ranches, including a 3,000 head buffalo ranch in Montana (Fig. 7-19). Another vision of a Buffalo Commons is presented in a book, *Bring Back the Buffalo!*, by Callenbach (1996). He proposed coupling restored prairie with wind-powered electricity generation (Fig. 7-20). Buffalo have been restored on many Indian reservations on the plains. Thus, the buffalo returns.

Bison are "keystone species" that have a disproportionately large impact on an ecosystem (Knapp et al., 1999). They preferentially graze on grasses, allowing forbs (broad-leaved plants, wild flowers) to flourish. Their grazing pattern leaves patches of heavily grazed and lightly grazed areas, enhancing biodiversity. They create mud wallows, which trap water in the spring that are suitable for ephemeral wetland species. Their grazing activities create a mosaic of patches, which influences fire. Fires burn on old dried grass in ungrazed patches. The following year, these burned areas become preferred bison grazing sites. These patches are spatially dynamic, moving across the landscape as old patches burn and burned areas are heavily grazed. Buffalo contribute to patchiness by defoliation, trampling, excretion, and dying. Elevated nitrogen in urine

patches increases rates of nutrient recycling and changes plant species composition (Damhoureyeh and Hartnett, 1997). Dead bison carcasses on the prairie decompose and provide a zone of high fertility, causing changes in plant composition and increased biodiversity (Knapp et al., 1999).

## African Grasslands and the Sahel

The Serengeti plains of east Africa provide a living laboratory of grassland:herbivore interactions (Fig. 7-21). The ecological relationships of Serengeti herbivores are described in detail by a number of authors, including Bell (1971) and McNaughton (1979). Observations of similar grasslands in Zimbabwe (formerly Rhodesia) led Savory (1988) to develop his HRM concept, which recommends that domestic livestock be managed in ways that mimic the grazing patterns of wild herbivores.

The **Sahel** is a region of semiarid land in northern Africa that lies between the Sahara to the north and savanna to the south (Fig. 7-22). It includes parts of numerous countries, including Ethiopia, Somalia, Sudan, Chad, Niger, Mali, Algeria, and Mauritania, stretching across the African continent. Since the 1960s, the Sahelian region has endured

**FIGURE 7-21**  The grasslands of east Africa are a living laboratory of grassland:herbivore interactions. (Courtesy of W. D. Hohenboken.)

drought, desertification with the Sahel expanding southward into the savanna, and famine (e.g., Ethiopia, Sudan, and Somalia). **Desertification** refers to long-lasting land degradation producing possibly irreversible desert conditions, due at least in part to human activities. While there is a tendency in these "Columbus was a scoundrel" days to blame disasters in Third World countries on European and western influences, there is a good case to be made that the problems of the Sahel are due to the social and political residues of colonialism, followed by misguided aid and development programs (Sinclair and Fryxell, 1985; Western and Finch, 1986; Orskov, 1987). The simple, but probably incorrect, explanation for the desertification of the Sahel is overgrazing by domestic livestock. In fact, **overgrazing** is the ultimate cause, but the reasons for it are complex. According to Sinclair and Fryxell (1985), historical evidence indicates that the people of the Sahel and their cattle have lived in reasonable ecological balance with their environment for the past 5,000 years, from when domestic cattle first appeared in this region. The degradation of the Sahel followed independence from colonialism and well-meant but misguided aid and development programs. In many cases, the best grazing land was converted to crop production, leaving graziers with the poorest and often the most easily damaged range.

**FIGURE 7-22**  (*Top*): Sand dunes invading a village in Sudan. It is not certain whether the movement of the Sahara Desert south into the Sahel is a natural climatic phenomenon, reversible when the current drought ends, or if it is a permanent desertification caused by human activities. (Courtesy of U. Hellden.)

(*Middle*): Camels transporting trees for use as fuelwood and charcoal production. The increasing pressure on shrubs and trees for fuel and forage is contributing to desertification in North Africa. (Courtesy of U. Hellden.)

(*Bottom*): Desertification in North Africa. Note the heavily browsed shrub, and the complete destruction of the grass cover. Overgrazing due to poor grazing management is contributing to this problem. (Courtesy of S. Louis.)

The livestock farmers of the Sahel have two main systems of life: transhumanance and nomadism. **Transhumanance** is a system in which some of the family members move with the livestock, while a permanent homestead is maintained, with some cropping. The livestock are mainly cattle, camels, sheep, goats, and donkeys. A **nomadic system** is one in which the people are constantly on the move, grazing their animals in a migratory pattern. Animal products (meat, milk, blood) are the principal source of food and wealth. The migration patterns of these pastoralists have traditionally been similar to those of the wild herbivores, moving with the rains to fresh feed. Transhumanance and nomadic livestock production in the Sahel functioned well until recently. Problems began with independence of African nations. National governments and international boundaries were imposed on nomadic peoples, who were discouraged or in some cases prevented from their migrations. The grasslands were grazed continuously with no opportunity for rest and recovery, as the nomadic peoples were forced to become sedentary. New governments encouraged cash crops such as peanuts on the traditional rangelands of the pastoralists (Franke and Chasin, 1981), depriving their herds of forage and further intensifying overgrazing of the remaining grassland. Western countries provided aid for economic development and encouraged crop production. At the same time, the overgrazing in areas populated by the former nomads was observed, so western aid was used to bore wells to provide better distribution of livestock. The people settled around the new wells, and "each bore hole became the center of its own little desert" (Wade, 1974, p. 236). The topsoil was blown away by the dry easterly winds of the dry season. The loss of surface vegetation changed the reflectance of light by the soil surface. The light-colored denuded soil had higher reflectance (**albedo**) than the vegetated land, causing the soil temperature to be cooler. Plant cover and litter absorb heat, leading to thermal currents, and plant transpiration pro-

vides moisture that condenses in the higher atmosphere to form rain. The net effect of the forced overgrazing was a change in the climate, with less rain, and development of prolonged drought conditions. The drought means less forage, still more overgrazing, more loss of vegetation and more denuded land, and still less rain. This feedback system is discussed at greater length by Sinclair and Fryxell (1985). Improved health care provided by the new governments to their people resulted in a rapid increase in the population of both humans and livestock. As desertification proceeded, the livestock stripped the leaves and bark off trees and shrubs to survive. The people, now sedentary, used the wood and any regenerating trees and shrubs for fuel, further intensifying the vegetative changes and soil erosion.

The problems of the Sahel follow a familiar pattern. "Intervention in Africa by the developed countries in the past 100 years has been the unwitting cause of each of the major ecological disasters on that continent" (Sinclair and Fryxell, 1985, p. 992). An ecologically stable nomadic system was destroyed and replaced with rapid population growth, overgrazing, soil erosion, desertification, and unstable cropping systems that emphasized short-term profits at the expense of sustainability. These problems, caused in general by well-intended advice and assistance, may be unsolvable, except by depopulating the Sahel (either voluntarily or involuntarily, as has been attempted in Ethiopia) and starting over to revegetate the land. Planting trees and shrubs, properly managed for forage and fuel, could facilitate the recovery process. Social factors, such as tribal rivalries, are a major impediment but a reality that must be factored into possible solutions.

There is some controversy as to whether desertification has actually occurred in the Sahel. There is evidence that the recent degradation was due primarily to a natural drought, which is a regular periodic occurrence in the Sahelian region (Mattsson and Rapp, 1991). Remote sensing observations from satellite photographs do not support the theory of the advancing desert and reveal rapid recovery and revegetation when the rains returned (Hellden, 1991). Forse (1989) reviewed the Sahelian situation in an article "The Myth of the Marching Desert" and suggested that United Nations efforts to halt desertification have been inappropriate.

Much of the land degradation in the Sahelian region has occurred following the conversion of the more productive rangelands to crop production. Nomadic herdsmen relied on these areas in dry years. Now, when dry years occur, the converted rangelands provide no sustenance for the human population either directly in the form of crops or indirectly in the form of animal products (Dodd, 1994). In the Sahel and other semiarid regions of the world, range livestock production is probably the only sustainable, ecologically sound system capable of producing human food. However, the human population that can be sustained in this manner is relatively low, as has been historically true with nomadic cultures. The human population explosion of the twentieth century has placed an impossible burden on these lands. Foreign aid programs should be designed to understand and enhance the positive attributes of pastoral cultures rather than seeking ways to replace them (Dodd, 1994). Unfortunately, livestock production has been perceived by foreign experts as the problem rather than the solution.

Oba et al. (2000) cite four false assumptions for the failure of rangeland development projects in sub-Saharan Africa:

1. Changes in grazing systems and in patterns of pastoralist land use will improve range productivity.

2. If grazing is rotated by developing water in specific areas of the range, degradation will be reversed.

3. Long-term grazing exclusion will reverse range degradation.

4. Sub-Saharan pastoralists have traditionally maintained more livestock than can be supported by the range.

Oba et al. (2000) maintain that traditional pastoralist land use is well adapted to arid land grazing. There is a high degree of livestock mobility (herding), which allows opportunistic coping with unpredictable rainfall and highly fluctuating forage distribution. Livestock mobility relieves areas of concentration and allows herds to exploit grazing resources that are unevenly distributed in time and space. Managing multiple livestock species—sheep, cattle (grazers), goats, and camels (browsers)—allows optimal use of highly variable grazing resources. These authors maintain that heavy grazing, rather than being destructive, is necessary for proper management of arid zone pastures. In essence, rangeland improvement projects based on results from more humid environments are doomed to fail. Over thousands of years, pastoralists have evolved the most efficient systems for their particular environments, and any tinkering by outsiders is likely to have negative rather than positive effects.

In East Africa, nomadic livestock herders have coexisted reasonably well with native herbivores (e.g., the abundant wildlife of the Serengeti). Lacking government programs for drought assistance, and the like, pastoralists in East Africa have been regulated by the same natural processes that govern wild herbivore populations (Skarpe, 1991). In contrast, European settlement of South Africa has resulted in the conversion of extensive native grasslands into shrub-dominated degraded landscapes (Moll and Gubb, 1989). The native savannah, a mixture of perennial grasses and shrubs, has been "selectively overgrazed and abused to the extent that the grasses have all but disappeared and only the more unpalatable shrubs remain" (Moll and Gubb, 1989, p. 152). Vast herds of migratory zebra and antelope were replaced first by cattle and then by sheep and goats after the grass disappeared. Along with shrubs, low-growing succulents dominate the Karoo (South African term for native grasslands). Rangeland degradation of this type is nearly irreversible. Loss of grass results in extensive erosion from infrequent but intense rains. In a visit to the Karoo region in 1992, I viewed extensive areas that were once grasslands but which are now degraded landscapes of rocks and low-growing succulents. I also observed an intense rainstorm on the Karoo, and the tremendous and rapid runoff of the water, further reducing the limited amounts of soil remaining. Reclamation of this type of degradation is either extremely long-term or impossible. As the grasslands of the Karoo region were degraded, cattle and horses were replaced by sheep and goats, and in many areas now, only goats can be raised (Fig. 7-23). Each cycle of degradation leaves the vegetation more sparse and the land more erodible and more difficult to reclaim (Milton et al., 1994). The last stage in this stepwise progression is a human-made desert (see p. 190).

## Desertification of Rangelands in Asia

In September, 1986, I visited China and first saw the Yangtze River at Nanjing. This huge river, at the end of a dry summer, was the color of mud. Near Shanghai, the brown river was too wide to see across. The amount of soil erosion that caused this huge river to appear as mud was staggering to contemplate.

Sheehy (1992) has reviewed the extent and causes of erosion and desertification of rangelands in north China. Desertification is occurring at the rate of over a million hectares per year, an unparalleled degradation of grasslands. Much of north China and Mongolia is arid or semiarid grasslands, traditionally used for livestock grazing. The bulk of the human population and intensive agriculture of China are in the coastal regions. Traditionally, the semiarid and arid steppe regions have been populated by nomadic tribes, who practiced extensive grazing of livestock, including sheep, goats, cattle, and horses. As is the case with the Sahel in Africa, the traditional system appeared to have been in reasonable balance with the grassland resources. Several factors have caused this to change in the last few decades. Since

**FIGURE 7-23** (*Top*): Goats can cause severe environmental damage and can even climb trees to strip the bark and foliage. (*Bottom*): In this South African former grassland, excessive livestock grazing has reduced the rangeland to acacia trees, which can support only browsing by Angora goats.

1949, the Chinese government has actively promoted the movement of the Han Chinese (the major ethnic group of temperate China) into the steppes of Central Asia, to assure its sovereignty of the area, particularly in areas near Mongolia and Russia. This policy has emphasized the conversion of the grasslands into production of grain and cash crops. As in the Sahel, the land is too arid for viable crop production, and the reduction in native grasslands and increase in sedentary farming populations have restricted the movements of nomadic herders. As a result, both the newly created croplands and the now overgrazed grasslands are subject to severe wind and water erosion.

While livestock grazing has a role in this desertification process, the problem is not livestock per se but political and economic events that have disrupted and destroyed traditional management practices that were in equilibrium with the environment (Frisina et al., 2001). Sheehy (1992) concluded that the political and socioeconomic situation is such that desertification in China will get considerably worse before it is controlled. Continued conversion of arid and semiarid steppe grasslands to crop production has the potential of causing political, social, and economic chaos in China (Sheehy, 1992). Some areas, such as the high altitude rangelands of Tibet (Fig. 7-24), are still in relatively good condition but face increasing pressure from an expanding human population and economic development (Miller and Schaller, 1996).

Asian rangelands are home to a number of endangered animal species, such as the blue sheep (*Pseudois schaeferi*). This endangered animal is threatened by poaching and competition with domestic livestock (Wang et al., 2000). Religion may play a role in wildlife conservation in Asia. The blue sheep occurs in Tibet (Fig. 7-25). The Buddhist religion emphasizes the sanctity of all life; educational programs invoking religious sentiment may help to reduce poaching of endangered species (Wang et al., 2000).

## Livestock Grazing and Australian Grasslands

The flora of Australia evolved in the absence of many large mammalian herbivores. As a result, Australian grass- and shrub-lands have been quite vulnerable to damage from heavy grazing by livestock and feral exotic animals (e.g., rabbits, camels, goats) introduced by Europeans. Additionally, the soils of Australia are of a chemical composition that makes them susceptible to "slipping" so that

**FIGURE 7-24**  High altitude rangelands in Tibet support nomadic herders and their livestock consisting mainly of yaks. (Courtesy of Winifred B. Kessler.)

FIGURE 7-25  The endangered blue sheep of Tibet. Buddhist emphasis on the sanctity of all life may help to reduce poaching of endangered species. (Courtesy of Daniel J. Miller.)

FIGURE 7-26  (*Top*): Aerial view of sheep on rangeland in Australia, showing overgrazing and formation of trails. (*Bottom*): The overgrazed land is highly susceptible to erosion. (Courtesy of G. Grigg.)

FIGURE 7-27  Many Eucalyptus forests in Australia suffer from die-back, a condition associated with overgrazing by livestock. (Courtesy of R. A. Leng.)

when grass cover is denuded, extensive gully erosion occurs (Fig. 7-26). Thus many areas of Australia show significant degradation resulting from livestock production. For both grazing and cropping purposes, much of the indigenous eucalyptus forest has been cleared. In Western Australia, the eucalyptus trees functioned as pumps by their transpiration of water. Following deforestation, the water table has risen to the surface, leaving mineral salts at the surface as the water evaporates, and thus large areas of Western Australia now are afflicted with severe soil **salination**. In other parts of Australia, eucalyptus trees in pastoral areas are in a declining state, with trees suffering from **die-back** (Fig. 7-27). This condition results from a complex interaction of factors associated with grazing, including changes in insect populations and soil grubs, soil compaction by cattle and sheep, destruction of seedlings to prevent replacement of aging trees by new growth, and pasture fertilization programs (Ohmart and Edwards, 1991).

As in many other areas of the world, predator control programs to protect livestock have severely impacted wildlife, including extinction of the top carnivore in Australia, the Tasmanian tiger. Efforts to reclaim Australia's degraded rangelands are in progress, including a massive tree-planting effort, the "Greening Australia" project, with a goal of planting 1 billion trees.

Extensive changes in rangeland characteristics have occurred in Australia as a result of development of artificial sources of water for livestock. Typically, water is obtained by "bores" or wells, and often the water flows unaided to the surface because of subterranean pressures (artesian wells). Most of the effects on native vegetation and wildlife are negative, as described in an extensive review by James et al. (1999). Negative effects of water development include overgrazing, subsequent loss of native vegetation, and extinction of small birds and mammals due to introduced predators (cats, foxes) that rely on drinking water.

Australia has abundant types of unique wildlife species such as kangaroos. In the past, the main efforts of stock raisers (graziers) have been to exterminate wildlife in the belief that they are competitors of sheep and cattle (Edwards et al., 1996). However, there is now interest in non-consumptive exploitation of Australian wildlife in endeavors such as ecotourism (Croft, 2000). Croft (2000) asserts that there are many such opportunities, if "entrepreneurs with the imagination to go beyond a 19th century view of wildlife exploitation (namely, eat it or wear it)" (p. 92) become involved. He recommends "in situ" commercial use of wildlife, such as wildlife tourism, in which nutrients are recycled within the system rather than being exported as components of meat or hides.

## Forage Trees and Shrubs on Rangelands

On arid and semiarid lands, the inclusion of trees, shrubs (brush), and other non-traditional forage plants in grazing ecosystems may be desirable. Many semiarid rangelands exist naturally in this form, as savanna. The foliage of trees and shrubs, especially leguminous ones, is often high in protein and other nutrients. *Acacia* spp. are important shrub legumes in the Sahelian region of Africa and are browsed upon by livestock, particularly goats and camels. Legume shrubs often contain high levels of tannins, which can reduce palatability and protein digestibility. However, some animal species have tannin-binding salivary proteins, which reduce the adverse effects of tannins (Mole et al., 1990). An intensive discussion of the biology and utilization of shrubs is provided by McKell (1989). Nutritional properties of forage trees and shrubs are reviewed by Topps (1992).

The prickly pear cactus (*Opuntia* spp.) can be utilized as a forage in arid areas. The spines are burned with a propane torch (Fig. 7-28) before use as a feed. This allows the cactus to be stored as an emergency feed and provides a means of controlling intake. Russell and Felker (1987) reviewed the characteristics of prickly pear and indicated that it deserves greater consideration as a forage crop in arid areas. They proposed that a combination of prickly pear with shrubby legumes such as *Prosopis* (mesquite) and *Acacia* spp. may be a productive forage system for arid environments, with the cactus providing energy and the legumes protein. In South Africa, spineless cactus (*Opuntia* spp.) is extensively grown as a drought-resistant emergency fodder crop. Plantations of spineless cactus are established in fenced paddocks from which livestock are excluded. During droughts, the cactus is manually or mechanically harvested to provide emergency feed. Guevara et al. (1999) found spineless cactus to be an economically viable forage in an arid area of Argentina. The spineless cactus is a recessive mutant of prickly pear and is destroyed by overgrazing if livestock are allowed free access to it. The American aloe (*Agave americana*) is also grown in South Africa as a drought fodder. Because of the fibrous nature of the leaves, they must be chopped or milled before animals will consume them. The aloe is also very effective in controlling soil erosion. When planted on slopes, it allows water from heavy rains to pass through, while retaining debris and silt. Erosion is a serious problem in arid areas such as South Africa and Australia; infrequent but heavy rains cause flash flooding and severe erosion of rangelands.

**FIGURE 7-28** (*Left*): Prickly pear cactus can be used as an emergency cattle feed by burning off the spines with a propane torch. (Courtesy of C. Russell.) (*Right*): Spineless cactus is widely grown in South Africa as emergency fodder for livestock during droughts. This is a fenced-off cactus plantation on a South African ranch.

Another drought-resistant fodder plant widely grown in South Africa is Oldman saltbush (*Atriplex nummularia*), a native of Australia. Saltbush is a **halophyte** (salt-tolerant plant) with a high concentration of salt in the leaves. *Atriplex* spp. concentrate salt in cells of the outer tissues of the leaves. Eventually, these cells burst, depositing crystalline salt on the leaf surface. The salt provides a reflective coating that shades the leaf from direct sunlight. The sharp salt crystals may also deter herbivores (Mares et al., 1997). Nevertheless, saltbush is readily consumed by livestock and is tolerant of heavy grazing pressure. Saltbush can even be established in arid coastal areas by irrigation with sea water. Another plant with forage potential for arid and semiarid lands is *Kochia scoparia* (kochia). Kochia is commonly viewed as a weed, but it produces about the same amount of digestible energy and protein as alfalfa with about half the water requirement. The nutritional and other properties of kochia as a forage have been reviewed by Rankins and Smith (1991). Tree lucerne (tree alfalfa) or tagasaste (*Chamaecytisus palmensis*) is a leguminous shrub used in parts of Australia as a forage, particularly under drought conditions (Borens and Poppi, 1990).

More than 60 million hectares of irrigated cropland worldwide have been damaged by salt. One method of reclaiming these lands for agricultural use would be by developing salt tolerant crops. Traditional plant breeding to accomplish this goal has been unsuccessful. Zhang and Blumwald (2001) developed transgenic tomato plants with an enhanced ability to sequester sodium in their vacuoles, allowing them to grow in saline soils. Sodium accumulated in the leaves but not in the fruit. Perhaps forage crops such as alfalfa could be genetically engineered in a similar manner. Biochemical mechanisms involved in salt tolerance of halophytes have been reviewed by Glenn and Brown (1999).

# LIVESTOCK–WILDLIFE INTERACTIONS

Livestock and wildlife interactions in riparian zones and wetlands have already been mentioned. There are numerous other interactions, such as competitive use of forage resources, predator control, disease transmission between wildlife and domestic animals, game ranching, and effects on endangered species. These will be briefly discussed.

## Habitat and Food Resource Interactions

Livestock and wildlife on rangelands share feed and habitat. Are their interactions detrimental or beneficial to themselves or the rangeland? Both effects can occur, or sometimes there may be no effect. With regard to feed, there may be **dietary overlap,** where domestic and wild species are competing for the same resources, or domestic livestock and **wild herbivores** may be utilizing different resources. The interactions of domestic livestock with free-ranging large herbivores have been reviewed by Vavra et al. (1989).

As discussed in Chapter 4, herbivores vary in feeding behavior and strategy and can be classified as grazers (roughage eaters), intermediate feeders, and concentrate selectors. Cattle, horses, elk, and bison are grazers; sheep and goats are intermediate feeders; and pronghorn and mule deer are concentrate selectors, to use examples from western U.S. rangelands. Domestic cattle, wild horses, and elk would thus be expected to overlap in their diets and compete for forage, whereas cattle and pronghorn would not be expected to have much dietary overlap. However, there are other factors involved. Large herbivores show plasticity in their diet preferences. Under some conditions, cattle may browse on trees and shrubs, and mule deer may graze on grass. Another factor besides dietary overlap is distribution overlap. For example, in Oregon research reviewed by Vavra et al. (1989), deer and cattle had a high distributional overlap (i.e., were found occupying the same habitat) but had little dietary overlap (they were eating different things). On the other hand, cattle and elk had a high probability of dietary overlap, but they occupied different habitats. Deer and elk had neither dietary nor distributional overlap. Thus the determination of potential competition between livestock and wild herbivores for feed and habitat is not easily assessed. As Vavra et al. (1989) point out, a simple animal unit equivalency (a cow eats 12 kg per day and a deer 2 kg, so one cow equals 6 deer for stocking rate determination) is clearly

unacceptable. Competition for forage is more likely between elk and cattle than between mule deer and cattle (Coe et al., 2001). Livestock can be used as management tools to improve habitat for big game species such as deer and elk, which often prefer to graze on the succulent regrowth of areas previously grazed by livestock, rather than stay in areas where livestock have been excluded. Carefully managed livestock grazing can improve forage quality on elk winter range (Clark et al., 2000).

Much study has been made of the interactions between herbivores on the Serengeti plains of East Africa. A large and diverse population of wild ruminants (antelope species, wildebeest, giraffe) and nonruminant herbivores (elephant, rhinoceros, hippopotamus, zebra) share the forage resources very efficiently and completely, all the way from aquatic vegetation (hippo) to foliage high in trees (elephant, giraffe). These diverse herbivores avoid competition by different feeding preferences and strategies, as well as by migration patterns (Bell, 1971). In Africa, selective harvesting of these wild herbivores for meat and other products is likely to be ecologically and economically more productive than moving domestic livestock into these grasslands and displacing or disrupting the wildlife. The presence of abundant wild herbivores on ranches in Africa has an additional benefit, in that predators (lions, leopards) selectively hunt wild game rather than livestock (Mizutani, 1999). There is increasing realization in Africa that the economic values of wildlife (meat, hides, curios, trophy hunting, photographic safaris) can exceed the returns from livestock and that optimal human economic development will be best served by the integration of sustainable wildlife utilization into multispecies animal production systems (Du Toit and Cumming, 1999). Emphasis on exports of beef to Europe from countries like Namibia and Botswana is diminishing, especially after the 2001 foot and mouth disease epidemic in Europe, which is believed to have originated in meat imported from Africa.

On western U.S. rangelands, interactions between livestock and wild herbivores are being modified by human activities. Traditional winter ranges of wild herbivores such as elk have been converted to ski areas and condominium sites. Cities, towns, and farms now occupy areas that once were winter ranges for wildlife. In some cases, this has forced wildlife and livestock into greater competition for feed and habitat. Elk may be as numerous as cattle at hay stacks on ranches. Ranchers rightly believe that they shouldn't bear the total cost of feeding the elk over the winter; perhaps the people who own the ski lodges and those who use them should pay a surcharge to cover the cost of winter feeding the wildlife that their activities have displaced.

Livestock production may sometimes increase **wildlife habitat.** Most farms with grazing livestock have fencerows with trees and shrubs, shelterbelts of trees, and other wooded areas for shade. The pastures are often somewhat poorly managed (from an agricultural standpoint) with weeds like thistles, wild rose bushes, and hawthorns, which provide nesting sites and food for birds. The fencerows and pastures provide habitat and food for many types of small mammals and songbirds, as well as foxes, raccoons, skunks, pheasants, and hawks. Earthworms, grubs, and insects abound in a pasture, providing food for songbirds. During the winter, grass seed in hay fed to livestock serves as food for songbirds. Contrast a cattle farm with intensive corn, soybean, or wheat production. With the intensive crop production, there are generally no fencerows at all, and the crops are grown in monoculture with virtually all weeds controlled by herbicides. Insects, earthworms, and grubs are killed by pesticides. Such environments do not provide much wildlife food or habitat and are basically year-round biological deserts. In Europe, it has been proposed that farmers should be paid for wild birds that nest on their property (Musters et al., 2001).

With appropriate management, livestock grazing can be used to enhance wildlife habitat (Severson, 1990). In wetlands, cattle grazing assists in suppressing growth of hydrophytes such as cattails (*Typha* spp.) and willows, which if undisturbed form a dense mass of vegetation, impeding water fowl nesting and most other

wildlife uses. In many areas of the great plains of western North America, the **prairie potholes,** the traditional nesting sites of migratory waterfowl, are becoming degraded by encroaching cattails and willows, severely reducing waterfowl reproduction. Cattle grazing fills the ecological role historically performed by bison (Fig. 7-29). Reduced livestock production in many grain-producing areas of the United States and Canada has led to degradation of the potholes as wildlife habitat.

Livestock production can benefit wildlife in other ways. Salt and mineral supplements supplied to grazing livestock are also used by deer and other wild herbivores, improving their nutritional status. Mineral deficiencies exist in many areas. For example, selenium deficiency is widespread, and wild herbivores benefit from selenium-containing supplements provided for livestock. Because of human invasion of wildlife habitat, wildlife are often more restricted in their range than was once the case, so if they live in an area characterized by a mineral deficiency, they may not have the opportunity to forage in other areas where the mineral status is adequate.

Livestock production may increasingly interact with preservation of **endangered species.** The Pacific salmon-riparian zone issue is an example. Reintroduction of wolves into various areas of the continental United States, as currently proposed by many environmentalists, would certainly interact with livestock production, as discussed in the next section (Predator and Pest Control Programs). In the U.S. Southwest, ranching has been criticized for having adverse effects on the survival of the endangered desert tortoise (Fig. 7-30). Berry (1986) reviewed some of the impacts of livestock grazing on desert tortoise populations. Sheep grazing results in reduced feed resources for the tortoises, as well as some physical damage to the animals and their burrows, causing increased tortoise mortality. Berry (1986) estimated that about 56 percent of crucial tortoise habitat was currently (1986) grazed by sheep and cattle, which is much less than historic figures. In Texas, Kazmaier et al. (2001) found no differences in numbers of

**FIGURE 7-29** (*Top*): A typical prairie pothole in North Dakota. These potholes are the major breeding areas of millions of North American waterfowl. Cattle, like the bison before them, help preserve the potholes by trampling vegetation such as cattails, which if undisturbed gradually fill in the pothole. (Courtesy of H. A. Kantrud.)

(*Bottom*): A pothole protected from cattle grazing. It has become overgrown with vegetation and ruined as waterfowl habitat. Introduction of cattle would restore the area to resemble the top figure. (Courtesy of H. A. Kantrud.)

**FIGURE 7-30** Environmentalists claim that the survival of the endangered desert tortoise in the U.S. Southwest is threatened by sheep grazing. (Courtesy of W. D. Edge.)

desert tortoises on grazed (by cattle) or ungrazed areas. They concluded that moderate grazing by cattle is compatible with maintenance of desert tortoise populations. Desert tortoises are by far the most abundant land animal ever listed as threatened or endangered in the United States (Freilich et al., 2000). However, their range is very large, so it is a problem for biologists to determine demographic trends in a large but sparsely distributed population. Freilich et al. (2000) found that desert tortoises are likely to be undercounted in dry years, as they remain in their burrows. Thus some estimates of low numbers may simply be artifacts due to environmental conditions.

The influences of livestock grazing on wildlife are contextual; in other words, they may be positive, negative, or no effect, depending on the context. For a given wildlife species, some influences of livestock may be positive and others negative. For example, the **sage grouse**, native to western North America, has been declining alarmingly. The sage grouse is dependent on large, continuous blocks of sagebrush habitat (Beck and Mitchell, 2000). Much of the destruction of sagebrush habitat has been by the ranching industry. Sagebrush range has been sprayed, burned, bulldozed, and so on in order to reseed with exotic forage grasses such as crested wheatgrass. Sagebrush is fire sensitive, and does not resprout after burning. Burned range encourages annuals such as cheatgrass, which has an explosive burning potential, reducing the likelihood of sagebrush recovery. On the other hand, high levels of livestock grazing can reduce competition between grasses and sagebrush, triggering increases in sagebrush density. Grazing reduces the buildup of fire fuels. Cattle grazing can stimulate forbs and insects that are sage grouse foods. Grazing provides bare areas (leks) that the birds use in mating displays. But grazing reduces grass cover that is needed for protection of nests. Certain species of ground squirrels, which are a major cause of egg predation, are favored by high levels of grazing. Thus some of the effects of livestock on sage grouse are negative, but with improved grazing management, livestock can actually be used to improve sage grouse habitat (Beck and Mitchell, 2000). Both prairie dogs and sage grouse have been proposed for listing under the Endangered Species Act (Frisina et al., 2001). Range management that favors recovery of prairie dogs is detrimental to sage grouse, and vice versa. Management to maintain these species, while allowing for range livestock production, requires achieving a balance. It may not always be possible to manage for maximum numbers of prairie dogs, sage grouse, and cattle on the same land base, but appropriate management can ensure a place for all of them (Frisina et al., 2001). Thus there is no simple answer.

Livestock grazing can influence populations of many types of birds. Some endangered birds, such as the mountain plover and McCown's longspur, use heavily grazed sites for nesting (Milchunas et al., 1998). The brown-headed cowbird (*Molothrus ater*) has expanded its range greatly in North America since the arrival of cattle. Prior to European settlement, cowbirds were confined to the grasslands of the Great Plains, where they had a nearly obligate commensalistic relationship with bison. Cowbirds typically feed in open habitats with short vegetation, such as those created and maintained by grazing ungulates. Cowbirds are brood-parasitic; they lay their eggs in other birds' nests. When the eggs hatch, the nestling cowbirds kill or outcompete the host's chicks. Increases in cowbird numbers are believed to be a significant factor in the decline of many North American songbirds. Management of cattle to rotate them away from songbird habitat during breeding season appears to be a useful strategy for reducing cowbird parasitism of songbirds (Goguen and Mathews, 2000).

Several endangered butterfly species depend upon livestock grazing to maintain their essential habitat of short grass swards (Weiss, 1999), such as the Bay checkerspot butterfly native to the San Francisco Bay area. In 1994 and 1995, severe population crashes of check-

erspots occurred after cattle were removed from two sites. Ironically, the biggest threat to the continued existence of these butterflies is the pressure for removal of cattle due to urbanization of the butterflies' habitat.

Cattle grazing on desert rangelands may impact the microhabitats of desert rodents, probably by affecting the amounts and types of vegetation. Jones and Longland (1999) showed that cattle, by preferentially feeding on certain plants, can create conditions that are more suitable for some species of rodents, while reducing microhabitat for other species. In California, Germano et al. (2001) observed that thick herbaceous cover had negative effects on threatened or endangered kangaroo rats, ground squirrels, and lizards. Cattle grazing, by removing dense plant cover, improved the habitat of these animals. They recommended that at least some of the conservation lands in the San Joaquin Valley should be grazed by well-managed herds of livestock.

It should be apparent from the preceding examples that there is no simple answer to the question of whether livestock grazing is good or bad for wildlife. It is not very useful for either pro- or anti-livestock people to pick out examples that support their point of view regarding livestock grazing. Each situation is different, and must be studied carefully; people with opposing views should not make rash decisions.[7] While farmers and ranchers are

---

[7] It is easy to pick out examples that support your point of view and ignore those that don't. There are numerous examples demonstrating that livestock grazing harms wildlife habitat. Environmentalists may (and often do) trumpet that information. There are other examples showing that livestock grazing improves wildlife habitat. Environmentalists may ignore that data, while ranchers trumpet it. The truth is that the effects of grazing on wildlife habitat are situational or contextual. There are situations, especially involving endangered species, where livestock grazing clearly should be curtailed. There are other situations where increased grazing is desirable. These decisions are best made by informed natural resource managers, on the scene, with local inputs.

often portrayed as "bad guys" by environmentalists, surveys indicate that the majority of American agricultural producers have favorable perceptions about wildlife on their farms and ranches (Conover, 1998) and are key players in wildlife conservation. Farmers and environmentalists are on the same team, although they are only beginning to realize it (Cheeke and Davis, 1997).

In the U.S., there is widespread opposition by livestock and forestry organizations to the **Endangered Species Act,** calling for relaxation of the guidelines for preservation of endangered species when such preservation would have negative economic impacts such as loss of jobs. The mood of society at large is not favorably inclined to allow other species to become extinct to preserve jobs or allow business as usual. Modifying the Endangered Species Act is not consistent with the image of farmers and ranchers as "the original environmentalists." This is another example of the agrarian value system (see Chapter 11) being opposed to the values of the larger society. Sometimes it will be necessary to "bite the bullet" and take an economic hit when species are threatened with extinction. Slogans like "humans are more important than fish" won't generate much sympathy.

## *Predator and Pest Control Programs*

Livestock production was in large part responsible for the complete elimination of wolves and the almost complete elimination of grizzly bears from the western United States. Except for a few wolves in Minnesota and Michigan, and the small populations of grizzly bears in National Parks, both the wolf and grizzly have been extirpated from the continental United States. A principal reason for the elimination of these animals was that they were **predators** of sheep and cattle. However, these animals would probably have been eliminated regardless of livestock losses. Humans have an innate fear of wolves, and the grizzly bear is a predator of people. It is very

**FIGURE 7-31** The wily coyote has thrived on ranchers' attempts to exterminate it. The public outcry against predator control programs will force ranchers to coexist with coyotes. Ranchers may discover that coexistence is mutually beneficial. (Courtesy of W. D. Edge.)

difficult for grizzlies and humans to coexist in Yellowstone National Park, for example. Numerous campers and hikers have been attacked and killed by grizzlies in Yellowstone.

In contrast to the wolf and grizzly, the coyote (Fig. 7-31) thrived on our efforts to eliminate it and has expanded its range as an animal of the western plains to cover the entire continental United States, in part because of the elimination of wolves. Coyotes generally do not remain in wolf habitat. Like the rat, the coyote thrives on association with humans. Our livestock and poultry and other farming activities provide it with plentiful food, and elimination of larger predators, which once kept coyotes in their place, has allowed the wily creature to occupy environments as diverse as western rangelands and the suburbs of Los Angeles (where it feeds on garbage, pets, and occasionally children). Recently (early 1990s), the cougar or **mountain lion** has increased in numbers in the western United States, largely because of legislation to protect the animal. Numerous cougar attacks on humans have occurred, and in some areas, such as Vancouver Island in British Columbia, the cougars are particularly aggressive towards people.

In the 1950s, the western livestock industry threw its entire arsenal at the **coyote** (*Canis latrans*). The animals were shot, trapped, poisoned, chased by helicopters, and their dens dynamited. The coyote thrived! In the 1960s and 70s, the developing environmentalist movement targeted predator control programs because poisoning and trapping were killing non-target species and protected species such as the golden eagle. Animal rights organizations were against trapping, poisoning, and denning (dynamiting coyote dens) for obvious reasons! Since that time, predator control has been an increasingly controversial issue.

Knowlton (1989), Wagner (1988), and Knowlton et al. (1999) reviewed the interactions of predators (mainly coyotes) and livestock (mainly sheep). There are many possibilities for helping to prevent losses of livestock to predators. Removal of predators should be selective. Not all coyotes kill lambs. The "killers" or "problem individuals" need to be removed (Linnell et al., 1999); the coyotes that feed only on mice and ground squirrels are actually beneficial to the rancher. Interestingly, both domestic and wild ungulates seem to be able to recognize which members of a coyote pack are a threat and act aggressively only towards certain individuals (Gese, 1999). Ungulates direct aggressive behavior towards the individuals (alpha coyotes) that have the greatest tendency to initiate attacks. The abundance of natural prey can influence coyote predation on sheep. For example, when jackrabbits are abundant, they are a buffer against coyote attacks on sheep (Stoddart et al., 2001). Often, a series of factors interact to cause a predator problem. Lambs are born in the spring, at the same time as coyotes produce their pups. Other food for coyotes is often in short supply in early spring, when the food requirements are greatest for feeding the pups. The situation suggests several solutions. Change the lambing time or move the sheep to a protected area. Use high-voltage electric fencing (New Zealand fences), which coyotes will not cross. Use guard dogs; there are several breeds that are very effective in keeping

## A DOG-EAT-DOG WORLD? BIG DOGS RUN THE SHOW!

*(Review of J. Cohn. 1998. "A Dog-Eat-Dog World?" BioSci. 48:430–434.)*

This paper is an interesting account of competition among North American canids, which range in size from the wolf on down to coyotes, foxes, and small endangered foxes such as the kit fox and swift fox. Domestic dogs of various sizes, depending upon breed, are also in the mix. As a general rule, larger canids beat up, harass, or kill smaller canids. At one time, coyotes were relatively uncommon in many parts of North America, due to the presence of wolves. Wolves kill coyotes, and coyotes, being intelligent, leave the country if wolves are present. The extirpation of the wolf from most of the United States cleared the way for coyotes, which now occur in all parts of the continental United States. Following the reintroduction of wolves into Yellowstone National Park, the number and size of coyote packs have decreased by about one third. Coyotes, in turn, attack red and gray foxes. Red foxes are a major cause of death of the small, endangered kit fox found in California. In some cases, competing canids can coexist. Kit foxes have multiple dens and use their dens year-round. They are usually not far from the safety of a burrow too small for a coyote or red fox to enter. Surprisingly, the presence of coyotes is beneficial to the kit fox, because they eliminate red foxes. Kit foxes can successfully share habitat with coyotes. Red foxes are major predators of nesting birds, some of which are endangered. On some California nature preserves, coyotes are beneficial to several species of endangered birds, by suppressing red foxes. In parts of southern California, conservationists regard coyotes as an important conservation tool.

predators at bay. These include Akbash, Great Pyrenee, Anatolian, Maremma, and Komondor breeds. The most important characteristics of guard dogs are high aggressiveness to predators, great attentiveness to sheep, and great trustworthiness (Andelt, 1999). Run some llamas, donkeys, or billy goats with the sheep; they are somewhat effective at keeping coyotes away from the flock. Use of guard llamas can reduce losses to coyotes and dogs (Meadows and Knowlton, 2000); large, alert llamas tend to be the most aggressive towards dogs (Cavalcanti and Knowlton, 1998). McAdoo and Glimp (2000) and Knowlton et al. (1999) offer a number of good sheep management techniques that can help reduce predation. These include selection of season of lambing, range-use timing, use of skilled herders, avoiding problem areas, and use of guard animals. The most bizarre (to a sheepman) idea is to feed the coyotes dog food during the critical whelping period! If the coyotes are killing lambs to feed their pups, perhaps feeding them during this period would be cheaper than losing lambs. Coyotes are territorial, and the same pair will occupy a den for several years. Perhaps it would be wise to feed them in the spring; after the pups are grown and gone to a different location, the parents will spend the rest of the year controlling mice, ground squirrels, and gophers. (The pups disperse because they must leave their parents' territory. If they don't find another unoccupied territory for their own, they will not survive.) If the original pair of coyotes is killed, they will be replaced by a new pair the next year. This cycle can go on endlessly. It would be better to keep one non-sheep-eating coyote pair on a given territory for several years, helping them through the critical period by feeding them so they don't kill lambs, than battling a new pair each spring, which are forced to kill lambs because they don't have enough other food. These ideas, of course, don't work as well in practice as in theory and are not likely to be received in a friendly fashion by most sheep producers! However, the livestock industry will be forced

to move in this direction; trapping and killing predators in large numbers does not meet with the public's approval, and the majority rules eventually in a democracy. The successful application of a "coyote-friendly" management system on an Oregon cattle ranch has been described by Hyde (1986). Hyde encourages coyotes to live on his ranch by friendly actions towards them and benefits from their control of ground squirrels and gophers, which reduce grass production. The cattle are managed so that calves are not exposed to potential coyote predation in their vulnerable first few weeks of life.

One of the reasons that coyotes thrived and expanded their range despite efforts to reduce their numbers is that they can adjust their reproduction in response to "being thinned out." When coyotes are heavily controlled, they respond with increased litter sizes and increased percentages of yearlings reproducing (Knowlton et al., 1999). The larger litters increase food requirements so that predation on livestock and wildlife can actually increase in response to predator control.

Farmers and ranchers have traditionally defended predator control programs on the grounds that it is more important to produce food for people to eat than to live with the predators. The general public is now saying (by their eating and buying habits) that it is more important to them to preserve wilderness than to eat lamb (which can be imported from New Zealand anyway). This argument is currently pertinent, because of the implementation of a program (discussed earlier) to reintroduce the **wolf** into Yellowstone National Park. This program was fought with all possible vigor by the sheep and cattle industries of Montana and Wyoming. The weight of public opinion prevailed and wolves have been released into the Park. Ironically, increased numbers of large predators such as the **grizzly bear** and **mountain lion** will likely result in human mortalities. This is already happening with mountain lions. The wolf, despite the folklore, rarely attacks people. The grizzly is notorious for its ferocity and unpredictability.

It should be appreciated that not all predator control programs are for the benefit of livestock producers. Control of coyotes, ravens, raccoons, skunks, and so on is widely used in wildlife refuges to protect nesting birds, particularly endangered species such as the whooping crane.

**Pest control programs** such as the poisoning of prairie dogs on rangeland are also controversial. The near-extinction of the **black footed ferret,** which lives exclusively in prairie dog colonies (towns), occurred because of the extensive poisoning of **prairie dogs.** This is another example of conflict between the interests of ranchers, farmers and environmentalists.

## Game Farming and Ranching

The raising of wild animals on game farms or ranches is a controversial subject in North America. Several species of deer (red deer, fallow deer, etc.) are raised commercially for meat, hides, and antlers. Deer farming is especially important in New Zealand and Germany. There is both strong interest and strong opposition to **game farming** in North America. Valerius Geist, of the University of Calgary, is an internationally prominent wildlife biologist who is vigorously opposed to game farming (Geist, 1985, 1989). Briefly, he proposes and provides evidence that native North American animals are less competitive than their counterparts in other parts of the world. Both the survivors of the megafauna extinctions of the last Ice Age and the "new" species that entered North America over the Siberian land bridge are less competitive than their Eurasian counterparts. The "old" megafauna that survived the Ice Age, like the white-tailed deer, black-tailed deer, and pronghorn, are species that deliberately avoid competition in favor of dispersal. The "new" North American animals that came from Siberia, such as the bighorn sheep, elk, moose, and caribou, are, according to Geist (1989), very susceptible to southern pathogens because of their evolu-

tionary history in the dry, cold climate of Siberia, which was inhospitable to pathogens and parasites. The "old" species, such as the native American deer, are not competitive with introduced species from Europe or Asia. Geist (1985, 1989) provides evidence that wherever these introduced species escape from game farms, as they inevitably do, and establish local populations, they drive out the native deer or hybridize with them, causing extinction by genetic pollution. The "new" species from Siberia, such as the bighorn sheep (Fig. 7-32), are very susceptible to the diseases and parasites of domestic sheep, to which they have very little resistance (Singer et al., 2001). They are unable to coexist with domestic sheep or, increasingly, escaped wild sheep from game farms (e.g., barbary sheep). The American Indians, also of Siberian ancestry, were similarly extremely vulnerable to the diseases of European immigrants and have not been competitive with immigrants from the Old World.

Fraser (1996) found that sika deer *(Cervus nippon)* from Japan when introduced into New Zealand steadily replaced the previously introduced red deer *(C. elaphus)*. Sika deer are grazers rather than browsers, giving them a competitive nutritional advantage in many environments, including parts of the United

**FIGURE 7-32** Bighorn sheep are highly susceptible to diseases of domestic sheep and are unable to survive in direct contact with domestic sheep. (Courtesy of Winifred B. Kessler.)

States where sika deer are out-competing the native white-tailed deer.

Thus game ranching in North America is controversial. However, in numerous other parts of the world, it offers considerable potential. In southern Africa, for example, game ranches have been successful in helping to preserve native species (Stanley-Price, 1985). The animals are managed for trophy hunting. It is in the best interests of the ranch owner to use management that produces sustainable populations, with a modest level of harvest (primarily of aged animals that have trophy-size horns or antlers). Integrated cattle and game ranching with a balance of grazing and browsing game species optimizes land use in many areas of southern Africa (Taylor and Walker, 1978). In South Africa, land owners are considered the owners of wildlife on their property, giving them an incentive for sustainable harvesting.

In some parts of the world, native herbivores that have coevolved with the grasslands may offer potential as replacements for traditional livestock. Sheep growers in Australia spend large amounts of money and effort to reduce kangaroo numbers. Grigg (1988) proposed that the best way to reduce desertification in Australia would be to develop a market for kangaroo meat and other products in order to give farmers an economic incentive to reduce their reliance on sheep, in favor of free-ranging kangaroos. Sheep production has caused tremendous land degradation in Australia, as well as contributing to the loss of native wildlife (e.g., extinction of the Tasmanian tiger because it killed sheep) and severe damage to native vegetation (e.g., eucalyptus dieback, associated with sheep production). Kangaroos have soft padded feet (Fig. 7-33), whereas the hard feet of sheep cause soil compaction and erosion. Because of the economic limitations of sheep production, Australian graziers have begun raising other livestock such as goats. Besides their hoof impact, goats cause severe damage to trees and shrubs. According to Grigg (1988), "If goat farming becomes widespread and intensified, we can expect sheep rangelands, already barren and

**FIGURE 7-33** The native marsupials of Australia such as the kangaroo have soft padded feet, whereas the hard feet of sheep cause soil compaction and erosion.

overgrazed, to take the final step into desert. If you doubt this, think of northern Africa" (p. 127). Other advantages of kangaroos over sheep and goats include their low water consumption (about half of the amount consumed by sheep), their low metabolic rate (about 30 percent less than for sheep), and hence their low maintenance energy requirements (Grice and Beck, 1994).

In North America, native herbivores such as deer, elk, and bison may warrant greater consideration as viable replacements for sheep and cattle on rangelands. These herbivores, and particularly deer, have evolved a seasonality of feeding behavior, whereby they store large amounts of body fat in summer, voluntarily reduce feed intake in the winter, and mobilize body fat to meet their energy needs. This seasonality in feed intake might be of advantage in reducing winter feeding requirements, if wild populations were managed for meat production. There is considerable interest in bison ranching on western rangelands. While cattle production is viewed negatively by many environmentalists, bison are in general held in high esteem, perhaps because of their earlier persecution to almost the point of extinction. While bison and cattle are closely related as bovids, there are some differences between them in grazing behavior that may cause cattle and bison to differ in their effects on grassland vegetation and biodiversity (Hartnett et al., 1996).

China is a special case as far as game farming is concerned (Fig. 7-34). As discussed previously (Chapter 3), traditional Chinese medicine utilizes various animal parts (paws, gall bladders, penises, antlers, horns, etc.) from a variety of animals (rhinoceros, tiger, bear, turtle, deer, etc.) With the high human population of more than 1 billion people in China, and their increasing affluence, the demand for these products is increasing and having a severe negative impact on the world's wildlife. Farming of animals such as bears, tigers, and turtles is taking place in China. While this is strongly opposed by animal rights advocates, it is probably preferable to meet the demand in this way than to have species such as tigers hunted to extinction in the wild. Because of a deeply ingrained belief in the medicinal pow-

ers of animals in China, it is unlikely that the demand for animal parts can be reduced in the near future.

## The Livestock–Wild Horse Controversy

Nostrils flared, mane and tail streaming in the wind, the wild stallion gallops effortlessly mile after mile across the dusty land (Fig. 7-35). This magnificent animal, and others of its kind, are being driven to extinction by the relentless greed of western cattlemen, who covet every last bit of forage on public lands of the American west. This may be the popular perception of the wild horse (mustang) on the western rangelands of North America. How valid is this perception?

Evolution of the horse and other equids has been extensively studied (Colbert, 1969; Simpson, 1950, 1953), drawing upon an abundant fossil record. The ancestors of the modern horse evolved in the Pleistocene epoch in North America, approximately 50 million years ago. These ancestral animals migrated from North America to South America and Europe by land bridges existing then. For reasons not fully known, the horse, along with numerous other

**FIGURE 7-34** The young antlers (velvet) of deer are a highly prized medicinal item in China. (Courtesy of R. Marten.)

**FIGURE 7-35** Born free: wild horses galloping across the dusty range, manes and tails streaming in the wind. Control of wild horse numbers has been one of the major environmental controversies of the western United States, pitting ranchers against environmentalists and animal rights activists. (Courtesy of J. C. Buckhouse.)

large herbivores, became extinct in North America following the last Ice Age. The so-called **wild horses** found today in North America are actually **feral** (domestic animals gone wild) **horses,** descended from three main origins: Spanish horses that escaped early in the Spanish conquest, escaped U.S. cavalry mounts, and cow horses and farm horses from farms and ranches in the settlement of the American west. Thus the "wild horse" is not a wild horse: it is a domestic animal that has reverted to a life in the wild. (This in no way diminishes its status as an animal deserving to be preserved). Wild donkeys in North America (burros) are also feral animals. There are small populations of true wild horses and wild asses in Asia (see Chapter 1).

The grasslands and prairies of most of North America coevolved with the bison, a bovid closely related to cattle. Cattle can fairly be regarded as domesticated replacements of bison. In contrast, because the horse did not coevolve with the grasslands that developed in the American west since the last Ice Age, the grasses are poorly equipped to withstand heavy grazing pressure by mustangs. The description by a pioneer Oregon rancher, Herman Oliver, of the impact of wild horses on the range is pertinent (Oliver, 1961):

The bands (of wild horses) were usually from about seven up to 25, each with its stallion leader, and the others mares. As is the case with the deer now, there was all kinds of summer range, but in winter the snows would force them down into the open grasslands on the ranches. There were far too many for the grass available. Occasionally a war somewhere in the world would cause a sudden demand for cavalry and artillery horses. The Boer War really was a bonanza for the cowboys, because without much finances or backing, a few men could go out and round up hundreds of these wild horses. Between wars the horses bred and multiplied enormously, except that an occasional hard winter would kill them off by the thousands, as is now the case with deer herds

that get too plentiful. For a long time we didn't pay so much attention to them, but finally came to see that wherever there were wild horses, there wasn't much range anymore. The bunch grass left and was replaced by sagebrush or weeds. They were especially hard on the grass in winter and spring, when they congregated by the hundreds on grassy hillsides, pawed away the snow, and dug out the grass—sometimes roots and all. If it happened to be a dry year, the grass on such a range never came back, so then the horses next year would ruin another hill. After this had gone on from 1870 to around 1910, forty years, there wasn't much grass left in entire counties. (pp. 144–145)

The wild horses did not coevolve with the grasslands, nor did they coevolve with predators (limited predation by mountain lions occurs in a few areas). The only major controls on their populations have been starvation (overpopulation and hard winters) and human predation—first by the American Indians, followed by the cowboys and ranchers. As described earlier, wild horses were caught and used by the U.S. cavalry and also sold to other nations at war. After the First World War, the horse was replaced by motorized vehicles. From then until the 1960s, large numbers of wild horses were rounded up and slaughtered for horse meat for export and for dog food. These practices ended in the 1960s, following the protests of private citizens concerned with the welfare of wild horses.

There are no significant natural controls on wild horse numbers except starvation. In common with cattle and other herbivores, they have a high reproductive rate, so populations increase rapidly. Their numbers are currently being controlled by round-ups, with the captured animals being sold on an **Adopt-a-Horse program,** or else being maintained in feedlots at taxpayer expense (Fig. 7-36). Thousands of unwanted "wild horses" are in feedlots in the western states, living out their lives on alfalfa hay and grain. These practices were

**FIGURE 7-36** A Bureau of Land Management (BLM) horse corral, with wild horses being kept on feed for the rest of their natural lives. Under the Wild Horse and Burro Protection Act, they cannot be slaughtered. (Courtesy of L. Coates-Markle.)

established by the 1971 Wild Horse and Burro Protection Act. This would seem to be an unwise use of resources, both financial and otherwise. According to Berger (1986, 1991), more than $121 million has been spent on wild horses since 1973, and from 1980 to 1987, the expenditure averaged $10.2 million per year; each of the threatened and endangered species of wildlife in the United States receives less than 6 cents for every dollar spent on wild horses. He claims, with good justification, that a major change in priorities is needed. In an update, Berger and Berger (2001) noted that between 1992 and 1995, there was a seven-fold increase in U.S. funding for threatened and endangered species (T and E species). More resources were devoted to salmon than to livestock and wild horses combined. They conclude that funding priorities (during the Clinton presidency) shifted towards an emphasis on biological diversity, and

state, "The shift away from increased spending on rangelands for livestock and toward biodiversity, has given the American public reason to be optimistic that 21st-century efforts to promote biological conservation and ecosystem health will exceed those of the past" (p. 593).

The wild horse issue in the United States is an emotional one, because of the unique role that horses have in our psyche. No one doubts the beauty and symbolism of freedom that the wild horses personify. Without human intervention, the populations of these feral animals would increase enormously, threatening rangeland stability, livestock production, and "real" wildlife. The livestock industry is not causing the extinction of the wild horse; there are many wild horse herds on western rangelands (Fig. 7-37). There is room on the range for horses, cattle and wildlife. Under prevailing conditions, all need to be managed, including wildlife. The natural controls have been eliminated, and humans have intruded into the ecology of range lands (ski resorts, snowmobiling, summer cabins, condominiums, cities and towns, farms and ranches). Of all of these intrusions, ranching probably has less undesirable impact than the other factors mentioned (which are largely the responsibility of non-rural and non-agricultural people). One form of human intervention to control wild horse numbers at a sustainable level is to treat horses with contraceptive devices. Some success with immunocontraceptive vaccines, delivered by dart guns, has been achieved (Gillis, 1994; Turner et al., 1997).

**FIGURE 7-37** A wild burro on an arid western range. These feral animals are descendants of prospectors' burros that escaped or were turned loose. (Courtesy of L. Coates-Markle.)

# STUDY QUESTIONS

1. Five hundred years ago, North America had vast grasslands. Now there are only small remnants of native prairie left. What human activities have been most responsible for the loss of most of the native North American grasslands?

2. What is a watershed? How does livestock grazing affect watersheds?

3. What are riparian areas? What is the influence of livestock grazing on the integrity of riparian zones?

4. Should livestock be managed or fenced so that they have no access to streams and rivers? Should this be a nationwide or global objective?

5. What is the "tragedy of the commons"? How do rangelands and ocean fisheries relate to this concept?

6. What is the postulated role of the microbiotic crust in rangeland soil fertility?

7. Why is the pristine state a moving target? Can we return rangelands or other ecosystems to their pristine state?

8. What are the benefits and disadvantages of having the government provide subsidies for emergency feed during times of drought?

9. What are the advantages and disadvantages of livestock grazing in reducing rangeland fires?

10. Compare the environmental effects of large wild ungulates (deer, elk, bison) in Yellowstone National Park with the effects of beef cattle on western rangelands.

11. What roles do beaver have as a keystone species? Why are beaver numbers much lower in Yellowstone National Park now than they were 100 years ago? What are the environmental consequences of this change in beaver numbers?

12. Why are many ranchers in Montana opposed to allowing bison to move out of Yellowstone National Park into adjacent rangelands?

13. What are some of the reasons for and against the reintroduction of wolves into some of their former territory in the "Lower 48" states?

14. What is the Sahel? Why is it that nomadic herders and their livestock have been able to live in the Sahelian region for hundreds or even thousands of years in reasonable ecological harmony, until the last 50 years?

15. How can changes in vegetation affect the climate of an area?

16. Goats have been blamed for causing desertification. Are goats really the problem?

17. How would you describe the effects of livestock grazing on rangeland health in North America, Africa, China, and Australia? Have the effects been positive, negative, or neutral? Explain.

18. Describe the competition for feed resources among deer, elk, cattle, and wild horses.

19. Discuss some positive impacts of livestock production on wildlife and wildlife habitat.

20. Do you favor modifying the Endangered Species Act to give greater consideration to economic impacts? Why?

21. This statement was a reader response in *Western Beef Producer:* "I'm all for it (changing the Endangered Species Act). If it doesn't get changed, people will become extinct. I don't believe in putting animals in front of the well-being of people." Do you agree? Write a response to this opinion.

22. Coyotes have tremendously expanded their range in the last 100 years and are now found in every U.S. state except Hawaii. Why have they expanded their range, in spite of extensive predator control programs?

23. Are you in favor of ranching or farming of wildlife? Why?

# REFERENCES

ALROY, J. 2001. A multispecies overkill simulation of the end-Pleistocene megafaunal mass extinction. *Science* 292:1893–1896.

ALT, K., and M. R. FRISINA. 2000. Natural regulation and Yellowstone National Park—Unanswered questions. *Rangelands* 22:3–6.

ANDELT, W. F. 1999. Relative effectiveness of guarding-dog breeds to deter predation on domestic sheep in Colorado. *Wildlife Soc. Bull.* 27:706–714.

ANONYMOUS. 1995. Sheep at home on the wild summer range. *Agr. Res.* (USDA) 43(6):4–8.

ARMOUR, C., D. DUFF, and W. ELMORE. 1994. The effects of livestock grazing on western riparian and stream ecosystem. *Fisheries* 19(9):9–12.

BANGSUND, D. A., D. J. NUDELL, R. S. SELL, and F. L. LEISTRITZ. 2001. Economic analysis of using sheep to control leafy spurge. *J. Range Manage.* 54:322–329.

BECK, J. L., and D. L. MITCHELL. 2000. Influences of livestock grazing on sage grouse habitat. *Wildlife Soc. Bull.* 28:993–1002.

BECK, S. 1997. Cows cause forest fires. *Oregon Beef Producer* 10(6):5.

BELL, R. H. V. 1971. A grazing ecosystem in the Serengeti. *Sci. Am.* 225:86–93.

BELSKY, A. J. 1996. Viewpoint: Western juniper expansion: Is it a threat to arid northwestern ecosystems? *J. Range Manage.* 49:53–59.

BELSKY, A. J., and D. M. BLUMENTHAL. 1997. Effects of livestock grazing on stand dynamics and soils in upland forests of the interior West. *Conserv. Biol.* 11:315–327.

BELSKY, A. J., W. P. CARSON, C. L. JENSEN, and G. A. FOX. 1993. Overcompensation by plants: Herbivore optimization or red herring? *Evolutionary Ecology* 7:109–121.

BELSKY, A. J., A. MATZKE, and S. USELMAN. 1999. Survey of livestock influences on stream and riparian ecosystems in the western United States. *J. Soil and Water Cons.* 54:419–431.

BERGER, J. 1986. *Wild horses of the Great Basin: Social competition and population size.* University of Chicago Press, Chicago.

BERGER, J. 1991. Funding asymmetries for endangered species, feral animals, and livestock. *BioScience* 41:105–106.

BERGER, J. 1999. Anthropogenic extinction of top carnivores and interspecific animal behaviour: implications of the rapid decoupling of a web involving wolves, bears, moose and ravens. *Proc. R. Soc. Lond.* B:2261–2267.

BERGER, J., and K. BERGER. 2001. Endangered species and the decline of America's western legacy: What do changes in funding reflect? *BioScience* 51:591–593.

BERRY, K. 1986. Desert tortoise (*Gopherus agassizii*) research in California, 1976–1985. *Herpetologica* 42:62–67.

BESCHTA, R. L. 1997. Riparian shade and stream temperature: An alternative perspective. *Rangelands* 19(2):25–28.

BEYMER, R. J., and J. M. KLOPATEK. 1992. Effects of grazing on cryptogamic crusts in Pinyon-juniper woodlands in Grand Canyon National Park. *Am. Midl. Nat.* 127:139–148.

BORENS, F. M. P., and D. P. POPPI. 1990. The nutritive value for ruminants of tagasaste (*Chamaecytisus palmensis*), a leguminous tree. *Anim. Feed Sci. Tech.* 28:275–292.

BOX, T. 2000. Public rangelands without cows? *Rangelands* 22:27–30.

BOYCE, M. S. 1998. Ecological-process management and ungulates: Yellowstone's conservation paradigm. *Wildlife Soc. Bull.* 26:391–398.

BRADY, W. W., M. R. STROMBERG, E. F. ALDON, C. D. BONHAM, and S. H. HENRY. 1989. Response of a semidesert grassland to 16 years of rest from grazing. *J. Range Management* 42:284–288.

BURKHARDT, J. W. 1996. Herbivory in the intermountain west. An overview of evolutionary history, historic cultural impacts and lessons from the past. *Station Bulletin 58*, Experiment Station, University of Idaho, Moscow.

BUTLER, P. J. 2000. Cattle distribution under intensive herded management. *Rangelands* 22:21–23.

CALLENBACH, E. 1996. *Bring back the buffalo! A sustainable future for America's Great Plains.* Island Press, Washington, D.C.

CAVALCANTI, S. M. C., and F. F. KNOWLTON. 1998. Evaluation of physical and behavioral traits of llamas associated with aggressiveness toward sheep-threatening canids. *Appl. Anim. Behav. Sci.* 61:143–158.

CEDERHOLM, C. J., M. D. KUNZE, T. MUROTA, and A. SIBATANI. 1999. Pacific salmon carcasses: Essential contributions of nutrients and energy for aquatic and terrestrial ecosystems. *Fisheries* 24:6–15.

CHAMBERLAIN, D. J., and M. S. DOVERSPIKE. 2001. Water tanks protect streambanks. *Rangelands* 23:3–5.

CHASE, A. 1987. *Playing God in Yellowstone. The destruction of America's first national park.* Harcourt Brace and Co., New York.

CHEEKE, P. R., and S. L. DAVIS. 1997. Possible impacts of industrialization and globalization of animal agriculture on cattle ranching in the American West (Can environmentalists save the ranch?). *Rangelands* 19:4–5.

CLARK, E. A. 1998. Landscape variables affecting livestock impacts on water quality in the humid temperate zone. *Can. J. Plant Sci.* 78:181–190.

CLARK, P. E., W. C. KRUEGER, L. D. BRYANT, and D. R. THOMAS. 2000. Livestock grazing effects on forage quality of elk winter range. *J. Range Manage.* 53:97–105.

COE, P. K., B. K. JOHNSON, J. W. KERN, S. L. FINDHOLT, J. G. KIE, and M. J. WISDOM. 2001. Responses of elk and mule deer to cattle in summer. *J. Range Manage.* 54:205.

COHN, J. 1998. A dog-eat-dog world? *BioScience* 48:430–434.

COLBERT, E. H. 1969. *The evolution of the vertebrates.* John Wiley and Sons, New York.

COLLINS, A. R., and F. H. OBERMILLER. 1992. Interdependence between public and private forage markets. *J. Range Manage.* 45:183–188.

CONOVER, M. R. 1998. Perceptions of American agricultural producers about wildlife on their farms and ranches. *Wildlife Soc. Bull.* 26:597–604.

CROFT, D. B. 2000. Sustainable use of wildlife in western New South Wales: Possibilities and problems. *Rangel. J.* 22:88–104.

DAMHOUREYEH, S. A., and D. C. HARTNETT. 1997. Effects of bison and cattle on growth, reproduction, and abundances of five tallgrass prairie forbs. *Am. J. Bot.* 84:1719–1728.

DAYTON, L. 2001. Mass extinctions pinned on Ice Age hunters. *Science* 292:1819.

DEVINE, R. 1993. The cheatgrass problem. *Atlantic Monthly,* May, pp. 40–48.

DOBKIN, D. S., A. C. RICH, and W. H. PYLE. 1998. Habitat and avifaunal recovery from livestock grazing in a riparian meadow system of the northwestern Great Basin. *Conserv. Biol.* 12:209–221.

DODD, J. L. 1994. Desertification and degradation in sub-Saharan Africa. *BioScience* 44:28–34.

DONAHUE, D. L. 1999. *The western range revisited, Removing livestock from public lands to conserve biodiversity.* University of Oklahoma Press, Norman, Oklahoma.

DU TOIT, J. T., and D. H. M. CUMMING. 1999. Functional significance of ungulate diversity in African savannas and the ecological implications of the spread of pastoralism. *Biodiversity and Conservation* 8:1643–1661.

EBERHARDT, L. L., and R. O. PETERSON. 1999. Predicting the wolf-prey equilibrium point. *Can. J. Zool.* 77:494–498.

EDWARDS, G. P., D. B. CROFT, and T. J. DAWSON. 1996. Competition between red kangaroos (*Macropus rufus*) and sheep (*Ovis aries*) in the arid rangelands of Australia. *Australian Journal of Ecology* 21:165–172.

ELMORE, W., and R. L. BESCHTA. 1987. Riparian areas: Perceptions in management. *Rangelands* 9:260–265.

EVANS, R. D., and J. R. JOHANSEN. 1999. Microbiotic crusts and ecosystem processes. *Crit. Rev. Plant Sci.* 18:183–225.

FERGUSON, D., and N. FERGUSON. 1983. *Sacred cows at the public trough.* Maverick Publications, Bend, Oregon.

FITCH, L., and B. W. ADAMS. 1998. Can cows and fish co-exist? *Can. J. Plant Sci.* 78:191–198.

FLEISCHNER, T. L. 1994. Ecological costs of livestock grazing in western North America. *Conserv. Biol.* 8:629–644.

FORSE, B. 1989. The myth of the marching desert. *New Scientist* No. 4:31–32.

FRANK, D. A., S. J. MCNAUGHTON, and B. F. TRACY. 1998. The ecology of the earth's grazing ecosystems. *BioScience* 48:513–521.

FRANKE, R. W., and B. H. CHASIN. 1981. Peasants, peanuts, profits and pastoralists. *Ecologist* 11:156–168.

FRASER, K. W. 1996. Comparative rumen morphology of sympatric sika deer (*Cervus nippon*) and red deer (*C. elaphus scoticus*) in the Ahimanawa and Kaweka Ranges, Central North Island, New Zealand. *Oecologia* 105:160–166.

FREILICH, J. E., K. P. BURNHAM, C. M. COLLINS, and C. A. GARRY. 2000. Factors affecting population assessments of desert tortoises. *Conserv. Biol.* 14:1479–1489.

FRISINA, M. R., C. L. WAMBOLT, B. SOWELL, S. J. KNAPP, M. SULLIVAN, and C. JOHNSON. 2001. Maintaining habitat for sage grouse, prairie dogs and livestock requires managing resources for a balance. *Rangelands* 23:17–19.

FRISINA, M. R., G. ANLIN, Y. JINFENG, and B. WEIDONG. 2001. Innovative land use practices are helping rehabilitate China's Ordos rangelands. *Rangelands* 23:10–15.

FRITTS, S. H., E. E. BANGS, J. A. FONTAINE, M. R. JOHNSON, M. K. PHILLIPS, E. D. KOCH, and J. R. GUNSON. 1997. Planning and implementing a reintroduction of wolves to Yellowstone National Park and central Idaho. *Restoration Ecol.* 5:7–27.

FUHLENDORF, S. D., and D. M. ENGLE. 2001. Restoring heterogeneity on rangelands: Ecosystem management based on evolutionary grazing patterns. *BioScience* 51:625–632.

GEIST, V. 1985. Game ranching: Threat to wildlife conservation in North America. *Wildl. Soc. Bull.* 13:594–598.

GEIST, V. 1989. Livestock/wildlife interactions: Behavioral and territorial relationships. *Proc. West. Sec. Am. Soc. Anim. Sci.* 40:496–499.

GERMANO, D. J., G. B. RATHBUN, and L. R. SASLAW. 2001. Managing exotic grasses and conserving declining species. *Wildlife Soc. Bull.* 29:551–559.

GESE, E. M. 1999. Threat of predation: do ungulates behave aggressively towards different members of a coyote pack? *Can. J. Zool.* 77:499–503.

GILLIS, A. M. 1991. Should cows chew cheatgrass on commonlands? *BioScience* 41:668–675.

GILLIS, A. M. 1994. Fiddling with foaling. *BioScience* 44:443–450.

GLENN, E. P., and J. J. BROWN. 1999. Salt tolerance and crop potential of halophytes. *Crit. Rev. Plant Sci.* 18:227–255.

GOGUEN, C. B., and N. E. MATHEWS. 2000. Local gradients of cowbird abundance and parasitism relative to livestock grazing in a western landscape. *Conserv. Biol.* 14:1862–1869.

GREEN, D. M., and J. B. KAUFFMAN. 1995. Succession and livestock grazing in a northeastern Oregon riparian ecosystem. *J. Range Manage.* 48:307–313.

GREENWOOD, K. L., and B. M. MCKENZIE. 2001. Grazing effects on soil physical properties and the consequences for pastures: A review. *Aust. J. Exp. Agric.* 41:1231–1250.

GRESH, T., J. LICHATOWICH, and P. SCHOONMAKER. 2000. An estimation of historic and current levels of salmon production in the northeast Pacific ecosystem: Evidence of a nutrient deficit in the freshwater systems of the Pacific Northwest. *Fisheries* 25:15–21.

GRICE, A. C., and R. F. BECK. 1994. Kangaroos in Australian rangelands. *Rangelands* 16:189–192.

GRIGG, G. 1988. Kangaroo harvesting and the conservation of the sheep rangelands. *Aust. Zool.* 24:124–128.

GUEVARA, J. C., O. R. ESTEVEZ, and C. R. STASI. 1999. Economic feasibility of cactus plantations for forage and fodder production in the Mendoza plains (Argentina). *J. Arid Environ.* 43:241–249.

HADLEY, C. J. 2000. The wild life of Allan Savory. *Rangelands* 22:6–10.

HANLEY, S., M. ARTHUR, and C. SERVHEEN. 1999. The importance of meat, particularly salmon, to body size, population productivity, and conservation of North American brown bears. *Can. J. Zool.* 77:132–138.

HARDIN, G. 1968. The tragedy of the commons. *Science* 162:1243–1248.

HARTNETT, D. C., K. R. HICKMAN, and L. E. FISCHER WALTER. 1996. Effects of bison grazing, fire, and topography on floristic diversity in tallgrass prairie. *J. Range Manage.* 49:413–420.

HEATH, M. E., and C. J. KAISER. 1985. Forages in a changing world. pp. 3–11 In: Heath, M. E., R. F. Barnes, and D. S. Metcalfe (Eds.). *Forages. The science of grassland agriculture.* Iowa State University Press, Ames, Iowa.

HELLDEN, U. 1991. Desertification—Time for an assessment? *Ambio* 20:372–383.

HILDERBRAND, G. V., C. C. SCHWARTZ, C. T. ROBBINS, M. E. JACOBY, T. A. HANLEY, S. M. ARTHUR, and C. SERVHEEN. 1999. The importance of meat, particularly salmon, to body size, population productivity, and conservation of North American brown bears. *Can. J. Zoo.* 77:132–138.

HOBBS, N. T. 1996. Modification of ecosystems by ungulates. *J. Wildl. Manage.* 60:695–713.

HODGINS, I. W., and R. W. ROGERS. 1997. Correlations of stocking with the cryptogamic soil crust of a semi-arid rangeland in Southwest Queensland. *Aust. J. Ecol.* 22:425–431.

HOLECHEK, J. L. 2001. Western ranching at the crossroads. *Rangelands* 23:17–21.

HOLECHEK, J. L., and K. HESS, JR. 1995. The emergency feed program. *Rangelands* 17:133–136.

HOLECHEK, J. L., H. GOMES, F. MOLINAR, D. GALT, and R. VALDEZ. 2000. Short-duration grazing: The facts in 1999. *Rangelands* 22:18–22.

HOLECHEK, J. L., M. THOMAS F. MOLINAR, and D. GALT. 1999. Stocking desert rangelands: What we've learned. *Rangelands* 21:8–12.

HUFF, D. E., and J. D. VARLEY. 1999. Natural regulation in Yellowstone National Park's northern range. *Ecol. Applic.* 9:17–29.

HUNTER, M., JR. 1996. Benchmarks for managing ecosystems: Are human activities natural? *Conserv. Biol.* 10:695–697.

HUNTSINGER, L., and P. HOPKINSON. 1996. Viewpoint: Sustaining rangeland landscapes: A social and ecological process. *J. Range Manage.* 49:167–173.

HYDE, D. O. 1986. *Don coyote. The good times and the bad times of a much maligned american original.* Arbor House Publ. Co., New York.

ISENBERG, A. C. 2000. *The destruction of the bison. An environmental history, 1750–1920.* Cambridge University Press, Cambridge, U.K.

JACOBS, B. F., J. D. KINGSTON, and L. L. JACOBS. 1999. The origin of grass-dominated ecosystems. *Annal. Mo. Bot. Garden* 86:590–643.

JAMES, C. D., J. LANDSBERG, and S. R. MORTON. 1999. Provision of watering points in the Australian arid zone: a review of effects on biota. *J. Arid Environ.* 41:87–121.

JENSEN, M. N. 2001. Can cows and conservation mix? *BioScience* 51:85–90.

JONES, A. 2000. Effects of cattle grazing on North American arid ecosystems: A quantitative review. *West. N. Amer. Natural.* 60:155–164.

JONES, A. L., and W. S. LONGLAND. 1999. Effects of cattle gazing on salt desert rodent communities. *The American Midland Naturalist* 141:1–11.

KAUFFMAN, J. B., and W. C. KRUEGER. 1984. Livestock impacts on riparian ecosystems and streamside management implications: A review. *J. Range Manage.* 37:430–437.

KAUFFMAN, J. B., and D. A. PYKE. 2001. Range ecology, Global livestock influences. pp. 33–52. In: Levin, S. A. (Ed.). *Encyclopedia of biodiversity,* Academic Press, San Diego.

KAUFFMAN, J. B., R. L. BESCHTA, N. OTTING, and D. LYTJEN. 1997. An ecological perspective of riparian and stream restoration in the western United States. *Fisheries* 22(5):12–24.

KAY, C. E. 1994a. Aboriginal overkill. The role of native Americans in structuring western ecosystems. *Human Nature* 5:359–398.

KAY, C. E. 1994b. The impact of native ungulates and beaver on riparian communities in the intermountain West. *Nat. Res. Environ. Issues* 1:23–44.

KAY, C. E. 1996. Viewpoint: Ungulate herbivory, willows and political ecology in Yellowstone. *J. Range Manage.* 50:139–145.

KAY, C. E. 1998. Are ecosystems structured from the top-down or bottom-up: A new look at an old debate. *Wildlife Soc. Bull.* 26:484–498.

KAY, C. E. and R. T. SIMMONS. (EDITORS). 2002. *Wilderness and political ecology aboriginal influences and the original state of nature.* The University of Utah Press, Salt Lake City, UT.

KAY, C. E., and J. W. WALKER. 1997. A comparison of sheep- and wildlife-grazed willow communities in the Greater Yellowstone ecosystem. *Sheep and Goat Res. J.* 13:6–14.

KAZMAIER, R. T., E. C. HELLGREN, D. C. RUTHVEN III, and D. R. SYNATZSKE. 2001. Effects of grazing on the demography and growth of the Texas tortoise. *Conserv. Biol.* 15:1091–1101.

KEIGLEY, R. B. 1997. An increase in herbivory of cottonwood in Yellowstone National Park. *Northwest Sci.* 71:127–136.

KEIGLEY, R. B. 2000. Head to head. Elk, beaver, and the persistence of willows in national parks: *Comment on Singer et al.* (1998). *Wildlife Soc. Bull.* 28:448–450.

KEITER, R. B. 1997. Greater Yellowstone's bison: Unraveling of an early American wildlife conservation achievement. *J. Wildl. Manage.* 61:1–11.

KEITER, R. B., and M. S. BOYCE. (Eds.). 1991. *The Greater Yellowstone ecosystem. Redefining America's wilderness heritage.* Yale University Press, New Haven, Connecticut.

KNAPP, A. K., and T. R. SEASTEDT. 1986. Detritus accumulation limits productivity of tallgrass prairie. *Bioscience* 36:662–668.

KNAPP, A. K., J. M. BLAIR, J. M. BRIGGS, S. L. COLLINS, D. C. HARTNETT, L. C. JOHNSON, and E. G. TOWNE. 1999. The keystone role of bison in North American tallgrass prairie. *BioScience* 49:39–50.

KNIGHT, R. L., W. C. GILGERT, and E. MARSTON. (Eds). 2002. *Ranching west of the 100th meridian: Culture, Ecology, and Economics.* Island Press, Washington, D. C.

KNIGHT, R. L., G. N. WALLACE, and W. E. RIEBSAME. 1995. Ranching the view: Subdivisions versus agriculture. *Conserv. Biol.* 9:459–461.

KNOWLTON, F. F. 1989. Predator biology and livestock depredation management. *Proc. West. Sec. Amer. Soc. Anim. Sci.* 40:504–509.

KNOWLTON, F. F., E. M. GESE, and M. M. JAEGER. 1999. Coyote depredation control: An interface between biology and management. *J. Range Manage.* 52:398–412.

LEOPOLD, A. 1944. Review of the wolves of North America by S. P. Young and E. H. Goldman. *J. Forestry* 42:928–929.

LEOPOLD, A. 1949. *A Sand County almanac.* Ballentine Books, New York. (Oxford University Press, 1970).

LIFFMANN, R. H., L. HUNTSINGER, and L. C. FORERO. 2000. To ranch or not to ranch: Home on the urban range? *J. Range Manage.* 53:362–370.

LINNELL, J. D. C., J. ODDEN, M. E. SMITH, R. AANES, and J. E. SWENSON. 1999. Large carnivores that kill livestock: Do "problem individuals" really exist? *Wildlife Soc. Bull.* 27:698–705.

MACFADDEN, B. J. 1997. Origin and evolution of the grazing guild in New World terrestrial mammals. *Trends in Ecology & Evolution* 12:182–187.

MACK, R. N., and J. N. THOMPSON. 1982. Evolution in steppe with few large, hooved mammals. *Amer. Nat.* 119:757–773.

MARES, M. A., R. A. OJEDA, C. E. BORGHI, S. M. GIANNONI, G. B. DIAZ, and J. K. BRAUN. 1997. How desert rodents overcome halophytic plant defenses. *BioScience* 47:699–704.

MARTIN, P. S., and C. R. SZUTER. 1999. War zones and game sinks in Lewis and Clark's West. *Conserv. Biol.* 13:36–45.

MATTHEWS, A. 1992. *Where the buffalo roam.* Grove Press, New York.

MATTSSON, J. O., and A. RAPP. 1991. The recent droughts in western Ethiopia and Sudan in a climatic context. *Ambio* 20:172–175.

MCADOO, J. K., and H. A. GLIMP. 2000. Sheep management as a deterrent to predation. *Rangelands* 22:21–24.

MCINNIS, M. L., and J. MCIVER. 2001. Influence of off-stream supplements on streambanks of riparian pastures. *J. Range Manage.* 54:648–652.

MCKELL, C. M. (Ed.). 1989. *The biology and utilization of shrubs.* Academic Press, San Diego.

MCLEOD, D. M., and L. W. VAN TASSELL. 1996. Economic and policy implications of brucellosis in the Greater Yellowstone Area. *Rangelands* 18:145–148.

MCNAUGHTON, S. J. 1979. Grazing as an optimization process: Grass-ungulate relationships in the Serengeti. *Amer. Nat.* 113:691–703.

MCNAUGHTON, S. J. 1993. Grasses and grazers, science and management. *Ecol. Appl.* 3:17–20.

MEADOWS, L. E., and F. F. KNOWLTON. 2000. Efficacy of guard llamas to reduce canine predation on domestic sheep. *Wildlife Soc. Bull.* 28:614–62.

MECH, L. D. 1995. The challenge and opportunity of recovering wolf populations. *Conserv. Biol.* 9:270–278.

MEYER, M. E., and M. MEAGHER. 1995. Brucellosis in free-ranging bison (*Bison bison*) in Yellowstone, Grand Teton, and Wood Buffalo National Parks: A review. *J. Wildlife Diseases* 31:579–598.

MILCHUNAS, D. G., W. K. LAUENROTH, and I. C. BURKE. 1998. Livestock grazing: Animal and plant biodiversity of shortgrass steppe and the relationship to ecosystem function. *OIKOS* 83:65–74.

MILLER, D. J., and G. B. SCHALLER. 1996. Rangelands of the Chang Tang wildlife reserve in Tibet. *Rangelands* 18:91–96.

MILTON, S. J., W. R. J. DEAN, M. A. DU PLESSIS, and W. R. SIEGFRIED. 1994. A conceptual model of arid rangeland degradation. *BioScience* 44:70–76.

MIZUTANI, F. 1999. Biomass density of wild and domestic herbivores and carrying capacity on a working ranch in Laikipia District, Kenya. *Afr. J. Ecol.* 37:226–240.

MOLE, S., L. G. BUTLER, and G. IASON. 1990. Defense against dietary tannin in herbivores: a survey for proline rich salivary proteins in mammals. *Biochem. Systemics Ecol.* 18:287–293.

MOLL, E. J., and A. A. GUBB. 1989. Southern African shrublands. pp. 145–175. In: McKell, C. M. (Ed.). *The biology and utilization of shrubs.* Academic Press, San Diego.

MUSTERS, C. J. M., M. KRUK, H. J. DEGRAAF, and W. J. TER KEURS. 2001. Breeding birds as a farm product. *Conserv. Biol.* 15:363–369.

OBA, G., N. C. STENSETH, and W. J. LUSIGI. 2000. New perspectives on sustainable grazing management in arid zones of sub-Saharan Africa. *BioScience* 50:35–51.

OHMART, C. P., and P. B. EDWARDS. 1991. Insect herbivory on Eucalyptus. *Annu. Rev. Entomol.* 36:637–657.

OLIVER, H. 1961. *Gold and cattle country.* Binfords and Mort Publishers, Portland, Oregon.

ORSKOV, E. R. 1987. The role of livestock in Africa: Are livestock occasionally contributing to famine? *Proc. Nutr. Soc.* 46:301–308.

PAINTER, E. L., and A. J. BELSKY. 1993. Application of herbivore optimization theory to rangelands of the western United States. *Ecol. Applic.* 3:2–9.

PEASE, C. M., and D. J. MATTSON. 1999. Demography of the Yellowstone grizzly bears. *Ecology* 80:957–975.

PEREVOLOTSKY, A., and N. G. SELIGMAN. 1998. Role of grazing in Mediterranean rangeland ecosystems. *BioScience* 48:1007–1017.

POPPER, D. E., and F. J. POPPER. 1999. The Buffalo Commons: Metaphor as method. *The Geographical Review* 89:491–510.

POWER, T. M. 1996. *Lost landscapes and failed economies. The search for a value of place.* Island Press, Washington, D.C.

RANKINS, D. L., JR., and G. S. SMITH. 1991. Nutritional and toxicological evaluations of kochia hay (*Kochia scoparia*) fed to lambs. *J. Anim. Sci.* 69:2925–2931.

RIPPLE, W. J., and E. J. LARSEN. 2000. Historic aspen recruitment, elk, and wolves in northern Yellowstone National Park, USA. *Biol. Conserv.* 95:361–370.

ROBERTS, R. G., T. F. FLANNERY, L. K. AYLIFFE, H. YOSHIDA, J. M. OLLEY, G. J. PRIDEAUX, G. M. LASLETT, A. BAYNES, M. A. SMITH, R. JONES, and B. L. SMITH. 2001. New ages for the last Australian megafauna: Continent-wide extinction about 46,000 years ago. *Science* 292:1888–1892.

ROMME, W. H., M. G. TURNER, L. L. WALLACE, and J. S. WALKER. 1995. Aspen, elk, and fire in northern Yellowstone National Park. *Ecology* 76:2097–2106.

ROWE, H. I., E. T. BARTLETT, and L. E. SWANSON, JR. 2001. Ranching motivations in 2 Colorado counties. *J. Range Manage.* 54:314–321.

RUSSELL, C. E., and P. FELKER. 1987. The prickly-pears (*Opuntia* spp., Cactaceae): A source of human and animal food in semiarid regions. *Econ. Bot.* 41:433–445.

SAVORY, A. 1988. *Holistic resource management.* Island Press, Washington, D.C.

SAVORY, A., and J. BUTTERFIELD. 1999. *Holistic management. A new framework for decision making.* Island Press, Washington, D.C.

SCHULLERY, P. 1999. *Searching for Yellowstone: Ecology and wonder in the last wilderness.* Houghton Mifflin Co., New York.

SEVERSON, K. E. (Ed.). 1990. Can livestock be used as a tool to enhance wildlife habitat? USDA Forest Service General Technical Report RM-194, Fort Collins, Colorado.

SHAW, J. H. 1995. How many bison originally populated western rangelands? *Rangelands* 17:148–150.

SHEEHY, D. P. 1992. A perspective on desertification of grazingland ecosystems in North China. *Ambio* 21:303–307.

SIMPSON, G. G. 1950. *Horses.* Oxford University Press, London.

SIMPSON, G. G. 1953. *The major features of evolution.* Columbia University Press, New York.

SINCLAIR, A. R. E., and J. M. FRYXELL. 1985. The Sahel of Africa: Ecology of a disaster. *Can. J. Zool.* 63:987–994.

SINGER, F. J., and R. G. CATES. 1995. Response to comment: Ungulate herbivory on willows on Yellowstone's northern winter range. *J. Range Manage.* 48:563–565.

SINGER, F. J., A. HARTING, K. K. SYMONDS, and M. B. COUGHENOUR. 1997. Density dependence, compensation, and environmental effects on elk calf mortality in Yellowstone National Park. *J. Wildl. Manage.* 61:12–25.

SINGER, F. J., D. M. SWIFT, M. B. COUGHENOUR, and J. D. VARLEY. 1998. Thunder on the Yellowstone revisited: An assessment of management of native ungulates by natural regulation, 1968–1993. *Wildlife Soc. Bull.* 26:375–390.

SINGER, F. J., L. C. ZEIGENFUSS, and D. T. BARNETT. 2000. Head to head. Elk, beaver, and the persistence of willows in national parks: Response to Keigley (2000). *Wildlife Soc. Bull.* 28:451–453.

SINGER, F. J., L. C. ZEIGENFUSS, R. G. CATES, and D. T. BARNETT. 1998. Elk, multiple factors and persistence of willows in national parks. *Wildlife Soc. Bull.* 26:419–428.

SINGER, F. J., L. C. ZEIGENFUSS, and L. SPICER. 2001. Role of patch size, disease, and movement in rapid extinction of bighorn sheep. *Conserv. Biol.* 15:1347–1354.

SKARPE, C. 1991. Impact of grazing in savanna ecosystems. *Ambio* 20:351–356.

SMITH, S. J. 1995. Viewpoint: Fuels management—Is livestock grazing the solution? *Rangelands* 17:97.

STANLEY-PRICE, M. R. 1985. Game domestication for animal production in Kenya: The nutritional ecology of oryx, zebu cattle and sheep under free-range conditions. *J. Agric. Sci.* 104:375–382.

STARRS, P. F. 1998. *Let the cowboy ride: Cattle ranching in the American West.* Johns Hopkins University Press, Baltimore.

ST. CLAIR, L. L., and J. R. JOHANSEN. 1993. Introduction to the symposium on soil crust communities. *Great Basin Naturalist* 53:1–4.

STEBBINS, G. L. 1981. Coevolution of grasses and herbivores. *Ann. Missouri Bot. Gard.* 68:75–86.

STODDART, L. C., R. E. GRIFFITHS, and F. F. KNOWLTON. 2001. Coyote responses to changing jackrabbit abundance affect sheep predation. *J. Range Mange.* 54:15–20.

TAYLOR, R. D., and B. H. WALKER. 1978. Comparisons of vegetation use and herbivore biomass on a Rhodesian game and cattle ranch. *J. Appl. Ecol.* 15:565–581.

TERWILLIGER, J., and J. PASTOR. 1999. Small mammals, ectomycorrhizae, and conifer succession in beaver meadows. *OIKOS* 85:83–94.

THIRGOOD, J. V. 1981. *Man and the Mediterranean forest. A history of resource depletion.* Academic Press, New York.

THUROW, T. L., and C. A. TAYLOR, JR. 1999. Viewpoint: The role of drought in range management. *J. Range Manage.* 52:413–419.

TOPPS, J. H. 1992. Potential, composition and use of legume shrubs and trees as fodders for livestock in the tropics. *J. Agric. Sci.* 118:1–8.

TRIMBLE, S. W., and A. C. MENDEL. 1995. The cow as a geomorphic agent—A critical review. *Geomorphology* 13:233–253.

TURNER, J. W., JR., A. T. RUTBERG, and J. F. KIRKPATRICK. 1997. Immunocontraception limits foal production in free-roaming feral horses in Nevada. *J. Wildl. Manage.* 61:873–880.

VAVRA, M. 1998. Public land and natural resource issues confronting animal scientists and livestock producers. *J. Anim. Sci.* 76:2340–2345.

VAVRA, M., M. MCINNIS, and D. SHEEHY. 1989. Implications of dietary overlap to management of free-ranging large herbivores. *Proc. West. Sec. Am. Soc. Anim. Sci.* 40:489–495.

WADE, N. 1974. Sahelian drought: No victory for western aid. *Science* 185:234–237.

WAGNER, F. H. 1988. *Predator control and the sheep industry.* Iowa State University Press, Ames.

WALLER, D. M., and W. S. ALVERSON. 1997. The white-tailed deer: A keystone herbivore. *Wildlife Soc. Bull.* 25:217–226.

WAMBOLT, C. L. 1998. Sagebrush and ungulate relationships on Yellowstone's northern range. *Wildlife Soc. Bull.* 26:429–437.

WANG, X.-M., J.-T. PENG, and H.-M. ZHOU. 2000. Preliminary observations on the distribution and status of dwarf blue sheep *Pseudois schaeferi.* *Oryx* 34:21–26.

WEISS, S. B. 1999. Cars, cows, and checkerspot butterflies: Nitrogen deposition and management of nutrient-poor grasslands for a threatened species. *Conserv. Biol.* 13:1476–1486.

WESTERN, D., and V. FINCH. 1986. Cattle and pastoralism: Survival and production in arid lands. *Human Ecology* 14:77–94.

WHITE, C. A., C. E. OLMSTED, and C. E. KAY. 1998. Aspen, elk, and fire in the Rocky Mountain national parks of North America. *Wildlife Soc. Bull.* 26:449–462.

WILLSON, M. F., S. M. GENDE, and B. H. MARSTON. 1998. Expanding perspectives on fish-wildlife interactions. *BioScience* 48:455–462.

WILSON, M. A. 1997. The wolf in Yellowstone: Science, symbol, or politics? Deconstructing the conflict between environmentalism and wise use. *Society & Natural Resources* 10:453–468.

WOLFF, W. J. 1992. The end of a tradition: 1000 years of embankment and reclamation of wetlands in the Netherlands. *Ambio* 21:287–291.

WUERTHNER, G. 1994. Subdivisions versus agriculture. *Conserv. Biol.* 8:905–908.

YOUNG, J. A., C. D. CLEMENTS, and G. NADER. 1999. Medusahead and clay: The rarity of perennial seedling establishment. *Rangelands* 21:19–23.

ZHANG, H.-X., and E. BLUMWALD. 2001. Transgenic salt-tolerant tomato plants accumulate salt in foliage but not in fruit. *Nature Biotech.* 19:765–768.

# CHAPTER 8

# Industrialization, Corporatization, and Globalization of Animal Agriculture

**CHAPTER OBJECTIVES**

1. To discuss the nature of intensive animal production

2. To discuss the industrialization, corporate control, and globalization of poultry production, and to consider whether other types of animal production may follow the lead of poultry to industrialization

3. To consider the risks and benefits of corporate swine production

4. To consider the validity of societal concerns about "factory farming"

## INTRODUCTION

Livestock production is increasingly under intense scrutiny by the general public of the United States, Europe, and most other industrialized countries. This is particularly true of young people, of high school, university, and the under-50 professional "white collar" status. Individuals in these groups are increasingly turning to a vegetarian or plant food-based diet for a variety of complex, interconnected reasons. In contrast to the days

when the author was a student, when vegetarian meals were practically unknown, college campuses, restaurants, meals served in association with meetings, symposia, and so on (even animal science meetings) routinely offer vegetarian options, and in many cases, meat entrees are regarded with the same enthusiasm as smoking sections. On numerous college campuses in the United States, student activists have pushed, sometimes successfully, for elimination of meat and other animal products from dormitory meals, and in some cases, from all food service institutions on the campus. What accounts for these trends? What are the potential impacts of these trends on the livestock industries? Do American high school and college students, in general, consider T-bone steaks and leg of lamb among their top dietary choices? If not, does it matter?

When I have raised some of these issues among animal science colleagues, a fairly typical response is that vegetarian lifestyles are not common and not increasing, that I am forming my opinions based on a small population in the "la-la land" on the U.S. West Coast, and that all is well in the animal production segment of agriculture. Another common response is the attitude that it is treasonous to even bring these issues up, with the implication that if we ignore what may be happening, maybe it will go away.

Many or most of the societal concerns impacting livestock production discussed in this chapter and Chapters 9 and 10 ultimately come down to a matter of personal opinions, ethics, and philosophies. If a person believes that raising animals for the purpose of killing and eating them is morally wrong, no amount of learned discussion from learned professors is likely to have the slightest impact on those opinions. If a person believes that keeping sows tethered in stalls or housing laying hens in small cages is morally wrong, no amount of physiological data from animal scientists showing that the animals are not under stress is likely to change any minds. Similarly, most people who are supportive of the production and use of animal products are unlikely to be swayed by emotional or other arguments from vegetarians and animal rights activists. Ultimately, the "correct" answer to these debates is a matter of personal choice. In our society, nobody has the right to dictate what others should eat. If most people decide that they do not wish to buy and consume avocados, the avocado growers will be in a bind, and many will have to change crops or occupations. The same is true of beef, lamb, or any other animal product. If public opinion is such that the demand

for these products decreases, there will be consequences for the producers of these commodities and their supporting infrastructure, including the animal science and veterinary medicine professions.

Therefore, these societal concerns are of vital interest to all segments of animal agriculture and are worthy of intensive examination and discussion. As is true with other publicly debated issues such as abortion, shouting and name calling are not very useful tactics in resolving issues related to animal agriculture but are the very tactics in common use (wacko environmentalists, tree huggers, corporate welfare ranchers, etc.).

The reasons for a considerable anti-animal agriculture attitude in postindustrial society are complex but in my opinion are intimately connected with the concepts of animal rights and animal welfare. My interpretation is that many if not most people who adopt a no-animal products lifestyle do so because they believe that modern, intensive, large-scale animal production systems are cruel and inhumane. Therefore, concerns about "factory farming," feed additives and food safety ("hormones and chemicals,") biotechnology techniques involving animals (e.g., cloning), and animal rights and welfare are all interrelated. Thus the order in which they are discussed in this book does not suggest an order of priority. Because I believe that public concern about the ethics of so-called "factory farming" is a precipitating factor in causing people to be "turned off" by animal agriculture, it is appropriate to discuss intensive animal agriculture before considering some of the other issues. The discussion presented is intended to be as open-minded and non-judgmental as possible, with the intent of providing a framework to promote discussion, debate, and the evolution of personal philosophies and ethics based on scientific knowledge rather than emotion. While that was the intention, I suspect that my attempts to be non-judgmental were not a total success in this chapter!

# INTENSIVE (INDUSTRIAL) ANIMAL PRODUCTION (FACTORY FARMING)

**Intensive livestock production** (Fig. 8-1) in confinement conditions (so-called **factory farming** by critics) has greatly increased the productivity and apparent efficiency of food production. It has also resulted in considerable criticism by various segments of society. Some of the concerns shared by many people include the perception that animals are viewed as machines, their welfare and well-being are compromised, they are kept under inhumane conditions, they are fed hormones, additives, and other chemicals that adversely affect their well-being and may affect the safety of the animal products to human health, multinational agribusiness companies have driven the family farmer out of business, and short-term profits are emphasized at the expense of animal welfare, soil fertility, and sustainability.

Until the twentieth century, the majority of people in most societies were agrarian—they were farmers and tillers of the soil. The Industrial Revolution created urban opportunities by providing employment, usually at a higher level of remuneration and with better working conditions than possible with subsistence farming. The Industrial Revolution led to advances in mechanization, so as farmers left the land, their labor contributions were replaced by machines. Before the mechanization of agriculture, farms were small by necessity, because the work was accomplished by the family members and draft animals. Most enterprises were "mixed" or "general" farms, with a variety of livestock and crops produced. A major goal was to be as self-sufficient as possible.

With the advent of farm tractors, draft horses were no longer required. Until that time, virtually all farms had to produce forage and grain for horses. With increasing mechanization, it became possible for each farmer to farm more land; in fact, it was a necessity to re-

**FIGURE 8-1** Intensive production of pigs (*top*) and broiler chickens (*bottom*). Critics claim that these confinement systems are factory farming and the animals are simply machines to produce meat. What is your opinion? (Bottom photo courtesy of R. A. Swick.)

coup the costs of the machinery. It soon became apparent that the costs of mechanization could be recovered most easily by specializing in a few or even just one type of crop or livestock species. Thus the old mixed farm with a few horses, dairy cows, beef cows,

Oklahoma, with a low human population and wide open spaces, was until recently vigorously courting corporate farms to encourage them to locate in Oklahoma. In the future, it is likely that intensive animal production in the United States will tend to be located in areas of low human population density and having soil and water characteristics that will minimize pollution concerns. Location in isolated areas will also provide better security against animal rights activism and bioterrorism. New Mexico has already become one of the major dairy states, shipping milk throughout the southwest and as far away as Iowa. Poultry production has long been important in the Delmarva peninsula (Delaware, Maryland, Virginia) but is rapidly being squeezed out by urbanization and environmental concerns. The Iowa legislature is considering legislation to establish livestock enterprise zones to promote intensive animal production and protect against "nuisance" lawsuits involving smell, noise, dust, and so on.

Another concern sometimes expressed is the lack of genetic variability in animals raised intensively. The poultry industry is based on a very small number of selected strains of birds, with the breeding controlled by industrial scale hybrid breeding centers. The dairy industry now depends largely on the high-producing Holstein cow. Critics are concerned that the control of plant and animal germ plasm will end up totally in the hands of large industrial companies. This is happening particularly quickly with poultry and large volume crops such as corn and soybeans. Large companies such as Monsanto are producing herbicide-resistant hybrid corn and soybeans. To obtain high yields and remain competitive, farmers may become tied to these companies (e.g., Monsanto is the sole producer of the herbicide Roundup [glyphosate], and has developed Roundup-resistant corn and soybeans. Thus the producer may become locked into reliance upon Monsanto for both seed and herbicide). Potential effects of herbicide-resistant crops have been discussed by CAST (1991) and Tauer and Love (1989). Many of

the same issues are involved in the control of animal germ plasm by large companies.

# INDUSTRIALIZATION, CORPORATIZATION, AND GLOBALIZATION OF POULTRY PRODUCTION

Poultry production, particularly for meat (broilers, fryers), has become an industrial process (Skinner, 1996). In the United States, a few companies control virtually 100 percent of meat bird production (Table 8-1). **Industrialization** occurred most rapidly in southeastern states (Table 8-2) such as Arkansas for reasons relating to inexpensive input costs (feed, labor), less than maximally stringent environmental regulations, and a friendly business climate (generous tax incentives). Poultry companies are vertically integrated, meaning that the company (the **integrator**) owns or controls most or all of the components of production from the breeder flocks (parent and grandparent stocks), all the way to the supermarket meat counter. Typically, the integrators own all components of the system (feed mills, slaughter plants, etc.) except for the actual facilities in which the birds are raised. The birds are produced by **growers** who raise birds under contract to the integrators. The growers provide the housing, equipment, labor, and management. The integrators own the birds and provide the chicks, feed, technical advice, and bird harvest.

Egg production is probably even more industrialized than broiler production. Layers are kept in small cages; feed and water are provided with automatic systems and eggs roll on to conveyors and are collected automatically. The major U.S. egg-producing state is Iowa (Table 8-3).

It is no coincidence that poultry production was the first segment of animal agriculture to become industrialized. Poultry, and

**TABLE 8-1**   Top broiler producing companies, mid-2001.
(Source: *Feedstuffs* Reference Issue, 2001.)

| Company | Average weekly production, million pounds[*] | Percent of market share |
|---|---|---|
| 1. Tyson Foods | 150 | 23.8 |
| 2. Gold Kist | 58 | 9.2 |
| 3. Pilgrim's Pride | 55 | 8.7 |
| 4. ConAgra Poultry | 52 | 8.3 |
| 5. Perdue Farms | 47 | 7.5 |
| 6. Wayne Poultry | 26 | 4.1 |
| 7. Sanderson Farms | 21 | 3.3 |
| 8. Cagle's | 19 | 3.0 |
| 9. Mountaire Farms | 17 | 2.7 |
| 10. Foster Farms | 15 | 2.4 |
| 11. Fieldale Farms Corporation | 13 | 2.1 |
| 12. O.K. Foods | 12 | 1.9 |
| 13. House of Raeford/Columbia Farms | 11 | 1.7 |
| 14. Peco Foods | 11 | 1.7 |
| 15. Allen Family Foods | 10 | 1.6 |
| 16. Choctaw Maid Farms | 10 | 1.6 |
| 17. Simmons Foods | 9 | 1.4 |
| 18. Townsends | 9 | 1.4 |
| 19. Case Farms | 7 | 1.1 |
| 20. George's | 7 | 1.1 |
| 21. B.C. Rogers Poultry | 5 | 0.8 |
| 22. Claxton Poultry farms | 5 | 0.8 |
| 23. Gold'n Plump Poultry | 5 | 0.8 |
| 24. Koch Foods | 5 | 0.8 |
| 25. Marshall Durbin Companies | 5 | 0.8 |
| 26. Peterson Farms | 5 | 0.8 |
| 27. Zacky foods | 5 | 0.8 |
| 28. Mar-jac Poultry | 4 | 0.6 |
| 29. Rocco | 4 | 0.6 |
| 30. Amick | 3.5 | 0.6 |

[*]In million pounds, ready-to-cook weight basis.

**TABLE 8-2** Top 15 broiler and turkey states, 2000.
(Source: *Feedstuffs* Reference Issue, 2001.)

| Rank | State | Broilers (million head) |
|---|---|---|
| 1 | Georgia | 1,229 |
| 2 | Arkansas | 1,191 |
| 3 | Alabama | 1,039 |
| 4 | Mississippi | 740 |
| 5 | North Carolina | 698 |
| 6 | Texas | 551 |
| 7 | Maryland | 283 |
| 8 | Virginia | 265 |
| 9 | Delaware | 248 |
| 10 | Missouri | 240 |
| 11 | Oklahoma | 223 |
| 12 | Kentucky | 208 |
| 13 | South Carolina | 197 |
| 14 | Tennessee | 151 |
| 15 | Pennsylvania | 133 |
| Total for 15 states: | | 7,396 |
| Percent of all broilers in 2000: | | 89.5% |
| **Rank** | **State** | **Turkeys (thousand)** |
| 1 | Minnesota | 43,500 |
| 2 | North Carolina | 41,000 |
| 3 | Arkansas | 28,000 |
| 4 | Virginia | 25,500 |
| 5 | Missouri | 23,000 |
| 6 | California | 18,000 |
| 7 | Indiana | 13,500 |
| 8 | South Carolina | 9,900 |
| 9 | Pennsylvania | 9,300 |
| 10 | Iowa | 7,100 |
| 11 | Ohio | 4,400 |
| 12 | South Dakota | 4,300 |
| 13 | West Virginia | 4,100 |
| 14 | Michigan | 3,500 |
| 15 | Illinois | 2,900 |
| Total for 15 states: | | 234,500 |
| Percent all turkeys in 2000: | | 86.8% |

**TABLE 8-3** Major egg producing states, 2000. (Source: *Feedstuffs* Reference Issue, 2001.)

| Rank—2000 (thousand) | State | Laying hens |
|---|---|---|
| 1 | Iowa | 31,063 |
| 2 | Ohio | 29,131 |
| 3 | California | 24,303 |
| 4 | Pennsylvania | 24,179 |
| 5 | Indiana | 23,038 |
| 6 | Georgia | 20,778 |
| 7 | Texas | 18,660 |
| 8 | Arkansas | 14,853 |
| 9 | Minnesota | 12,480 |
| 10 | Nebraska | 11,840 |
| 11 | North Carolina | 11,022 |
| 12 | Florida | 10,738 |
| 13 | Alabama | 10,183 |
| 14 | Missouri | 6,665 |
| 15 | Mississippi | 6,601 |
| Total for 15 states: | | 255,534 |
| Percent of all laying hens in 2000: | | 76.9% |

**FIGURE 8-2** An advantage of poultry over other livestock is that large numbers of birds of the same genetics and age can be produced by setting eggs in incubators. Thus the reproductive process of chickens is very easy to automate. (Courtesy of Robert A. Swick.)

particularly chickens, have a number of biological attributes that allow them to be raised industrially. Some of these are the following:

1. Reproduction
   (a) Poultry are reproduced by setting eggs for hatching in incubators (Fig. 8-2). The entire reproductive process can be controlled. There is virtually no practical limit to the number of chicks that can be hatched at one time; this is regulated simply by the number of eggs of known hatchability set in the incubator.
   (b) Birds can be hatched on a set schedule as needed. The integrator can establish schedules with a large number of growers, a year or more in advance. On Jan. 2, for example, the integrator can plan for Grower A to receive 200,000 day-old chicks on July 15 and Grower B to receive 150,000 chicks on July 16. Thus the reproductive attributes are also a tremendous advantage for management and scheduling production in an industrial mode.
   (c) In contrast to the situation with birds, mammalian reproduction cannot be so easily automated. It is not usually technically feasible to set up production schedules so that an integrator is assured of having 50,000 21-day old pigs on May 25 or 100,000 day-old calves on Oct. 15. Thus it is difficult for integrators to manage mammalian species to produce large numbers of similar-aged offspring on a precise, set schedule.

2. Genetics and Selection
   Poultry are **short-cycle animals.** They reach reproductive maturity quickly. Because of this, the large number of eggs per bird, and the ease of storing and then incubating large numbers of eggs, it is feasible in poultry

breeding centers to make rapid and continual progress in increasing growth rate and feed efficiency. Broilers used to reach market weight in 8 to 10 weeks; now they are marketed at 5 weeks of age. Over the last several decades, the average growth rate of broilers has increased by 1 gram per bird per day, each year. Thus a broiler grown to 45 days of age in 1997 would weigh approximately 1 pound more than a broiler grown to the same age in 1987 (10 years $\times$ 45 g/year).

This rapid increase in bird performance has not been without its problems. There are numerous "**diseases of industrialization**" that are directly or indirectly attributable to selection for maximum growth rate, including ascites, spiking mortality, sudden death syndrome, and so on.

3. Nutrition

The digestive physiology and dietary strategy (feeding behavior) of chickens are important attributes for industrialization. Chickens are gramnivores (seed eaters). They are ideally suited for high energy, grain-based diets. Chickens are the most efficient of the livestock species in converting corn-soy diets into meat. Poultry nutritionists have developed high energy, low fiber diets that give a high feed conversion efficiency. One of the goals of poultry nutritionists is to lower dietary fiber in broiler diets to as close to zero as possible. This is not possible in swine, cattle, rabbits, and so on. In these animals, dietary fiber has important roles in maintaining normal gastrointestinal tract health and motility. For now and the foreseeable future, corn (a high energy grain) is and will be the cheapest source of metabolizable energy (ME) for livestock. Poultry are able to utilize this feedstuff and its ME more efficiently than can any other livestock species.

Another aspect of digestive physiology that is an advantage for poultry is that they excrete uric acid instead of urea and produce a very concentrated urine. Thus poultry excreta is very dry. Large numbers of birds can be kept on litter with a high stocking density, without the litter becoming wet and foul-smelling. With adequate ventilation, a broiler house with 100,000 "wall to wall" broilers can be dry and without excessive levels of ammonia and humidity in the air. This is not possible with pigs or cattle, because of the much higher volume of urine excreted. The **high stocking density** possible with poultry also means lower building costs relative to other types of intensive production, when a greater area per pound of animal is required. The excretion of a concentrated urine also means that poultry have low water requirements compared with other livestock.

4. Management and Production Flexibility

As discussed previously, the production of chicks can be geared exactly to anticipated need. The production cycle is short—five weeks and counting down. If there is a poor corn harvest, an increase in inflation resulting in reduced consumer spending, and so on, the poultry industry can adjust very quickly to reduce production. This is a major advantage over cattle and swine. Conversely, if there is a sudden increase in meat demand, the poultry industry can gear up in a matter of a few weeks to increase production.

5. Animal Behavior

Broilers can be raised in large numbers at a **high stocking density.** The lights are kept dim so the birds can see to eat, but they don't engage in much other activity. In the author's opinion and judgment, broiler chickens are "content" under these conditions. They generally do not exhibit signs of distress. I own a small farm and have "free range" chickens around the farm. These birds engage in many activities, including wrecking my vegetable and flower gardens! I once had some broiler chicks raised under these conditions. The chicks were raised by a bantam hen and had free range. They looked absolutely pathetic! They seemed to have no idea where they were, looking randomly into the sky and stumbling over each other. My opinion, after studying these birds, is that the selection of broilers for rapid growth and feed efficiency has indirectly selected for birds that are behaviorly adapted to confinement and high stocking density. It is my opinion that intensive production of broiler chickens is not cruel or inhumane. This is both a matter of opinion and an ethical question; ethical concerns are discussed in more detail in Chapter 10.

At any rate, it is a fact that broiler chickens can be successfully raised at very high stocking densities, which would appear difficult to achieve with other types of animals.

6. Low Labor Requirements

Another advantage of broiler production is that it is highly automated and mechanized. Automatic feeders and waterers are used, and the bird harvesting can be mechanized. The primary responsibilities of the grower are to check daily to ensure that all automatic systems are working, to observe the birds for any problems, and to remove any dead birds. The other major activity is to prepare the litter by rototilling, for example, between production cycles. Thus, compared to raising other livestock, poultry production is highly mechanized and automated; as a result, labor requirements are low.

7. Marketing Advantages

Poultry meat is white and has real and perceived human health benefits over red meat (see Chapter 2). There are no religious restrictions against poultry meat, as there are for beef and pork. Poultry meat is much cheaper in the supermarket than other types of meat. This is clearly due to the economic efficiencies and subsidies associated with the industrial production system.

A negative factor in the marketing of industrially produced broilers is that the meat is widely perceived as lacking taste and other desirable eating qualities. Broilers are being marketed at an increasingly young age, now down to about five weeks. A common perception with meat animals is that when the animals are killed very young, the meat is bland and lacks taste. I have had the experience of eating chicken strips in a fast food restaurant and not being able to ascertain if they were chicken or fish. I finally decided that the amorphous, tasteless white slab of tissue was probably chicken, but it could have gone either way! Perhaps the day will come when the average consumer decides that chicken meat is so tasteless that it is not worth buying. More likely, they will probably think that that's the way it is supposed to taste!

The amorphous, innocuous looking structure of poultry meat may even be a marketing advantage compared with red meats. It distances the consumer from thinking about where it came from. Red meat such as beef is unpalatable to some people because they make a connection between the red color, blood, and the origin of the product from a live animal.

8. Uniformity of Product

Through genetic selection (see 2. on p. 261), modern strains of broilers are highly uniform in conformation, growth rate, and feed efficiency. The high degree of uniformity in size and conformation produces a final product that is highly uniform in composition and taste. This is a major marketing advan-

who have experienced the loss of their department and who likely will not be replaced when they retire.

Over the years, numerous reports and articles dealing with the status of departments of poultry science have been published. One of the earliest was the paper of Sunde et al. (1972) entitled "Problem of Disappearing Poultry Science Departments." They believed that the major reason for the decrease from 44 departments in 1960 to 21 in 1971 was the need by universities to eliminate classes with small enrollments. They concluded, "The long term effects of the mergers of the 1960's will not be known for some time. The effects are just beginning to show and only time will tell the story as it should be told" (p. 1087). Thirty years later, the story would seem to be that the mergers have not had a serious impact on the poultry industry. In 1987, a symposium on Poultry Programs of the Future was held at the Poultry Science Association (PSA) annual meeting in Corvallis, Oregon (Reynnells, 1988). At this symposium, Cook (1988) documented a further decline to 16 departments by 1987. He suggested that poultry programs could remain relevant by moving research programs into molecular biology, to keep poultry research on the cutting edge. Beck (1992) documented the further attrition of poultry departments to only 14. She concluded, "Vertical integration in the industry has led to specialization and decreased job availability, both in reality and in student perception" (p. 1330).

Beck (1992) asked two pertinent questions: "Why should a heavily integrated industry support scientific endeavors at universities? Why should (university) administrations support programs that attract few students?" (p. 1331). Her answers were as follows: "Universities have the capability to address long-term questions, animal welfare questions, and concerns of food quality and safety. Is it not short-sighted for industry to take the stance that university programs are obsolete and for universities to abandon such an important industry?" (p. 1331) Ort (1994) suggests that agriculture today is at the interface between society and its natural resource base and that poultry science can survive as a discipline if it increases its relevance to issues of global development and environmental preservation.

Pardue (1997) conducted a survey of poultry meat producers to evaluate concern over loss of poultry science departments and to assess future needs. While 44 percent of industry respondents noted "extreme concern" about loss of poultry programs, they ranked communication and business skills as much more important than poultry background as desired skills and training in prospective employees. Pardue (1997) observed that a fundamental shift has occurred in types of positions typically filled by poultry science graduates, with a change in emphasis from production to food science and technology. Another area where there has been increased demand is for veterinarians with expertise in avian pathology to identify and control disease outbreaks in birds kept under intensive conditions.

## Future Trends for Industrial Poultry Production

One concept being discussed is **supply chain optimization,** the next level of integration in the poultry industry. An article in *Feedstuffs* (Jan. 20, 1997), entitled "Poultry Industry Must Look to Next Level of Integration," begins with this paragraph: "Is your company making the most money possible? Are you creating the maximum value for your shareholders? Are you generating the maximum return possible on your investment? Integrated poultry producers can significantly improve their business results in each of these areas by improving the level of integration among individual operating departments."

Aspects of supply chain optimization include the control of all inputs from their source. The major cost in producing chickens is feed (it's not chicken feed, accounting for over 70 percent of the cost of production). Supply chain optimization suggests that instead of buying feed ingredients on the open market, integrators can derive more profit by setting up contract growers for corn and other commodities. This concept implies that wherever there is a profit to be made in the process, it should be made by the **single profit center,** or corporate headquarters. In the future, grain farmers will likely be tied into the process, in the same way that contract poultry growers are now. It could be (and is) argued that grain producers will be better off if they have a contract and produce grain as specified by the integrator rather than operating as independent farmers and gambling in the marketplace.

Another motivation for the poultry industry to lock in grain production is to avoid competition on the global market for feed grains. A potential positive impact of coordination of grain and broiler production is that it might facilitate recycling of poultry manure as a fertilizer. Control of grain production would give the integrators the opportunity to direct that the manure be used in crop fertilization. A recent development is genetically altered **value-added grains,** such as a high oil corn developed by DuPont. Optimal utilization of such grains necessitates that their identity be preserved from the field to the feed mill (**identity-preserved value-added grain**). This might best be achieved by integrating crop and bird production in a vertical integration enterprise.

In another *Feedstuffs* (Jan. 27, 1997), an article entitled "Tyson Sends Message about Chicken, and Next Millennium," one of the major poultry integrators, Tyson Foods, outlined its plans for the next millennium (perhaps they meant the next century rather than the next 1,000 years!) The first paragraph: "In a bold, confident and detailed fiscal report, Tyson Foods has left the clearest message possible to beef, pork and poultry producers: Chicken will lead them into the coming millennium, and Tyson will lead chicken." In the same article, the CEO of Tyson Foods also commented, "The company is positioned stronger than any other producer to make high-quality chicken products accessible to the world, noting that the company's production is 2.5 times its nearest competitor and that the only entities

producing more chicken than Tyson are the countries of Brazil and China" (p. 1).

This article and others like it are notable in that the main points are how we, the company, can maximize profits. The live chicken is simply an "input." Growers and their concerns are not mentioned. Many people can accept industrial production of nuts and bolts, computer chips, and so on but have difficulty applying with much enthusiasm the industrial concept to living, breathing animals, and thus may lack the same intensity of enthusiasm as the CEO of Tyson Foods for the industrial model.

Besides being the self-proclaimed leader in industrial poultry production, Tyson Foods has a seafood division involved in **industrial fishing.** Despite the well-documented depletion of West Coast fishing stocks, leading to the almost unimaginable situation where several species of salmon are now listed as endangered species, industrial fishing pushes on (Fig. 8-6). The Tyson message for the next millennium includes a brief statement on the seafoods division, which described the past year as being full of obstacles, including biomass replenishment, fishing quotas, and low prices (*Feedstuffs,* Jan. 27, 1997). Tyson Foods is now the nation's leading surimi producer (surimi is deboned fish tissue, which can be

**FIGURE 8-6** Industrial fishing has led to rapid depletion of the world's fisheries resources. (Courtesy of Anders Skrede.)

used to make fish sticks, synthetic "crab meat," etc.). Fish for surimi production are harvested using drift nets, which indiscriminately mine the sea, contributing to the loss of endangered salmon species. As in the rest of the Tyson comments on the Next Millennium (*Feedstuffs*, Jan. 27, 1997), no mention is made of any aspect except the corporate bottom line and any nuisance factors, such as fishing quotas, that negatively affect profits. The history of industrial agriculture, fishing and forestry in the twentieth century in North America is not one that inspires much confidence in the sustainability of these activities. Industrial fishing, for example, has brought the cod of the east coast and the salmon of the west coast of North America to the brink of extinction. Now they're after the so-called "trash fish," which can be made into surimi.

The February 17, 1997, issue of *Feedstuffs* had two interesting articles. On page 1, an article began with the following sentence: "The U.S. broiler industry, as a gesture of goodwill, said Feb. 7 that it plans a joint venture project in Russia intended to help put poultry production on a firmer basis in Russia." On page 4 of the same issue is an article entitled "USDA Seeks Comments on Contract Poultry Grower Concerns," discussing how the USDA is seeking public comment on concerns that contract poultry growers have raised about their relationships with poultry companies. Three areas of concern are (1) grower payments tied to the performance of other growers; (2) the accuracy of feed weights and feed delivery as well as pick-up procedures and (3) procedures for weighing live birds. The agency said it has heard repeated concerns from poultry growers who say they are treated unfairly by the poultry companies with whom they contract. "The comments we receive will enable us to determine whether and what type of additional regulatory action may be necessary to protect contract growers in this era of increasing concentration in agriculture," said Dan Glickman, secretary of agriculture.

In an earlier article (*Feedstuffs*, August 7, 1995), growers complained of unfair practices such as feed theft, underweighing of

birds, and targeting of poor quality chicks (dwarfs, which grow slowly) to outspoken growers and grower-union organizers. It is not inevitable that there should be an adversarial relationship between integrators and growers, but the impression seems to be that the integrators' major concern is with their profit, regardless of social costs. Cole (2000) in a book *Communication in Poultry Grower Relations* provided guidelines for reducing integrator-grower antagonism.

Not surprisingly, following the "goodwill to Russia" announcement, the Feb. 24, 1997, issue of *Feedstuffs* had an article "Contract Poultry Growers Opposed to Russian Venture," beginning with: "The Arkansas Contract Poultry Growers Assn. said it is opposed to spending tax money to help finance a new U.S.-Russia poultry venture. The association criticized a proposed joint venture that was recently announced by the governments of the two countries and U.S. poultry interests" (*Feedstuffs*, Feb. 17, 1997).

This article raises another societal concern about industrial poultry production: **political influence and clout.** The dominance of Arkansas-based poultry companies, the relationship of the former president of the United States to Arkansas, the firing of Mr. Clinton's first secretary of agriculture, Mr. Mike Espy, for inappropriately receiving gifts from a major Arkansas-based poultry firm (Tyson Foods target of corruption case, *Feedstuffs*, June 30, 1997), and the $70 million U.S.-taxpayer funded airport in Arkansas (whose major function appears to be for the transport of Arkansas-raised chicken to Asia and on the return flights bringing Asian-made electronic goods and clothing to Wal-Mart, a major Arkansas-based chain store) seem to form a pattern suggesting unusual political influence and clout of the poultry industry. On Dec. 29, 1997, Tyson Foods pleaded guilty to federal gratuities statutes in the Mike Espy case and paid $4 million in fines and $2 million investigation costs (*Feedstuffs*, Jan. 5, 1998, p. 1). Under the plea bargain agreement, Tyson Foods will not be barred from continuing millions of dollars of sales to the U.S. military and school lunch programs.

Another example of political clout is that the U.S. government has passed legislation allowing chicken meat chilled to a temperature of 26° F to be labeled and sold as fresh chicken (despite the fact that a broiler chilled to 26° F is frozen hard as a rock!). This provides a nationwide marketing advantage for Arkansas chicken, which can be sold throughout the United States as "fresh" rather than "frozen."

Political influence is also being exerted with land grant universities. The University of Arkansas Poultry Science Center of Excellence is heavily funded by the vertically integrated poultry companies. Kansas State University has formed a "strategic alliance" with Cargill, Inc. (see Chapter 10). What do these companies expect in return for their underwriting of university programs?

The dominance and political strength of the industrial poultry industry seems to raise legitimate societal concerns. Will meat in the supermarket be so cheap after the competitors are gone?

The poultry industry industrialized in the United States without the general public realizing it. Basically, it is a done deal. The major poultry integrators in the United States now have so many financial resources and so much political influence that the industry is unassailable. There's probably only one thing that could end the industrial poultry business in the United States, and that is if it moves to Thailand, Brazil, Russia, China, and other countries that have cheaper inputs. Shoes, electronics, chicken—why not produce them where the costs are lowest and export the product to the United States? In the global food system, both capital and technology are highly mobile and will move to wherever input costs are least expensive (the Nike shoe analogy). An additional advantage to the integrators in moving to other countries is that in many countries, the concept of animal rights is almost non-existent. An old poultry integrator's joke is that there is good news and bad news in Russia. The bad news is that animal rights activists are coming to protest the industrial pro-

duction of chickens; the good news is that they won't be here for a hundred years!

As a final comment on this subject, in *Feedstuffs* (May 5, 1997), Tyson Foods announced 10 proposed poultry complexes in China, including feed mills, processing plants, and so on. Each complex will be capable of producing and processing one-half million broilers per week. Local farmers will be sought to be growers.

# INDUSTRIAL SWINE PRODUCTION

As discussed previously, the poultry industry industrialized in the United States without the general public realizing it. The swine industry has had a much rockier road. Corporate swine mega-farms with over 100,000 sows, producing over 2 million pigs per year, are constantly in the news (e.g., virtually every week in *Feedstuffs* magazine). The principal reason is the air and water pollution associated with large-scale swine production. **Swine odor** contains over 200 compounds, including ammonia, hydrogen sulfide, mercaptans, phenols, indoles, skatoles, and numerous other nitrogen and sulfur containing compounds, as well as organic acids, aldehydes, and ketones. Odors emanating from swine facilities are objectionable for a number of reasons. They are unpleasant and "cling" or stick to skin, hair, and clothing for a long time. In simple English, swine manure stinks! Numerous studies have indicated that exposure to swine odor can cause negative mood changes including tension, depression, anger, fatigue, and confusion (Schiffman et al., 1995). Changes in immune system function have been noted.

The industrialized swine industry has developed most rapidly in North Carolina, which is now the second leading pork-producing state (Table 8-4). The advantages of industrial pork production relate to efficiency and production of a product of low cost to the consumer in the supermarket. Objections to the industry are summarized by Thu and Durrenberger (1994)

**TABLE 8-4**  Swine numbers in leading U.S. states, 2000.
(Source: *Feedstuffs* Reference Issue, 2001.)

| Rank | State | Inventory (1,000 head) |
|---|---|---|
| 1 | Iowa | 15,400 |
| 2 | North Carolina | 9,400 |
| 3 | Minnesota | 5,800 |
| 4 | Illinois | 4,200 |
| 5 | Indiana | 3,400 |
| 6 | Nebraska | 3,100 |
| 7 | Missouri | 2,900 |
| 8 | Oklahoma | 2,340 |
| 9 | Kansas | 1,570 |
| 10 | Ohio | 1,510 |
| 11 | South Dakota | 1,360 |
| 12 | Pennsylvania | 1,040 |
| 13 | Michigan | 950 |
| 14 | Texas | 920 |
| 15 | Colorado | 840 |
| Total for 15 states: | | 54,730 |
| Percent of U.S. Inventory | | 91.4% |

in an article: "North Carolina's Hog Industry: The Rest of the Story" and in their book *Pigs, Profits, and Rural Communities* (Thu and Durrenberger, 1998). Some of the negative points are the following:

1. Total disruption of communities and rural residents by the overpowering stench of swine manure, preventing any pleasurable outside activities.

2. Drastic drop in property values after a swine mega-farm is built in an area. Homes become unsalable, except at give-away prices.

3. Squeezing out of local hog farmers, and thus the infrastructure of family farm-based agriculture.

4. Loss of jobs and local income. As with intensive poultry production, much of the process is automated, including automatic feeders, waterers, waste removal systems, and so on. There is a net loss of jobs when family hog farms are replaced by corporate farms.

5. **Political gerrymandering** is often involved in creating a climate suitable for corporate farm development. Murphy Farms, Inc., is one of the largest corporate swine farms. According to Thu and Durrenberger (1994), Wendell Murphy, founder of Murphy Farms, Inc., was investigated by the N.C. Bureau of Investigation for possible illegal financial dealings while he was a state senator. While a senator, he played a pivotal role in developing legislation to assist his hog production company, including having intensive hog farms exempt from conducting environmental impact studies or public hearings. Residents received no notice of the construction of these facilities in their neighborhoods and had little or no opportunity to raise questions or voice concerns. Perhaps it is not inappropriate to mention that in 1997 Wendell Murphy joined the *Forbes* 400 list of the 400 richest people in America, with a net worth of $1 billion. Don Tyson also joined the *Forbes* 400 list in 1997, with a net worth of $1.2 billion. In 1999, Murphy Family Farms was purchased by Smithfield Foods, which now controls most of the pig production in the United States (Table 8-5). Mr. Murphy was fortunate in selling out just before Hurricane Floyd slammed into North Carolina, causing extensive environmental damage from flooded pig farms. Thus he was spared the trauma of spending some of his billion dollars on environmental cleanup!

6. Hog production in North Carolina has caused immense pollution. The failure of hog lagoons has caused rivers to become swine manure sewers. In one well-publicized lagoon failure, the New River was essentially depopulated of fish and other aquatic organisms all the way to the Atlantic Ocean, as described in *Feedstuffs* (July 3, 1995):

"Anti-corporate farming activists have the smoking gun they've been looking for. On June 21, North Carolina, the nation's second-

**TABLE 8-5** Leading U.S. pig producing companies, 2000.
(Source: *Feedstuffs* Reference Issue, 2001.)

| Rank | Company |
|------|---------|
| 1 | Smithfield Foods |
| 2 | Premium Standard Farms |
| 3 | Seaboard Farms |
| 4 | Prestage Farms |
| 5 | Tyson Foods |
| 6 | Cargill |
| 7 | Iowa Select |
| 8 | Christensen Farms |
| 9 | Purina Mills |
| 10 | Goldsboro Hog Farm |
| 11 | Land O'Lakes |
| 12 | The Hanor Co. |
| 13 | Heartland Pork Enterprises |
| 14 | Pipestone System |
| 15 | Sands System |

"ambush-predator" form, which attacks fish and immobilizes them with a highly poisonous neurotoxin (Burkholder and Glasgow, 1997).

The swine industry in North Carolina was, in 1997, the subject of intense political debate. North Carolina Governor Jim Hunt proposed a two-year moratorium on hog farm construction, representing a two-year "time out" to seek solutions to environmental concerns (*Feedstuffs*, April 14, 1997, p. 1). The moratorium is supposed to give N.C. State University researchers time to develop alternatives to lagoon systems and solutions to odor and water quality concerns that have aroused the public. Although the two-year moratorium seems to have general public support, the corporate swine industry is pulling no punches. According to *Feedstuffs* (April 14, 1997), Murphy Family Farms and other corporate producers are "working around-the-clock to counter further restrictions on the pork industry. 'They haven't seen anything yet,' exclaimed Jeff Turner, a Murphy farms representative. 'We will put enough people in there (the State Capitol building) to bust the walls out of the building. Maybe then they will pay us some attention'" (p. 1).

Corporate farms seem to like to fall back on the family farm image when things get sticky. Does Murphy Family Farms, with assets of several hundred million dollars, conjure up images of your typical family farm? The president of the North Carolina Pork Council is quoted (*Feedstuffs*, April 14, 1997, p. 1) as observing, "Farmers have been the state's lifeblood since its inception. We are good stewards of our environment, and we have been willing to do what is necessary to protect the water, air and land." This would appear to be a self-serving attempt to exploit the public's positive image of real family farms in defense of corporate mega-farms. Another interesting statement in the *Feedstuffs'* article: "The pork industry fears the bill (the moratorium proposed by the Governor) could precipitate widespread local control of hog production" (p. 1). This is apparently a terrifying prospect to the corporate swine industry!

largest hog-producing state, registered the largest documented hog waste spill in recent history when 25 million gallons of slurry flowed from an eight-acre lagoon into streams that feed New River, approximately 30 miles upstream from the Atlantic Ocean" (p. 1).

Pollution of estuaries with swine and poultry manure is causing biological changes and the emergence of new fish diseases due to polluted water. One of them is a toxic algae, *Pfiesteria piscicida*, that has caused extensive fish kills (Noga et al., 1996). This organism is also toxic to humans, causing open sores, nausea, memory loss, fatigue, disorientation, and incapacitation (Gratlan et al., 1998). The toxic algae is a dinoflagellate, a class of single-celled aquatic organisms that exhibits both plant and animal characteristics (see Chapter 6). It has a complicated life cycle involving toxic and nontoxic forms. Water pollution is one of the factors causing the formation of an

The intensive swine and poultry industries of North Carolina suffered a severe blow (no pun intended) when Hurricane Floyd, a so-called 500-year event, slammed into the state in 1999. Hog and poultry farms on flood-plains were flooded, swine lagoons flooded and failed, and dead pigs, poultry, and tons of animal waste contaminated rivers. Fortunately for Wendell Murphy, Murphy Family Farms was purchased by another corporate entity (Smithfield Foods) shortly before Hurricane Floyd hit North Carolina. The cleanup costs were shouldered by taxpayers, in the form of emergency state and federal disaster aid.

One solution to the environmental problems posed by swine mega-farms is to locate them in isolated areas such as in the Great Plains region. This is now a trend. Wyoming, Utah, Colorado, Oklahoma, and other wide open states are developing swine industries. While circumventing most of the air pollution concerns, this development raises other issues. Much of the western United States is semiarid, and water is a valuable commodity. Swine farms use large quantities of water. There is also the specter of pollution of groundwater and aquifers with swine manure from leaking lagoons. In terms of resource utilization, is it sustainable or wise to ship feed grains from the Midwest to Wyoming and Utah to raise huge numbers of pigs? This totally uncouples crop and livestock production. What are the implications of this 100 years from now, in terms of soil erosion, soil fertility, and so on in the Midwest where the crops are grown? These are not the sort of questions that Americans are accustomed to asking. Over the long term, a **sustainable swine production system** should maintain or enhance the environment and the resource base (land, water, air, human); the quality of life for producers, pork consumers, and society as a whole; the profit level of producers; and the quality of pork produced (Honeyman, 1996).

Corporate swine farms have not escaped controversy in the West. In 1997, Governor Frank Keating of Oklahoma declared an emergency measure to regulate corporate swine farming, in response to public demands that

politicians curb the state's exploding sow herd (*Feedstuffs*, June 23, 1997, p. 5). The Oklahoma swine population has quadrupled since 1991. Murphy Farms, Land O'Lakes, Seaboard Corp., Dekalb Swine Breeders, Inc., Cargill Swine Products, Inc., and Tyson Foods are some of the companies seeking the "wide open spaces" of Oklahoma. Even though Oklahoma is a state that has traditionally not been receptive to government regulations and "big government," citizens are clamoring for control of corporate swine farming because of odor problems and water quality concerns. After being rejected from locating an integrated swine program with 250 growers in Hyde County in South Dakota, Tyson Foods is considering putting hogs on South Dakota Indian reservations (*Feedstuffs*, June 16, 1997, p. 15): "The state has eight reservations, representing millions of acres. Some are near the Missouri River, which bisects the state, and they offer isolation, grain and water access. They might also provide protection from county zoning and state corporate farming restrictions." Unlike the apparent success of industrial poultry production, the jury is still out on industrial pork production. There is widespread and deep public antipathy, distrust, and suspicion. Most of the advantages of poultry for intensive production, described in the previous section, don't apply to pigs. The poultry industry can ratchet down the price of chicken meat to make pork terminally non-competitive. Leaders of the industrial poultry industry, such as Tyson Foods, have made no secret of the fact that their goal is to relentlessly increase **market share,** at the expense of beef and pork. It is not human nature for individuals in such organizations to conclude "OK, we're big enough. We're a 7 billion dollar company. We'll just stop where we are, and leave some of the meat business for other commodities."

Another negative aspect of industrial poultry and swine production is the effects on **human physical and mental health** (Gustafsson, 1997). Intensive swine units can be hazardous to human health. Numerous deaths of workers from hydrogen sulfide exposure have occurred in swine confinement buildings

(Donham and Thu, 1993). Hydrogen sulfide is of particular concern with slurry systems of manure handling (Gustafsson, 1997). Levels as high as 1500–2000 ppm in the air can be reached during slurry pumping. Levels above 200 ppm can be lethal to humans. As many as 30 percent of workers in intensive livestock units develop bronchitis, asthma, and other respiratory problems. A variety of airborne contaminants from swine confinement buildings can cause respiratory problems in workers (Asmar et al., 2001). They include various gases (methane, carbon dioxide, ammonia, hydrogen sulfide), swine odor component chemicals (acids, phenols, cresols, indoles), and particulate matter such as animal dander, hair, fecal solids, feed dust particles, fungal spores, fungal hyphae, and bacterial endotoxins. Repeated exposure to these airborne contaminants can cause permanent lesions of the epithelial cells in the respiratory tract, chronic respiratory disease, and death (Asmar et al., 2001). Volatile organic acids have the greatest potential to lower air quality near swine facilities, because of their ease of atmospheric transport (Zahn et al., 1997).

Residents who live near industrial swine facilities report a decreased quality of life (Wing and Wolf, 2000), a fact that will not be doubted by anyone who has been in the vicinity of a large hog facility. Specific health concerns are increased incidence of headaches, runny nose, sore throat, coughing, diarrhea, and burning eyes (Wing and Wolf, 2000).

In North Carolina, industrial hog production has been concentrated on the coastal plain region, an area that is poor economically and has a large proportion of African Americans in the population. The siting of hog facilities in regions inhabited by the poor, minorities, and generally powerless people has been termed **environmental injustice** (Wing et al., 2000). In other words, corporate swine farms have been foisted on people the least likely to have the financial, educational, and legal means at their disposal to fight back, such as rural African Americans and Native Americans (see previous comments regarding South Dakota Indian reservations). Environ-

mental injustice is the disproportionate burden of pollution on people of color and the poor (Wing et al., 2000). Intensive swine production is concentrated in areas of North Carolina that have the highest disease rates, least access to medical care, and the greatest need for positive economic development and educational systems (Wing et al., 2000). Instead, they get stinking, water-polluting pig farms. Is this something that animal scientists, the developers of the technology, can look at with pride? The presence of these swine facilities in poor and underdeveloped regions may preclude other types of more positive development, scaring off more desirable businesses (Cole et al., 2000).

Manure from industrial swine farms is applied to adjacent crop lands, often owned by independent farmers. Because of transport costs, hog manure is often applied at excessive rates, causing air and water pollution. In a study in Iowa, Jackson et al. (2000) found that large hog farms minimized the area used for manure spreading by underestimating manure nitrogen content, projecting above-average crop yields, and applying manure to crops (e.g., soybeans) that don't require additional nitrogen.

Intensive poultry and swine production are highly automated. Most physical work in these facilities is unskilled. For example, on an egg farm, the birds are kept in cages. Feeding and watering are automatic. The eggs automatically roll on to a moving belt to be taken to be washed and packaged. The physical work in a layer facility involves workers who walk through the facility daily, checking to make sure that the automatic systems are working and removing dead birds. This work is mainly low wage and often done by non-English speaking farmworkers who receive the wages and substandard benefits typical of disenfranchised and powerless people. Workers in poultry processing plants and livestock slaughter houses similarly receive low wages, few benefits, and little respect. The trend in hourly earnings for poultry processing plant workers, corrected for inflation, is negative. Real earnings per hour were lower in the early 1990s than in the late 1970s and most of the 1980s

(Schrimper, 1997). The average hourly wage of a poultry processing plant worker in 1995 was $5.27 (Schrimper, 1997). Meanwhile, Donald Tyson of Tyson Foods, Inc., joined the *Forbes* 1997 list of the 400 richest people in the United States, with a fortune of $1.2 billion (*Feedstuffs*, Oct. 6, 1997, p. 2).

Especially in the swine area, the emergence of industrialized mega-farms has great implications on **mental health of family farmers.** To strive to compete with the industrialists, many farmers have borrowed heavily to build new and larger facilities. This has led many to bankruptcy and to mental health problems, including suicides arising from the stress of economic uncertainty. This can be dismissed by some as the price paid for progress. To others, it is another example of the unpaid social costs of **"slash and burn" factory farming.** The exodus of family hog farms was accelerated in 1998, when hog prices dipped to as low as $10 per 100 lb. *Feedstuffs* magazine described this as "chicken time for pigs" in an article entitled "Independent Pork Production May Be in Last Desperate Year" (*Feedstuffs*, Dec. 21, 1998, p. 1).

Various segments of agribusiness and industrialized animal production have been very effective not only at exploiting the favorable public image of family farmers, but also at staying in the background and letting farmers carry the ball for them. As Caneff (1993) pointed out:

These groups (agribusiness companies, organizations of large-scale farmers, and commodity and livestock promotion groups) need farmers as foot soldiers in their war against any social movement that threatens their primacy in the U.S. food system, including the animal protection movement. The farm and commodity organizations rail against the animal protection movement to distract farmer-members from recognizing the real threats facing them, many of which those organizations helped create—like the continuing industrialization of animal production. Factory-style livestock production has little to recommend it for family farmers. As farmer-essayist Gene Logsdon wrote, "Which is more of a threat

to your independent business as a family livestock farmer: animal rights or animal mega-factories?"

Caneff (1993) concluded that the animal rights "threat" may be the salvation of family farming. In a similar vein, Cheeke and Davis (1997) discussed how environmentalists might save the ranching industry. Caneff (1993) described a number of ways that farmers can remain competitive with industrialized animal production. These include forming alliances with the environmental and animal protection movements and putting a stop to being the "shock troops" for the pharmaceutical industries, swine conglomerates, and poultry integrators. The same could be said for many animal scientists. Do I have any obligation to promote the injection of dairy cows with bovine somatotropin just because Monsanto is investing $500 million in developing it?

I'll conclude this section with a quotation from *Feedstuffs* (Aug. 15, 1994):

"Huge specialized pork factories will be a passing fancy in the long-term history of U.S. hog production," predicted Harold Breimyer, University of Missouri agricultural economist and professor. "Because they've staked their future, rigidly and wholeheartedly, on technology, they are bound to implode. Such production methods, ultimately, will collide with high-priced energy. Held hostage to rigid, systems-oriented production schemes, meat factories will fold like a house of cards during the next major energy crunch. Cheap energy is a massive force in the way we live and work," Breimyer noted. He said reliance on cheap fuel dictates long commuting distances and serves as an underpinning for the nation's cheap food policy, but it won't always be that way. Within 10–15 years, an economic shock wave, prompted by an energy crisis, will turn mega pork farms into dinosaurs, he predicted.

"Hog factories are a passing event. They rest solely on cheap labor and cheap energy. When the cost of motor fuel doubles and triples and quadruples, with electric power rates following suit, those big hog installations, empty of hogs, will be good shelter during a rainstorm, if lingering odor is not too strong."[2]

If the preceding scenario comes to pass, the rise and fall of the swine mega-farms will devastate many investors, but more importantly, will devastate rural communities. The family farms and their associated infrastructure will be gone, replaced by the rubble of failed pig factories.

There is an additional reason why mega-swine farms are likely to fail. They are based on technological solutions to biological problems. **Technological solutions** are based on human control. Nuclear power plants, for example, are built with many safeguards, which theoretically make failure impossible. But failures occur, due to human error. Swine waste lagoons are designed so they can't fail, but they do. An example is Circle 4 Farms, a mega-swine farm in Utah. It has plans for 120,000 sows, a feedmill, processing plant, and so on. The farm is located in an arid area, and the waste is pumped into lined lagoons (Fig. 8-7), which theoretically cannot contaminate groundwater. However, in 1996, workers accidentally siphoned swine waste slurry into the underground aquifer. Local residents, who had previously been supportive of Circle 4 Farms, were angered about the introduction of swine waste into their water supply and further angered by the fact that the incident was not reported to state authorities for several months. Local health authorities were not informed for several more months. According to *Feedstuffs* (May 26, 1997), state

**FIGURE 8-7**  A lined lagoon on a swine mega-farm in Utah. Theoretically, these lagoons cannot leak and contaminate groundwater. However, in 1996, workers accidentally siphoned swine waste slurry from one of these lagoons into the underground aquifer. Industrial swine production has been plagued with lagoon failures and water contamination, leading to widespread public opposition to industrial pig production.

officials claimed that the waste was not toxic, the site was isolated, and the slurry was a drop in the bucket in the aquifer. Enthusiasm for Circle 4 Farms in the closest town, Milford, Utah, was initially high but has waned considerably following this incident. The U.S. public, with good reason, is increasingly skeptical of assurances from local officials and corporate farms that there is nothing to worry about ("just trust us"). Residents of Milford, Utah, have rates of respiratory and diarrheal diseases several times the state of Utah average and higher than in comparably sized towns (*Feedstuffs*, March 20, 2000, p. 16). Residents believe that these problems began to develop after the arrival of Circle 4 Farms. As far as contamination of groundwater is concerned, DNA finger printing techniques should allow for tracing the movement of pathogens and their point of origin.

Independent hog farmers claim that they can compete with industrial pork production if they were "playing on a level field." At various times in the twentieth century, American farm policy was designed to reduce the number of farmers. For example, former Secretary of Agriculture Earl Butz, an

[2] When will the oil run out? K.S. Deffeyes (2001) reviewed this issue in *Hubbert's Peak. The Impending World Oil Shortage*. M. King Hubbert was an American geophysicist working for Shell Oil. He calculated that world oil production would hit a peak in the 1970s (hence, Hubbert's Peak) and decline thereafter. Deffeyes (2001) updated Hubbert's predictions, and presents evidence indicating that oil production is on its way down and severe shortages will occur by 2025.

agricultural economist from Purdue University, was notable for his blunt statement to farmers "get big or get out." Clearly, farm policy under his jurisdiction would not have emphasized assistance for small farmers.

Integrated pork producers are increasingly looking at Latin America as a region for expansion. Seaboard Corporation, frustrated by attempts to build a large hog operation in Oklahoma, has purchased one-half million acres (700 square miles) in Argentina. The American Soybean Association (ASA) is promoting swine production in Brazil, in an attempt to reduce competition in the world market for soybeans. If Brazilians produce and consume more pork, they might export fewer soybeans. Of course, Brazil might wind up exporting pork (a value-added product) instead of soybeans, which eventually would reduce demand for soybeans in the United States.

In an address to the 2001 annual meeting of the American Society of Animal Science, ag economist R. L. Plain of the University of Missouri provided 10 trends that are shaping the U.S. swine industry (Plain, 2001):

1. Improved herd performance. Over the last 20 years, hog farmers have produced 3% more pork per breeding animal per year.

2. Fewer and bigger hog farms. In 2000, 235 hog farms owned 52% of U.S. pigs.

3. Specialization. In 1920, 75% of U.S. farms had pigs. Today, 5% have pigs.

4. Fewer and bigger packing plants.

5. Geographic shift in production, to grain-deficit regions such as western states.

6. Integration of production and packing (killing). Four major packers are on the list of the 6 largest hog producers.

7. Integration of packing and processing.

8. Contracting. Over 32% of hogs in 1999 were finished under production contracts.

9. Globalization. World trade in pork is increasing by 8% per year.

10. Not in my backyard (NIMBY). There is increasing societal aversion to pig production "in my neighborhood."

John Ikerd, another ag economist at the University of Missouri, listed 10 top reasons for rural communities to be concerned about large-scale, corporate hog operations (Ikerd, 1998, unpublished paper). Starting at the bottom and working up:

10. Hogs stink.

9. The work is bad for people.

8. Putting too much stuff (swine manure) in one place causes problems.

7. Consumers have little if anything to gain.

6. Continuing regulatory problems are inevitable.

5. Hog factories destroy public confidence in agriculture. The wholesome image of family farmers is being rapidly destroyed by the bad publicity of corporate hog farms.

4. Future of the community is turned over to outside interests.

3. The decision-making process can rip communities apart. The introduction of large-scale corporate hog farms is invariably accompanied by wrenching community debates as people choose up sides.

2. Hog factories degrade the productive capacities of rural people. Industrial hog operations destroy more jobs than they create. Hog factories are looking for people who are dependable, who know how to carry out orders, and will work hard for little money.

1. Tomorrow's problems are disguised as today's solution.

Ikerd concludes:

Hog factories are a short-run solution, at best, that may create more long-run problems than they solve today. Sooner or later, non-thinking jobs will be done elsewhere on the globe, where people will work harder for less money, and are accustomed to doing whatever they are told. Hog factories may keep rural people from doing the things that need to be done today to ensure the future of their communities. Hog factories will not create communities where our children and their children will choose to live and grow. Communities with a future must take positive actions today to ensure a desirable quality of life for themselves, their children, and rural children of future generations.

A leading critic of industrial swine farms is Robert F. Kennedy, Jr., who heads an organization called the Water Keepers Alliance (WKA). The WKA claims that "factory hog farmers are outlaws and bullies who have destroyed thousands of miles of public waterways and aquifers, shattered the lives of many rural Americans and treated millions of animals with unspeakable cruelty" (*Feedstuffs*, July 15, 2002, p. 15).

# INDUSTRIALIZATION OF OTHER TYPES OF ANIMAL PRODUCTION

To a large degree, **dairy cattle** production can be intensified in much the same way as poultry and swine production. In the United States, dairy farms are becoming fewer and larger, with some having more than 10,000 cows. Advantages of large size include the usual economies of scale, such as computerized, highly automated milking facilities, bulk purchase of commodity feed ingredients, automation of feeding systems, and so on. Subjectively, it would seem that animal welfare for dairy cows in confinement is not as severely compromised as in the case of intensive poultry and swine production. In some areas, the dairy industry is returning to its grazing roots, and intensive grazing of pastures is being used to reduce costs (Hanson et al., 1998). Leading dairy states are shown in Table 8-6.

**Beef cattle** production involves a mix of intensive and extensive systems. Cow-calf pro-

**TABLE 8-6** Inventory of milk cows, top dairy states (1999–2000). (Source: *Feedstuffs* Reference Issue, 2001.)

| Rank | State | Milk cows (1,000 head) |
|---|---|---|
| 1 | California | 1,466 |
| 2 | Wisconsin | 1,365 |
| 3 | New York | 701 |
| 4 | Pennsylvania | 616 |
| 5 | Minnesota | 545 |
| 6 | Texas | 345 |
| 7 | Idaho | 318 |
| 8 | Michigan | 299 |
| 9 | Ohio | 260 |
| 10 | Washington | 247 |
| 11 | New Mexico | 232 |
| 12 | Iowa | 217 |
| 13 | Vermont | 160 |
| 14 | Missouri | 159 |
| 15 | Florida | 158 |
| Total for 15 states: | | 7,088 |
| % of all milk cows in U.S.: | | 77.4% |

duction systems generally involve animals on range or pasture, while the feedlot finishing phase is an intensive system. The welfare of both dairy and beef cattle under intensive production is probably less adversely affected by confinement than for layers kept in small cages or sows tethered in stalls. In a beef feedlot, for example, the animals have considerable freedom of movement and opportunity for normal activity as compared with broilers in a broiler house or pigs in a confinement swine facility. **Sheep production** is largely an extensive, pasture and range-based enterprise (Fig. 8-8).

Beef producers have traditionally valued their independence and way of life. However, the declining market share of beef compared with poultry and lack of profitability has forced U.S. beef producers to consider various

**FIGURE 8-8** Cattle in feedlots (*top*) have considerable freedom of movement compared with broilers and swine in industrial systems (Fig. 8-1). The same is true for dairy cattle in confinement (*middle*). Sheep production is largely an extensive, pasture-based enterprise (*bottom*). (Middle photo courtesy of T. A. Schultz.)

types of alliances. Of major concern is the lack of uniformity of beef cattle, with many different breeds and crosses. This has made it difficult for meat packers to market a uniform, quality product. Vertical integration, with defined breeding and management systems, could enhance the competitiveness of the beef industry. Farmer- and rancher-owned cooper-

atives might be one type of organization that could produce more consistency in beef production while allowing some degree of control and independence for cattle producers rather than simply being contract growers. One farmer-owned cooperative, Farmland Industries, has merged with the fourth-largest beef packer in the United States to create a "huge, ranch to retail, vertically integrated beef production system" ("Farmland, USPB Form Beef Company," *Feedstuffs*, Aug. 4, 1997, p. 1). In 2001, Tyson Foods, the poultry giant, acquired the largest U.S. beef processor, IBP. Tyson Foods has assured the beef industry that it will not vertically integrate beef production. It refers to the merged company as being a "protein provider" rather than a meat company.

Leading beef cattle-producing U.S. states are shown in Table 8-7. Leading feedlot companies are listed in Table 8-8.

In an unpublished paper, Swick and Cremer (2000) (American Soybean Association) concisely described the current state of the U.S. beef industry compared with poultry production:

In the early 1970's, beef was king in America. Today, beef represents an unbranded and non-integrated protein source with falling market share. The industry largely ignores its consumers and continues to offer raw pieces of beef flesh described anatomically using an antiquated grading system that favors a high fat product. Each segment of the industry acts as if it were an island. The cow-calf breeding operation, feedlot, meat packing plant, truck lines, merchandising, and retailing are all independent businesses, each responsible for making as much profit as possible. Once the product is off the truck or out the door and the money changes hands, all care and much of the responsibility is finished.

In contrast, since the 1970s, large chicken companies such as Tyson Foods, Charoen Pokphand (in Thailand), ConAgra and Perdue Farms have become relentless in developing products such as chicken nuggets, processed meat, pre-

**TABLE 8-7** All cattle and calves on feed, top 15 states, 1999–2000. (Source: *Feedstuffs* Reference Issue, 2001.)

| Rank 2000 | State | Inventory (1,000 head) |
|---|---|---|
| 1 | Texas | 13,700 |
| 2 | Kansas | 6,700 |
| 3 | Nebraska | 6,600 |
| 4 | California | 5,150 |
| 5 | Oklahoma | 5,050 |
| 6 | Missouri | 4,250 |
| 7 | South Dakota | 4,050 |
| 8 | Iowa | 3,650 |
| 9 | Wisconsin | 3,350 |
| 10 | Colorado | 3,150 |
| 11 | Minnesota | 2,550 |
| 12 | Montana | 2,250 |
| 13 | Kentucky | 2,260 |
| 14 | North Dakota | 1,980 |
| 15 | Idaho | 1,970 |
| Total in top 15 states: | | 60,260 |
| Percent of U.S. total: | | 61.9% |

**TABLE 8-8** Top U.S. beef feedlot operations, 2000. (Source: *Feedstuffs* Reference Issue, 2001.)

| Rank | Company |
|---|---|
| 1 | Cactus Feeders Inc. |
| 2 | ContiBeef LLC |
| 3 | ConAgra Cattle Feeding Co. |
| 4 | Caprock Industries |
| 5 | National Farms Inc. |
| 6 | J.R. Simplot Co. |
| 7 | Cattlco Inc./Liberal Feeders |
| 8 | Friona Industries |
| 9 | Agri Beef Co. |
| 10 | AzTx Cattle Co. |
| 11 | Irsik & Doll |
| 12 | Hitch Enterprises Inc. |
| 13 | Four States Feedyard Inc. |
| 14 | Gottsch Feeding Corp. |
| 15 | Barrett-Crofoot Inc. |

cooked turkey ham, and other specialties. Innovative ideas such as chicken shops, franchise restaurants, and retail outlets have become part of these integrated companies. Branding and brand recognition of poultry products has been an essential element allowing these companies to gain control of the entire business from grain plow to dinner plate. Consequently, more consumers have switched to chicken and decided to eat less beef. The swine industry in the U.S. in the past 5–8 years has begun to follow the lead of the poultry industry. Major acquisition and consolidation has taken place similar to what happened in the poultry industry 15 to 20 years ago. Consumer preference for lean meat and creative processing is being heard by pork producers and consumption is on the upswing.

Do you find anything wrong with these observations? What is the future of the U.S. beef industry?

Fish farming, or **aquaculture,** is also being industrialized. Aquaculture has been a traditional activity in many parts of Asia, and particularly in China. Fish culture was often integrated with other forms of agriculture, with animal wastes and crop by-products used to feed and fertilize fish ponds (Little and Edwards, 1999). Herbivorous species such as carp are used, usually in a polyculture with several species of fish occupying different trophic levels (bottom feeders, filter feeders, etc., feeding at different levels of the food chain).

In many countries, aquaculture is being industrialized, especially with high-value species such as salmon, trout, shrimp, and prawns. Norway, Chile, Canada, and Scotland are major producers of salmon, which are raised intensively in net pens (Fig. 8-9).

**FIGURE 8-9** Net pens for farming Atlantic salmon in Norway. (Courtesy of A. Skrede).

Shrimp and prawn culture is important in tropical countries in Southeast Asia (e.g., Thailand, Indonesia), China, Australia, and Ecuador. Aquaculture is largely in the form of industrialized production and is basically an aquatic version of broiler chicken production. There are various environmental concerns with aquaculture, especially with carnivorous species such as salmon. Salmon are fed diets based on fish meal, often derived from oceanic fisheries. Salmon are raised in high density in net pens, producing large amounts of excreta and waste feed. The net pens are usually placed in protected areas such as fjords, where water flow may be restricted, causing the wastes to accumulate. The environment is conducive to the development of diseases, which may spread to native fish in the polluted water. Another form of pollution with aquaculture is genetic pollution. Atlantic salmon is the major farmed salmonid. There are often escapees from net pens, during storms or when the nets are damaged by seals or sea lions. The escaped fish may reproduce in the wild or hybridize with native salmon stocks. In British Columbia, for example, DNA evidence confirms that escaped Atlantic salmon are successfully reproducing in the wild (Volpe et al., 2000).

Shrimp and prawn farming in the tropics has caused environmental problems (Naylor et al., 2000). These farms are established by digging ponds near the sea, often by clearing coastal mangrove forests. Mangrove destruction can cause a decline in local fisheries that actually exceeds the gains from shrimp production. Effluent water from shrimp ponds is highly polluted.

On the positive side, intensive aquaculture has the potential for reducing pressures on wild fish through overfishing. Especially in Asia, rational development of intensive aquaculture should make a major contribution to food production.

# ALTERNATIVES TO FACTORY FARMING

Not all countries have embraced industrial farming. Several European countries are going in a different direction. For example, Sweden has developed animal production guidelines that emphasize food safety (for humans and animals—thus antibiotics in feeds are not permitted), concern for animal suffering and welfare, no use of growth promotants or hormones, and sustainability. Sustainability in this model includes preserving Swedish farmland and natural areas, preserving farmers by striving for economic sustainability, preserving clean air and water, minimizing use of agricultural chemicals, minimizing greenhouse gas emissions, and recycling plant nutrients from the city to the countryside. The Swedish Animal Protection Act of 1988 states, "Animals which are bred and kept for production of food, wool, skins, or furs shall be kept and handled in a good environment for animals and in such a way as to promote their health and allow natural behavior." Of course, this presupposes that the natural behavior is known. For swine, Swedish behavioral scientists have found that pigs live naturally in maternal groups of a few sows, some juveniles, and a boar that shows up for mating and then lives independently on his own. The pigs spend most of the day exploring their environment, searching for food sources and consuming foods. When parturition nears, a sow withdraws from the group and builds a nest in an isolated area. About 10 days after giving birth, the sow leads the piglets to rejoin the group, where the piglets integrate socially with other litters, with very little aggression or fighting. Older sows help "educate" young sows in acquiring piglet-rearing skills. Sows wean the piglets gradually, with a minimum of postweaning stress and enteric disorders.

The Swedish philosophy is to raise animals as closely as possible to their natural behavior, which means building housing structures that allow the animals to engage in their normal behavior. Similar strategies are encouraged for other types of livestock. Swedish law requires that layers be maintained in cages that satisfy the hens' need of nesting boxes, perches, and dust baths. Foxes must be kept so that they can socialize with other foxes, dig their own burrows, and have environmental enrichment. Dairy cattle must be on pasture during the summer. Beef cattle must be raised on pasture. Pigs must have access to outdoor areas and pasture during the summer. There are many other regulations pertaining to transportation, feeding, surgical procedures, slaughter, animal shows, exhibitions and sporting events, and so on. Certain breeds are prohibited, such as Belgian Blue cattle, which have a gene for double muscling. This produces more meat but also a smaller pelvis, causing calving difficulties. Most Belgian Blue calves are delivered by caesarean. In Sweden, this type of animal is considered unacceptable because it is not normal and negatively affects animal welfare.

The Swedish regulations and guidelines are markedly different from those in the United States. In Sweden, a very high priority is placed on raising animals in ways that allow them to engage in their natural behaviors and in systems that preserve the integrity of "old-fashioned" family farm-oriented agriculture. Americans would probably never accept such massive government involvement in regulating every facet of animal production, even including what breeds are allowed. If the U.S. government passed a law prohibiting the raising of Belgian Blue cattle, for example, there would probably be a rush on breeding stock by ranchers who aren't about to let the government tell them that they can't raise Belgian Blues. American society is highly individualistic; hence there is a great lack of appreciation by Americans that other societies may not share the same values. However, a gradual change is occurring, and a few American animal scientists are daring to study alternative methods to the industrial model. For example, at Iowa State University, a "hoop system" of housing is being investigated, using inexpensive plastic hoops as the frame for plastic-covered loose housing of pigs (Honeyman, 1996). In 2000, over 1 million pigs were raised in straw-bedded hoop structures in Iowa. John McGlone of Texas Tech University, Lubbock, has a program researching sustainable swine production (*www.pii.ttu.edu*).

Sweden has joined the European Economic Union (EU). In the EU, consisting of virtually all Western European and some Middle European countries, an EU-wide regulator framework exists. Sweden's regulations are being integrated into those of the EU. Thus European animal production will increasingly diverge from the American model. This process is also being fueled by public perception that major European agricultural disasters (mad cow disease, foot and mouth disease epidemics, dioxin contamination of feed, etc.) are a direct result of factory or industrialized farming. The divergent European and American views on agriculture spill over into trade issues The EU has banned importation of meat from animals treated with growth hormones, effectively blocking the export of U.S. beef to Europe. Rather than accepting that European consumers might not want beef from animals treated with hormones, the American position is that this is a phoney way of erecting barriers to free trade, and represents a form of subsidy to European farmers. A similar reaction has occurred to the European ban on GMO crops. The American position is that we're going to produce GMO crops, and by golly the rest of the world is going to buy our grains whether they like it or not!

Most alternatives in the United States to factory farming seem to have limited economic viability. There is some interest in free-range layers, for example. Raising free-range chickens may work as a hobby but would be a tough way to earn a living. An average middle-class American would like to earn a net annual income of at least $20,000. The selling price per dozen eggs would need to be very high to generate this kind of income for a poultry farmer producing free-range eggs.

Less intensive systems of animal production are not automatically more humane. One of the original reasons for intensive confinement systems was to reduce animal exposure to inclement weather, predators, and so on. Extensive production systems have their own types of problems. For example, free-range hens are susceptible to grass impaction, in which the digestive tract becomes impacted with a mess of tangled, rope-like grass (Christensen, 1998).

# INDUSTRIAL LIVESTOCK PRODUCTION, BIOTERRORISM, AND NATURAL DISASTERS

Large concentrations of animals in one place are tempting targets for bioterrorists. This is another disadvantage of industrial raising of livestock and poultry. The year 2001—the year of the World Trade Center and Pentagon bombings, and a foot and mouth disease (FMD) epidemic in Great Britain—exemplifies the perverse possibilities. Animal rights extremists have threatened to unleash FMD and other diseases on corporate farms, to attempt to drive them out of business. Political terrorists threaten to use biological weapons, such as anthrax, plague, brucellosis, and so on against the American public. These organisms, when introduced into livestock, can spread to humans. A case can be made that mega-swine farms and other confinement animal facilities are "sitting ducks" for terrorists. On the other hand, a case can be made that corporate farms might be better equipped to provide biosecurity than traditional farms. At any rate, livestock production

could emerge on the front lines of terrorist activity, of both the animal rights and political-religious varieties. Many authorities believe the United States is poorly equipped to respond to "agriterrorism."

Besides being susceptible to terrorism, large animal facilities are susceptible to natural disasters. On July 29, 2001, a sow barn at Circle 4 Farms in Utah was destroyed by fire, killing 5,000 sows and 11,000 young pigs (*Feedstuffs*, August 6, 2001, p. 3).

# GLOBALIZATION: THE END OF AMERICAN AGRICULTURE?

American farmers like to say, "We can produce food more efficiently than any other country in the world." It is ironic, then, that a University of California–Davis agricultural economist predicts that American agriculture will fade away. In a provocative book, *The End of Agriculture in the American Portfolio*, Steven Blank makes an excellent case that the United States no longer wants or needs farmers. The crux of his argument is that because of globalization, prices paid for farm products are determined globally, while costs are determined locally. If China and Chile can produce apples cheaper than they can be produced in Washington state, then supermarkets will purchase apples from China and Chile. This is exactly what has happened; China is now the world's number one producer of apples, and Washington farmers are bulldozing out their apple orchards. Large supermarket chains now purchase commodities on a global basis; they buy from wherever they can get products the cheapest.

Blank (1998) has postulated an **economic food chain** (Fig. 8-10). Societies begin at the base level, where most of the population is involved in agriculture and subsistence. As societies progress, they begin to produce goods from natural resources. For example, a country in stage 1 might mine iron ore, while a country in stage 2 makes the steel. The United States entered stage 2 during the Industrial

| Development Stage | Economic Activities | Resources Emphasized |
| --- | --- | --- |
| 4 | Information production (research, computers, pharmaceuticals, etc.) | Capital and management |
| 3 | Hi-tech manufacturing (electronics, automobiles, consumer goods) | Land, capital and management |
| 2 | Base manufacturing (steel, heavy industry, industrial revolution industries) | Land and capital |
| 1 | Exploit natural or human resources (mining, forestry, agriculture) | Land and labor |
| Base | Food production (peasants) | Land and labor |

**FIGURE 8-10**  The Economic Food Chain, as proposed by Blank (1998).

Revolution of the eighteenth and nineteenth centuries. The fallout from the end of stage 2 in the United States is the "rust belt" of the east, with idled steel mills and factories. Stage 3 is triggered by the development of a skilled labor force, enabling a country to go into high-tech manufacturing. Countries of southeast Asia are in this stage, making TVs, VCRs, cameras, and other electronic goods. The fourth stage requires a highly educated population; people are paid for what they know. Thus the focus is on knowledge and information production. The United States, Japan, and some European countries are moving into stage 4.

Blank argues that the United States has moved beyond the agricultural stage. Most commodities that can be produced by American agriculture can be produced more cheaply elsewhere (Fig. 8-11). Most Americans don't

care where their car, camera, or TV are made; they seek the best buy. The same thing applies to raspberries and beef. As a country moves up the economic food chain, land used by agriculture becomes more valuable for other purposes. A farm near a city is used with greater economic return if it is converted to a housing development. As Blank (1998) states: "Some of the best farmland in the country is disappearing because the most profitable use of that land is no longer farming" (p. 121). He maintains that the truly sustainable forms of agriculture that will survive in the United States are golf courses, nurseries, and turf farms. Equine facilities could be included also. It's hard for many of us to think of golf courses as agriculture, but the economic returns from a 500 acre golf course greatly exceed the returns from the same land used as a traditional farm. In Japan, for instance, businessmen will pay as much as one half million dollars for a golf club membership. Golf courses, nurseries, turf farms and equine facilities do not yet have to compete on the world market.

The closer a business is to the food distribution system, the more likely it will survive in the United States. Some jobs can be done far away from the consumer, while others involving perishable products must be nearby. "No one wants to eat a tuna fish sandwich that was made a month ago in another country" (p. 112). Blank (1998) uses the hamburger to illustrate this. Americans don't want to travel far to get a

**FIGURE 8-11**  An abandoned grape plantation (vineyard) in Arizona. American agriculture is increasingly non-competitive with various foreign suppliers of fruits and vegetables such as Mexico and Chile.

hamburger. Thus fast-food restaurants are located near consumers. The hamburger bun is not baked at the fast-food outlet. Buns are perishable, but not as perishable as hot burgers. Therefore, buns can be baked at central bakeries located within a reasonable trucking distance from fast-food restaurants. Thus bakeries and fast-food outlets are in the United States to stay. The flour used in the bun is readily stored and transported and could come from a flour mill in another country, if cheaper. The wheat is extremely storable, and could come from wherever wheat is cheap (Argentina, Canada, Australia?). The beef in the hamburger could be imported (Argentina, Brazil, Australia, etc.). The cheese might come most cheaply from New Zealand. The lettuce, onion, and tomato might come from Mexico or Chile. The pickle could come from anywhere. The average American teenager gulping down a hamburger neither knows nor cares where its components came from; all that matters is that it tastes good and is cheap.

Globalization of world trade will accelerate these changes. The United States will produce unique knowledge, information, and high-tech pharmaceuticals and other products, which will be traded to less developed nations for food. Developing countries like Mexico, Chile, Thailand, and so on have lower labor and other in-put costs and can produce basic food commodities—meat, milk, vegetables, fruit—more cheaply than American farmers can produce them. The only way that American agriculture can be competitive is with massive subsidies. We don't subsidize production of TVs, VCRs, textiles, cameras, and so on. That is why they are imported from other countries. American agriculture is heavily subsidized "to keep the family farm in business." Increasingly, when American consumers realize how much extra they pay for food indirectly (taxes) in the form of subsidies, they will demand changes. Blank claims that family farms are an expensive lifestyle that the United States cannot afford; it is an inefficient use of resources that leads to bankruptcy. Subsidies merely postpone the inevitable.

These are not ideas that American farmers and agriculturalists enjoy hearing. However, it is becoming patently obvious that most types of American agriculture, including livestock and poultry production, will become increasingly marginal. Farming requires large amounts of capital (money). Increasingly, lenders (banks) will become less interested in taking the risk of agricultural lending, when other industries offer less risk and higher returns.

Are Americans willing to place their food security in the hands of other countries? Events such as the 2001 World Trade Center and Pentagon bombings might cause a reappraisal of the wisdom of a reliance on imported food. However, if we want to retain agricultural and food production capabilities, continued subsidization is a given.

One means of avoiding Blank's grim scenario is the development of niche markets. A small segment of society is opposed to globalization and supports local production. These markets could be encouraged. "Buy local" is their slogan.

This is a brief review of Blank's book. His ideas are provocative. Current events in agriculture seem to support the scenario he has outlined.

The time for the United States to be the world's major food-producing area may have come and gone (or it may be going). Production of corn, soybeans and other grains is expanding at a tremendous rate in Argentina and Brazil. Russia and Ukraine have great grain-producing potential. Chile has developed a huge fruit and vegetable export trade. Turkey has recently completed a massive irrigation project and has essentially created a new California. The world is awash in food production capacity. Global demand will be met by those able to meet it at the lowest cost. This is particularly true as trade barriers come down. European and American taxpayers won't be willing to subsidize their farmers forever. The handwriting is on the wall.

## AMISH, MENNONITES, AND HUTTERITES: THE LAST OF THE MOHICANS OF NORTH AMERICAN AGRICULTURE?

While Steven Blank (1998) has predicted the end of American agriculture, there are some North American farmers who are thriving. These include the Anabaptists (the Amish, Mennonites, and Hutterites). These societies have been described in detail by Kraybill and Bowman (2001). Amish and Mennonites live mainly in eastern North America, especially in Ohio, Pennsylvania, Indiana, and the Canadian province of Ontario. Hutterites live in communal groups (colonies) mainly in the prairie provinces of Canada (Alberta, Saskatchewan, Manitoba) and the U.S. states of Washington, Montana, North and South Dakota, and Minnesota. Hutterites and Amish are probably the extremes in Anabaptist lifestyles. Amish live on individual farms, and farm primarily with horse-drawn equipment, using horse-drawn buggies instead of automobiles. Hutterites own no private property. They live in colonies of about 100 people. In contrast to the Amish, they use the latest in tractors and equipment, including computers. Livestock production is an important component of Amish, Hutterite, and Mennonite farms. In their own ways, these groups are very competent and effective farmers. Their numbers are increasing rapidly. New Hutterite colonies spring up regularly in western North America. In the east, Amish farmers are on the move, establishing new settlements. Because of their modest lifestyles, using few consumer products, these hard-working farmers will continue to thrive and present one form of viable alternative to corporate, industrial agriculture.

## CONCLUSIONS

The world is moving towards a vast, interconnected global network, fostered by reduction and elimination of tariffs. Agriculture, like other industries, is part of this trend. In rapidly expanding economies, such as those in many Asian and Latin American countries, the demand for meat and other animal products is increasing rapidly. Much of this demand will be met by chicken meat, and to a lesser extent, pork. As discussed in this chapter, these short-cycle animals, which can be raised intensively in confinement, have various characteristics that allow them to be produced on an industrial scale. There are several billion people in the world who are either undernourished or consume a largely uninspiring plant-based diet. They represent a huge potential market for industrially produced meat. Concerns about animal welfare, preservation of family farms, and so on are indulged in by well-fed societies but are issues of little importance to the majority of the world's population. While there likely will be continuing negative perceptions of industrial animal production in North America and western Europe, in the rest of the world it will be "full speed ahead." Future growth of animal production and opportunities for animal and poultry scientists will occur largely in those parts of the world that are presently termed "underdeveloped," but which in many cases are now industrializing rapidly. Multinational poultry and swine companies based in North America and Europe will play a major role in food production in Asia. These companies will be a major source of opportunity for animal science students in North America and Europe. Not coincidentally, industrial animal production dovetails very nicely with the fast-food restaurant system, which is also of global extent and expanding rapidly (Schlosser, 2001). While there is significant opposition (Kneen, 1995a, b) to the transnational corporations that are rapidly moving to control world food production and distribution, they are likely to prevail.

## THE SKY IS THE LIMIT

*(Editorial. 2000. Feed Tech. Vol. 4, No. 8.)*

"Every once in a while, futuristic ideas emerge from the surface about how our agricultural world should be designed for the coming century. One country currently facing future dilemmas is the Netherlands, where agriculture and nature fight for the limited space available outside the cities. The Dutch Minister of Agriculture is supporting the idea of creating so-called 'agroproduction parks,' where 300,000 pigs are kept in a high rise building, together with a quarter of a million layers, a million broilers and a lot of salmon, and hot houses on top of the building to grow vegetables. Next to this building, which has an estimated size of 400,000 m², abattoirs, processing plants and a feedmill should be built. One complete unit could provide food for a community of one million people.

When you think about it, it is not such a bad idea. Of course, there are disadvantages of putting together so many animals of different species, but in terms of efficiency it could be worthwhile. Savings on costs are imaginable as less labor is required, less transport is needed when located close to a harbor, and production efficiency can be improved. The environment will also profit from larger units, because it is more economical to invest in (more expensive) environmental friendly equipment.

On the other hand, there are some negative thoughts: Feedmills have to close down, family farms run out of business, and (intensive) agriculture will lose its fragile image. Agriculture becomes an industrial activity, so there will be no further need for a ministry of agriculture. The minister signs his own dismissal. But as a minister, one should think ahead, and doing so . . . the sky is the limit."

Despite some of the negative aspects of "factory farming" discussed in this chapter, it is likely that the mass production of poultry and/or pigs in China and other Asian countries will result in less environmental degradation and destruction of wildlife habitat than alternatives such as slash and burn crop production, deforestation for livestock grazing, and so on. On a long-term basis, beyond the comprehension of most of us, industrial animal production may well turn out to be a temporary blip involving the last 50 years of the twentieth century and part or all of the twenty-first century. When human population growth has stabilized, petroleum reserves have been depleted, and new processes developed for producing palatable meat substitutes from plant products, our descendants might look back a couple of hundred years from now and ask, "How could they do that to animals?"

## STUDY QUESTIONS

1. Discuss whether the term "factory farming" is a valid and legitimate description of intensive swine and poultry production, or is it a "value-laden" animal rights term that should be avoided by animal scientists?

2. Some critics of industrial swine and poultry production claim that the high efficiency of these enterprises is partly due to the externalization of some costs, which are paid for by society at large. Explain.

3. What is vertical integration in the poultry industry?

4. Some critics of industrial poultry production claim that the growers are "the new sharecroppers." Discuss.

5. What advantages are there for poultry over beef cattle, in terms of sustainability, for industrial production?

6. What are short-cycle animals? What advantages are there for these types of animals in meat production?

7. Why are chickens more efficient than cattle or pigs in converting corn-soy diets to meat? (or are they?)

8. Factory farming is said to be a "value-laden" term. Define a value-laden term.

9. Critics of intensive animal production are sometimes called Luddites. What are Luddites?

10. Former U.S. Secretary of Agriculture, Earl Butz, often said "We can go back to organic agriculture, but who wants to pick the 50 million Americans who will starve to death?" Is this a valid or correct point of view?

11. Is industrial production of chickens and pigs cruel and inhumane, or is the animals' welfare better than it was under traditional systems? Discuss.

12. Consult some recent issues of *Feedstuffs*. What evidence do you find that industrial production of pigs is a controversial issue?

13. What are some ideas for the beef industry to remain competitive, when faced with inexpensive, industrially produced chicken meat and pork?

14. Why have departments of poultry science at North American universities decreased in number over the past 50 years? What do you see as future opportunities for university graduates in the poultry industry?

15. What is supply chain optimization? Is it a good or bad concept for family farmers?

16. What is the connotation of a "single profit center" in the industrialized poultry industry?

17. What is surimi? What is the relevance of surimi to natural resource utilization?

18. What two U.S. states are numbers one and two in pig production? Where might pig production expand greatly in the next 25 years?

19. What does "globalization of animal agriculture" mean? What effect will this have on job opportunities for animal science graduates? Does industrial globalization have any implications for elementary and high school programs?

20. Is there any point in trying to "save the family farm"? Defend your viewpoint.

21. Industrially produced poultry meat is a highly uniform product. Beef is quite a variable product, with thousands of individual farmers and ranchers using their favorite breeds, crossbreeding systems, and so on to produce cattle lacking in uniformity. What are the disadvantages of lack of uniformity? Are there any advantages?

22. In 1997, the National Academy of Sciences recommended an increase in the daily calcium intake of Americans, equivalent to one more glass of milk per person per day. This would amount to a 19 percent increase in annual fluid milk sales. In a report of this in *Feedstuffs* (Sept. 8, 1997), a dairy economist stated, "Imagine what that would do to the farm price of milk" (p. 17). OK, imagine what it would do to the farm price of milk. What are some possible and probable responses of the dairy industry to a 19 percent increase in annual milk demand?

# REFERENCES

ASMAR, S., J. A. PICKRELL, and F. W. OEHME. 2001. Pulmonary diseases caused by airborne contaminants in swine confinement buildings. *Vet. Human Toxicol.* 43:48–53.

BECK, M. M. 1992. Status of Poultry Science Departments and poultry research within combined departments. *Poult. Sci.* 71:1328–1331.

BLANK, S. C. 1998. *The end of agriculture in the American portfolio.* Quorum Books, Westport, Connecticut.

BURKHOLDER, J. M., and H. B. GLASGOW, JR. 1997. Trophic controls on stage transformations of a toxic ambush-predator dinoflagellate. *J. Eukaryotic Microbiol.* 44:200–205.

CANEFF, D. 1993. Family farmers should move toward the animal welfare movement. *Amer. J. Alternative Agric.* 8:4,46.

CAST. 1991. *Herbicide resistant crops.* Council for Agricultural Science and Technology, Ames, Iowa.

CHEEKE, P. R., and S. L. DAVIS. 1997. Possible impacts of industrialization and globalization of animal agriculture on cattle ranching in the American West (Can environmentalists save the ranch?). *Rangelands* 19:4–5.

CHRISTENSEN, N. H. 1998. Alleviation of grass impaction in a flock of free-range hens. *Vet. Record* 143:397.

COLE, D., L. TODD, and S. WING. 2000. Concentrated swine feeding operations and public health: A review of occupational and community health effects. *Environ. Health Perspect.* 108:685–699.

COLE, L. 2000. *Communication in Poultry Grower Relations. A Blueprint to Success.* Iowa State University Press, Ames, Iowa.

COOK, R. E. 1988. Poultry research programs in the future. *Poult. Sci.* 67:890–896.

DEFFEYES, K. S. 2001. *Hubbert's Peak. The impending world oil shortage.* Princeton University Press, Princeton, New Jersey.

DONHAM, K. J., and K. M. THU. 1993. Relationships of agricultural and economic policy to the health of farm families, livestock, and the environment. *J. Amer. Vet. Med. Assoc.* 202:1084–1091.

GRATTAN, L. M., D. OLDACH, T. M. PERL, M. H. LOWITT, D. L. MATUSZAK, C. DICKSON, C. PARROTT, R. C. SHOEMAKER, C. L. KAUFFMAN, M. P. WASSERMAN, J. R. HEBEL, P. CHARACHE, and J. G. MORRIS, JR. 1998. Learning and memory difficulties after environmental exposure to waterways containing toxin-producing *Pfiesteria* or *Pfiesteria*-like dinoflagellates. *Lancet* 352:532–539.

GUSTAFSSON, B. 1997. The health and safety of workers in a confined animal system. *Livestock Prod. Sci.* 49:191–202.

HANSON, G. D., L. C. CUNNINGHAM, M. J. MOREHART, and R. L. PARSONS. 1998. Profitability of moderate intensive grazing of dairy cows in the Northeast. *J. Dairy Sci.* 81:821–829.

HONEYMAN, M. S. 1996. Sustainability issues of U.S. swine production. *J. Anim. Sci.* 74:1410–1417.

JACKSON, L. L., D. R. KEENEY, and E. M. GILBERT. 2000. Swine manure management plans in north-central Iowa: Nutrient loading and policy implications. *J. Soil and Water Cons.* 55:205–212.

KNEEN, B. 1995a. The invisible giant. Cargill and its transnational strategies. *The Ecologist* 25:195–199.

KNEEN, B. 1995b. *Invisible giant. Cargill and its transnational strategies.* Fernwood Publishing, Halifax, Nova Scotia; Pluto Press, London and East Haven, Connecticut.

KRAYBILL, D. B., and C. F. BOWMAN. 2001. *On the backroad to heaven. Old order Hutterites, Mennonites, Amish, and Brethren.* The Johns Hopkins University Press, Baltimore.

LITTLE, D. C., and P. EDWARDS. 1999. Alternative strategies for livestock-fish integration with emphasis on Asia. *Ambio* 28:118–124.

NAYLOR, R. L., R. J. GOLDBURG, J. H. PRIMAVERA, N. KAUTSKY, M. C. M. BEVERIDGE, J. CLAY, C. FOLKE, J. LUBCHENCO, H. MOONEY, and M. TROELL. 2000. Effect of aquaculture on world fish supplies. *Nature* 405:1017–1024.

NOGA, E. J., L. KHOO, J. B. STEVENS, Z. FAN, and J. M. BURKHOLDER. 1996. Novel toxic dinoflagellate causes epidemic disease in estuarine fish. *Marine Pollution Bull.* 32:219–224.

ORT, J. F. 1994. Positioning poultry science for the twenty-first century: A commitment to people. *Poult. Sci.* 73:215–223.

PARDUE, S. L. 1997. Educational opportunities and challenges in Poultry Science: Impact of resource allocation and industry needs. *Poult. Sci.* 76:938–943.

PLAIN, R. L. 2001. The U.S. swine industry. Where we are and how we got here. *J. Anim. Sci.* 79 (Suppl. 1): 98(Abst.).

RANDALL, A. 1998. The "lessons" of Luddism. *Endeavour* 22:152–155.

REYNNELLS, R. 1988. Symposium: Poultry programs of the future. *Poult. Sci.* 67:878.

SCHIFFMAN, S. S., E. A. SATTELY-MILLER, M. S. SUGGS, and B. G. GRAHAM. 1995. The effect of environmental odors emanating from commercial swine operations on the mood of nearby residents. *Brain Res. Bull.* 37:369–375.

SCHLOSSER, E. 2001. *Fast food nation: The dark side of the all-American meal.* Houghton Mifflin Co., New York.

SCHRIMPER, R. A. 1997. U.S. Poultry processing employment and hourly earnings. *J. Appl. Poult. Res.* 6:81–89.

SKINNER, J. L. 1996. *American poultry history. 1974–1993.* Watt Publishing Co., Mt. Morris, Ilinois.

SUNDE, M. L., T. E. HARTUNG, and L. S. JENSEN. 1972. Problem of disappearing Poultry Science Departments. *Poult. Sci.* 51:1079–1087.

SWICK, R. A., and M. C. CREMER. 2000. Livestock production: A model for aquaculture? Conference on Aquaculture in the Third Millennium, 20–25 February 2000, Bangkok, Thailand.

TAUER, L. W., and J. LOVE. 1989. The potential economic impact of herbicide-resistant corn in the USA. *J. Prod. Agric.* 2:202–207.

THU, K., and E. P. DURRENBERGER. 1994. North Carolina's hog industry: The rest of the story. *Culture and Agriculture* 49:20–23.

THU, K. M., and E. P. DURRENBERGER (Eds.). 1998. *Pigs, profits, and rural communities.* State University of New York Press, Allbany.

VOLPE, J. P., E. B. TAYLOR, D. W. RIMMER, and B. W. GLICKMAN. 2000. Evidence of natural reproduction of aquaculture-escaped Atlantic salmon in a coastal British Columbia River. *Conserv. Biology* 14:899–903.

WING, S., and S. WOLF. 2000. Intensive livestock operations, health, and quality of life among eastern North Carolina residents. *Environ. Health Perspect.* 108:233–238.

WING, S., D. COLE, and G. GRANT. 2000. Environmental injustice in North Carolina's hog industry. *Environ. Health Perspect.* 108:225–231.

ZAHN, J. A., J. L. HATFIELD, Y. S. DO, A. A. DISPIRITO, D. A. LAIRD, and R. L. PFEIFFER. 1997. Characterization of volatile organic emissions and wastes from a swine production facility. *J. Environ. Qual.* 26:1687–1696.

# CHAPTER 9

# Food Quality and Safety Issues

## MEAT QUALITY AND TASTE

There seems to be a widespread perception that food quality is not what it used to be. The tomatoes in the supermarket these days are said to be tasteless and the "agribusiness chicken" is as bland and tasteless as cardboard. The agricultural scientists and "agribiz" have bred the taste out. Bread isn't as good as it used to be; they take all the goodness out of it and sell it back to you as additives in fortified bread. There's nothing in the supermarket that is comparable to a vine-ripened tomato out of the back garden, and Grandma's roast chicken on Sunday had a taste and aroma totally lacking in today's supermarket and fast-food chicken. Are these valid perceptions? If it is true that today's supermarket tomatoes and chickens are not as tasty as in "the good old days," does this mean that there is something wrong with how these products are produced? The need for extensive postharvest storage and transportation in the modern world is a critical factor. Grandpa did not walk out his backdoor to pick a vine-ripened tomato in North Dakota in December. He lived without eating tomatoes for most of the year. Grandma (or Grandpa) raised, killed, eviscerated, and prepared the chicken for roasting. Are today's consumers prepared to do that? Ironically, the "putting the taste back in" is being accomplished by biotechnology and gene transfer, as discussed later in this chapter.

The taste and eating qualities of meat (**organoleptic properties**) can be influenced by the feeding and management systems under which the animals are produced. In general, the meat of young animals is more tender and more bland in taste than meat of older animals of the same species. As the growth performance of poultry and livestock has been increased by modern production techniques, the age at slaughter weight has been steadily decreasing. The growth rate of the broiler chicken has increased by about 1 g per bird per day each year for the last several decades, so a broiler grown to 45 days in 1991 would weigh approximately 1 pound more than a broiler grown to the same age in 1981. There is no sign that this improved production rate is leveling off. Maybe it should not be surprising if the taste of the meat from such a young bird is less pronounced than of meat from an older bird. Because it is easy to measure growth rate and feed efficiency, perhaps poultry scientists have exerted greater effort on production-related research than research on eating quality of the final product. On the other hand, perhaps it is not possible to achieve both goals. Maybe it is not possible to make meat from a young broiler taste like that from an older bird. The consumers ultimately will decide which is more important: cost or taste? If they find the mass-produced, but very inexpensive, modern broiler undesirable, consumption of poultry meat will decline, and the poultry industry will adjust to reclaim lost markets. It is difficult to conclude that general consumer dissatisfaction with the product exists today when the per capita consumption of poultry meat is increasing rapidly and linearly in the United States. Sixty years ago, a political

slogan was "a chicken in every pot." That referred to chicken in every family's diet once per week. The modern poultry industry has delivered, and then some, on that promise!

Another aspect of the taste of current animal products is that somehow confinement rearing will give a different product than traditional systems. There is a market for eggs from **free-range chickens.** Besides the interest in these from an animal rights perspective, there is also the common belief that eggs from free-ranging birds taste better. Perhaps the more varied diet, including earthworms and insects, will contribute to the taste. Again, the consumer is faced with a choice. Eggs from free-ranging hens cost more to produce than those from layers kept in cages (Fig. 9-1). Those people who believe that the eggs are better from free-ranging birds must be prepared to pay more for them. For those who detect no difference in taste and have no animal rights concerns regarding caged layers, the cost per dozen eggs is the bottom line.

The beef industry has had to deal with consumer views on the quality of beef. The fat, saturated fatty acid, and cholesterol contents have had a major impact on consumer perception of "red meat." Other issues such as the "hamburger connection," grazing on public lands, competition with humans for grains,

**FIGURE 9-1** Free-range chickens. Eggs and broilers produced under free-range conditions are more expensive than products from conventionally reared chickens because of higher land and labor costs and lower bird performance. (Courtesy of C. S. Winstead.)

and so on have also impacted the beef industry. In response to these consumer concerns, the industry has made numerous changes and is continuing to change. Competition for the consumer dollar drives these changes.

As discussed in other chapters, the fat content of beef can be reduced by changes in the feeding system used (Chapter 2), genetic selection, and use of products of biotechnology such as growth hormone and repartitioning agents (see Chapter 5). Taste and texture, including tenderness, are influenced by fat content. It is possible to produce beef with a very low fat content and with a strong taste. Cattle raised to market weight on semiarid rangelands and tropical grass pastures, such as in Australia and South America, may be several years old before they reach market weight (literally teenagers in some cases!). The beef is very lean, tasty, and as tough as leather. However, this is an extreme case. The fat content can be reduced by using a longer feeding period of forage, with a shorter feedlot period of grain feeding. In areas with abundant high quality forage, it is possible to produce beef of good eating quality and low fat content on a forage feeding system entirely. However, because grain is higher in energy and a cheaper source of net energy for cattle than forages in the United States, and grain-fed animals grow faster, **forage-finished beef** will generally be higher in price to the consumer than feedlot beef, reflecting the higher production costs. Among other considerations, the forage-raised animal is older at slaughter than the feedlot-finished animal, so a greater proportion of its lifetime nutrient intake has been used for maintenance. Allen et al. (1992a, b) have developed and evaluated forage systems for beef production from conception to slaughter. Griebenow et al. (1997) have reviewed forage-based beef finishing systems.

One of the fast-food restaurant chains, McDonald's, introduced a low fat ground beef hamburger containing carrageenan, an extract of seaweed. This material replaced fat in providing functional properties such as texture and eating quality. Other materials such as an oat bran product to replace fat are also being developed. Thus food technology research provides another means of lowering the fat content of animal products without necessarily eliminating their basic appeal to the human palate (although the "seaweed hamburger" did not succeed and was discontinued, because of lack of consumer demand).

The flavor of lamb and mutton is unappealing to many people. This aversion is largely due to substances associated with the fat that are given off as the meat is cooked. Australian scientists have developed a feed additive, based on high linoleic acid vegetable oils (e.g., sunflower, safflower oils) protected against rumen fermentation by encapsulation with formaldehyde-treated casein, which can be fed as a supplement to market lambs to produce a product with greater consumer acceptance. As discussed in Chapter 2, it is possible to modify the fatty acid composition of ruminant body fat by dietary means.

Ultimately, whether the meat and other products of today's intensive animal production taste different than when livestock and poultry were produced under traditional systems is a matter of opinion. Some people will be forever convinced that there is a difference. In part, this may reflect how we reminisce about things in general; nostalgia always distorts reality. Taste perceptions are more acute in young children than in older people. Perhaps the memories of the special taste of Grandma's chicken have been distorted by time. Perhaps not! As time marches on and the baby boomers age, perhaps older folks will yearn for the days when Chicken McNuggets were really tasty!

## MEAT AND FOOD SAFETY

**Food safety** is a significant concern of American consumers. Ironically, there is an inverse relationship between food safety and consumer concern about food safety. As the food supply has become increasingly safe, people have become suspicious and cynical, and sometimes hysterical, about what "agribusiness" is doing to their health. Cynicism is expressed in the feeling that multinational food corporations will do anything they can get away with to make a dollar. Hysteria occurs regularly. In 1989, there was the Alar scare. Alar is a chemical that was sprayed on apples to slow ripening to give a more desirable color, retard fruit drop from trees, and maintain fruit crispness after harvest. The CBS television network ran an "exposé" on Alar, with a backdrop of a giant apple marked with a skull and crossbones. Actress Meryl Streep gave press conferences on the dangers of apples. The public panicked. Apples were taken out of school lunches. Apple sales slumped. It was all a great fuss about nothing; there was no health hazard (Marshall, 1991). At about the same time (1988), the British public panicked about salmonella in eggs. Again, despite the very slight hazard (Humphrey, 1990), paranoia reigned and egg consumption in Britain fell precipitously. The British had barely got over the salmonella scare when they were hit with another one, bovine spongiform encephalopathy (BSE, mad cow disease). Eating beef was like playing Russian roulette—it might cause your brain to turn into a sponge in a few years! (BSE is discussed in more detail in a later section).

Surveys conducted by the Food Marketing Institute indicate that chemical residues on foods and antibiotics and hormones used in livestock production are the two greatest food safety concerns of American consumers (Table 9-1). In contrast to these perceived health threats, the FDA considers microbial contamination (food poisoning) to be the major health hazard associated with food, followed by malnutrition and diet-related factors involved in degenerative diseases like coronary heart disease and osteoporosis. Pesticide residues and food additives are below natural toxins in foods as very minor or slight hazards, according to the FDA. Antibiotics and hormones used in livestock pose no detectable risk to human health, or at least it is so slight that it has not been detected yet.

Senauer et al. (1991) discuss some of the reasons why people become alarmed about issues that are not important, while dismissing real risks. Some of the factors influencing the perception of risk are shown in Table 9-2. Risks borne voluntarily, such as cigarette smoking or driving automobiles, are perceived to be of less concern than those imposed on one involuntarily (e.g., pesticide residues in food). The perception of benefit is also a factor. The lower the perceived benefit, the lower the tolerance for the risk. These factors help explain why some seemingly minor issues cause hysteria, while major issues are virtually ignored. With the Alar incident, the implication of children, apples, and cancer was a potent mix. Besides, the benefit from Alar use was perceived to be very minor. The salmonella incident in Britain was similar. Health warnings for many years about the links between egg consumption, blood cholesterol level, and coronary heart disease were largely ignored. Publicity about a low incidence of salmonella in eggs caused a sud-

TABLE 9-1   Food safety concerns of American consumers (Adapted from Senauer et al., 1991).

| Factor | % Ranking of Serious Hazard | | |
| --- | --- | --- | --- |
| | 1987 | 1988 | 1989 |
| Pesticide and herbicide residues | 76 | 75 | 82 |
| Antibiotics and hormones | 61 | 61 | 61 |
| Nitrates in foods | 38 | 44 | 44 |
| Additives and preservatives | 36 | 29 | 30 |
| Artificial food colors | 24 | 21 | 28 |

**TABLE 9-2** Factors influencing
the perception of risk
(Adapted from Senauer et al., 1991.)

| Decreased Perception of Risk | Increased Perception of Risk |
|---|---|
| Risk assumed voluntarily | Risk borne involuntarily |
| Effect immediate | Effect delayed |
| No alternatives available | Many alternatives available |
| Risk known with certainty | Risk not known |
| Exposure is an essential | Exposure is a luxury |
| Encountered occupationally | Encountered nonoccupationally |
| Common hazard | "Dread" hazard |
| Affects average people | Affects especially sensitive people |
| Will be used as intended | Likely to be misused |
| Consequences reversible | Consequences irreversible |

**TABLE 9-3** Consumer concerns about
food safety
(Source: National Cattlemen's Beef
Association, 2001.)

| Item | % Extremely Concerned |
|---|---|
| Bacterial contamination | 39% |
| Pesticide residues | 38% |
| Chemical additives | 33% |
| Hormones | 30% |
| Genetically modified foods (GMO) | 26% |
| Antibiotics | 23% |
| Irradiated foods | 21% |
| **Food Safety Concerns** | |
| **Food Type** | **% Concerned** |
| Fish | 26% |
| Chicken | 22% |
| Prepared foods | 21% |
| Beef | 14% |
| Pork | 12% |
| Fruits and vegetables | 4% |

den and dramatic drop in egg consumption, nearly destroying the British poultry industry. Not only was the number of contaminated eggs low, but the problem can be virtually eliminated by thorough cooking. Salmonella on poultry meat can be killed by irradiation of the meat. However, distrust of irradiated foods, despite repeated tests confirming the effectiveness and safety of food irradiation, kept this technique from commercial application in the United States until 1992.

A more recent survey on consumers' concerns about food safety is given in Table 9-3, conducted in 2001 by the National Cattlemen's Beef Association. In agreement with the survey reported by Senauer et al. (1991), consumers continue to regard pesticides, chemical additives, and hormones as major food safety concerns, despite scientific evidence to the contrary. Probably reflecting a number of bacterial contamination problems in foods during the 1990s, bacteria were the number

one concern, whereas they didn't even hit the radar screen in 1991 (Table 9-1).

The principal concerns of the general public with use of feed additives, growth promotants, and other "chemicals" in current livestock production cannot, in my opinion, be adequately dealt with by data and scientific interpretations. The overwhelming preponderance of evidence and hard data indicates that these substances are safe, effective (increase performance as claimed—i.e., efficacy), improve the economic efficiency of food production, do not adversely affect the animal's welfare or well-being, and represent a high ratio of benefit to risk. They are subjected to extensive testing for safety and efficacy before approval for their commercial use is granted. However, there is a significant proportion of the general public that does not trust the regulatory, scientific, and business establishment. They do not believe the data, its interpretation, or the messengers. They do not trust the government to set or enforce safety standards, nor do they trust farmers. Current controversy surrounding the approval of bST for administration to dairy cattle is a case in point. As discussed in Chapter 5, the safety and efficacy of bST are virtually impossible to refute. The bottom line is that people have a vague sense of uneasiness about injecting cows with hormones to increase milk production. They perceive the possibility of some risk and don't perceive a benefit. However, after bST for dairy cattle was approved, the issue has largely faded away from public concern, although there is residual "background noise" about hormones in food.

## Bovine Spongiform Encephalopathy: Will Eating Beef Kill You?

In the 1980s, a new cattle disease (**mad cow disease,** raging cow disease) appeared in Great Britain, which has had drastic effects on the British beef industry and consumers. Affected animals develop various neurological symptoms, including exaggerated limb movements, muscular jerking, anxiety, and frenzied movements, culminating in death (Fig. 9-2). The brain shows extensive damage, with degeneration of neurons, hypertrophy of connective tissue, and a sponge-like appearance (spongiosis). The condition is called **bovine spongiform encephalopathy (BSE)** (Fig. 9-3) and is apparently one of a series of diseases in humans and animals known as **transmissible spongiform encephalopathy (TSE).** Although the etiology is not certain, it is generally believed by scientists investigating it that BSE was introduced into British cattle by the feeding of meat meal prepared from sheep infected with scrapie, a TSE. **Scrapie** causes sheep to act in a frenzied manner, with signs similar to those of BSE. Scrapie has been recognized in sheep for many years. It has a long incubation period of several years and appears to be transmitted to the offspring of affected animals.

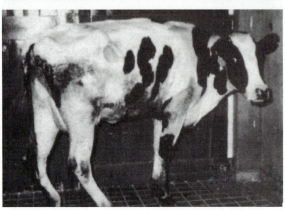

**FIGURE 9-2** Holstein cows exhibiting signs of BSE. Typically, animals show bulging eyes *(top)* and emaciation, weakness, and staggering *(bottom).* (Courtesy of S. Franklin.)

When scrapie is identified in a flock of sheep, the entire flock is slaughtered. The disease also resembles some human conditions such as **Creutzfeldt-Jakob disease (CJD)** and **Kuru.** The CJD disease occurs in the general human population at a low incidence and causes fatal brain degeneration. Kuru is a disease of a stone-age tribe in the remote highlands of Papau New Guinea (PNG). It is believed that the mode of transmission was cannibalism, including eating the brain in funeral rituals. The disease is now rare, because of educational programs to eliminate cannibalism. Cannibalism ended in PNG in 1959, and the last recorded case of Kuru was in 2000, which illustrates the long lag time possible between exposure and disease in the prion-induced

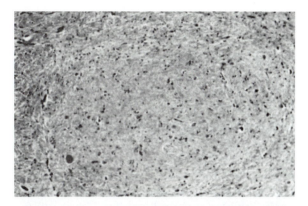

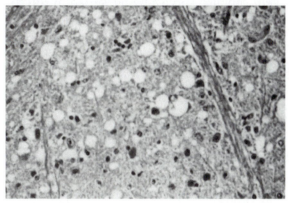

**FIGURE 9-3** Normal bovine brain tissue *(top)* and vacuolated spongiosis of brain tissue from a cow affected with BSE *(bottom)*. (Courtesy of S. Franklin.)

diseases. An outbreak of CJD occurred in France in the early 1990s, in patients suffering from congenital dwarfism who were given human growth hormone extracted from the pituitary glands of cadavers. Apparently one or more of the pituitary glands from which growth hormone was extracted was infected with CJD (Aldhous, 1992).

Other TSE diseases in humans are Gestmann-Straussler-Scheinker syndrome, a form of dementia, and fatal familial insomnia (Donnelly et al., 1999). TSE diseases in the United States include **scrapie** of sheep, **chronic wasting disease** (CWD) of deer and elk, and **transmissible mink encephalopathy** (TME). The CWD occurs in several western states (Colorado, Wyoming) in wild elk and deer, and in deer and elk herds on several game farms (Spraker et al., 1997). Hunters are ad-

vised not to eat the meat from any deer or elk that exhibit abnormal behavior or neurological signs. The TME disease has occurred on Wisconsin mink farms and has been circumstantially linked to the feeding of the meat of "downer" dairy cows to mink (Hadlow, 1999). (Downer cows are those in untreatable recumbancy, for a variety of reasons such as milk fever.) Some have speculated that the downer cows may have BSE, but there is no evidence to support this idea.

**Scrapie** has been known in sheep for over 200 years (Donnelly et al., 1999). Some animal genotypes and breeds of sheep are resistant to scrapie (Balter, 2000). Variations in susceptibility to scrapie involve polymorphisms or variations in the nucleotide sequence of the gene coding for PrP, the normal protein that becomes converted to the prion form (see below). Some polymorphisms make sheep nearly impervious to scrapie, while others make them highly susceptible. There is no evidence that scrapie can be directly transferred to humans.

The BSE disease has been thoroughly reviewed by Dealler and Lacey (1990), Wills (1991), and more recently by Brown et al. (2001). Among the major concerns are a long incubation period of several years so it is difficult to identify affected animals, and the extreme resistance of the causative agent to destruction. Disinfectants, ionizing radiation, and cooking do not destroy infectivity. The only methods that are effective are strong sodium hypochlorite solutions (bleach) and hot solutions of sodium hydroxide (Taylor, 2000). Although the causative agent is not conclusively known, most authorities now believe that an infectious agent known as a **prion** is involved. The term prion was coined by Prusiner[1] (1982); for many years the existence of prions was hotly debated, but they are now generally recognized as the causative

_____

[1] Stanley B. Prusiner was awarded the 1997 Nobel Prize for medicine for his discovery and elucidation of prions and prion diseases. His acceptance speech on prions was published in 1998 (Prusiner, 1998).

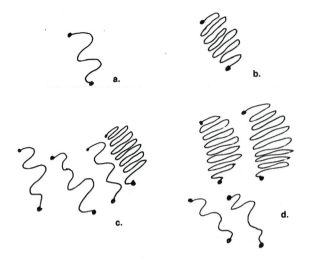

**FIGURE 9-4** Conversion of normal proteins into prions. (a) Normal PrP protein in a nerve cell. (b) The same protein, twisted into a stable configuration called a prion. (c) A prion touches another normal PrP protein and converts it to a prion. (d) Normal PrP protein that has been "prionized" by contact with a prion.

agents of TSE diseases. A prion is an unusually folded variant of a normal body protein, **PrP,** which occurs mainly in nerve cells. A prion has the same amino acid sequence as the normal PrP protein but has been perturbed so as to cause it to fold in an abnormal manner, making it very resistant to degradation. The conformation change in the PrP protein changes it from an α-helical form to a β-flat sheet. The factor that causes a normal PrP protein to convert to a prion is physical contact with another prion. Thus once a prion enters the body, it can eventually initiate a chain reaction of prion formation (Fig. 9-4). Nerve cells store prions because they are resistant to lysosomal enzymes that normally break down unneeded proteins. Prions are stored until the nerve cell bursts, exposing adjacent cells. Thus a "circle of death" spreads throughout the brain, causing the small holes that create the spongiform appearance. The folded structure of a prion protein gives it a great ability to survive such treatments as high temperature, disinfectants, and protein-digesting enzymes in the gut and tissue cells. In fact, the resistance

of prions to inactivation is almost unbelievable. Brown et al. (2000) found that the hamster-adapted scrapie agent retained infectivity even after the tissue has heated at 600° C. This is the temperature used for ashing feeds in the Proximate Analysis system of feed analysis. Brown et al. (2001) speculated that during ashing, an inorganic replica of the prion's molecular geometry was made, that retained an ability to change the PrP precursor protein to its infectious, β-pleated isoform. The scrapie agent can also survive being buried in soil for 3 years (Brown and Gadjusek, 1991).

How are orally ingested prions absorbed and transported to the brain? It is commonly assumed that intact proteins are not absorbed. However, there is a subclass of enterocytes (intestinal lining cells), the M-cells, whose function is to sample gut antigens and relay them to the immune system for inspection (Nagler-Anderson et al., 2001). Antigen-processing cells differentiate between those absorbed proteins that are dangerous antigens and those that are harmless. Prions are apparently absorbed by M-cells located in lymphoid tissue called Peyer's patches (Beekes and McBride, 2000). Because they closely resemble normal PrP, prions do not elicit an immune response and are released into the blood where they are carried passively to the brain.

Prions probably originally arose as a result of mutations in the PrP protein and then spread at a low rate in infected populations. Prions can be transferred from the mother to a fetus. The existence of CJD in humans, which occurs in a very small number of people, could be due to mutation of the gene coding for the normal PrP protein. Recent cases in France (see the preceding) have involved direct transfer of infected human tissue, while the "mad cow disease" situation, while not proven, might be another route of exposure.

Can mad cow disease be acquired by people from eating infected beef? The worst fears of the British government (which had assured the public that beef was safe) and beef producers were realized in 1996, when cases of CJD were linked to consumption of beef (Will et al., 1996). Mad cow disease became a

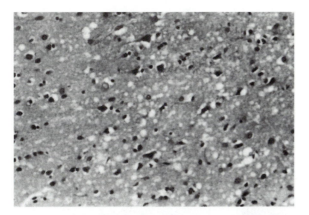

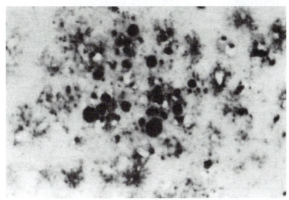

**FIGURE 9-5** *(Top):* The cerebral cortex in Creutzfeldt-Jakob disease (CJD) in humans shows characteristic spongiform change with numerous small vacuoles. *(Bottom):* Brain tissue from a human with new variant CJD showing large round vacuoles. (Courtesy of J. W. Ironside.)

worldwide number one public issue. What could be more damaging to a food industry than the news that eating a food commodity (beef) could destroy your brain and kill you? While it has not been conclusively proven that the new variant of CJD in Britain, which afflicts people much younger than was noted previously, is linked to mad cow disease, the evidence is apparently fairly convincing to medical scientists. Direct proof of the ability of BSE to infect humans is unlikely to be obtained in a controlled experiment involving human subjects.

The new variant strain of CJD (Fig. 9-5), referred to as vCJD, is distinct from other forms of CJD that occur sporadically at low incidence (spCJD). The distinguishing features

of vCJD include a rapid onset in young patients, extensive deposition of PrP amyloid in the brain as large plaques, and a prominent involvement of the cerebellum. These diseases, and BSE, produce a characteristic "signature" of lesions when infective tissue is administered to mice. Infective tissue from cows with BSE produces a signature identical to that of vCJD but different from that of spCJD (Bruce et al., 1997). Bruce et al. (1997) concluded, "Our transmission studies, in combination with the surveillance data, provide compelling evidence of a link between BSE and vCJD" (p. 498). These results were substantiated by Scott et al. (1999).

In a similar study, Hill et al. (1997) demonstrated identical physical and chemical characteristics of prion protein (PrP) in vCJD and BSE. Hill et al. (1997) concluded, "We now find that the biological and molecular transmission characteristics of vCJD are consistent with it being the human counterpart of BSE" (p. 448). In other words, mad cow disease can infect humans, and probably has.

The publicity surrounding mad cow disease and its possible transfer to humans has at least temporarily wrecked the British beef industry. An initial response by the British government was to propose that *all* cattle in Britain be slaughtered and burned. Less drastic measures were taken, but thousands of potentially infected cattle were shot and incinerated. No cases of bovine BSE-linked CJD have been reported in the United States. It is recognized by the U.S. cattle industry that a single case, or even a well-placed but misinformed rumor, could virtually destroy the U.S. beef industry.

The number of cattle with BSE in the U.K. is declining (Donnelly et al., 1999) but the number of human cases of vCJD continues to increase (Andrews et al., 2000), because of the long incubation period from initial exposure. People at risk are those who consumed British beef in the 1980s, including tourists. The United States now bans blood donations from people who spent more than six months in the U.K. in that period. BSE can be transmitted via blood among primates (Lasmezas et al.,

2001). Non-human primates have been infected in zoos via consumption of contaminated meat (Bons et al., 1999).

BSE can be transmitted from cattle to sheep, causing a scrapie-like syndrome (Hope et al., 1999). Although no drug treatments for TSE diseases yet exist, limited success in increasing the survival time of scrapie-infected mice has been achieved with certain cyclic tetrapyrroles (Priola et al., 2000). A possible risk is the use of cosmetics and pharmaceutical products containing derivatives from beef. Interestingly, there is a genetic basis for susceptibility in humans, as there is for scrapie in sheep. The vCJD has occurred only in people who are homozygous for methionine at codon 129 of the PrP gene (Hillerton, 1998).

The outbreak of BSE in British cattle apparently stemmed from changes by the rendering industry in the processing of animal offal and carcasses into meat and bone meal. A key factor was the elimination of a solvent extraction step that had previously been used and that apparently inactivated the BSE infective agent in scrapie-infected sheep carcasses (Taylor et al., 1995, 1998). As a result of the mad cow episode, many countries have now banned the feeding of ruminant-derived meat meal to ruminants, and several countries have banned all use of meat meal in feeds. The use of meat meal in any animal diets will likely come under increased scrutiny.

The mad cow disease event in Britain reinforced the public perception that governments are more interested in protecting private industry than public health, after years of pronouncements from both the British government and the beef industry that British beef was perfectly safe (Shaoul, 1997).

Current information on BSE is available from a number of websites:

*www.maff.gov.uk/animalh/bse/index.html*

*www.maff.gov.uk/maffhome.html*

*www.oie.int.eng/info/en_esbmonde.htm*

*www.cjdsurveillance.com/*

*www.aphis.usda.gov/oa/bse/*

*www.mad-cow.org*

## Microbial Contamination of Animal Products

### Salmonellosis

The main concerns with microbial contamination of meat and eggs have involved Salmonella bacteria (Fig. 9-6). **Salmonellosis** in humans (food poisoning) causes diarrhea, fever, headache, and other symptoms typical of flu, generally lasting 3–5 days. Poultry products have been of particular concern as sources of Salmonella infection through contamination of broilers during processing and the presence of the organisms on or in eggs. Raw egg products (e.g., mayonnaise) are a common source of infection. The word *Salmonella* is derived from the name of a USDA scientist, D. E. Salmon, who studied the role of this organism in food poisoning over 100 years ago.

There are many strains of Salmonella, and numerous species. Over 2,000 serotypes are known, and new ones are discovered annually (Lax et al., 1995). Individual serotypes can be identified by differences in surface antigens. Serotyping is an important means of tracing the origin of a *Salmonella* outbreak in humans or animals. Some species, such as *S. pullorum* and *S. gallinarium*, cause clinical disease in poultry. The major concern with contamination of broilers and eggs is with *S. typhimurium* and *S. enteritidis*. There are many strains (serotypes) of these species. *S. enteritidis* is a common inhabitant of the chicken digestive tract. Most flocks of laying hens are infected but without clinical signs. They periodically shed the organism in the excreta, from which other birds may be infected. A small percentage (about 0.5%) of the eggs laid by an infected hen may contain the organism (Humphrey, 1990). Because of the ubiquitous nature of *S. enteritidis* in poultry, proper han-

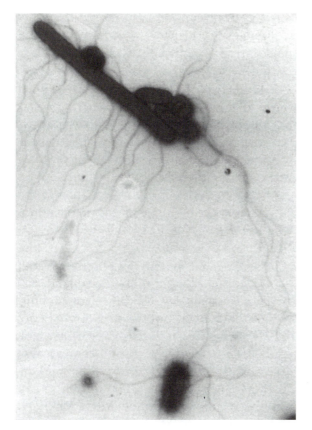

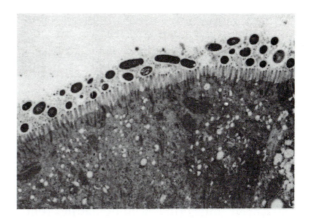

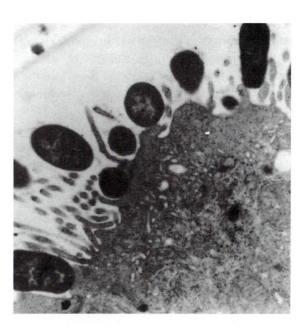

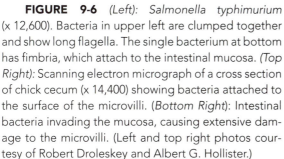

**FIGURE 9-6** *(Left): Salmonella typhimurium* (x 12,600). Bacteria in upper left are clumped together and show long flagella. The single bacterium at bottom has fimbria, which attach to the intestinal mucosa. *(Top Right):* Scanning electron micrograph of a cross section of chick cecum (x 14,400) showing bacteria attached to the surface of the microvilli. (*Bottom Right*): Intestinal bacteria invading the mucosa, causing extensive damage to the microvilli. (Left and top right photos courtesy of Robert Droleskey and Albert G. Hollister.)

dling, and cooking of eggs is important. Freshly laid eggs are unlikely to represent a public health hazard (Humphrey, 1990). They should be cooled and refrigerated as soon as possible to prevent multiplication of bacteria in any infected eggs. Eggs should be thoroughly cooked; pasteurized eggs should be used in recipes calling for raw eggs.

In the late 1980s, salmonella poisoning became a major public health issue in Great Britain and much of western Europe, caused by *S. enteritidis* phage-type 4 (PT4). This organism can infect both layers and broilers without producing clinical signs. It can be shed within the egg, so new-laid eggs can be contaminated. This makes *S. enteritidis* PT4 a greater public health threat than other poultry-associated salmonellas, which are not found within the intact egg. Eradication of infected birds and flocks is being undertaken in Britain (Humphrey, 1990). Apparently, the emergence of PT4 as a problem came about because of successful efforts to rid poultry of *S. pullorum* (pullorum disease) and *S. gallinarum* (fowl typhoid). The PT4 strain of *S. enteridis* filled the ecological niche in the gut created by the elimination of the two pathogens (Baumler et al., 2000). This is an example of how an apparently wise idea (to eliminate two poultry diseases) can have unexpected consequences (human disease).

Birds are most susceptible to being infected with *Salmonella* during the first few days of life. After they acquire a normal gut microflora, they are resistant to colonization with *Salmonella*. This has led to a technique to increase the resistance of baby chicks to *Salmonella* infection by inoculating them orally with adult bird intestinal contents (Schleifer, 1985). The procedure is known as **competitive exclusion.** The bacteria in the inocula occupy the attachment sites on the intestinal and cecal mucosa, thereby reducing the available sites for *Salmonella* colonization.

Certain carbohydrates in the feed or water have also been shown to reduce *Salmonella* infection. Administration of lactose and mannose to chicks provides protection (Hinton et al., 1990; Oyofo et al., 1989). **Oligosaccharides** (short chains of sugar molecules) such as fructooligosaccharide (FOS) also have some protective activity, particularly when given in conjunction with a competitive exclusion inocula (Bailey et al., 1991). These sugars and oligosaccharides are not digested by the birds but serve as growth-stimulating factors for beneficial microbes in the gut, thus inhibiting pathogens by competitive inhibition. Some oligosaccharides serve as binding sites for pathogenic bacteria. It is postulated that the bacteria "confuse" binding sites in the intestine (these sites contain sugar molecules) with the oligosaccharides, so they bind to the oligosaccharides and are excreted (flushed) from the gut in this manner. The "hot" compound in chili peppers, capsaicin, when fed at 20 ppm in the diet of broiler chicks, suppresses the growth of *Salmonella* without affecting taste of the meat (Sams et al., 1995).

Transmission of *Salmonella* to a farm occurs via contaminated feed, carrier animals, and other vectors. Continual vigilance is required by the poultry industry to minimize this problem. Free-range systems of poultry production greatly increase the risk of infection of birds with *Salmonella*. "Free-range" eggs should never be consumed raw or partially cooked (this recommendation should be observed for all eggs, regardless of source).

Irradiation of food products such as poultry meat kills *Salmonella*. As with any new technology, irradiation of foods has its critics, particularly because it involves radioactive isotopes. Following exhaustive testing for safety, irradiation was approved for commercial use for poultry products in the United States in 1992. To increase consumer confidence, the poultry industry now refers to irradiation as "cold pasteurization."

*Salmonella* infection is also a problem in cattle (Lax et al., 1995), especially calves. A particular feature of salmonellosis in calves has been the development of antibiotic-resistant strains. This has some implications in human health. Drug-resistant *Salmonella* organisms of animal origin have been observed. Holmberg et al. (1984) reported the occurrence of antibiotic-resistant *S. newport* in people in four midwestern U.S. states, which was traced to hamburger originating from a beef herd in South Dakota where subtherapeutic levels of chlortetracycline were fed as a growth promotant. The affected individuals were hospitalized with severe, bloody diarrhea, abdominal cramps, nausea, and vomiting. Patients who had been taking antibiotic medications for unrelated reasons prior to consuming the contaminated hamburger were the most severely affected. More recently, White et al. (2001) documented a high level (20%) of a sample of retail ground meats contained antibiotic-resistant *Salmonella*. Another means of infection is in farmers or animal caretakers working with sick calves that have been treated with antibiotics. Wall et al. (1995) documented a number of such cases in Britain. Livestock producers should be aware that they can contact numerous diseases from livestock and should observe good hygiene. This is especially true when handling calves and other animals with diarrhea. Exposure of children to calves and other animals in petting zoos and fairs is another risk source.

A new strain of *Salmonella*, *S. typhimurium* DT 104, has emerged in livestock and poultry. It has multidrug resistance, being resistant to ampicillan, chloramphenicol, streptomycin, sulfonamides, and tetracycline (Poppe et al., 1998). It is primarily a pathogen of cattle. Humans may be infected from contaminated foods of animal origin, especially from drinking unpasteurized milk. Symtoms in humans include diarrhea, fever, headache, nausea, abdominal pain, and vomiting. The symptoms in cattle are watery to bloody diarrhea, decreased milk production of dairy cattle, fever, anorexia, dehydration, and depression. Contact with affected animals, such as calves, dogs, and cats, can lead to human infections. The DT104 strain has an enhanced ability to acquire antibiotic resistance. There is concern that use of a new class of antibiotics, fluoroquinolones, for control of salmonellosis in humans may be compromised by its use in livestock and poultry (Poppe et al., 1998).

Development of antibiotic resistance by bacteria has been reviewed by Davies (1994) and Khachatourians (1998). Commonly, genes coding for resistance to antibiotics are carried by transmissible genetic units called resistance plasmids (R factors). **Plasmids** can be transmitted quite readily among bacteria. Thus there is concern that plasmids from antibiotic-resistant bacteria in livestock could be transferred to human pathogens, complicating the treatment of human disease (Witte, 1998).

## *Campylobacter* Contamination of Poultry Meat

*Campylobacter* spp. (mainly *C. jejuni*) contamination of poultry carcasses is of public health concern. *Campylobacter* and *Salmonella* are responsible for the majority of acute cases of human gastroenteritis (food poisoning). Another organism contaminating poultry that causes food poisoning in humans is *Arcobacter* spp., formerly classified as *Campylobacter butzleri* (Wesley and Baetz, 1999). The *Campylobacter* bacteria do not

cause disease in poultry but are prevalent in the digestive tract and contaminate the carcasses during processing. The stress associated with feed withdrawal and confinement during transport to the processing facility causes shedding of the bacteria, which then contaminate the skin and feathers (Stern et al., 1995). Carcasses become contaminated during processing (defeathering, evisceration, chilling). Control of poultry carcass contamination with *Salmonella* and *Campylobacter* is a continuing, unsolved problem for the poultry industry and for consumers. Swine are also commonly infected with *C. jejuni* and *C. coli* (Harvey et al., 1999).

*Campylobacter* contamination of poultry meat is emerging as a major public health issue. *Campylobacter* infection of humans has been linked to **Guillain-Barre,** a rare and potentially fatal nerve-damaging disease. Antibiotic-resistant strains of *Campylobacter* have developed, with the implication that the use of antibiotics (specifically the **fluoroquinolone antibiotics**) in poultry has caused the emergence of antibiotic-resistant strains of the bacteria. Fluoroquinolones are not used as growth promotants and are not continuously in poultry diets. They are used specifically to treat disease, including respiratory diseases of chickens and turkeys. They are prescription products that must be prescribed by a veterinarian. In humans, the fluoroquinolones are used to treat food-borne illnesses. Quinolone-resistant *C. jejuni* infections in humans are an increasing problem and are largely due to the acquisition of resistant strains from infected poultry (Aarestrup et al., 2000; Smith et al., 1999).

In Europe, enterococci bacteria resistant to the antibiotic vancomycin (vancomycin-resistant enterococci—VRE) have become a public health problem. The source of the resistance was linked to the use of avoparcin as a growth promoter in livestock and poultry (Bager et al., 2000). Avoparcin is a structural analog of vancomycin. Use of avoparcin as a feed antibiotic has been banned in Europe. A new drug, synercid, is effective against VRE. It is an analog of virginiamycin, an antibiotic that has been used as a feed additive for many years. Virginiamycin has also been banned for animal use in Europe (Bager et al., 2000). Another antibiotic used to treat salmonellosis in children is ceftriaxone, an expanded spectrum cephalosporin antibiotic. Ceftriaxone-resistant *Salmonella* have now transferred from cattle to humans (Fey et al., 2000).

Increasing foreign travel and the internationalization of the food trade make the use of antibiotics in food production a public health issue of global dimensions (Wegener, 1999). *Camphlobacter* is one of the most common causes of "traveler's diarrhea" and infections with quinolone-resistant strains is increasingly occurring. A major source for Americans is travel to Mexico, where quinolones are widely used in poultry production (Wegener, 1999).

It is worth pointing out that livestock can acquire pathogens from humans. The odds that a swine herd could have a high prevalence of *Salmonella* infection were 11 times as high if there were no toilet facilities on the premises as if there were (Barber, 2001). There are numerous incidents of bacterial and tapeworm infections in livestock linked to contamination of the feed with human feces (Barber, 2001). On many industrial animal production facilities, the employees come largely from developing countries such as Mexico, where antibiotic-resistant bacteria and internal parasites may be more frequent in humans. In another instance, a cattle owner transmitted asymptomatic renal tuberculosis (*Mycobacterium bovis*) to a herd of cattle by urinating on the straw (Barber, 2001).

The use of antibiotics as feed additives has environmental effects. Chee-Sanford et al. (2001) reported that antibiotic resistance was tranferred from swine manure to soil bacteria and groundwater adjacent to swine waste lagoons. Osterblad et al. (2001) found that wild rodents in areas where livestock are raised and other human activity occurs have high levels of antibiotic-resistant *E. coli*, while animals in areas free of human activity had no antibiotic-resistant *E. coli*.

It seems apparent that the use of antibiotics as feed additives has a number of undesirable public health and environmental consequences. This conclusion has yet to sink in with many peo-

ple involved in livestock and poultry production. Antibiotic use has been sharply reduced in Europe. It is likely that the United States will eventually follow the European decisions to ban feed use of antibiotics, despite the pleas of the industry, including many animal scientists and veterinarians. Antibiotics are too useful in human medicine to warrant jeopardizing their effectiveness by feeding them routinely to animals. The entire blame for antibiotic resistance can't be placed on livestock use; there is widespread misuse of antibiotics in human medicine (Goldmann, 1999). Even routine and justified use of antibiotics may lead to resistance. Hoiby et al. (2000) found that excretion of penicillin in human sweat led to multidrug resistance of skin bacteria such as *Staphylococcus aureus*.

The United States has in place a National Antimicrobial Resistance Monitoring System to monitor antimicrobial resistance in humans and animals. Information can be obtained from the following websites:

animals: *www.fda.gov/cvm*

humans: *www.cdc.gov*

## *E. coli*—Associated Illnesses

*Escherichia coli* (*E. coli*) bacteria are common inhabitants of the gastrointestinal tract and are responsible for a variety of human and animal diseases. They have been grouped into various categories on the basis of clinical syndromes produced. Some of the more common categories are the following:

1. Enteropathogenic *E. coli*

   • cause infantile diarrhea

2. Enterotoxigenic *E. coli*

   • cause neonatal diarrhea in pigs, calves and lambs

3. Enterohemorrhagic *E. coli*

   • cause **hemorrhagic colitis (HC)** and **hemolytic uremic syndrome (HUS)** in humans

Of major concern in the public health area is a toxin-producing strain of enterohemorrhagic *E. coli*, **E. coli 0157:H7**. This organism has caused numerous outbreaks of HC and HUS around the world. The toxin is called Shiga-like toxin or verotoxin; these toxins bind to cell membranes and block protein synthesis (Whipp et al., 1994). **Hemorrhagic colitis (HC)** is characterized by abdominal cramps and watery diarrhea, followed by gastrointestinal bleeding (Dorn, 1993). The site of infection with serotype 0157:H7 is the large intestine, in contrast to most other *E. coli* infections, which occur in the small intestine. **Hemolytic uremic syndrome (HUS)** is characterized by acute kidney failure.

Most outbreaks of 0157:H7 have been directly or indirectly traced to animal products, especially undercooked beef and raw milk. A major outbreak occurred in Washington state in 1993, involving undercooked hamburgers at a fast-food chain (Bell et al., 1994). Over 500 cases, with several deaths, occurred. The incubation period is 4–8 days before clinical signs are seen. The elderly, children, and those with compromised immune systems are most susceptible. Because of the association of 0157:H7 *E. coli* with animal products, and its high virulence, the potential acquisition of antibiotic resistance by this organism would be an alarming situation (Dorn, 1993).

Numerous outbreaks have been caused by consumption of 0157:H7-contaminated raw milk (Dorn, 1993). In 1996, over 9,000 people in Japan suffered from *E. coli* poisoning linked to contaminated radish sprouts. Also in 1996, unpasteurized apple juice tainted with *E. coli* 0157:H7 caused an outbreak of poisoning in several western U.S. states. These cases involving plant products are likely due to contamination of the plants with animal manure. Although cattle seem to be the major host of serotype 0157:H7, deer can also be infected (Dorn, 1993; Keene et al., 1997), and consumption of deer meat in western Oregon (Corvallis) has caused poisoning (Keene et al., 1997).

Only a small proportion of cattle harbor the 0157:H7 organism (Whipp et al., 1994). Contamination of carcasses with fecal matter is the major route of entrance into the food chain. Hamburger is the major culprit, because meat from one infected carcass can contaminate a large amount of product when ground beef is prepared. Dairy cattle seem to be more likely to harbor the organism than are beef cattle. Much of the meat in hamburger is derived from cull dairy cows. Hamburger that is not thoroughly cooked can cause infection, because the bacteria survive in the middle of the hamburger patty. Fast-food chains are a particular risk, because many restaurants within a company may receive meat from the same source and use a uniform cooking technique (Bell et al., 1994). Hundreds of diners could be at risk from a single batch of meat. Other cuts of beef like steaks and roasts are unlikely to cause *E. coli* problems, because any bacterial contamination would be on the surface only and would be killed during cooking.

While high temperature cooking will kill bacteria such as *E. coli*, the negative side is that the formation of carcinogenic **heterocyclic amines** is increased as cooking temperature and time increase (see Chapter 2).

Rasmussen et al. (1993) demonstrated that normal rumen fermentation inhibits growth of toxic *E. coli* and suggested that stress and fasting before slaughter might increase infection rates. The rumen protozoa may serve as a reservoir for toxic *E. coli*; killing the rumen protozoa prior to slaughter with defaunating agents (e.g., yucca extract saponins) may "flush" *E. coli* from cattle (M. A. Rasmussen, personal communication). Thus there may be some management and feeding procedures for cattle that might minimize risks of contamination. Management practices that contribute to dietary stress during marketing and transport can increase pathogen loads of cattle at slaughter (Rasmussen et al., 1993).

Cattle may harbor 0157:H7 in their large intestine and shed the bacteria into the environment via their feces. Thus waterborne outbreaks of human poisoning may occur from contaminated water. For example, about 30 percent of the people in Walkerton, Ontario, Canada, contracted 0157:H7 from the municipal water supply that had been contaminated with cattle manure. Diez-Gonzalez et al. (1998) reported that grain-fed cattle had a lower colon pH and more acid-resistant *E. coli* than cattle fed only hay. They suggested that if grain-fed feedlot cattle were switched to a hay diet shortly before slaughter, the acid-resistant *E. coli* would be shed, thus reducing carcass contamination with the bacteria. The significance of the acid resistance is that the human stomach normally kills ingested bacteria, but acid-resistant *E. coli* may pass through the stomach to infect the intestine. In contast, Hovde et al. (1999) found that *E. coli* were equally acid resistant in grain-fed and hay-fed cattle. The effects of diet management on *E. coli* in cattle have been reviewed by Russell et al. (2000).

Although adult cattle can harbor 0157:H7, they are not adversely affected by it. This is because they lack cellular receptors for the 0157:H7 shiga toxins (Pruimboom-Brees et al., 2000). Newborn calves are susceptible. In humans, 0157:H7 can attach to the intestinal mucosa via a protein called intimin and can secrete highly virulent shiga toxins into the mucosal cells (Russell et al., 2000).

Cattle are readily infected with 0157:H7 by contamination of feed with manure (Lynn et al., 1998). The organism also survives extensive periods in contaminated water troughs (Rice and Johnson, 2000). Pasture-fed livestock can be contaminated by *E. coli* shed by deer (Sargeant et al., 1999). Humans, and especially children, may be exposed to *E. coli* on farm visits, in petting zoos, country fairs, and so on.

In August, 1997, a recall of 25 million pounds of hamburger was made by Hudson Foods, Inc. Some of the hamburger was contaminated with *E. coli*. Although there was no evidence linking this outbreak with the feeding of by-products to cattle, the use of **broiler litter** (dried poultry waste) as a feedstuff received a lot of television and print media attention. For example, an article "The Next Bad Beef Scandal?" in *U.S. News and World Report* (Sept. 1, 1997) was subtitled "Cattle Feed Now Contains Things Like Chicken Manure and Dead Cats." (The reference to dead cats refers to meat meal; cats and dogs euthanized

by veterinarians and animal shelters are sent to rendering plants rather than landfills.) Broiler litter has been used for many years as a **non-protein nitrogen** (NPN) source for ruminants. While when properly processed there is no reason why DPW is not a completely satisfactory feedstuff, it has an obvious public relations problem. The association of mad cow disease with the feeding of meat meal has focused public attention on the feeding of by-products to livestock, and especially to cattle. A short time ago, animal scientists proudly emphasized how livestock can convert by-products such as DPW, waste paper, bakery waste, restaurant grease, and so on into high quality human food. Now the use of these materials is being viewed by the public as undesirable, public health-threatening, and an example of corporate greed!

## Other Food-Borne Pathogens

The major food-borne pathogens associated with meat or other animal products are *Salmonella*, *Campylobacter*, and *E. coli*, with the exception of shellfish and other seafoods, which are not relevant to this book (although I suppose they could be with reference to aquaculture-raised organisms.) The nematode *Trichinella spiralis* causes the disease **trichinosis**, associated with uncooked pork from pigs fed garbage (restaurant scraps, etc.). This was once a significant problem, but with dissemination of information that pork should be well cooked, and the intensification of the swine industry leading to a cessation of garbage feeding, trichinosis now rarely occurs.

Enterococci (*Enterococcus faecalis* and *E. faecium*) are human gut pathogens. Vancomycin-resistant *E. faecium* are an important threat to public health in Europe, apparently related to the use of avoparcin as a feed additive. Avoparcin, a glycopeptide antibiotic, is structurally similar to vancomycin. Another class of antibiotics is the streptogramins. Virginiamycin, used as a growth promotant in the United States since 1974, is a streptogramin, as is a new antibiotic, quinupristin-dalfopristin. McDonald et al. (2001) found that at least 17 percent of chickens sampled in supermarkets in four U.S. states were contaminated with streptogramin-resistant *E. faecium*. Continued use of virginiamycin in poultry diets will likely increase the incidence of antibiotic resistance. Sorensen et al. (2001) demonstrated that antibiotic-resistant *E. faecium* isolated from supermarket chicken and pork could survive and reproduce in the human digestive tract. These findings prompted an editorial in the *New England Journal of Medicine* calling for an end to the use of antibiotics as growth promotants in livestock and poultry (Gorbach, 2001).

## Waterborne Diseases in Humans Linked to Livestock

**Protozoan parasites** in water can cause disease in humans. *Giardia lamblia* and *Cryptosporidium parvum* are the two most important ones in the United States (Smith, 1993). Pathogenic protozoa have a life cycle involving more than one host. The organisms produce oocytes that are excreted by a host animal. The oocytes contaminate water and may infect humans when the water is consumed. The oocytes sporulate in the gastrointestinal tract, releasing sporozoites, which invade the intestinal cells and differentiate into trophozoites. The trophozoites undergo asexual reproduction to form merozoites that invade new tissues. Both *Cryptosporidium* and *Giardia* can cause severe diarrhea in humans. Several outbreaks of these diseases have been linked to contamination of water supplies by cattle feces. Olson et al. (1997) found that *Giardia* infection rate is high in dairy calves and suggested that *Giardia* control should be considered as a routine part of a dairy herd health program to avoid water contamination and human disease.

Livestock infected with "crypto" may experience brief diarrheal disease but generally are not severely affected. Newborn calves are often infected, suffering transient diarrhea (Anderson, 1998). The main concern with *Cryptosporidium* in cattle is the potential for contamination of water with infected cattle feces, leading to outbreaks of human cryptosporidiosis.

A massive outbreak of ***Cryptosporidium*** infection occurred in Milwaukee, Wisconsin, in 1993, caused by contamination of the water supply with oocysts. It was estimated that over 400,000 people had watery diarrhea attributable to *Cryptosporidium* infection (MacKenzie et al., 1994). The source of the oocysts was never conclusively identified, but possible sources included cattle pastures adjacent to rivers in the watershed, slaughterhouses, and human sewage. Newspaper and TV coverage prominently suggested cattle as the source of the protozoa, probably leaving a widespread public perception that this was true, even though the source was never conclusively identified. Both *Giardia* and *Cryptosporidia* cause diarrhea in livestock, especially in young calves (Atwill, 1996). *Giardia* infections are very common in domestic ruminants and adversely affect animal performance (Olson et al., 1995). Livestock producers and animal caretakers should be aware of the potential for being infected from exposure to livestock feces, especially when working with sick animals with diarrhea. Contamination of streams and other water sources with livestock wastes is undesirable for a variety of reasons, including the risk of exposing humans to protozoal diseases.

Bacteria such as *Salmonella* and *E. coli* that may cause food-borne diseases may also be of concern as sources of waterborne disease associated with water contamination by animal excreta. ***Paratuberculosis*** or **Johne's disease** in cattle is caused by the bacterium *Mycobacterium paratuberculosis*. It causes a chronic, contagious inflammation of the gut (enteritis) characterized by persistent and progressive diarrhea, weight loss, debilitation, and eventual death. The organism is present in the milk and feces of infected cows. This organism has been implicated in **Crohn's disease** in humans, in which there is intestinal inflammation. The evidence of an involvement of *M. paratuberculosis* in Crohn's disease is inconclusive (Pell, 1997; Chiodini and Rossiter, 1996), but the potential relationship is of concern to the dairy industry (Pell, 1997; Stabel, 1998). The association of *M. paratuberculosis* with Crohn's disease stems from work by Chiodini in which the organism was isolated from patients with Crohn's disease but not from an equal number of control subjects. Chiodini and Rossiter (1996) concluded, "The hypothesis that *M. paratuberculosis* may play an etiologic role in the pathogenesis of Crohn's disease is a valid hypothesis that should continue to be investigated" (p. 466). It has not been conclusively determined if normal pasteurization will completely eliminate *M. paratuberculosis* from milk (Collins, 1997).

*Listeria monocytogenes* is a bacterium occurring in the gut of ruminants and poorly fermented silage. It can cause severe, fatal neurological symptoms in humans and has been involved in several incidences involving contamination of vegetables with manure, and with unpasteurized dairy products (Pell, 1997; Vasavada, 1988).

Clearly, animal waste management must consider not only the effects of excess nutrients such as nitrogen and phosphorus, but also the possibility of food-borne and water-borne diseases transferrable to humans. It is also evident that as livestock and poultry production becomes more intensive, with large numbers of animals on small land areas, these problems are likely to increase.

# FEED ADDITIVES, HORMONES, AND CHEMICALS

As discussed in Chapter 5, many feed additives, growth promotants, and substances that influence animal metabolism are used in modern systems of livestock production. Are these additives safe? Do **residues** exist in milk, meat, or eggs, and if so, are they hazardous to humans?

These substances are used because they affect animals in ways that increase their productivity. It is unlikely that very toxic, cancer-causing chemicals that are dangerous

to humans would be beneficial to animals. The fact that they have positive, rather than negative, effects on animals suggests that they are not intrinsically bad for people. On the other hand, they could conceivably induce changes (e.g., antibiotic-resistant bacteria, estrogenic effects) that are or could be detrimental to some people.

It cannot be stated with absolute certainty that the use of antibiotics and other feed additives and growth promotant implants is completely without risk to humans. However, regulatory agencies perform exhaustive analyses of the risks vs. benefits and always err on the side of safety when establishing guidelines and residue tolerances. Despite their protestations, the U.S. public is extremely well protected by regulatory agencies. Although it has become a cliche, it is undeniably true that the United States and other developed countries currently have the safest food supply that has ever existed. Our principal food-related problems are "super-nutrition" or overconsumption, as discussed in Chapter 2.

The estrogenic compound **diethylstilbestrol (DES)** was used in human medicine and as a feed additive at very low levels. DES became implicated as a carcinogen because massive doses (50 mg/day) of DES given to pregnant women caused an increased incidence of cervical cancer in their daughters (0.14–1.4 cases per thousand exposures). No case of cancer or other pathology was linked to DES residues in meat. When DES was a controversial topic in animal production, animal scientists delighted in making calculations that showed that one would have to eat huge quantities of liver to consume an adverse dose of DES. One calculation indicated that one would have to eat 5,500 pounds of beef liver per day to consume a daily therapeutic dose of 5 mg of DES.

Another calculation by Thomas Jukes, a renowned skeptic of the environmental activist and "health food" movements, will be briefly presented. The calculations are reasonably on target and useful in illustrating how potential hazards of "chemicals" can be magnified out of proportion. On the other hand, such calculations seem to do little to reassure the public, coming as they do from industry advocates. The example (Jukes, 1978) compares the potential daily intake of DES from consumption of beef liver from animals implanted with DES to the doses implicated in provoking cancer in humans:

(a) Per capita annual consumption of liver = 0.7 kg. Assume a generous value of 1.0 kg.

(b) Residual DES in liver from implanted beef animal = 0.11 ppb (0.11 µg/kg).

(c) Yearly intake of DES = 1.0 kg × .11 µg/kg = .110 µg/yr = 110 ng/yr = 0.3 ng/day.

(d) Administration of DES to pregnant women was 50 mg/day = 50,000,000 ng/day.

Thus there is a difference of 165 million fold between the DES dose given to pregnant women that caused a low incidence of cancer and the amount of DES consumed by an individual eating 1 kg of beef liver per year.

Jukes (1978) further calculated, using an estimate of 50 million women of childbearing age in the United States, that the number of cancers attributable to DES in beef would be from 0.14–1.4 cases per 1,000 years. This is likely an overestimate, because DNA repair mechanisms provide protection against low levels of carcinogens. Jukes (1978) concluded, probably validly, that

> The entire campaign carried out by the FDA towards a ban on DES in beef production, as far as any real protection of the public is concerned, is a massive exercise in futility, a travesty of science, and an affront to the intelligence. It is a diversionary movement to shift the blame for use of DES to the farmer, when it should be placed on the misuse of DES by the medical profession. Worst of all, the attack by the FDA is calculated to foster completely unnecessary alarm among consumers of beef. (p. 89)

Despite a vigorous defense of DES by the animal industry, its use in animal production in the United States was banned by the FDA in 1979. Dire predictions on the adverse effects that the DES ban would have on livestock production proved incorrect. Alternative and probably safer, better products have replaced it. Probably the main source of exogenous estrogens in the human diet is from phytoestrogens in plants. Alfalfa sprouts and soybeans are two common sources of plant estrogens. Zearalenone, a mycotoxin found in corn and other grains, has potent estrogenic effects and is likely ingested in small amounts by humans.

A book, with the blatantly misleading title of *Cancer from Beef. DES, Federal Food Regulation, and Consumer Confidence*, by Marcus (1994) suggests that use of DES as a feed additive or implant for beef cattle resulted in production of beef that was carcinogenic to humans. Unfortunately, public opinion can be influenced by erroneous but inflammatory books such as this one. As discussed previously, there is no evidence that use of DES in cattle caused any negative human health effects.

Hormones are administered to cattle in the United States and Canada in the form of implants. **Implants** are small pellets containing hormones that are administered subcutaneously in the ear. The pellet releases the hormones slowly and continuously. The ear, with any remaining implant, is discarded at slaughter. Some of the commercial products available are Compudose (estradiol-17-β), Synovex (progesterone and estradiol benzoate), and Ralgro (zeranol, an estrogenic substance derived from fungi). Their use in cattle production has been reviewed by Duckett et al. (1996). Implants improve growth rate, feed efficiency, and beef tenderness. There is relatively little information on tissue levels of these hormones in meat (Andersson and Skakkebaek, 1999). These authors conclude, "based on our current knowledge possible adverse effects on human health by consumption of meat from hormone-treated animals cannot be excluded" (p. 477), although there is no evidence apparently that there is a health risk. European governments have used the precautionary principle to ban the importation of hormone-treated beef

concern about the possibility of harmful effects of residues in foods, which have yet to occur, our ancestors had to cope daily with the real possibility of being killed by what they ate.

Of more concern than pesticide residues in the food are the natural toxins freely sold in "health food" stores. Herbal products such as comfrey tea are potentially hazardous. Comfrey contains a number of carcinogenic pyrrolizidine alkaloids (Culvenor et al., 1980) and causes liver cancer when fed to rats (Hirono et al., 1978). These "natural" products are unregulated. Imagine the public reaction if a multinational food company tried to market a product known to contain carcinogens!

## The Validity of Rodent Testing for Risk Assessment

It is of vital importance to know if feed additives and other chemicals used in livestock production cause cancer. Any product that represents a significant risk or hazard should not be used. No livestock producer would want to be responsible for inflicting cancer on anyone. How can the potential risk of these substances be assessed? Traditionally, it has been accomplished by administering graded levels of the test substance to laboratory rats or mice and carefully evaluating them for the presence of tumors. A common criticism of this testing procedure is that the doses of the test compound may be tremendously greater than what humans could possibly be exposed to. Why are such massive doses used? A simple example will illustrate the dilemma. Suppose a substance causes 15 cases of cancer per 100,000 people. We would all agree that this is an unacceptable hazard. Assume that at similar levels of exposure, laboratory rats would have the same cancer incidence as humans. How could we test this substance in rats, at levels comparable to normal human exposure? For adequate statistical analysis, we need to use a large number of animals. We decide to use 1,000 rats on the control diet, and 1,000 rats on the control + additive

diet. This is a huge number of animals. We feed the two diets for two years and then kill all 2,000 rats and examine them for presence of tumors. Unfortunately, in the test group we would likely have only 0.15 tumors, that is, none. If we had 10,000 rats per treatment, we'd expect 1.5 tumors. There would likely be a few spontaneous tumors in both groups, not caused by the additive. Assume there were 3 tumors in the control rats and 4 in the treatment group. Can we conclude that this is a treatment effect?

Obviously, unless a substance is wildly carcinogenic, testing it at normal levels of human exposure is not feasible. To resolve this dilemma, standardized testing procedures have been developed using higher dose levels. Commonly, the **Maximum Tolerated Dose (MTD)** is used; it is the largest estimated dose that the animals can tolerate without a 10 percent weight loss compared to controls. The MTD in most cases is still hundreds of times greater than normal levels of exposure.

The interpretation of rodent test results is complicated by the relationship between dose and response. At low levels of exposure, DNA repair mechanisms in animals may repair the mutations caused by a carcinogen. Thus it is not valid to extrapolate tumor incidence linearly back to a threshold value. Doing this may overestimate the threshold by as much as 15 times (Kraybill and Flynn, 1990). According to Ames et al. (1990a), about 99.99 percent of the pesticides in the American diet are natural compounds (nature's pesticides) produced by plants as chemical defenses. A high proportion of these natural chemicals are carcinogens, mutagens, teratogens, and clastogens (substances that can break chromosomes) in standard tests (Ames et al., 1990b), leading Ames and Gold (1990) to conclude that the hazards to humans of rodent carcinogens may be much lower than often assumed. If current means of risk assessment do not accurately predict hazard, they err on the side of safety. Consumers are well-protected. The major disadvantage of current procedures is that it is possible that some safe and useful agricultural chemicals are not approved for use.

**TABLE 9-4** Comparison of average daily exposures to synthetic (in bold) and natural chemicals in foods (Adapted from Gold et al., 1992).

| HERP (%)* Index | Food Item or Pesticide Residue, Daily Exposure | Carcinogen Present in Food or Pesticide |
|---|---|---|
| 0.1 | Coffee (3 cups) | caffeic acid+ |
| 0.04 | Lettuce (1/67 head) | caffeic acid |
| 0.03 | Orange juice (4/5 glass) | d-limonene |
| 0.03 | Black pepper (446 mg) | d-limonene |
| 0.004 | Potato (55 g) | caffeic acid |
| 0.003 | Carrot (12 g) | caffeic acid |
| 0.001 | Plum (2 g) | caffeic acid |
| 0.001 | Tap water (1 liter) | chloroform |
| 0.0002 | **Toxaphene** (food residue) | toxaphene |
| 0.0002 | Apple (1, whole; Alar residue) | from Alar, 1988 |
| 0.000001 | **Lindane** (residue) | lindane |
| 0.0000001 | **Chlorobenzilate** (residue) | chlorobenzilate |
| 0.000000006 | **Captan** (residue) | captan |

*HERP Index = human exposure/rodent potency. The larger the number, the greater the potential hazard or risk.

+ Caffeic acid is a phenolic compound or tannin and is chemically unrelated to caffeine.

Ames and coworkers have developed an index to rank carcinogenic chemicals as possible hazards to human health (Gold et al., 1992). The **HERP index** (human exposure/rodent potency) is calculated by using estimates of average daily human consumption of a carcinogen (mg per kg body weight per day) divided by the rodent potency ($TD_{50}$ in mg per kg body weight per day [$TD_{50}$ = dose toxic to 50 percent of test animals]). Representative values are shown in Table 9–4. When viewed against the large background of naturally occurring carcinogens in typical portions of common foods, the residues of synthetic pesticides and environmental pollutants rank very low. Gold et al. (1992) emphasize that the HERP values are not necessarily of much relevance to human cancer and do not imply that coffee consumption, for example, is a grave threat to one's health. Their point is to demonstrate that traditional means of testing synthetic chemicals for carcinogenicity are not of much relevance either. HERP values are a step towards putting potential risk in perspective. Excessive concern about pesticide residues, for example, can lead to regulatory actions banning or restricting use of agricultural chemicals when the residues do not present significant hazard. Restriction of pesticide use could increase the cost of fruits and vegetables that contain substances such as sulforaphane and β-carotene, which reduce the risk of cancer and thus actually increase cancer risk. According to Ames and Gold (1997), "There is no convincing evidence that synthetic chemical pollutants are an important source of human cancer. Regulations targeted to eliminate minuscule levels of synthetic chemicals are enormously expensive: the Environmental Protection Agency has estimated that environmental regulations cost society $140 billion/year" (p. 1041).

## The HACCP Program and Food Safety

A good discussion of the methods to ensure safety of animal-origin food products, on a global basis, is provided by Hubbert et al. (1996). This comprehensive book covers human health hazards of concern in foods of animal origin, principles of safe food production, processing and handling, and the roles of local, national, and international organizations in ensuring quality and safety of food involved in the global trade of food products. The USDA has introduced a **Hazard Analysis and Critical Control Points (HACCP)** program as a means of improving food safety. In the case of animal products, each segment of the process from farm to table is responsible for identifying and preventing food safety hazards. Traditionally, meat inspection has been a visual process, but the main hazards—microbes and chemical contaminants—are not detectable by inspectors. The HACCP program (pronounced *has cep*) is intended to address potential pathogen problems by management practices at the farm and processor level, rather than relying on examination of the end product.

The seven basic principles of the HACCP system include the following:

1. Identify hazards.
2. Find critical control points in the process.
3. Establish critical limits for each critical control point.
4. Monitor.
5. Take corrective action if monitoring shows there are deviations outside the limits of a critical control point.
6. Keep records on each critical control point.
7. Verify that the HACCP plan is working correctly.

To get an idea of how HACCP works, imagine cooking a pork loin roast.[2] The recipe says to cook this particular size of roast at 325° F for one hour. Doing this will allow the meat's internal temperature to reach 160° F. Here is how a HACCP plan would make sure this was done:

1. **Identify hazards:** Proper cooking will prevent a microbial food safety problem.

2. **Find critical control points in the process:** It is critical that the internal meat temperature reaches 160° F to ensure the meat is safe to eat. The cooking step, with a certain oven temperature for a specified time, is therefore a critical control point.

3. **Establish critical limits for each critical control point:** If the oven target temperature is 325° F and the oven temperature varies by 5° F., 320° F will be the minimum acceptable temperature.

4. **Monitor:** A thermometer in the oven will show the actual oven temperature.

5. **Take corrective action if monitoring shows there are deviations outside the limits of a critical control point:** The HACCP plan would say the temperature of the oven should be raised or lowered to keep it in the acceptable range. Extending the cooking time could also be used to reach the critical internal meat temperature.

6. **Keep records on each critical control point:** The oven temperature needs to be recorded at periodic intervals to show that the oven did not go out of the acceptable range or that corrective action was taken.

7. **Verify that the HACCP system is working correctly:** The actual internal temperature

---

[2] This example is adapted from the Pork Quality Assurance booklet published by the National Pork Producers Council, P.O. Box 10383, Des Moines, IA 50306.

of the cooked meat is taken and recorded. The monitoring records show that the temperature of the oven did not go out of the acceptable oven temperature range, or if it did that the appropriate corrective action was taken to ensure product safety.

The USDA has informed meat packers they must follow a HACCP system in their plants. Groups of hazards that can affect the safety of the meat have been identified as:

- microbial contamination
- chemical hazards such as antimicrobial and chemical tissue residues
- physical hazards like the accidental presence of broken needles or other metals

The packers will initially address microbial contamination through an in-plant HACCP process, but they will ask producers to help control antimicrobial and chemical residues and physical hazards. The producer will be responsible for properly observing withdrawal times to ensure there are no violative antimicrobial tissue residues. This is because a packer is not able to hold the animals for withdrawal once the producer brings them to the plant.

Although various segments of the meat industry may have resisted changes needed to assure the safety of meat (Gill, 1995), public concern following widely publicized incidents of *E. coli* contamination in 1996 and 1997 will likely force implementation of HACCP systems. Of course, irradiation of meat would eliminate microbial contamination as an issue, but public suspicion of food irradiation has limited the adoption of this technique. Another major problem has been the decline in public knowledge of personal responsibility in hygiene and food preparation, and the acceptance of the concept of restaurant food preparation being minimum-wage employment. You get what you pay for.

A central feature of the HACCP program is that responsibility for food safety will be everybody's concern, including the livestock producers, auction yards, feedlots, and so on. The days of sending mud- and manure-caked cattle to market will soon be over. Food safety begins at the farm level. Many farmers will have to clean up their act. There are animal welfare considerations as well. Improvement in animal management and hygiene will not only reduce microbial contamination but will also improve the welfare of livestock. One manure-covered steer entering a packing plant will jeopardize the entire lot by contamination. As is generally the case, these changes will probably be seen by many as "government intrusion," excess bureaucracy, and so on but in fact are long overdue. Food safety is everybody's business, and responsibility for it starts on the farm. Up to now, we have concentrated only on the end product in meat inspection to attempt to ensure wholesomeness of the product. It is now recognized that all steps in the "agri-food chain" are important, and the process begins with a healthy, clean animal when it leaves the farm.

# GLOBALIZATION OF AGRICULTURE AND FOOD SAFETY

In a somewhat fatuously titled book *Food, Sex and Salmonella. The Risks of Environmental Intimacy*, Waltner-Toews (1992) asserts that "What sex is to inter-personal relationships, eating is to the human-environment interrelationship, a daily consummation of our *de facto* marriage to the living biosphere" (cited by Baker [1993] in a review, p. 693). His point is that with increasing globalization of the food industry, we have less and less awareness of where our food comes from and what it is exposed to before it reaches our plates. Within

North America, an armada of trucks daily transports our food supply to and from various parts of Canada, the United States, and Mexico. In 1997, an outbreak of hepatitis A in Michigan was traced to contaminated strawberries grown in Mexico. With the continent-wide movement of food, contamination of a food product at one location can affect millions of people. The 1996 report that BSE (mad cow disease) might have been transferred to humans sent shudders through the world beef trade. How many countries had received British beef and beef cattle breeding stock in the years since BSE was first identified in British cattle? Use of pesticides or feed additives in one country might expose people to residues in many countries, even if their use is banned in the receiving countries. Bacteria such as *E. coli* 0157:H7 in hamburger could be destined for worldwide distribution in beef used by a fast-food chain. Continued global integration of food production and marketing is likely to increase. In the long term, this will be beneficial to many people. Waltner-

Toews (1992) suggests, for example, that restructuring the global economy so that Brazil and Mexico and Nigeria (for example) have the money and skills to put in good water and sewage treatment plants may equally benefit the developed countries that obtain food from the lesser developed nations. Perhaps the stimulus of self-interest will prod the industrialized countries into funding improvements in the global food network.

Niche markets for agricultural products exist for locally produced food. "Eating locally" might provide consumers with greater safeguards concerning food quality and safety. Locally grown food will probably remain a niche market, at least until the end of the era of cheap fossil fuel to maintain the armada of trucks on the North American freeways. On the other hand, "eating locally" sometimes involves an anti-establishment mind-set that includes such things as consuming raw milk, with numerous recent cases on record of *E. coli* and *Salmonella* poisoning as a result.

# STUDY QUESTIONS

1. Many people believe that they remember chicken meat tasting better "in the old days" than it does now. What are some possible reasons for this perception?

2. What are the major human health hazards associated with meat, milk, and eggs?

3. What was the cause of the outbreak of mad cow disease in Great Britain? Is mad cow disease a threat to humans?

4. Discuss the postulated roles of prions in the etiology of BSE. What is etiology?

5. What dangers, if any, are associated with eating raw (uncooked) eggs and drinking raw (unpasteurized) milk?

6. Discuss the use of the competitive exclusion technique. How do oligosaccharides function?

7. What is *E. coli* 0157:H7, and what is its significance to the beef industry?

8. How has the development of the fast-food restaurant business influenced the potential for human health problems associated with animal production?

9. Medical authorities suggest that when meat is barbecued, it should not be charred or overcooked. Why? Neither should it be undercooked. How would you define "properly cooked" meat?

10. What is the role, if any, of livestock in such diseases as giardiasis and cryptosporidiosis?

11. Discuss the current and past use of DES in animal production.

12. What are environmental estrogens? What are dioxins?

13. Why are eagles and hawks more likely than sparrows and starlings to suffer ill effects of environmental estrogens?

14. In an article in *Feedstuffs* (July 14, 1997), it is stated that 1 ppt is equivalent to less than one second per 25,000 years. Verify this statement by doing the calculation.

15. What is organic beef?

16. Why do plants contain toxic substances? Why are there such things as poisonous plants?

17. Nature's pesticides—what are they?

18. Are pesticides manufactured by Dow Chemical or Monsanto (for example) natural? If not, why not?

19. Does coevolution sound like a plausible theory to you? Explain your viewpoint.

20. During the French Revolution, the peasants revolted. What relationship might this event have had with LSD?

21. How has globalization of agriculture and the food industry affected food safety? Are you in favor of or opposed to free trade and a global food industry? Explain.

22. What is the HACCP program?

# REFERENCES

AARESTRUP, F. M., N. E. JENSEN, S. E. JORSAL, and T. K. NIELSEN. 2000. Emergence of resistance to fluoroquinolones among bacteria causing infections in food animals in Denmark. *Vet. Rec.* 146:76–78.

ALDHOUS, P. 1992. French officials panic over rare brain disease outbreak. *Science* 258:1571–1572.

ALLEN, V. G., J. P. FONTENOT, and D. R. NOTTER. 1992b. Forage systems for beef production from conception to slaughter: II. Stocker systems. *J. Anim. Sci.* 70:588–596.

ALLEN, V. G., J. P. FONTENOT, D. R. NOTTER, and R. C. HAMMES, JR. 1992a. Forage systems for beef production from conception to slaughter: I. Cow-calf production. *J. Anim. Sci.* 70:576–587.

AMES, B. N., and L. S. GOLD. 1990. Chemical carcinogenesis: Too many rodent carcinogens. *Proc. Natl. Acad. Sci.* USA 87:7772–7776.

AMES, B. N., and L. S. GOLD. 1997. Environmental pollution, pesticides, and the prevention of cancer: Misconceptions. *The FASEB Journal* 11:1041–1052.

AMES, B. N., M. PROFET, and L. S. GOLD. 1990a. Dietary pesticides (99.99% all natural). *Proc. Natl. Acad. Sci.* USA 87:7777–7781.

AMES, B. N., M. PROFET, and L. S. GOLD. 1990b. Nature's chemicals and synthetic chemicals: Comparative toxicology. *Proc. Natl. Acad. Sci.* USA 87:7782–7786.

ANDERSON, B. C. 1998. Cryptosporidiosis in bovine and human health. *J. Dairy Sci.* 81:3036–3041.

ANDERSSON, A. -M., and N. E. SKAKKEBAEK. 1999. Exposure to exogenous estrogens in food: Possible impact on human development and health. *Europ. J. Endocrinol.* 140:477–485.

ANDREWS, N. J., C. P. FARRINGTON, S. N. COUSENS, P. G. SMITH, H. WARD, R. S. G. KNIGHT, J. W. IRONSIDE, and R. G. WILL. 2000. Incidence of varient Creutzfeldt-Jakob disease in the UK. *The Lancet* 356:481–483.

ATWILL, E. R. 1996. Assessing the link between rangeland cattle and water-borne *Cryptosporidium parvum* infection in humans. *Rangelands* 18:48–51.

BAGER, F., F. M. AARESTRUP, and H. C. WEGENER. 2000. Dealing with antimicrobial resistance—the Danish experience. *Can. J. Anim. Sci.* 80:223–228.

BAILEY, J. S., L. C. BLANKENSHIP, and N. A. COX. 1991. Effect of fructooligosaccharide on *Salmonella* colonization of the chicken intestine. *Poul. Sci.* 70:2433–2438.

BAKER, T. 1993. Review of: Walter-Toews, D. Food, Sex and Salmonella: The Risks of Environmental Intimacy. *Can. Vet. J.* 34:693–694.

BALTER, M. 1999. Scientific cross-claims fly in continuing beef war. *Science* 284:1453–1455.

BALTER, M. 2000. On the hunt for a wolf in sheep's clothing. Science 287:1906–1908.

BARBER, D. A. 2001. New perspectives on transmission of foodborne pathogens and antimicrobial resistance. *J. Am. Vet. Med. Assoc.* 218:1559–1561.

BAUMLER, A. J., B. M. HARGIS, and R. M. TSOLIS. 2000. Tracing the origins of *Salmonella* outbreaks. *Science* 287:50–52.

BEEKES, M., and P. A. MCBRIDE. 2000. Early accumulation of pathological PrP in the enteric nervous system and gut-associated lymphoid tissue of hamsters orally infected with scrapie. *Neuroscience Letters* 278:181–184.

BELL, B. P., M. GOLDOFT, P. M. GRIFFIN, M. A. DAVIS, D. C. GORDON, P. I. TARR, C. A. BARTLESON, J. H. LEWIS, T. J. BARRETT, J. G. WELLS, R. BARON, and J. KOBAYASHI. 1994. A multistate outbreak of *Escherichia coli* 0157:H7-associated bloody diarrhea and hemolytic uremic syndrome from hamburgers. *J. Am. Med. Assoc.* 272:1349–1353.

BONS, N., N. MESTRE-FRANCES, P. BELLI, F. CATHALA, D. C. GAJDUSEK, and P. BROWN. 1999. Natural and experimental oral infection of nonhuman primates by bovine spongiform encephalopathy agents. *Proc. Natl. Acad. Sci.* 96:4046–4051.

BROWN, J. P., R. G. WILL, R. BRADLEY, D. M. ASHER, and L. DETWILER. 2001. Bovine spongiform encephalopathy and variant Creutzfeldt-Jakob disease: Background, evolution, and current concerns. *Emerg. Infect. Dis.* 7:6–16.

BROWN, P., and D. C. GADJUSEK. 1991. Survival of scrapie virus after 3 years' interment. *Lancet* 337:269–270.

BROWN, P., E. H. RAU, B. K. JOHNSON, A. E. BACOTE, C. J. GIBBS, JR., and D. C. GAJDUSEK. 2000. New studies on the heat resistance of hamster-adapted scrapie agent: Threshold survival after ashing at 600°C suggests an inorganic template of replication. *Proc. Natl. Acad. Sci.* 97:3418–3421.

BRUCE, M. E., R. G. WILL, J. W. IRONSIDE, I. MCCONNELL, D. DRUMMOND, A. SUTTIE, L. MCCARDLE, A. CHREE, J. HOPE, C. BIRKETT, S. COUSENS, H. FRASER, and C. J. BOSTOCK. 1997. Transmissions to mice indicate that 'new variant' CJD is caused by the BSE agent. *Nature* 389:498–501.

CAST. 1990. *Pesticides and safety of fruits and vegetables.* Council for Agricultural Science and Technology, Ames, Iowa.

CHEEKE, P. R. 1998. *Natural toxicants in feeds, forages, and poisonous plants.* Interstate Publishers, Danville, Illinois.

CHEE-SANFORD, J. C., R. I. AMINOV, I. J. KRAPAC, N. GARRIGUES-JEANJEAN, and R. I. MACKIE. 2001. Occurrence and diversity of tetracycline resistance genes in lagoons and groundwater underlying two swine production facilities. *Appl. Environ. Microbiol.* 67:1494–1502.

CHIODINI, R. J., and C. A. ROSSITER. 1996. Paratuberculosis: A potential zoonosis? *Vet. Clin. North Amer. Food Animal Pract.* 12:457–467.

COLBORN, T., and C. CLEMENT. (EDS.). 1992. *Chemically induced alterations in sexual and functional development: The wildlife/human connection.* Princeton Scientific Publishing, Princeton, New Jersey.

COLBORN, T., F. S. VOM SAAL, and A. M. SOTO. 1993. Developmental effects of endocrine-disrupting chemicals in wildlife and humans. *Environ. Health Perspect.* 101:378–384.

COLLINS, M. T. 1997. *Mycobacterium paratuberculosis:* A potential food-borne pathogen? *J. Dairy Sci.* 80:3445–3448.

CULVENOR, C. C. J., M. CLARKE, J. A. EDGAR, J. L. FRAHN, M. V. JAGO, J. E. PETERSON, and L. W. SMITH. 1980. Structure and toxicity of the alkaloids of Russian comfrey (*Symphytum x uplandicum Nyman*), a medicinal herb and item of the human diet. *Experientia* 36:377–379.

DAVIES, J. 1994. Inactivation of antibiotics and the dissemination of resistance genes. *Science* 264:375–382.

DEALLER, S. F., and R. W. LACEY. 1990. Transmissible spongiform encephalopathies: The threat of BSE to man. *Food Microbiol.* 7:253–279.

DIEZ-GONZALEZ, F., T. R. CALLAWAY, M. G. KIZOULIS, and J. B. RUSSELL. 1998. Grain feeding and the dissemination of acid-resistant *Escherichia coli* from cattle. *Science* 281:1666–1668.

DONNELLY, C. A., S. MAWHINNEY, and R. M. ANDERSON. 1999. A review of the BSE epidemic in British cattle. *Ecosystem Health* 5:164–173.

DORN, C. R. 1993. Review of foodborne outbreak of *Escherichia coli* 0157:H7 infection in the western United States. *J. Amer. Vet. Med. Assoc.* 203:1583–1587.

DUCKETT, S. K., D. G. WAGNER, F. N. OWENS, H. G. DOLEZAL, and D. R. GILL. 1996. Effects of estrogenic and androgenic implants on performance, carcass traits, and meat tenderness in feedlot steers: A review. *Prof. Anim. Scient.* 12:205–214.

EATON, D. L., and J. D. GROOPMAN. (EDS.). 1994. *The toxicology of aflatoxins.* Academic Press, San Diego.

ESPINOZA, O. B., M. PEREZ, and M. S. RAMIREZ. 1992. Bitter cassava poisoning in eight children: A case report. *Vet. Human Toxicol.* 34:65.

FEY, P. D., T. J. SAFRANEK, M. E. RUPP, E. F. DUNNE, E. RIBOT, P. C. IWEN, P. A. BRADFORD, F. J. ANGULO, and S. H. HINRICHS. 2000. Ceftriaxone-resistant salmonella infection acquired by a child from cattle. *N. Engl. J. Med.* 342:1242–1249.

FOSTER, K. R., P. VECCHIA, and M. H. REPACHOLI. 2000. Science and the precautionary principle. *Science* 288:979–981.

FRIES, G. 1995. A review of the significance of animal food products and potential pathways of human exposure to dioxins. *J. Anim. Sci.* 73:1639–1650.

GIBBONS, A. 1991. Moths take the field against biopesticide. *Science* 254:646.

GILL, C. O. 1995. Current and emerging approaches to assuring the hygienic condition of red meats. *Can. J. Anim. Sci.* 75:1–13.

GOLD, L. S., T. H. SLONE, B. R. STERN, N. B. MANLEY, and B. N. AMES. 1992. Rodent carcinogens: Setting priorities. *Science* 258:261–265.

GOLDMAN, D. A. 1999. The epidemiology of antimicrobial resistance. *Ecosystem Health* 5:158–163.

GORBACH, S. L. 2001. Antimicrobial use in animal feed—Time to stop. *N. Engl. J. Med.* 345:1202–1203.

GRIEBENOW, R. L., F. A. MARTZ, and R. E. MORROW. 1997. Forage-based beef finishing systems: A review. *J. Prod. Agric.* 10:84–91.

HADLOW, W. J. 1999. Reflections on the transmissible spongiform encephalopathies. *Vet. Pathol.* 36:523–529.

HARBORNE, J. G. 1993. *Introduction to ecological biochemistry.* Academic Press, Inc., San Diego.

HARVEY, R. B., C. R. YOUNG, R. L. ZIPRIN, M. E. HUME, K. J. GENOVESE, R. C. ANDERSON, R. E. DROLESKEY, L. H. STANKER, and D. J. NISBET. 1999. Prevalence of *Campylobacter* spp isolated from the intestinal tract of pigs raised in an integrated swine production system. *J. Am. Vet. Med. Assoc.* 215:1601–1604.

HILL, A. F., M. DESBRUSLAIS, S. JOINER, K. C. L. SIDLE, I. GOWLAND, J. COLLINGE, L. J. DOEY, and P. LANTOS. 1997. The same prion strain causes vCJD and BSE. *Nature* 389:448–450.

HILLERTON, J. E. 1998. Bovine spongiform encephalopathy: Current status and possible impacts. *J. Dairy Sci.* 81:3042–3048.

HINTON, A., JR., D. E. CORRIER, G. E. SPATES, J. O. NORMAN, R. L. ZIPRIN, R. C. BEIER and J. R. DE-LOACH. 1990. Biological control of *Salmonella typhimurium* in young chickens. *Avian Diseases* 34:626–633.

HIRONO, I., H. MORI, and M. HAGA. 1978. Carcinogenic activity of *Symphytum officinale. J. Nat. Cancer Inst.* 61:865–868.

HOIBY, N., C. PERS, H. K. JOHANSEN, H. HANSEN, and THE COPENHAGEN STUDY GROUP ON ANTI-BIOTICS IN SWEAT. 2000. Excretion of β-lactam antibiotics in sweat—A neglected mechanism for development of antibiotic resistance? *Antimicrob. Agents Chemother.* 44:2855–2857.

HOLMBERG, S. D, M. T. OSTERHOLM, K. A. SENGER, and M. L. COHEN. 1984. Drug-resistant *Salmonella* from animals fed antimicrobials. *N. Engl. J. Med.* 311:617–622.

HOPE, J., S. C. E. R. WOOD, C. R. BIRKETT, A. CHONG, M. E. BRUCE, D. CAIRNS, W. GOLDMANN, N. HUNTER, and C. J. BOSTOCK. 1999. Molecular analysis of ovine prion protein identifies similarities between BSE and an experimental isolate of natural scrapie, CH1641. *J. General Virology* 80:1–4.

HOVDE, C. J., P. R. AUSTIN, K. A. CLOUD, C. J. WILLIAMS, and C. W. HUNT. 1999. Effect of cattle diet on *Escherichia coli* 0157:H7 acid resistance. *Appl. Environ. Microbiol.* 65:3233–3235.

HUBBERT, W. T., H. V. HAGSTAD, E. SPANGLER, M. H. HINTON, and K. L. HUGHES. 1996. *Food safety and quality assurance: Foods of animal origin.* 2nd ed., Iowa State University Press, Ames, Iowa.

HUMPHREY, T. J. 1990. Public health implications of the infection of egg-laying hens with *Salmonella enteritidis* phage type 4. *World's Poult. Sci. J.* 46:5–13.

JOBLING, S., T. REYNOLDS, R. WHITE, M. G. PARKER, and J. P. SUMPTER. 1995. A variety of environmentally persistent chemicals, including some phthalate plasticizers, are weakly estrogenic. *Environ. Health Perspect.* 103:582–587.

JUKES, T. H. 1978. Scientific agriculture: The choice of activism or extinction. *Proc. Pacific Northwest Anim. Nutr. Conf.* 86–91.

KEENE, W. E., E. SAZIE, J. KOK, D. H. RICE, D. D. HANCOCK, V. K. BALAN, T. ZHAO, and M. P. DOYLE. 1997. An outbreak of *Escherichia coli* 0157:H7 infections traced to jerky made from deer meat. *J. Amer. Med. Assoc.* 277:1229–1231.

KHACHATOURIANS, G. G. 1998. Agricultural use of antibiotics and the evolution and transfer of antibiotic-resistant bacteria. *Can. Med. Assoc. J.* 159:1129–1136.

**KRAYBILL, H. F., and L. T. FLYNN.** 1990. *From mice to men: The benefits and limitations of animal testing in predicting human cancer risk.* Amer. Council on Science and Health, New York.

**LAMBERT, B., and M. PEFEROEN.** 1992. Insecticidal promise of *Bacillus thuringiensis.* Facts and mysteries about a successful biopesticide. *BioScience* 42:112–122.

**LASMEZAS, C. I., J. -G. FOURNIER, V. NOUVEL, H. BOE, D. MARCE, F. LAMOURY, N. KOPP, J. -J. HAUW, J. IRONSIDE, M. BRUCE, D. DORMONT, and J. P. DESLYS.** 2001. Adaptation of the bovine spongiform encephalopathy agent to primates and comparison with Creutzfeldt-Jakob disease. Implications for human health. *Proc. Nat. Acad. Sci.* 98:4142–4147.

**LAX, A. J., P. A. BARROW, P. W. JONES, and T. S. WALLIS.** 1995. Current perspectives in salmonellosis. *Br. Vet. J.* 151:351–377.

**LYNN, T. V., D. D. HANCOCK, T. E. BESSER, J. H. HARRISON, D. H. RICE, N. T. STEWART, and L. L. ROWAN.** 1998. The occurrence and replication of *Escherichia coli* in cattle feeds. *J. Dairy Sci.* 81:1102–1108.

**MACKENZIE, W. R., N. J. HOXIE, M. E. PROCTOR, M. S. GRADUS, K. A. BLAIR, D. E. PETERSON, J. J. KAZMIERCZAK, D. G. ADDISS, K. R. FOX, J. B. ROSE, and J. P. DAVIS.** 1994. A massive outbreak in Milwaukee of Cryptosporidium infection transmitted through the public water supply. *N. Engl. J. Med.* 331:161–167.

**MARCUS, A. I.** 1994. *Cancer from beef. DES, federal food regulation, and consumer confidence.* The John Hopkins University Press, Baltimore.

**MARSHALL, E.** 1991. A is for Apple, Alar, and . . . Alarmist? *Science* 254:20–22.

**MATTOCKS, A. R.** 1986. *Chemistry and toxicology of pyrrolizidine alkaloids.* Academic Press, San Diego.

**MCDONALD, L. C., S. ROSSITER, C. MACKINSON, Y. Y. WANG, S. JOHNSON, M. SULLIVAN, R. SOKOLOW, E. DEBESS, L. GILBERT, J. A. BENSON, B. HILL, and F. J. ANGULO.** 2001. Quinupristin-Dalfopristin-resistant *Enterococcus faecium* on chicken and in human stool specimens. *N. Engl. J. Med.* 345:1155.

**MCGAUGHEY, W. H., and M. E. WHALON.** 1992. Managing insect resistance to *Bacillus thuringiensis* toxins. *Science* 258:1451–1455.

**NAGLER-ANDERSON, C., C. TERHORST, A. K. BHAN, and D. K. PODOLSKY.** 2001. Mucosal antigen presentation and the control of tolerance and immunity. *Trends in Immunol.* 22:120–122.

**OLSON, M. E., T. A. MCALLISTER, L. DESELLIERS, D. W. MORCK, K. -J. CHENG, A. G. BURET, and H. CERI.** 1995. Effects of giardiasis on production in a domestic ruminant (lamb) model. *Am. J. Vet. Res.* 56:1470–1474.

**OLSON, M. E., N. J. GUSELLE, R. M. O'HANDLEY, M. L. SWIFT, T. A. MCALLISTER, M. D. JELINSKI, and D. W. MORCK.** 1997. *Giardia* and *Cryptosporidium* in dairy calves in British Columbia. *Can Vet. J.* 38:703–706.

**OSTERBLAD, M., K. NORRDAHL, E. KORPIMAKI, and P. HUOVINEN.** 2001. How wild are wild mammals? *Nature* 409:37–38.

**OYOFO, B. A., J. R. DELOACH, D. E. CORRIER, J. O. NORMAN, R. L. ZIPRIN, and H. H. MOLLENHAUER.** 1989. Effect of carbohydrates on *Salmonella typhimurium* colonization in broiler chickens. *Avian Diseases* 33:531–534.

**PELL, A.** 1997. Manure and microbes: Public and animal health problem? *J. Dairy Sci.* 80:2673–2681.

**POPPE, C., N. SMART, R. KHAKHRIA, W. JOHNSON, J. SPIKA, and J. PRESCOTT.** 1998. *Salmonella typhimurium* DT104: A virulent and drug-resistant pathogen. *Can. Vet. J.* 39:559–565.

**PRIOLA, S. A., A. RAINES, and W. S. CAUGHEY.** 2000. Porphyrin and phthalocyanine antiscrapie compounds. *Science* 287:1503–1506.

PRUIMBOOM-BREES, I. M., T. W. MORGAN, M. R. ACKERMANN, E. D. NYSTROM, J. E. SAMUEL, N. A. CORNICK, and H. W. MOON. 2000. Cattle lack vascular receptors for *Escherichia coli* 0157:H7 shiga toxins. *Proc. Nat. Acad. Sci.* 97:10325–10329.

PRUSINER, S. B. 1982. Novel proteinaceous infectious particles cause scrapie. *Science* 216:136–144.

PRUSINER, S. B. 1998. *Prions. Proc. Natl. Acad. Sci.* USA 95:13363–13383.

RASMUSSEN, M. A., W. C. CRAY, JR., T. A. CASEY, and S. C. WHIPP. 1993. Rumen contents as a reservoir of enterohemorrhagic *Escherichia coli. Microbiol. Lett.* 114:79–84.

RHEEDER, J. P., W. F. O. MARASAS, P. G. THIEL, E. W. SYDENHAM, G. S. SHEPHARD, and D. J. VAN SCHALKWYK. 1992. *Fusarium moniliforme* and fumonisins in corn in relation to human esophageal cancer in Transkei. *Phytopath.* 82:353–357.

RICE, E. W., and C. H. JOHNSON. 2000. Short communication: Survival of *Escherichia coli* 0157:H7 in dairy cattle drinking water. *J. Dairy Sci.* 83:2021–2023.

RUSSELL, J. B., F. DIEZ-GONZALEZ, and G. N. JARVIS. 2000. Invited Review: Effects of diet shifts on *Escherichia coli* in cattle. *J. Dairy Sci.* 83:863–873.

SAFE, S. H. 1995. Environmental and dietary estrogens and human health: Is there a problem? *Environ. Health Perspect.* 103:346–351.

SAMS, A. R., E. M. HIRSCHLER, A. P. MCELROY, J. G. MANNING, and B. M. HARGIS. 1995. Flavor evaluation of light and dark meat from broilers fed capsaicin. *Poul. Sci.* 74:205–207.

SARGENT, J. M., D. J. HAFER, J. R. GILLESPIE, R. D. OBERST, and S. J. A. FLOOD. 1999. Prevalence of *Escherichia coli* 0157:H7 in white-tailed deer sharing rangeland with cattle. *J. Am. Vet. Med. Assoc.* 215:792–794.

SCHLEIFER, J. H. 1985. A review of the efficacy and mechanism of competitive exclusion for the control of *Salmonella* in poultry. *World Poult. Sci. J.* 41:72–83.

SCOTT, M. R., R. WILL, J. IRONSIDE, H. -O. B. NGUYEN, P. TREMBLAY, S. J. DEARMOND, and S. B. PRUSINER. 1999. Compelling transgenetic evidence for transmission of bovine spongiform encephalopathy prions to humans. *Proc. Nat. Acad. Sci.* 96:15137–15142.

SENAUER, B., E. ASP, and J. KINSEY. 1991. *Food trends and the changing consumer.* Eagen Press, St. Paul, Minnesota.

SHARPE, R. M., and N. E. SKAKKEBAEK. 1993. Are oestrogens involved in falling sperm counts and disorders of the male reproductive tract? *Lancet* 341:1392–1395.

SHAOUL, J. 1997. Mad cow disease; the meat industry is out of control. *The Ecologist* 27:182–187.

SMITH, D. E., J. R. SKALNIK, and P. C. SKALNIK. 1997. The bST debate: The relationship between awareness and acceptance of technological advances. *Agric. Human Values* 14:59–66.

SMITH, J. L. 1993. *Cryptosporidium* and *Giardia* as agents of foodborne disease. *J. Food Protec.* 56:451–461.

SMITH, K. E., J. M. BESSER, C. W. HEDBERG, F. T. LEANO, J. B. BENDER, J. H. WICKLUND, B. P. JOHNSON, K. A. MOORE, and M. T. OSTERHOLM. 1999. Quinolone-resistant *Campylobacter jejuni* infections in Minnesota, 1992–1998. *New Engl. J. Med.* 340:1525–1532.

SORENSEN, T. L., M. BLOM, D. L. MONNET, N. FRIMODT-MOLLER, R. L. POULSEN, and F. ESPERSEN. 2001. Transient intestinal carriage after ingestion of antibiotic-resistant *Enterococcus faecium* from chicken and pork. *N. Engl. J. Med.* 345:1161.

SPENCER, K. C. (ED.). 1988. *Chemical mediation of coevolution.* Academic Press, San Diego.

SPRAKER, T. R., M. W. MILLER, E. S. WILLIAMS, D. M. GETZY, W. J. ADRIAN, G. G. SCHOONVELD, R. A. SPOWART, K. I. O'ROURKE, J. M. MILLER, and P. A. MERZ. 1997. Spongiform encephalopathy in free-ranging mule deer (*Odocoileus hemionus*), white-tailed deer (*Odocoileus virginianus*), and Rocky Mountain elk (*Cervus elaphus nelsoni*) in northcentral Colorado. *J. Wildlife Dis.* 33:1–6.

STABEL, J. R. 1998. Johne's disease: A hidden threat. *J. Dairy Sci.* 81:283–288.

STERN, N. J., M. R. S. CLAVERO, J. S. BAILEY, N. A. COX and M. C. ROBACH. 1995. *Campylobacter* spp. in broilers on the farm and after transport. *Poult. Sci.* 74:937–941.

TAYLOR, D. M. 2000. Inactivation of transmissible degenerative encephalopathy agents: A review. *Vet. J.* 159:10–17.

TAYLOR, D. M., K. FERNIE, I. MCCONNELL, C. E. FERGUSON, and P. J. STEELE. 1998. Solvent extraction as an adjunct to rendering: the effect on BSE and scrapie agents of hot solvents followed by dry heat and steam. *Vet. Record* 143:6–9.

TAYLOR, D. M., S. L. WOODGATE, and M. J. ATKINSON. 1995. Inactivation of the bovine spongiform encephalopathy agent by rendering procedures. *Vet. Record* 137:605–610.

TYLER, C. R., S. JOBLING, and J. P. SUMPTER. 1998. Endocrine disruption in wildife: A critical review of the evidence. *Crit. Rev. Toxicol.* 28:319–361.

VASAVADA, P. C. 1988. Pathogenic bacteria in milk—A review. *J. Dairy Sci.* 71:2809–2816.

WALL, P. G., D. MORGAN, K. LAMDEN, M. GRIFFIN, E. J. THRELFALL, L. R. WARD, and B. ROWE. 1995. Transmission of multi-resistant strains of *Salmonella typhimurium* from cattle to man. *Vet. Record* 136:591–592.

WALTNER-TOEWS, D. 1992. *Food, sex and salmonella: The risks of environmental intimacy.* N. C. Press Ltd., Toronto.

WEGENER, H. C. 1999. The consequences for food safety of the use of fluoroquinolones in food animals. *New Engl. J. Med.* 340:1581–1582.

WESLEY, I. V., and A. I. BAETZ. 1999. Natural and experimental infections of *Arcobacter* in poultry. *Poult. Sci.* 78:536–545.

WHIPP, S. C., M. A. RASMUSSEN, and W. C. CRAY, JR. 1994. Animals as a source of *Escherichia coli* pathogenic for human beings. *J. Amer. Vet. Med. Assoc.* 204:1168–1175.

WHITE, D. G., S. ZHAO, R. SUDLER, S. AYERS, S. FRIEDMAN, S. CHEN, P. F. MCDERMOTT, S. MCDERMOTT, D. D. WAGNER, and J. MENG. 2001. The isolation of antibiotic-resistant salmonella from retail ground meats. *N. Engl. J. Med.* 345:1147.

WILL, R. G., J. W. IRONSIDE, M. ZEIDLER, S. N. COUSENS, K. ESTIBEIRO, A. ALPEROVITCH, S. POSER, M. POCCHIARI, A. HOFMAN, and P. G. SMITH. 1996. A new variant of Creutzfeldt-Jakob disease in the UK. *Lancet* 347:921–925.

WILLS, P. R. 1991. Prion diseases and the frame-shifting hypothesis. *New Zealand Vet. J.* 39:41–45.

WITTE, W. 1998. Medical consequences of antibiotic use in agriculture. *Science* 279:996–997.

to free people's minds from ignorance, prejudice, and provincialism and to stimulate a lasting attitude of inquiry" (Oregon State University Bulletin, 1994–95).

The killing and eating of other sentient animals[1] by humans is an ethical and moral issue worthy of consideration, especially by the people directly involved in the raising/slaughter of animals. Most people who eat meat don't think too deeply about all the processes involved in converting a living animal to meat on their plate. The farther one is removed from agriculture, the easier it is not to think about this issue. One of the best things modern animal agriculture has going for it is that most people in the developed countries are several generations removed from the farm and haven't a clue how animals are raised and "processed." In my opinion, if most urban meat eaters were to visit an industrial broiler house to see how the birds are raised and could see the birds being "harvested" and then being "processed" in a poultry processing plant, they would not be impressed, and some, perhaps many, of them would swear off eating chicken and perhaps all meat. For modern animal agriculture, the less the consumer knows about what's happening before the meat hits the plate, the better. If true, is this an ethical situation? Should we be reluctant to let people know what really goes on, because we're really not proud of it and concerned that it might turn them to vegetarianism?

Webster (1994) has put it well:

> I believe that it is in the long-term interest of animal farmers to contribute to the proper education of consumers by demonstrating quite openly where their meat and milk is coming from and how it is produced. An excellent test of animal welfare is to discover whether their owner can display his animals with pride to any fair-minded observer . . . The special pleading required to suggest that the welfare of broiler fowls or laying hens is satisfactory,

despite their appearance, is deeply unconvincing to almost any unbiased observer. (p. 268)

Assuming that the welfare of animals during the production stage, transport to slaughter, and the slaughtering process is considered acceptable, the ethical question remains: are we morally justified in killing and eating animals, especially when we have no compelling need to do so except that we like to eat meat?

The answer to this question has little or nothing to do with ecological and environmental considerations, such as the oft-cited ones that livestock can convert by-products into human food, can be raised in a manner non-competitive with human food production, and so on. It has little to do with animal welfare. It has everything to do with animal and human rights. Do we, as humans, having an ability to reason and to communicate abstract ideas orally and in writing and to form ethical and moral judgements using the accumulated knowledge of the ages, have the right to take the lives of other sentient organisms, particularly when we are not forced to do so by hunger or dietary need, but rather to do so for the somewhat frivolous reason that we like the taste of meat? In essence, should we know better?

Ethics in animal agriculture is closely tied to the subjects of animal welfare and animal rights. These issues will be discussed separately, with the recognition that ethical or moral behavior regarding human use of domestic animals is influenced by welfare and animal rights considerations. It is intuitively obvious and substantiated by research that animals can experience pain and suffering as well as feelings of well-being. For these reasons, it does not seem unreasonable that animal production practices that cause **pain** or **suffering** or that negatively impact an animal's well-being are morally questionable. In a democratic society, people who are not directly involved in animal production nevertheless have ethical and moral rights and obligations and can legitimately express their opinions on animal welfare issues (Hurnik, 1993). In an advanced society, the extension

---

[1] A **sentient being** is one that has the power of perception and awareness.

**FIGURE 10-1** Interest in animal rights and the ethical treatment of animals shows cultural differences. It is unlikely that the method of restraining and transporting pigs to market in Vietnam, shown here, would be acceptable in most industrialized societies. (Courtesy of Robert A. Swick.)

of moral interest to other sentient organisms besides humans is ethically valid and is of course what is happening in our society. Interest in animal rights and **ethical treatment of animals** is much less pronounced in societies in which there is not strong respect for and recognition of human rights (Fig. 10-1) and in countries where food is not in abundant supply (Hurnik, 1993). An interest in the ethical treatment of animals is a natural evolution in a democratic society, as basic human needs are met, and people are encouraged to think for themselves and express themselves in public. Many livestock production practices in the past have been brutal, painful, traumatic to human and beast, and unnecessary. I have participated in dehorning parties that were all of

the above! It has long been considered a sign of "toughness" and "machismo" to be able to perform castrations, dehorning, cut up calves *in utero* if they were too large to be born, cut decaying, stinking eyes out of cows with cancer eye, and so on. These types of macho "cowboying" are no longer considered admirable or acceptable by society, and for good reason.

What do we or should we do when ethics and economics collide? Livestock producers and their defenders sometimes justify brutal or painful practices because of economics: the animal is not worth the expense. Why should I call a veterinarian and pay a $60 vet bill for a $30 sheep? What about a pregnant cow that breaks her leg? Should you keep her until she calves, so you can get a live calf, and then shoot the cow? This has been known to me to happen, both on privately owned and university-run farms. Rollin (1991) described the case of a sow with a broken leg, which was kept alive until she was able to deliver her litter. Is it ethically correct to keep an animal alive in pain and discomfort to have it give birth some time in the future but withhold veterinary treatment of the animal because "it won't pencil out"? Most people in the United States today would be appalled that such a question even arises. Further discussion of these sorts of ethical issues is provided by Rollin (1995a). The book *Issues in Agricultural Bioethics* by Mepham et al. (1995) also considers ethical issues in modern agriculture, with an emphasis on bioethical issues in crop production and in biotechnology. Ethical issues in the plant area are in many ways less contentious than with animals, because animals are sentient beings while it is generally recognized that plants are not. Lehman (1995) also has discussed ethical issues relating to animal production.

Some of the ethical issues involved in animal production include the following examples:

(a)  Induction of nutrient deficiency for economic gain: Is it ethical to feed a veal calf an iron deficient diet to create iron-deficiency anemia, so that the meat is white?

(b)  Is it unethical to keep laying hens in very small cages (Fig. 10-2), in which many of the birds will have broken bones before they are culled as "spent hens"?

(c)  Is it ethical to select broiler chickens for very rapid growth rate, with the result that **"diseases of industrialization"** occur, with birds dying of heart attacks or lung failure because their rate of weight gain has exceeded the ability of the cardiac and pulmonary systems to support it?

(d)  Is it ethical for a farmer to load old or weak cows on a truck to take them to an auction yard or slaughter plant, knowing that they are very likely to fall down en route and never be able to stand again (downer cows)?

(e)  Is it ethical to slaughter animals in view of other animals, or does it not matter because the others "are just going to die soon anyway"?

(f)  Is it ethical to withhold veterinary care or otherwise provide inadequate medical treatment of animals, because their economic value is less than the cost of treatment?

(g)  Is it unethical to show disrespect for animals? For example, in a sheep experiment, sheep were killed humanely for collection of organs for mineral analysis. A graduate student participating in this study took a sheep head, stuck it on his hand, and pretended to charge other students with it. With his other hand, he made the eyes blink. Is this unethical, immoral, adolescent behavior or just plain stupid? (This is a true story.)

(h)  Turkeys have been selected for large breast muscles and rapid growth, with the result that modern breeds and strains of turkeys are incapable of natural reproduction. They must be mated by artificial insemination. Is this development ethically questionable?

**FIGURE 10-2** (*Top*): A modern egg farm, with birds kept in small cages. The eggs roll to a conveyer belt in front, to be moved automatically to the egg processing and packaging facility. Feed and water are provided automatically. There is little human labor, except for daily checking for dead birds. (*Bottom*): The crowding of laying hens in small cages has been a controversial animal welfare issue. (Bottom photo courtesy of Robert A. Swick.)

(i) Assume that it would be possible to develop a strain of chickens that were almost brain-dead. They could eat, drink, and grow very rapidly, but would not be capable of much else. They could be raised at very high densities. They would essentially be meat machines. Is it morally wrong to produce such birds, or is this actually a positive development because the birds are virtually unaware of their environment and thus probably experience no deprivation of normal stimuli?

These are just a few of many ethical issues that livestock producers and animal scientists must deal with today. Not too many years ago, you could simply dismiss anyone who asked such questions and go about your business. Those days are over, probably forever.

An ethical question that has arisen is "how valuable is an animal's life compared to a human life?" This question has been most hotly debated in terms of laboratory animals used in **biomedical research.** At one extreme are animal rights proponents who claim that animals are equivalent to humans in value and that there is no more moral justification for using rats in experiments than there would be in using children. The founder of PETA (People for the Ethical Treatment of Animals) is quoted as asserting, "A rat is a pig is a dog is a boy; they're all equal." (Kertz, 1996, p. 257). W. Ray Stricklin (personal communication) points out that, on a scientific basis, this statement is basically correct; the rat, pig, dog, and boy are mammals with similar physiology. The difference between the boy and the other animals listed cannot be defined scientifically but only by application of ethical principles. It seems obvious that most people would consider a human life more valuable than an animal's life. Dennis (1997) has considered these questions, in identifying morally relevant differences between animals and humans, in support of the use of animals in biomedical research. In contrast to many such writings, the paper by Dennis (1997) seems to be a

non-ranting, reasonable description of the differences between humans and other animals, which leads to his conclusion, "In the strict biological sense, human beings are animals too, but in the broader sense, human beings are much more than animals. The life of a man, woman or a child is worth far more than the life of a mouse, rat, dog, or monkey" (p. 618).

How do humans differ from other animals in a morally relevant sense? The following is a partial list selected from a longer list proposed by Dennis (1997):

- ability to appreciate order or beauty
- ability to apply moral judgments
- ability to deliberate on consequences prior to an action
- ability to love
- ability to recognize pure truth, as in mathematics
- desire to search for truth
- culture
- imagination
- introspective understanding
- knowledge of one's own mental state
- language
- ability to mentally juxtapose oneself into alternate circumstances
- logic
- reason
- spirituality
- time consciousness
- understanding of cause and effect
- ability to experience nostalgia

While some animals may possess some of the abilities from this list, only humans "put it all together." Domestication of animals has produced animals whose behaviors are compatible with life in human society, which may lead some people to attribute greater awareness to these animals than is truly the case. Many people today treat their pets (especially dogs) as members of the family and converse with them as if the pet really understood what they are talking about. In many cases, they truly believe that their pets understand them.

The most obvious difference between humans and other animals is **language.** Language is probably a prerequisite for intelligent thought and reasoning ability. Our thoughts are framed by words and language, which allow us to communicate with others and ourselves. Try to imagine how you would think if you had no language. Our minds are engaged with thoughts almost constantly, all expressed to our inner selves in the form of our language. Dennis (1997) concludes that human life has special value because of our unique mental and language abilities. "The activities and experiences that we most highly value, whether they are intellectual, cultural, relational or achievement related, are all distinctively human and require, as a minimum, a human brain" (p. 615). It is not necessary to invoke religious beliefs, including the intervention of a deity and the hypothesis of a soul, to differentiate moral distinctions between humans and other animals. Rollin (1995a) provides a good summary of our ethical history concerning farm animals:

> Animal agriculture is as old as civilization, and our ancient contract with animals, for all its flaws, has been a model of natural justice and fairness. We have, in most cases, coexisted with our domestic animals better than we have with one another. Their presence has been a manifest and positive one, reflected and extolled in our arts, crafts, literature, mythology, song, and story. Our children still sing of Old MacDonald's farm. No one sings of Old MacDonald's factory. We must do what needs to be done to preserve that ancient contract, else we diminish not only the animals but ourselves as well. (p. 141)

A further discussion of ethical issues will be given following discussion of animal welfare, animal rights, and biotechnology in animal production.

## Terminology of Bioethics

Basic philosophical theory with reference to animal agriculture has been presented by Kunkel (2000) and Shapiro (1999). A brief discussion of some of the philosophical background to current ethic issues will be given here.

Concern for humane treatment of animals is relatively recent. The "Father of Modern Philosophy," the French philosopher Rene Descartes, believed that animals were merely machines, incapable of consciousness or feeling pain. This callous attitude was prevalent for many years among scientists, veterinarians, agriculturalists, and so on, and justified what are now considered cruel or inhumane practices. Even today, there are a few people who deny that animals can feel pain or suffer and so believe we are justified in doing whatever we want to them.

Hurnik (1993) reviewed ethical theories used in consideration of farm animal welfare. Some of them are briefly discussed here.

The **divine command** position is derived from a supernatural authority, as expressed in and understood from scriptures: "God gave us dominion over animals." There are obvious weaknesses in this theory: whose "divine command" is to be followed? The theory is rejected by non-believers in supreme beings, and so on.

**Kantian philosophy** derives from the writings of the German philosopher Immanual Kant. Kant believed it was wrong to mistreat animals, not because of any harm done to the animals, but because doing so would be detrimental to the character of the abuser. One might become so habituated to abuse that one would be tempted to treat humans in the same way (Thompson, 1997).

**Utilitarianism** is a belief that the proper course of action is the one that provides the greatest good for the greatest number. It is a form of **consequentialism** theory, which evaluates morality in terms of the outcome. The right action is the one that benefits the most individuals. The well-being of humans is weighted more highly than that of animals, but utilitarians still must balance benefits to humans against the harm that may be done to other sentient creatures. Utilitarianism may lead to ridiculous conclusions, such as it is better to murder 7 million people than 7 million plus 1.

Modern utilitarianism theory applied to domestic animals began with the book *Animal Liberation*, by Peter Singer (1975). He asserted that contemporary industrialized animal production causes suffering to animals and that this suffering outweighs the benefits to humans of cheaper animal products. He concluded that consumers are morally obligated to become vegetarians (**ethical vegetarianism**) to alleviate animal suffering.[2]

**Communitarian ethics** and care-based ethics refer to the concept that we do not treat all people equally but have a sense of community and shared relationships with others (Fraser, 1999). We have responsibilities to immediate family members that we do not have to strangers. We see ourselves as members of a number of different communities that create different ethical requirements. These communities often include animals we treat as family members (pets), farm animals, and those whose lives we affect in other ways. To each of these, we feel quite different kinds and degrees of responsibility (Fraser, 1999). For example, most poultry farmers probably have different degrees of ethical concern for the family dog and an industrial broiler chicken.

Modern **animal rights theory** stems from the book *The Case for Animal Rights*, by Tom Regan (1983). Regan argues that animals have moral rights based on the concept of "inherent value" (Shapiro, 1999). According to Regan's theory, animals are subjects of a life with in-

terests that deserve equal consideration to the interests of humans. A problem with Regan's theory is deciding how to evaluate these interests. In the lifeboat scenario described by Regan (1983), if a human and a dog were adrift at sea in a lifeboat with enough food for only one of them to survive, it would be morally permissible to discard the dog because the interests of the human are greater than those of the dog. Regan reasons that death to a normal human being is incomparably worse than death to any non-human animal, because a normal human's capacity to form and satisfy desires is much greater (Kunkel, 2000). (However, Regan also suggested that if the child were retarded and the dog normal, we should save the dog!) Regan recommended adoption of a vegetarian/vegan diet because it minimizes the harm done to other animals. Davis (2000) takes issue with this position in a paper, "What is the Morally Relevant Difference between the Mouse and the Pig?" According to Davis, the production of a vegetarian diet inevitably results in the death of "animals of the field" such as field mice. How can we justify the killing of field mice to produce grains for the human diet, but not the killing of pigs to produce human food? The utilitarian answer is to use the method of food production that causes the least amount of suffering (**the least harm principle**). How can we evaluate the comparative suffering of a field mouse and a pig? Actually if we compare an equal amount of weight of field mice with an equal weight of pig, many more lives are lost and suffering is presumably greater when dozens of mice instead of one pig are killed. Following the least harm principle, perhaps we should raise elephants or draft horses as meat animals (Fig. 10-3). Killing one large draft horse for meat would result in the loss of less "subjects of a life" than killing a dozen or more sheep or pigs. Davis (2001) argued that a diet containing the meat of grazing animals is a more ethical choice than a vegan diet, because more subjects of a life (animals of the field) are killed when food crops are produced. Also, there is less wildlife habitat in a cultivated field for crop production than in a pasture for grazing animals.

---

[2] Peter Singer is now a bioethicist at Princeton University, where his appointment has been controversial, especially with regard to euthanasia. In a 1979 book *Practical Ethics*, he suggested that children less than a month old have no human consciousness and that parents should be allowed to kill a severely disabled infant to end its suffering and to increase the family's happiness. As a critic of animal agriculture, he carries a lot of baggage.

**FIGURE 10-3** Draft horses in Hungary being raised for meat production. If one accepts the least harm principle, then it would be more acceptable to slaughter one large animal than an equivalent weight of smaller animals. For example, one 1,500 lb draft horse represents one subject of a life, while the same weight of chickens represents about 375 subjects of a life.

A milestone in the development of the animal rights story was the publication of Ruth Harrison's book *Animal Machines* (Harrison, 1964), in which the term **"factory farming"** was coined to describe the negative effects of intensive confinement systems for livestock and poultry. Harrison (1993) states that an advertisement in a farm magazine prompted her to write the book: "It stated authoritatively: 'The modern pig farmer sees the pig itself as merely a cog in a machine for converting feedingstuffs into cash at the bank.' That philosophy, and the practices it encapsulates, are what led me to write *Animal Machines* (1964) nearly thirty years ago, and to pursue change ever since" (p. 5).

The publication of *Animal Machines* prompted the British government to set up a commission, the Brambell Committee, to "enquire into the welfare of animals kept under intensive husbandry systems." The committee concluded, "In principle we disapprove of a degree of confinement of an animal which necessarily frustrates most of its major activities which make up its natural behaviour." Following the Brambell Committee recommendations, a Farm Animal Welfare Advisory Committee was established. Its first public statement was the setting out of the "five freedoms" which animals are entitled to (see Animal Rights—this chapter). Other recommendations of this committee are reviewed by Harrison (1993).

Bernard Rollin of Colorado State University has been an outspoken critic of intensive animal production, on ethical and philosophical grounds. He has articulated a new social ethic for animals (Rollin, 1995a), which will be discussed later in this chapter under Animal Rights.

Bioethics is an important issue for animal scientists, if for no other reason than because vocal, and sometimes dangerous, critics of modern animal production practices are publically attacking animal agriculture. Two of the best known organizations are People for the Ethical Treatment of Animals (PETA) and the Animal Liberation Front (ALF). The ALF is a violent fringe element that has attacked and burned intensive animal facilities on commercial farms and at universities and research organizations. PETA has assumed the moral high ground by virtue of its name; who would want to be a member of PUTA (People for the Unethical Treatment of Animals)?! Thus whether you like it or not, animal agriculture and its defenders are in a contest for the public's good will. Ignorant or superficial responses merely put points on the board for PETA.

Loew (1993) has discussed how modern society views livestock production and how animal agriculture might respond. He points out that most animal science and veterinary students now come from urban areas, and bring with them "urban values" with respect to animal life. "Most of their teachers, however, represent the values of a former, more pastoral, United States. Many professors have great difficulty dealing with students whose attitudes are really quite different from the ones they themselves held when they were students" (p. 105). Loew (1993) refers to the phenomenon of an "urban prism" characterized by, among other things, an elevated moral status for animals. He states, "There is a pastoral myth in our urban book of myths that somehow the family farmer is more virtuous than the so-called corporate farmer, that the family farmer cuts the chicken's head off in a way that is somehow better than the way a factory farmer does it"! (p. 106).

**TABLE 10-1** Types of views about animals
(Adpated from Loew, 1993.)

| Attitude Toward Animals | Primary Animal Interest | U.S. Regional Occurrence |
|---|---|---|
| Naturalistic | Wildlife and the outdoors | General |
| Ecologistic | Ecosystems and natural habitats | Northwest |
| Humanistic | Individual animals (e.g., companion animals) | Northeast Northwest |
| Moralistic | Animal rights, welfare, and exploitation | Northeast |
| Scientific | Physical attributes and biological functioning | General |
| Aesthetic | Artistic and symbolic implications of animals | General |
| Utilitarian | Practical and material values of animals | South |
| Dominionistic | Mastery and control of animals (sporting activities) | South |
| Negative | Dislike or avoid animals | General |
| Neutral | No interest in or views about animals | General |

Loew (1993) points out that attitudes towards animals differ in different parts of the United States. A summary is given in Table 10-1. For example, in the South, there is a very high ranking on the utilitarianism and dominionistic side (the good ol' boy hunting mentality), which may have contributed to the rapid development of intensive poultry and swine production in that region. On the West Coast (bean sprouts and sandals country), attitudes toward animals are generally much different. Older people tend to be much higher on the utilitarian and dominionistic scales than younger people. Loew (1993) wonders whether these young people will turn into old people or whether we are watching a dying generation whose attitudes towards animals will disappear as the current younger generation marches on: "If the latter is true, it means that we live in a country that has changed profoundly with regard to its attitudes towards animals." Loew (1993) concludes: "A new urban view of animal life is emerging in the United States . . . If agriculture wishes to respond strategically to this view, it needs to understand that it isn't simply a matter of convincing people to think the way they once did. People apparently aren't going to think the way they used to and it is encumbent upon agriculture to respect these profound, and apparently lasting, demographic and attitudinal changes that have occurred" (p. 108).

Fraser (1999) has suggested that animal ethicists (philosophers) and animal welfarists (scientists) live in two different worlds or cultures, neither communicating with or understanding the other. Philosophers talk about rights and social contracts, while scientists talk about stereotypies, cortisol levels and productivity, or, as Fraser (1999) puts it, "as if one group speaks English and the other Tibetan" (p. 176). Fraser (1999) suggests that scientists can grasp the telos concept, and can take up the task of attaching "some biological flesh to this philosophical bone" (p. 177). Scientists can measure the key features that make up the telos (see Animal Rights, this chapter) of an animal. Animals should feel well, function well, and lead natural lives (Fraser, 1999). They should feel well by being free from prolonged or intense fear, pain, and other unpleasant states, and by experiencing normal pleasures. Animals should function well in the sense of satisfactory health, growth and normal behavioral, and physiological functioning. They should lead natural lives through the development and use of their natural adaptations.

# ANIMAL BEHAVIOR AND WELFARE

An important and expanding area of animal science is **ethology,** the study of animal behavior. Knowledge of behavioral characteristics of animals has several important implications in livestock production (Grandin, 1997a). Animal handling and management are facilitated by knowing an animal's behavioral patterns and preferences (Houpt, 1998). Sheep instinctively climb and can be driven more easily to the highest point of a pasture than to the lowest point. Thus it would make little sense to locate the handling facilities at the lowest point of a sheep pasture (unless unavoidable for logistical reasons) and then to rant and shout how stupid sheep are because they won't go into the pen! By making use of the animals' preferred behavioral patterns when designing animal handling equipment such as corrals, loading chutes, and so on, stress on both the animals and the operators can be reduced.

Animal management techniques that reduce or eliminate stress are an important component of **animal welfare.** The term welfare refers to the state of an individual in relation to its environment (Broom, 1991; Dawkins, 1998; Fraser and Broom, 1990). Welfare status can be measured by a number of techniques, including behavioral measures of responsiveness, stereotypies and preferences, and biochemical measurements of hormones associated with stress responses. These have been reviewed in detail by Fraser and Broom (1990).

Environmental factors that adversely affect an animal's welfare include pain, injury, fear, frustration, absence of normal stimuli, sensory deprivation, and overstimulation (Broom, 1991). These will be briefly discussed. **Pain** is a biological response to stimulation of pain receptors. It can be eliminated by the use of analgesics in operations such as dehorning and castration. It is also of interest that animals produce their own painkillers (analgesics). These are substances known as **endorphins,** which are peptides (opioids) produced in the pituitary. Endorphins bind to the same central nervous system receptors as do the morphine opiates and seem to play a role in controlling pain perception. Endogenous endorphins have as high as 30 times the analgesic potency of morphine. Secretion of endorphins may limit the perception of pain by animals undergoing procedures such as dehorning and may modulate the fear of animals approaching death in a slaughter house. There are marked differences in the perception of and relief of pain between individual animals (Cook, 1997), including sex-related differences in response to analgesia. **Fear** is an aversive response dependent on an animal's perception of its environment, such as during handling, transportation, preslaughter period, surgery, and so on. Fear can be measured by behavioral and physiological responses (escape attempts, aggression, adrenal cortex activity, heart rate, etc.). Both acute and chronic fear can seriously harm the performance and welfare of livestock and poultry (Jones, 1996). **Frustration** occurs when animals wish to respond appropriately to their environment but are prevented from doing so. For example, lack of sufficient feeder space in a group-feeding situation may prevent some animals from feeding. They may respond with stereotyped behaviors or aggression towards other individuals. Absence of normal stimuli that are important to animal behavior may result in abnormal behaviors. Suckling animals deprived of the normal stimulation of the mother's teats, as in early weaning, may exhibit teat-seeking behavior by abnormal means involving the navel, ears, penis, or scrotum of their penmates. **Sensory deprivation** occurs when animals with elaborate behavioral characteristics are kept in a pen with nothing to do. Humans and domestic animals kept under such circumstances exhibit a number of behavioral abnormalities. Stimulation may be provided by social interaction with other animals and by the provision of physical variety in the environment. Pigs are often provided with toys (e.g., old tires, opportunities to root), which sometimes reduce incidence of abnormal behaviors such as tail biting. Thus with knowledge of an animal's behavioral needs, animal housing and management can be designed to improve animal welfare.

**Poor welfare** can be assessed in a number of ways. Evidence of physical damage is one of the most obvious, such as broken bones of layers kept in cages, cannibalism, pecking, wounds, tail biting, and so on. This damage may occur from misdirected behaviors among individuals (e.g., tail biting, pecking) or from physical inadequacy of the environment. For example, **spent layers** (laying hens at the end of their egg production span) that have been kept in battery cages often are found to have a high incidence of broken bones (e.g., tibia) when sent to slaughter (Gregory and Wilkins, 1989). This is due to reduction in bone strength associated with lack of exercise (Knowles and Broom, 1990). Modified cages containing perches for the birds to roost on reduce this problem (Hughes and Appleby, 1989). Applications of knowledge of behavior to poultry management have been reviewed by Mauldin (1992), Mench (1992), Craig and Swanson (1994), and Barnett and Newman (1997).

Another example of physical expression of poor welfare is a high incidence of stomach ulcers in pigs kept under crowded conditions. Disease problems are greater when the welfare of animals is poor, due to inhibitory effects of high adrenal cortex activity on the immune system (Broom, 1991). High serum glucocorticoid levels occur with stressful conditions. However, assessment of animal welfare by physiological measurements such as blood corticosteroid levels is not yet feasible (Rushen, 1991). Much more basic knowledge of the pituitary-adrenocortical axis and its response to stress is needed before the fundamental question "At what level of change is welfare at risk?" can be validly answered (Barnett and Hemsworth, 1990; Rushen, 1991).

Broom (1991) considers that reduced life expectancy of livestock kept under intensive conditions is an indication of poor animal welfare. **Short life expectancy** is characteristic of intensive livestock production. This is in part a management decision; laying hens are normally kept only for one laying cycle. If allowed to molt and come back into production, they could live several years. However, it is economically more efficient to replace the "spent hens" with pullets ready to lay. Today it is common to use "forced molting." Hens are deprived of food and water for several days to induce molting. Dairy cattle are biologically capable of producing 10–15 calves in their lifetime, whereas normally they may last only 3–5 lactations in modern dairy systems. High milk production is associated with an increased incidence of metabolic diseases, lameness (laminitis), mastitis, damaged udder ligaments (udder breakdown), and other problems. Again, economic factors are involved. Because of continual selection for increased genetic potential for high milk production, a cow is culled after 3–5 lactations and replaced with a heifer with a greater production potential. This is a major factor explaining the continual increase in annual milk production per cow in the U.S. dairy industry.

Animals kept under confinement conditions may exhibit abnormal physical activity or behavior. A common type is stereotypy. **Stereotypies** are repeated sequences of movements that serve no obvious purpose (Broom, 1991). These include route tracing (pacing up and down a fenceline), bar biting, tongue rolling, and so on. They are common in confined animals, such as those in zoos (e.g., the constant swaying of an elephant chained to a post), in sows in stalls or tethered, calves in crates, and so on. The occurrence of stereotypy is an indication of poor animal welfare (Lawrence and Terlouw, 1993), but lack of stereotypies is not necessarily an indicator of good welfare.

Improvement in animal welfare can be achieved by having accurate measures of welfare, finding out what the preferences of the animal are, and designing facilities and handling techniques that minimize adverse effects on welfare. The big challenges are to precisely define welfare and to decide how to measure it (Dawkins, 1998). **Animal preferences** can be tested by giving them a choice of types of cages, for example, and observing which they use when given a choice. A considerable amount of work has been conducted in Europe on alternatives to cages for housing of poultry. The European Union (EU) has established regulations for animal housing, based in part on experimental work and in part on emotion and political considerations. Numerous alternative systems that can improve poultry welfare have been developed, such as free-range, deep litter floor system, multi-tiered house with perches, aviary system, perchery system, get-away cages, and numerous others (Appleby and Hughes, 1990, 1991; Wegner, 1990). When properly designed and used, these alternative systems improve bird welfare but also increase costs. When not properly used, they may be worse for bird welfare than are cages. For example, free-range systems often degenerate into a muddy mess of droppings and parasite-infected soil, with unhealthy, disease-ridden birds.

Poultry production has been at the forefront of animal welfare concerns, particularly with egg producing hens (layers) and meat chickens (broilers). A good source of information is the proceedings of the First North American Symposium on Poultry Welfare (*Poultry Science* 77:1763–1845, 1998).

For **laying hens** (layers), confinement in **battery cages** is a major welfare issue (Knowles and Wilkins, 1998), both from the point of view of the prevention of normal behavior and with the problems of broken bones due to bone fragility from lack of exercise. Keeping hens in small cages has facilitated development of industrial-style egg production. Advantages include reduced feather picking and cannibalism, elimination of coccidiosis (a protozoal disease whose transmission is facilitated when birds are kept "on the floor" on litter), and reduction of exposure to dust and ammonia. Tauson (1998) and Appleby (1998) reviewed various types of cage modifications, including installation of perches, which can improve bird welfare. Some of these improvements could be made at negligible expense to producers (Appleby, 1998).

Another issue with layers is **forced molting.** Hens exhibit an egg production cycle. Many weeks after they begin laying eggs, production gradually falls off. When it reaches an uneconomic level, one of two things is done. Either the hens are sold as "spent hens," or

they are force-molted. Molting is a process by which the hens cease egg laying, lose most of their feathers (molt), and undergo metabolic rejuvenation, which results in the initiation of a new laying cycle. Hens molt naturally, but to get an entire flock to molt at the same time, hens are force-molted by withdrawing feed and/or water. Feed may be withheld for as long as 21 days. Animal rights groups claim, with justification, that to deny food for three weeks to a bird that normally eats every day is tantamount to forced starvation and causes animal suffering. On the other hand, Ruszler (1998) points out that the natural behavior of the wild hen is to go off feed for 21–28 days during natural molting. Nevertheless, it is unlikely that public perception will ever be influenced to accept forced molting as natural. Poultry scientists and the poultry industry tend to look at this issue strictly in terms of economics. In a study of why egg producers decide to force-molt their flocks, McDaniel and Aske (2000) concluded: "If the egg industry is expected to operate in an atmosphere of free enterprise in which supply and demand establish the market price, then egg producers should be allowed to use molting to alter the supply of eggs in the short term" (p. 1245). Obviously, welfare doesn't figure prominently in this view. Similarly, in a study of forced molting by feed withdrawal, Webster (2000) stated, "One can conclude at least that an induced molt that uses a long period of feed withdrawal need not cause harm to hens, and it may even improve their survivability" (p. 199).

The public is unlikely to believe such fatuous statements by scientists, and neither do all scientists. In a letter to the editor of *Poultry Science*, Duncan and Mench (2001), with reference to the Webster study, state, "We dispute the conclusions drawn by Webster in his paper on molting. This paper describes a badly designed experiment with a flawed discussion . . . That Webster was able to deprive hens of feed for 21 days without killing any does not prove absence of suffering" (p. 934).

Dawkins (1999) states that apart from physical health, which is the cornerstone of all good welfare, the most important additional component of poultry welfare is psychological health (contentment), which can be most reliably assessed through the birds' own choice of behavior. Chickens have a suite of responses (biochemical, physiological, and behavioral) to various kinds of dangers that evolved in their jungle fowl ancestors and persist today. In the highly artificial conditions in which chickens are kept today, these natural responses may be inappropriate or even lethal. **Hysteria,** in which birds "go berserk," may be lethal in modern systems of poultry production. In the jungle, where chickens evolved, flapping and squawking when a predator approached may have been a highly effective means of scaring off the predator (Dawkins, 1999). Birds still retain most of their evolutionary heritage, based on millions of years of dealing with environmental threats. Dawkins (1999) states, "They have evolved to respond to each of these dangers in completely different ways. They do not respond to a water deficit by getting hysterical any more than they respond to the presence of a predator by drinking" (p. 297). Extreme or inappropriate expression of fear-related behavior (fearfulness) may have negative effects on productivity and welfare of chickens (Jones and Hocking, 1999; Minvielle et al., 2002).

**Feather picking** and **cannibalism** are major problems in poultry, especially in layers (Jones and Hocking, 1999). Certain birds become targets, with other birds pecking at their feathers. This leads to bleeding and intensified attacks, often causing death. Feather picking has been controlled by **beak trimming** of chicks, thus reducing the ability of the mature birds to effectively attack each other. Beak trimming is opposed by animal welfare advocates. Ironically, cannibalism is more severe in free-range situations and is a major factor in limiting more widespread adoption of free-range production of chickens (Jones and Hocking, 1999). Jones and Hocking (1999) indicate that there are genotype differences in cannibalism and that selection can be used to reduce feather picking. Many people question the morality of

altering the genetic makeup of animals, either by natural selection or genetic engineering, to better fit them for what they see as the unnatural environments of intensive production systems (Jones and Hocking, 1999).

Other concerns expressed by animal rights advocates with respect to caged layers are the inabilities of the birds to engage in various normal behaviors such as nest building, dust bathing, perching, and wing flapping (Wall et al., 2002). In the European Union countries, new regulations require that cage size be large enough for birds to engage in these activities.

There are a number of welfare and animal rights issues with the raising of **broiler chickens.** An obvious public concern is the high bird density; broiler houses have birds "wall to wall." Selection for rapid growth has led to increased problems with metabolic diseases such as ascites (fluid accumulation due to pulmonary and cardiac insufficiency, resulting in sudden death). Rapid growth induced by high nutrient intake can cause severe lameness, bone defects, and deformity (Julian, 1998).

Many people find the mechanization of broiler production to be objectionable—the factory farming of chickens. One aspect that has been difficult to automate is the catching and loading process, in which birds are caught by hand and loaded on trucks. Mechanical broiler harvesters, using rubber fingers to gently guide the birds on to conveyors have been developed (Lacy and Czarick, 1998) (Fig. 10-4). It might seem that this is just one more example of how the modern poultry industry abuses chickens. However, catching chickens by hand is dirty, unpleasant, low-paying work and is also not pleasant for the chickens. Bird welfare is probably improved by the use of mechanical catchers (Lacy and Czarick, 1998).

For **swine,** a major welfare question is the keeping of sows in gestation stalls (Fig. 10-5). The animals are confined so that about all they can do is stand up, lie down, eat, drink, and excrete. Various efforts have been made to design sow facilities that give the animals more freedom of movement while retaining the production efficiency and low labor costs

with animal rights activists. Battling with opponents can be an all-embracing task, with no apparent end or favorable outcome. Many cattle ranchers in the American West, for example, spend almost as much time in disputes with environmentalists as they do ranching. It's a war of attrition. Perhaps the best long-term solution is to raise animals in such a way that societal concerns are defused.

Having knowledge of animal welfare is important to animal scientists and livestock producers in responding to animal rights activists. Under what circumstances are livestock subjected to inhumane or cruel treatment? Is keeping laying hens in small cages cruel or not? To some extent, this is an ethical or philosophical question that can never be satisfactorily answered. No matter how many measures of animal welfare are made that indicate that the bird or animal in confinement is content, some people would never agree that anything other than complete freedom is acceptable.

It seems intuitively obvious that animals raised under good animal welfare standards and with gentle handling will be more productive and easier to manage than those kept under poor conditions and/or mistreated by rough handling. Fear of people from past rough handling can make animal handling difficult and sometimes dangerous, especially with large animals like cattle and horses. As reviewed by Munksgaard et al. (1997), gentle handling of dairy cows increases milk production, and the cows are less likely to kick or otherwise adversely respond. Dairy cows can recognize individual people and attempt to avoid people who have treated them aversively (Rushen and de Passille, 1998). Increased stress can lead to physiological responses that initiate disease such as gastrointestinal disturbance (e.g., enteritis). Many of these issues are common sense and well-known to farmers. Good farmers have always known that livestock respond favorably to good treatment. Many animal rights activists have had little or no experience with animals other than pets and may be unaware of the covenant between most farmers and their livestock. Not withstanding this fact, it is also true

that there are legitimate concerns with animal welfare in modern animal production systems.

An interesting application of animal behavior response is the use of aversion techniques to protect livestock from consuming poisonous plants. Researchers at the USDA Poisonous Plants Laboratory in Logan, Utah, have successfully used aversion techniques to make cattle avoid consuming larkspur (*Delphinium*), a poisonous plant. This technique involves feeding the poisonous plant with lithium chloride, a substance that causes temporary illness. The animals associate the sensations of illness with consumption of larkspur and thereafter avoid the plant (Ralphs and Olsen, 1992).

# ANIMAL RIGHTS

The animal rights movement has developed rapidly in Europe and North America (Fig. 10-9). It is one of the most important issues facing the livestock industries. Ultimately, it is an issue that could end the commercial production of domestic animals. In fact, if the aims of animal rights extremists were achieved, all use of animals by humans for any purpose would end, including the keeping of pets, zoo animals, and so on. Domestic animals would become extinct (the ultimate violation of their rights!).

It is important to differentiate between animal welfare and animal rights, but also important to note that they are very closely linked (Gonyou, 1994). **Animal welfare,** as previously discussed, relates to providing animals with adequate environments and management to meet their intrinsic physiological and behavioral needs. **Animal rights** refers to a belief system that animals intrinsically have the same rights to life and liberty as afforded to humans (Getz and Baker, 1990). From the acceptance of the concept that animals possess these rights follows the logical belief that use of animals by humans represents an infringement on their "God-given" rights and that killing an animal for meat is no different

**FIGURE 10-9** How times change! The top photo shows a rabbit drive in the western United States in the 1890s to kill jack rabbits that were destroying crops. Rabbit drives were held frequently. The animals were driven into a fenced area by people spread out in a long line, and once in the enclosure, the animals were clubbed to death. The rabbit drive was a festive occasion. One hundred years later, animal rights activists protest against a circus parade. (Bottom photo courtesy of W. Jamison.)

than murder and cannibalism. In my opinion, this is not an amusing, fanciful or way-out idea. If one accepts the philosophical concept of animal rights, one necessarily then equates slaughter of livestock with murder. Thus the intensity of animal rights activists is legitimate. If you sincerely believe that killing a cow is murder, you have a moral responsibility to be intense in your opposition to it. The animal rights issue can be compared to the abortion issue in this regard. No amount of "educating them" by "pro-choice" advocates will change the minds or intensity of "pro-

life" advocates, or vice versa. Animal scientists and livestock producers are naive if they think they can "educate" the animal rights activists. In fact, as discussed later, animal rights activists in general are far better educated than those opposed to them (Jamison, 1992), not only with reference to general level of education, but also in terms of their knowledge of animal production practices. They know what is going on and don't like it.

The philosophical basis of the animal rights movement has been very well described by Rollin (1990). When taken to its logical intellectual conclusion, an animal rights ethic would lead to no use of animals. It is hoped that (from an animal scientist's perspective) there is a middle ground. Rollin (1990) states that animal scientists have by training and inclination the general belief that their mandate is to develop the means to produce greater amounts of food more efficiently and more cheaply. We have operated on the premise that maximum yield and efficiency are the goals. Animal breeding and genetics, reproductive physiology, and animal nutrition—the three major animal science disciplines—have focused on increasing yields and efficiency. Genetic selection, artificial insemination, embryo cloning and transfer, fine-tuning diets to improve feed efficiency, use of feed additives or implants to stimulate growth rate—these have been the foundation of animal science research. These disciplines have recently been joined by the current attraction, biotechnology, to speed up these processes in order to produce even higher yields with ever-increasing efficiency. Perhaps it is time to back off and introduce some other variables such as animal welfare and rural sociology. This is not always easy to do. Rollin (1990) states, "To many animal scientists, this new ethic (backing off from the search for ever-increasing efficiency) is thus tantamount to a repudiation of their life's work" (p. 3461). It appears to be true that eggs can be produced most efficiently and cheaply by keeping hens in very small battery cages. By the end of their production period, they will have a frazzled appearance with many of the feathers broken, a bare neck, and probably some broken bones (Gregory and Wilkins, 1989). Animal rights adherents, and American and European society as a whole, are rejecting this maximum efficiency at any cost ethic and demanding that moral concern for livestock and poultry and for the farmers themselves be considered. They are willing to pay for it. **Cheap food, at any moral cost, may not be worth the price.** Perceptive animal scientists have responded to this changing tide of human values. One might capriciously think that an article in the *Journal of Animal Science* entitled "Educational Methodology in Dealing with Animal Rights and Welfare in Public Service" (Getz and Baker, 1990) might consist of a diatribe on how to "educate" the animal rights enthusiasts. Instead, their concluding statement is this: "Animal rights and welfare groups must be recognized as organizations involved in the animal industry of the future. They are likely to influence future policies. In fact, they ultimately may improve animal agriculture programs because challenging current methods, procedures and assumptions usually leads to improvements" (p. 3474). Similarly, Friend (1990) in an article "Teaching Animal Welfare in the Land Grant Universities" argues that animal science students should be exposed to the philosophical basis of the animal rights movement, in part to encourage intellectual activity to form their own values. Inflammatory and polarizing name-calling, by either side, has little if any redeeming value. How many people in the abortion debate have changed their minds because of shouting, screaming, and name-calling?

A common perception of livestock people is that animal rights activists don't understand the livestock industry (they "don't get it," in current terminology) because of their urban backgrounds, and they just need to be educated (Jamison and Lunch, 1992). Jamison (1992) has pointed out that the activists do "get it," they know what is going on, and they don't like it. In a study of a large group of animal rights activists, he found that the "proto-

typical activist" was a Caucasian female in her thirties, well-educated, frequently with one or more graduate degrees, working in a professional position with a high income and living in a large metropolitan area, with a high degree of political activism in environmental and feminist issues. These activists are often suspicious of agriculture, science, and business, and have a profound aversion towards industrialized agriculture. Jamison (1992) measured attitudes of animal rights activists by the use of "feeling thermometers," with a scale ranging from 0 to 100, with 0 being a very cool or negative response and 100 being a very warm or positive response. His results showed a very warm response to environmentalists, feminists, and veterinarians, and a very negative response to farmers and ranchers, scientists, politicians, and business people. Their views on the farm treatment of various livestock are interesting:

| Animal | Mean Score[*] |
|---|---|
| Horses | 36 |
| Sheep | 29 |
| Dairy cows | 22 |
| Beef cows | 15 |
| Pigs | 13 |
| Turkeys | 13 |
| Layers | 11 |
| Broilers | 7 |
| Mink | 4 |
| Veal calves | 2 |

[*]100 = very positive; 0 = very negative

It is evident that the greater the perception of intensity and industrialization of the production, the more negative the perception. Jamison (1992) concludes:

Animal rights activists are demographically much more mainstream than previously anticipated. They are not marginal to the political system, and their political values are based on classic American ideals of equality. Similarly, they are urban dwellers whose experience with the life and death processes inherent to animal production are severely limited. Ultimately, the debate over the rights of farm animals has little to do with the reality of their treatment. Instead the debate is about the perception of what is real, and in public policy, perception becomes reality. Agriculturalists and animal rights activists have different realities. (p. 136)

It seems obvious that as the size of farms increases and the number of farmers decreases, agriculture will have less and less political influence at the voting booth in the United States. By wholeheartedly supporting research and extension programs to maximize efficiency and productivity, which has led to fewer farms, agricultural scientists appear to have actively contributed to reducing their base of support. A common complaint by professors in land grant universities is that many of the students in agriculture don't have a farm background. What else could be expected, when we have directed our research and extension to programs that lead to fewer and fewer farmers? The ultimate irony is the case of poultry scientists. They have been so successful in developing the technology that allowed the industrialization of poultry production that they have now largely become irrelevant. As the former department head of a former department of poultry science said to me, "We've been so effective we've worked ourselves out of our jobs." There is, in my opinion, a direct cause and effect relationship between the industrialization of poultry production and the elimination of Departments of Poultry Science (see Chapter 8).

The concept of animal rights is closely linked with animal welfare (Gonyou, 1994). Most people probably associate humane treatment in part with their perception of the animal's **intelligence** (Allen, 1998). Dogs and pigs are commonly perceived as more intelligent than sheep, chickens, and turkeys (Davis and Cheeke, 1998). If pigs are indeed more intelligent than turkeys, does this have an influence on how they should be raised? Do pigs need a more enriched environment than turkeys or broiler chickens? The types of cognitive abilities animals have provide clues as

to the types of situations in which they might suffer (Nicol, 1996). How do you compare intelligence among different species? Nicol (1996) believes that the question of animal awareness is critical to animal welfare considerations. She states, "We should give animals the benefit of the doubt that awareness, and hence suffering, is a likely experience for many animals. . . . Although morally satisfactory, such an approach is not very helpful. Where, if anywhere, do we draw the line? At the chimpanzee, the hen, and flatworm, the amoeba?" (p. 388). Such are the issues of which bioethics is made.

Some behavioral scientists have argued that the mental abilities of domestic animals are less well-developed than those of their wild counterparts (Nicol, 1996), on the basis that selection pressure in the wild would favor quick rather than dull-witted individuals. Some psychologists have even claimed that domestic animals are not "real animals" because their behavior has been so modified by humans. Nicol (1996) concluded that there is no strong evidence that domestic animals have fewer cognitive abilities than have wild animals. The subject of animal cognition is of more than just academic interest. Whether or not modern broiler production is humane or not, for example, depends in part on the awareness of the birds of their environment and their responses to it.

Grazing animals have especially well-developed spatial abilities and can form spacial maps of their environment. Cattle and horses on rangelands can be aware of fine details of an extensive area. A humbling experience for me was working as a sheepherder's assistant on mountainous rangeland. After getting hopelessly lost numerous times, I could simply give my horse free rein and it would unerringly return to camp, by the most direct route possible. It knew the country intimately. In terms of spatial ability, I admit to being less intelligent than a horse. Thus our assessment of intelligence depends on the basis of measurement.

**Anthropomorphism** is the belief that animals have the same feelings and emotions as humans. Often animal rights concerns are raised by people who mentally place themselves in the position of an animal (how would you like to be penned up in a little cage like that?). Animal welfare should be assessed mainly by objective measurements rather than by anthropomorphic ideas. Another factor influencing people's viewpoint is the "cute and cuddly" syndrome. Many people who fully accept the use of beef, chicken, and pork in their diet are aghast at the thought of eating rabbit meat. "How could you kill such a cute, fuzzy, pink-eyed creature?" (My suggestion of a breeding program to develop an ugly rabbit was not well received!) Following a field trip to a veal farm, I was told by a student, "I'll never eat veal again—they're just babies!" The perception that only "non-cute" animals should be used for meat is illogical but widespread. As mentioned earlier, humans may be "hard-wired" to respond favorably to the facial features of infant animals (Fig. 10-7).

If animals have rights, what rights do they have? In the UK, the following "Five Freedoms" are recognized as an operating standard by the UK Farm Animal Welfare Council (Webster, 1994, 2000):

1.  freedom from thirst, hunger, and malnutrition—by ready access to freshwater and a diet to maintain full health and vigor

2.  freedom from discomfort—by providing a suitable environment including shelter and a comfortable resting area

3.  freedom from pain, injury, and disease—by prevention or rapid diagnosis and treatment

4.  freedom to express normal behavior—by providing sufficient space, proper facilities, and company of the animal's own kind

5.  freedom from fear and distress—by ensuring conditions that avoid mental suffering

At first glance, most people would probably agree that these are reasonable "rights" of

animals. Application in practice is not so simple. This can be illustrated by comparing battery cages vs. free range for laying hens. Freedoms 1–3 deal with production issues, while Freedoms 4 and 5 deal with ethical issues. In many respects, hens kept in battery cages may have Freedoms 1–3 met in a more satisfactory manner than those on free range. Free-range hens are subject to cannibalism associated with the "pecking order," whereby a bird low on the pecking order is attacked and injured or killed by the other hens. Birds on range may be subject to predation and inclement weather. In terms of a controlled, hassle-free environment, battery hens may be better off than those on range. On the other hand, they are certainly unable to engage in many normal behaviors, such as nest-building and egg-laying rituals, dust-bathing, and so on. Aspects of "**environmental enrichment**" (Fig. 10-10) such as provision of nest boxes raises the philosophical issue of whether animals can be frustrated or experience a sense of deprivation by not having certain resources that they have never experienced (Barnett and Newman, 1997). There is no black or white, right or wrong answer to these questions. They are ethical questions, which, presumably, can be reduced to components that can be addressed individually. Ultimately, however, the decisions reached in terms of farm animal welfare legislation will likely rest on personal beliefs and emotions, over the spectrum "I don't care what the data indicates; chickens should not be kept in little cages like that" to "Oh heck, they're only chickens. What's all the fuss about?" As is usual in decisions reached by political consensus, the optimal housing for layers will probably involve compromises on both sides, with enriched modified cages containing perches for roosting, and perhaps nest boxes and dust baths (Barnett and Newman, 1997). The cost of these modifications will be borne by the consumer. Surveys suggest that many consumers are willing to pay more for meat, milk, and eggs that they perceive are produced in an animal-friendly manner (Richardson, 1994). The majority of vegetarians (81%) as well as those considering

**FIGURE 10-10** An example of environmental enrichment. The calf has been provided with "toys" to play with. (Courtesy of Julie Morrow-Tesch.)

**FIGURE 10-11** Fast-food restaurants have become leading players in animal welfare issues. McDonald's, Burger King, and Wendy's have established standards for humane treatment of animals. McDonald's, the world's largest purchaser of animal protein, thus has an important role in livestock and poultry production, by taking the lead in forcing changes in the ways that food animals are raised.

becoming vegetarians or reducing meat consumption cite concerns about animal suffering and cruelty as the major determinant of this dietary choice (Richardson, 1994). In other words, a loss in market demand for animal products is driven mainly by concerns regarding the welfare of food animals, particularly those raised in "factory farms." Thus animal rights is one of the most important issues faced by the livestock industries.

Improved animal welfare can have unexpected consequences. For example, improvement of animal (pig) welfare by environmental enrichment can lead to beneficial economic effects by increasing pork meat quality (Klont et al., 2001).

How can livestock producers be convinced that it is in their interests to seek high standards of animal welfare? Webster (2001) discussed this issue in depth. Farmers respond

to an economic incentive. In Europe, legislation mandating certain minimum standards of animal welfare has been adopted. Verification that the standards are adhered to is achieved through an independent audit of quality standards. If all farmers are required to meet a minimum standard, this extra cost of doing business is absorbed by the marketplace (consumers). Radical improvements to the welfare of meat animals could be achieved at very little cost to consumers (Webster, 2001). In the United States, several large **fast-food restaurant chains** (e.g., McDonald's, Burger King, Wendy's) have voluntarily set up standards for humane treatment of animals, in consultation with Animal Scientists. This is a perceived pro-active approach by these businesses; they detect that the American public soon will demand evidence that animal products (meat, milk, eggs) used in the fast-food industry have been produced in a humane manner (Fig. 10-11). An organization of U.S. egg producers, the United Egg Producers (UEP), has established welfare guidelines, including minimum cage sizes and feed and water guidelines. This action was partially an effort to catch up with McDonald's and Burger King, which had earlier announced animal welfare guidelines for suppliers of eggs to their restaurants. Obviously, animal welfare issues are beginning to enter the U.S. mainstream, and the driving force may be fast-food restaurant chains. Unfortunately, some animal industry groups regard the adoption of standards of animal care by fast-food chains as a sell-out to animal rights activists.

The aim of the "Five Freedoms" is not to eliminate stress but to prevent suffering (Webster, 2001). Suffering may occur when an animal has difficulty coping with stress, either because the stress is too severe or because the animal is unable to take action to relieve the stress. Competing stresses may occur. For example, a high-producing dairy cow may simultaneously experience the motivation to eat in order to relieve her metabolic hunger for nutrients to support lactation, a desire not to eat in order to relieve the discomfort of excessive rumen fill, and a powerful desire to lie down and rest (Webster, 2001).

Webster (2001) described some elements of poor animal welfare that are encountered in modern production systems:

1. hunger or acute metabolic disease, through improper feeding and/or breeding, e.g., the high-producing dairy cow

2. chronic discomfort, through bad housing, loss of condition, etc., e.g., pigs on concrete floors

3. chronic pain or restricted movement, due to distortion of body shape or function, e.g., lameness in broiler chickens, chronically lame dairy cows (laminitis)

4. increased disease, through overwhelming exposure to pathogens, pollutants, etc., e.g., postweaning enteritis in pigs.

5. chronic anxiety or frustration, through improper housing, bad management, or restricted social contact between animals, e.g., stereotypies

6. metabolic exhaustion, due to prolonged, excessive productivity, e.g., "spent" layers, downer cows

In the preceding list, the concept of **metabolic exhaustion** is introduced. In this case, animals may suffer because they are not killed; instead, they are worked until they "collapse in the harness" (an expression meaning that a draft horse has been worked until it dies on the job). This is the fate of a lot of dairy cows; they are kept in production until they are "burned out."

Mellor and Stafford (2001) suggested that the "Five Freedoms" could be expressed as five domains of animal welfare compromise (Table 10-2). They propose that legislative action is necessary to establish minimum standards of animal care and welfare. Higher than minimum standards should be encouraged. In the United States, the development of higher standards is being driven by fast-food restaurant chains, which in turn are being driven by consumer action.

Ironically, beef production is viewed unfavorably by many consumers in the United States, who see beef production systems as the epitome of human exploitation of animals. There are a couple of ironies here. First, of all the important livestock species, beef cattle are raised most closely under the conditions in which they evolved. Of further irony is that beef cattle ranching is perceived as especially undesirable, and yet range cattle live under the most natural conditions of any livestock. "Of all production systems, beef production most closely approximates the social ethic of husbandry" (Rollin, 1995a, p. 55). The final irony is that beef is rapidly losing market share to poultry meat, which is produced entirely un-

**TABLE 10-2**　Five domains of potential animal welfare compromise and how problems may be prevented or minimized
(Adapted from Mellor and Stafford, 2001.)

| Domain | Description | How Prevented or Corrected |
|---|---|---|
| 1 | Water deprivation, food deprivation, malnutrition | Ready access to freshwater and an appropriate diet in sufficient quantities and with a composition that maintains full health and vigor |
| 2 | Environmental challenge | Providing a suitable environment including shelter and a comfortable resting area, whether outdoors or indoors |
| 3 | Disease, injury, functional impairment | Prevention or rapid diagnosis and treatment |
| 4 | Behavioral or interactive restriction | Providing sufficient space, proper facilities, and the company of the animal's own kind |
| 5 | Mental (and physical) suffering | Minimizing the conditions that produce unacceptable levels of anxiety, fear, distress, boredom, sickness, pain, thirst, hunger, and so on |

der "factory farm" conditions. For a variety of reasons (many of them unwarranted or unfair), beef has shouldered the bulk of the criticism of animal products, while poultry meat has escaped largely unscathed. Webster (2001) puts it this way: "We eat less red meat (beef and lamb) but more poultry, which is, to say the least, inconsistent with our perceptions of the relative welfare problems of lambs and broiler chickens" (p. 234).

Rollin (1995a) has adapted (from Aristotle) the word **telos** to refer to the essential nature of animals. In other words, while it is absurd to think of animal rights in terms of free speech, freedom of the press, freedom of religion or freedom to own property, it is not absurd that animals have rights to engage in the basic behaviors that define them—that is, as Rollin (1995a) puts it, "fish gotta swim and birds gotta fly" and they suffer if they don't. The telos is those characteristics that make a pig a pig, or a chicken a chicken. Do ducks have a right to swim? (See Fig. 10-12.) Perhaps without realizing it, society is moving towards a new social ethic for animals, which does not

preclude utilization of animals for food, but which demands that food animal production systems be humane and sensitive to the animals' basic needs and natures. This is a logical extension of the ever-expanding sphere of societal concern for sentient beings.

Rollin (1995a) has in fact proposed a **new social ethic for animals.** He has distinguished two sets of ethics, which he calls Ethics$_1$ and Ethics$_2$. **Ethics$_1$** is the set of rules, principles, and beliefs about right and wrong, good and bad, justice and injustice, duty and obligation that governs people's behavior, or that they believe ought to govern their behavior. **Ethics$_2$** is the study, critique, analysis, and criticism of Ethics$_1$—and examines the logic and coherence of fundamental concepts and beliefs. The traditional social ethic for the treatment of animals has been one that forbids cruelty to animals—that is, deliberate, sadistic, useless, unnecessary infliction of pain, suffering, and neglect. Rollin (1995a) maintains that this traditional social ethic is not adequate to deal with ethical concerns about intensive, industrialized agriculture. He believes that society is attempting to

**FIGURE 10-12** (*Top*): Ducklings raised in total confinement for meat production, without access to water for swimming. (*Bottom*): Traditional Long Island duckling production, where the birds have access to water. As waterfowl, do ducks have a "right" to swim? (Courtesy of Albert G. Hollister.)

extend the moral machinery it has developed for dealing with people (e.g., concepts of rights) and modifying it appropriately for use with animals. The demand for animal rights fills the gap left by the loss of traditional husbandry agriculture and its built-in guarantee of protection of fundamental animal interests. There is some evidence to support Rollin's position, including the work of Davis and Cheeke (1998) and Schneider (2001). Schneider (2001), in a study in Canada, concluded that the dominant view in North American society of the moral status of animals is not the Judeo-Christian position but is the new animal ethic described by

Rollin (1995a). Fraser (2001) also believes that there is a new societal perception of animal agriculture. Fraser (2001) summarizes the views of critics of animal agriculture, which he calls "The New Perception of Animal Agriculture":

1. Animal agriculture is highly detrimental to animal welfare.
2. Animal agriculture is mainly controlled by large corporations rather than by individuals or families.
3. Animal producers are motivated purely by profit, with no compassion for the animals or traditional ethic of animal care.
4. Animal agriculture increases world hunger by using grain and land to produce animal products for the wealthy instead of providing basic necessities for the hungry.
5. Animal agriculture produces unhealthy food.
6. Animal agriculture is harmful to the environment. (p. 635–636)

Fraser (2001) states that the animal industries and animal scientists have not responded well intellectually, ethically or politically to these challenges. He concludes: "There is an urgent need for scientists and ethicists to avoid simply aligning themselves with advocacy positions and instead to provide knowledgeable research and analysis of the issues" (p. 640).

Fraser and Duncan (1998) discussed the concept of **motivational affective states** (MAS), which can be either positive or negative. These are adaptations that motivate certain types of behavior. Negative MASs (e.g., thirst, fear) evolved in response to "need situations," because the action benefits fitness. Positive MASs (e.g., playing, exploring) evolved in "opportunity situations"; the pleasure inherent in the behavior motivates the animal to perform it. The MAS hypothesis provides a basis for predicting the effects of animal management on welfare. The public sets ethical priorities related to the perceived subjective experiences of animals.

Minimizing suffering (strong negative MASs) such as severe hunger, pain, or fear) is of primary concern, while allowing animals to pursue normal pleasures of life (positive MASs) is considered relevant to welfare but of lower priority than prevention of suffering (Fraser and Duncan, 1998). The telos concept relates to positive MASs, as does Rollin's Ethics$_2$ concept.

Finally, the livestock industry and animal scientists could enhance (or at least not further diminish) their public image by avoiding the use of disrespectful or derogatory terms for animals. For example, the term "spent hens" for end-of-lay chickens implies that we exploit chickens and as soon as we're done with them, they are discarded like trash. Andy Rooney of the CBS program *Sixty Minutes* made a big issue of "spent hens," which was followed by an outpouring of public revulsion about "throw away" hens. Another example is "trash fish," such as Pacific hake and carp, which are used for animal feed because they are not sought after for human consumption. Referring to another form of life as "trash" is not the best way to maintain public support.

Some practices invite criticism. A new industry, which has developed in Australia, is the shipping of live sheep to the Middle East, in "sheep ships" loaded with 120,000 or more sheep (Fig. 10-13). High mortality of sheep on the high seas, with dead sheep thrown overboard, and the perceived "cramming" of sheep into tight quarters, has led to a perceptible change in the traditionally positive Australian attitude towards animal agriculture. One of the reasons for the high mortality on the ships was that the sheep were brought from rangelands and offered a concentrate diet on the ships. Since the animals had never seen grain before, many of them did not recognize it as feed and starved to death. New regulations now require a feedlot period to adapt the sheep to grain-based diets before shipping. However, the bad publicity generated by the sheep ships will live on, in the form of new converts to the animal rights movement. The sheep ships validate for many people that modern industrial animal agriculture is devoid of feeling and concern for

**FIGURE 10-13** (*Top*): Loading a sheep ship, for shipping live sheep from Australia to the Middle East. (*Bottom*): This ship is loaded with 120,000 sheep. (Courtesy of Jan Z. Foot.)

the welfare of animals—they are just commodities to be used "efficiently." Australians have also developed a live cattle export trade. The loss of a cattle ship in a storm, drowning hundreds of cattle, stirred Australian public opinion against the live animal export trade. The trade in live animals may occur globally. The world's largest feedlot company, Cactus Feeders, Inc., of Amarillo, Texas, has applied for a permit to import live cattle from Australia (*Feedstuffs*, November 26, 2001, p. 1). This action is a predictable consequence of the globalization of agriculture and will likely increase worldwide concern for animal welfare. Also predictable is the likelihood of more shipwrecks and animal drownings, boosting the numbers of animal rights activists.

358 CONTEMPORARY ISSUES IN ANIMAL AGRICULTURE

Genetic manipulation through animal breeding techniques may be successful in alleviating some of the welfare problems associated with intensive animal production (Jones and Hocking, 1999; Swanson, 1995), such as hysteria, feather picking, and cannibalism in poultry. However, Swanson (1995) concluded, "The aesthetics of placing animals in barren environments in general may be a 'hard sell' to the public even if science produces evidence to the contrary. The visual impacts of animals kept in barren environments coupled with anthropomorphic intuitions prompt people to infer animal suffering" (p. 2750).

An example of genetic manipulation that could alleviate animal suffering is the production of **blind hens.** Commercial layers may peck the feathers of their cagemates, causing bleeding and further attacks (cannibalism). If their vision is blocked (by use of plastic spectacles, for example), cannibalism is reduced. Ali and Cheng (1985) developed a line of

**FIGURE 10-14** A featherless broiler chicken, with potential for efficient production in hot climates. How do you regard this development? How do you think that animal rights groups will respond? (Courtesy of A. Cahaner.)

blind hens. The blind hens were less active, had better feathers, showed less cannibalism, and were more productive than normal layers. They suggested that blind hens would be less stressed in cage systems than normal hens. Sandoe et al. (1999) discussed this issue and concluded that by utilitarian standards the blind hens had a better life than their sighted counterparts but that this idea is almost universally condemned by the general public.

A genetic manipulation achieved through conventional breeding techniques is a featherless chicken (Fig 10-14). These birds, produced by Israeli geneticist Avigdar Cahaner of the Hebrew University of Jerusalem, originated from a spontaneous natural mutation called "scaleless," which was reported by Abbott and Asmundson, 1957. The birds produced by Cahaner are featherless broilers, which he asserts are environmentally friendly and well suited for production in warm and subtropical regions (Cahaner, A. Personal Communication, Nov. 12, 2002). They are less susceptible to heat stress than conventional feathered broilers. They have lower production costs in terms of reduced ventilation requirements and greates feed conversion efficiency. The absence of feathers reduces water needs for processing (plucking), and eliminates the need for disposal of feathers. The featherless chicken is by definition not susceptible to feather picking.

The concept of rights is an ever-widening circle (Fig. 10-15). When the United States was founded, rights were extended only to white male landowners. Gradually, full human rights have been extended to all peoples in the United States (Fig. 10-15). Interest in the rights of animals is a natural extension of the expanding concept of rights. Many people accept the concept of some rights for some animals. For example, try beating your dog with a big stick on a busy downtown sidewalk. You'll quickly discover that most people believe in certain rights for dogs. However, you might well be able to stomp on a rat or a cockroach and not elicit much reaction. Clearly, the public is extending the concept of rights to an ever-

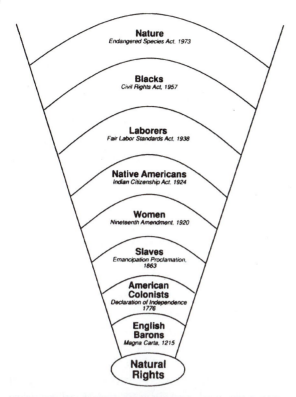

**FIGURE 10-16**  A fanciful view of the emerging field of animal law! (Courtesy of J. Cleland.)

**FIGURE 10-15**  (*Top*): The expanding concept of rights, entering the animal domain as a natural consequence of extension of egalitarianism to the natural world. (*Bottom*): The expanding moral covenant to the natural world began with concern about domesticated animals, and especially companion animals, and is progressing to encompass an ever-expanding menagerie. (Courtesy of Wesley V. Jamison.)

expanding circle of life, beginning with animals that are "most like us," such as dogs (Fig. 10-15). Other domesticated animals such as farm livestock are close behind. Wildlife, particularly large ungulates (e.g., bison), top-of-the-food-chain carnivores (wolves, eagles, lions, etc.), and other appealing animals (e.g., whales) are increasingly within the limits of our moral covenant. Clearly, the animal rights movement represents a natural extension of society's concern for fair play. The use of animals as "meat machines" on "factory farms" owned by extremely wealthy corporations clearly does not resonate very well with many people. The animal rights movement is here to stay and has been provided with plenty of ammunition by modern animal agriculture.

Concerns about animal welfare and animal rights have led to the formation of a new field of study, **animal law.** The Animal Legal Defense Fund, formerly the Attorneys for Animal Rights, works within the legal system to promote "liberation" of animals from human exploitation. A fanciful view of an "animal lawyer" in action is given in Fig. 10-16.

## The Good Old Days of Animal Welfare: Rollin, Reality, and Horses

Rollin (1995a) believes strongly that before the development of industrial agriculture, farmers and their animals lived in peaceful, symbiotic coexistence. What was good for the animals was good for the farmer. It is easy to romanticize the good old days. My own experience, at the tail end of the pre-industrial era, is not totally supportive of this view. Animals were often kept in muddy, primitive conditions. The farmwork was done by horses. They were frequently overworked and abused. I have personal knowledge of situations in which work horses died because of the ignorance of and abuse by their owners. The situation for horses was made tremendously worse than for other livestock because they were used for all transportation needs by all segments of society. Millions of military horses died in wars, after enduring starvation, pain, suffering, and cruelty. In the Alaska-Yukon gold rush, pack horses were used until they died of exhaustion on the trail. The trail from Skagway to the Yukon was littered with dead horses. Goldseekers coming behind simply walked over them; in places, the trail was several layers deep in dead horses. The best thing that ever happened to horses was the invention of the internal combustion engine!

# RELIGION AND THE USE OF ANIMALS

And God blessed them,
and God said unto them,
Be fruitful, and multiply,
and replenish the earth,
and subdue it: and have dominion
over the fish of the sea,
and over the fowl of the air,
and over every living thing
that moveth upon the earth.

Genesis 1:28

Do we humans have dominion over every living thing that moves upon the earth? We in western societies often seem to think so. In fact, we do hold dominion over many forms of life, and their survival as species depends on our goodwill. There are a relatively small number of animals that we have been unsuccessful in having dominion over, such as rats and cockroaches. At the microscopic level, we are largely at the mercy of bacteria and viruses, which can mutate rapidly and plague us with new diseases.

The issue of animal rights is heavily impacted by our **religious beliefs.** If you believe literally in the preceding quotation from Genesis, you can believe that you have been given a mandate to exploit all other forms of life more or less indiscriminately. You can believe that humans are special products of creation, having a soul which no other creature possesses. Thus you could in your mind-set divorce yourself from the biological reality of evolution and consider yourself apart from natural ecosystems. To a considerable extent, this has been the history of Western civilization for the past 2000 years, which has brought us to where we are today. In many respects, it is not a pretty sight!

This is a subject "where angels fear to tread" (and fools rush in!) Suffice it to state that a significant aspect of the animal rights movements is the rejection of the concepts that humans have dominion over other forms of life and that we "have been put here" to run the show as we see fit to meet our needs.

Shapiro (1999) has provided a good review of the views of several religions concerning treatment of animals. In general, Western philosophies tend to teach that humans are dominant and are a special creation of God, while Eastern philosophies (Buddhism, Hinduism, etc.) tend to believe that humans are equal to other animals. Animal rights philosophies more closely resemble those of the Eastern religions. Millions or perhaps billions of Buddhists and Hindus are vegetarians and oppose the killing of animals.

Allendorf (1997) discussed the similarities between Zen Buddhist teachings and modern ecology and suggested that ecologists, conservation practitioners, and the public could benefit greatly from an improved understanding of Zen Buddhist teachings, which emphasize the interconnectedness of individuals with their surroundings. "A bison cannot be understood in isolation from the prairie; understanding requires study of the bison-prairie unit" (p. 1045). Knowledge (science) is not equivalent to understanding, and understanding cannot be achieved through intellect alone (Regosin and Frankel, 2000).

Religious leaders are beginning to "step up to the plate" concerning environmental issues, some of which involve livestock production. For example, in 2001, 12 Roman Catholic bishops from the U.S. Pacific Northwest and British Columbia issued a pastoral letter, "The Columbia River Watershed: Caring for Creation and the Common Good" (*www.columbiariver.org*). They urged changes in priorities in resource exploitation, including changes in timber harvesting practices and provision of low-interest agricultural loans tied to land, water, and energy conservation practices. The pope has issued statements addressing the world's "ecological crisis." He has called for the concerted efforts of individuals, peoples, nations, and the international community for the defense and preservation of the natural and human environments, which are not safeguarded solely by market forces.

Religion influences livestock production in terms of **food taboos**. Some religions prohibit the consumption of pork, for example. Hinduism prohibits the killing and eating of cattle. These types of food taboos may have evolved as a means of fostering cultural and religious identity. "What distinguishes human groups are cultural practices—our customs and beliefs regarding the proper foods, the proper clothes, the proper adornments, the proper behavior" (Milton, 1997, p. 53). Having food taboos helps in establishing "us" vs. "them." Such tribalism seems fundamental to human behavior. Milton (1997) described how people in one Brazilian tribe eat monkeys, whereas an adjacent tribe considers eating monkey abhorrent. Needless to say, the monkey-eaters and the monkey-noneaters were bitter enemies, and killed each other on sight. She cited numerous other examples, including one tribe that didn't eat deer meat (hence the title of the article "Real Men Don't Eat Deer"). North Americans tend to react similarly concerning the consumption of horse, dog, cat, and rat meat, all of which are delicacies in some other parts of the world. Religious taboos against consuming certain types of meat may have developed in some cases because the animals performed especially valuable services while alive. Cattle in India were more valuable for work, fuel, and milk production than as a meat source and came to be viewed as sacred animals.

# BIOTECHNOLOGY: SCIENCE, GOD, AND GENE JOCKIES

**Biotechnology** deals with applications of technology to biology and generally refers to the manipulation of living cells and their components. Examples of biotechnology processes of current interest in animal science include embryo manipulation and gene transfer (transgenic animals), cloning, and the production of biologically active substances (e.g., bST) by genetically modifying microbes to cause them to produce the desired substance. A few examples will be briefly discussed to indicate the scope and potential of biotechnology applications in animal agriculture. A thorough review of biotechnological production and potentials of genetically modified livestock is provided in a series of papers edited by Hafs (1993).

Some of the biological processes that might be amenable to biotechnological modification (genetic engineering) include the endocrine system, biochemical pathways of cellular metabolism, structural proteins of wool and milk, and the immune system. In the endocrine area, attention has focused mainly on growth hormone (GH). Transgenic livestock with increased secretion of GH have

resistance of livestock to various diseases (Muller and Brem, 1991). Genetic manipulation of rumen and intestinal microflora is another possible application of biotechnology. The production of transgenic mice that produce cellulase activity in the pancreas has been achieved (Robinson and McEvoy, 1993). Transgenic pigs possessing phytase activity to enhance phosphorus digestion have been developed (see Chapter 6). Genetic engineering may have potential in combating mastitis, the most prevalent disease of dairy cattle. Kerr et al. (2001) reported the production of transgenic mice that secrete a potent anti-staphlococcal protein into milk. *Staphylococcus aureus* is the major contagious mastitis pathogen. Kerr et al. concluded, "These results clearly demonstrate the potential of genetic engineering to combat the most prevalent disease of dairy cattle" (p. 66).

There is considerable interest in using lactating animals as bioreactors to produce various proteins useful in human medicine, such as tissue plasminogen activator (Wilmut et al., 1991; Houdebine, 1995; Meisel, 1997; Velander et al., 1997). Use of animals to produce pharmaceuticals in their milk or other tissues raises some ethical issues. In some cases, the animals are subject to prolonged restraint and immobility. For example, **pregnant mare urine** is used as a source of estrogenic hormones (Freeman, 2000). Pregnant mares are restrained in stalls and equipped with urine collection bags. Although provision for exercise is usually made, there have been animal rights concerns with this activity. The mares must be pregnant to produce the desired hormones. The foals, especially the males, are in some cases raised as feedlot animals and slaughtered, with the meat exported from North America to Europe. This aspect is also criticized by animal rights organizations. In China and several other Asian countries, **bear farming** has been developed. Bear bile is especially valued as a component of traditional Chinese medicaments and is in high demand in China. As an alternative to killing wild bears for their gall bladders and bile, bears equipped with bile

duct cannulas are kept in cages and used as sources of bile. This activity raises ethical issues. Given that for the foreseeable future there will be a high demand for bear bile in China, which is the preferable choice: killing wild bears or having bear farms with cannulated animals?

On a more high-technology level, the production and cloning of transgenic animals with human genes for the production of milk containing pharmaceuticals raises ethical issues. These may involve animal rights issues such as animal confinement, moral issues as to how far humans should go in exploitation of animals, and risk factors such as the inadvertent introduction of infectious agents. Prion-caused diseases such as spongiform encephalopathies are obvious possibilities.

**Antisense technology** is the use of a reverse sequence of the nucleotides of a gene to repress or inhibit expression of that gene. This can be useful in crop and livestock production. For example, the ripening of fruit such as tomatoes can be reversibly inhibited by antisense RNA introduced into transgenic plants (Oeller et al., 1991). This technology could reduce losses due to overripening of fruits and vegetables during transportation or because of lack of refrigeration. Tomatoes could be picked when vine-ripened, restoring the widely perceived lack of taste of supermarket tomatoes.

These are very exciting developments. They are also very frightening to many people. There are serious religious overtones. If you believe, as many people do, that a God created everything in its present form, then altering the genetic makeup of an animal and transferring genes from one species to another present very serious ethical concerns. Are these scientists "playing God"? This issue also relates to education and **scientific literacy.** Major advances in biotechnology are taking place and will take place whether we like it or not—the genie is out of the bottle. These advances have major implications for agriculture and medicine. Unfortunately, the scientific expertise of the American public is apparently shrinking rather than expanding. The ethical

## ATTACK OF THE ANTHRAX "VIRUS"

*(R. Lewis. Attack of the Anthrax "Virus" The Scientist, November 12, 2001, p. 42.)*

In 2001, after the September 11 bombing of the World Trade Center towers, the U.S. was beseiged by an outbreak of bioterrorism. The following is a quotation from the article by Lewis:

"Americans are getting a crash course in microbiology. While the press and public, not to mention various government officials, struggle to distinguish bacteria from viruses, antibiotics from antibodies, and viruses from vaccines, an underlying message is emerging: As a nation, our science illiteracy has gone from mere embarrassment to a life-threatening problem. In the days following the first cases, poor *Bacillus anthracis* suffered a major identity crisis; first dubbed 'a dangerous chemical,' the pathogen was soon designated 'the anthrax virus' by a government spokesperson whose error was picked up and parroted by many in the media. Moving up the evolutionary ladder, a London tabloid vividly described 'fungal spores of anthrax.' "

What does it matter if many people think that anthrax is caused by a virus producing fungal spores? It matters because public perceptions influence public health responses. One response was the widespread use of powerful antibiotics to "build up resistance" to anthrax. The extensive consumption of powerful antibiotics by people who are not sick is an excellent method of creating antibiotic-resistant bacteria via the process of microevolution (microevolution is a change in gene frequency, such as when a drug kills off susceptible bacteria, leaving resistant variants the opportunity to multiply).

Responses to threats of bioterrorism and emerging diseases are best made by government officials and citizens who are knowledgeable about the nature of the threats and the appropriate (and non-appropriate) means of defending against them. This is why greater scientific literacy of the American public is important.

and legal decisions that will be forced on society by advances in biotechnology should be made by a society with some understanding of science. The unraveling of the genetic code, and the ability of scientists to make genetic modifications, hit at the very core of the value systems of many people. The more they can understand biological science, the less ominous and frightening will be the scientific advances. The time frame for public acceptance of scientific findings can be virtually glacial, almost on a geologic time scale. It is now universally accepted that the Earth is not flat. However, the Darwinian principle of **evolution** on which biotechnology is based (genetic engineering is artificially accelerated evolution) is still hotly debated publicly. The battle is raging on, while the war was won (or lost, depending on one's viewpoint) over one hundred years ago.

An interesting ethical concern with biotechnology is that of vegetarians who object to the transferring of animal genes into plants, which has been done. If you consume a tomato that contains chicken or fish genes, are you still a vegetarian? Or is it cannibalism if you consume a transgenic plant or animal containing human genes? We live in interesting times!

Public concern about the implications of biotechnological developments is certainly appropriate (Mepham, 1993). There have been many examples of unexpected and unwanted consequences of new technology since the Industrial Revolution began. The fact remains that a scientifically literate populace is necessary to evaluate the potential consequences, both advantageous and detrimental, of the development and application of new technologies. Throwing up our hands and yearning for the old days when these choices and concerns didn't exist is not a viable option. Genetic engineering of animals raises special concerns, not the least of which is that whatever tinkering with the genetic code of animals is possible could also be done with humans. There is a widespread belief, resting on a foundation of religion, that the human genome is inviolate. Human eugenics is widely considered to be morally wrong.

Some of the potential concerns or unexpected consequences of agricultural biotechnology relate to food safety. Are there risks of pathogenic agents in milk from transgenic animals used to produce pharmaceuticals? For example, prions that cause BSE (mad cow disease) are an obvious concern (Houdebine, 1995; Wills, 1996). **Food allergens** are proteins. The occurrence of proteins in unexpected places, such as peanut proteins in transgenic corn, might create problems for people allergic to peanuts. Potential benefits and problems associated with transgenic livestock have been reviewed by Dziadek (1996). Several countries now require labeling of foods derived from **genetically modified organisms** (GMO products).

The granting of **patents** for genetically modified animals has been very controversial in the United States. There are religious undertones. If one believes that only God can create life, patenting new life forms raises moral and spiritual issues that are difficult to reconcile in scientific or legal terms. The objective of patenting genetically engineered animals is so that the companies involved can protect their proprietary rights to recover their costs of research and development and make a profit. These seem to be legitimate objectives. In the case of plants, the Plant Variety Protection Act of the U.S. allows plant breeders to patent new varieties. This has encouraged commercial development of new varieties of crop and forage plants, including those involving gene transfer. In numerous cases, such as new varieties with increased pest resistance, there are favorable environmental and ecological effects.

While there are major concerns about the ethical and religious aspects of tinkering with the genetic code, in 1997 there was a firestorm of public concern about the consequences of not tinkering with the genetic code—that is, by **cloning** animals. Wilmut et al. (1997) announced that a sheep (Dolly) was a clone of another sheep (its mother?), obtained by introducing cells from the mammary gland of an adult sheep into an embryo, resulting in a new animal genetically identical to the animal from which the DNA was derived. Public concern was with the possible application of this technology to humans. Clearly, we will not lack ethical dilemmas resulting from scientific and technological advances. There is a real need for a scientifically literate public and for better education for scientists in ethics and philosophy. It has been a widely held view of scientists that science is "value-free" and "ethics-free." Scientists claimed that they don't make moral or ethical judgements. Animal scientists are often bewildered and annoyed when the public, or segments of it, reacts negatively to technological developments that improve efficiency, such as development of BST to increase milk production. The unspoken words are "Here we've busted our backsides to increase efficiency and give the American consumer the cheapest and safest food the world has ever known, and what sort of gratitude do we get? They are biting the hand that feeds them." A mandatory course in **agricultural ethics** in every ag school would seem to be a good place to start in bridging this gap.

Bernard Rollin, a bioethicist with an understanding and positive appreciation of animal agriculture, has addressed ethical issues in medical and agricultural sciences in *The Frankenstein Syndrome: Ethical and Social Issues in the Genetic Engineering of Animals* (Rollin, 1995b). Briefly, the **Frankenstein syndrome** is the public perception that science and technology are running amok, with potentially hellish consequences of a wrecked planet populated by genetically engineered monsters. People don't understand the new technologies in biology, but they are deeply disturbed by them, as the reactions to Dolly's birth announcement indicate. Animal scientists are often at the forefront of these developments, as in Dolly's case. Are we prepared to deal with the ethical consequences of our findings, or do we just dump them on the politicians and lawyers? If we take the latter approach, we must be prepared to accept the consequences.

According to Rollin (1996), **genetic engineering** is probably the most powerful technology ever devised by humans. He and many others blame the scientific community for much of the public mistrust of biotechnology. Scientists in general have distanced themselves from ethical and moral concerns, with such statements as "science cannot make value judgements," "science is value-free," and "genetic engineering should not be hampered by ethical considerations." By default, they have ceded these concerns to lawyers and politicians.

Rollin (1996) suggests that one common ethical issue raised by genetic engineering, that scientists should not "play God," is a spurious one and does not represent a genuine ethical issue. Humans have been participating in altering evolutionary processes from our earliest history, by eliminating megafauna (e.g., mammoths, mastadons) by hunting, and by domestication and selection of animals and plants. Genetic engineering is simply a means of accelerating this process. Thus he concludes that genetic engineering is not intrinsically wrong. There are genuine issues of risk in modern technology, and increasingly, the general public does not have confidence in the ability of the scientific community to identify these risks or to manage them. As Rollin (1996) suggests, the public has endured too many technology failures that technologists said couldn't or wouldn't happen (e.g., Chernobyl nuclear plant meltdown, explosion of the Challenger space shuttle, escape of the killer bees, spillage of millions of gallons of swine manure into rivers, contamination of aquifers with swine waste, adverse environmental effects from DDT, beached oil tankers, etc.). Scientists tend to minimize the likelihood of unanticipated consequences of new technologies or think they can fix them with even better technology. Failure to involve the public in the application of biotechnology leads to cynicism and rejection of new technologies. The public no longer responds well to "trust us, we know what we're doing."

Rollin (1996) has identified a number of potential risks in the genetic engineering of livestock; these are summarized in the following:

1. Unexpected problems may not be detected until the genetically altered organism has been widely disseminated. Genetic engineering is basically selective breeding "in the fast lane," without the safeguard of an enforced waiting period to fully assess the results.

2. Genetic engineering of animals for productive traits may have unsuspected harmful consequences for humans who consume the resultant animals or their products (e.g., milk).

3. Accelerated loss of genetic diversity by speeding up animal selection could lead to greater problems with harmful recessive genes, loss of hybrid vigor, and greater susceptibility to pathogens.

4. Creation of genetically engineered animals may alter the pathogens to which they are host. For example, in a mouse model for AIDS research, the AIDS virus might become more virulent and infectious by interacting with natural mouse viruses to create an airborne AIDS virus.

5. Ecological damage could result from radically altering an animal and then having it escape into an uncontrolled environment. This could readily happen with genetically engineered fish (Berkowitz and Kryspin-Sorensen, 1994). There are many examples of unforeseen consequences when animals are released into a new environment (e.g., rabbits in Australia).

6. Genetically engineered animals may accelerate environmental degradation. For example, large areas of equatorial Africa are presently unsuitable for cattle production because of the presence of the tsetse fly, which is a vector for the protozoan (*Trypanosoma*) causing sleeping sickness (Matzke, 1983). Production of *Trypanosoma*-resistant cattle could result in ecological devastation of these areas.

7. There are potential military applications such as using genetically engineered animals as carriers to infect populations with human pathogens.

8. Socioeconomic effects may occur such as allowing multinational corporations to monopolize and control the global food supply.

9. Adverse effects on animal welfare may occur. Transgenic pigs containing the gene for bovine somatotropin exhibited increased growth rate but also suffered pathological effects and pain (Pursel et al., 1989).

Rollin (1995b) proposed a **"principle of welfare conservation,"** meaning that animals created by genetic engineering should not be worse off than they are now. In other words, animal well-being should be considered and must not be compromised.

The well-demonstrated willingness of industrialized agriculture to sacrifice animal welfare for increased productivity suggests that the industry itself cannot be relied upon to adequately address these issues. Expanding technologies necessitate new rules and regulations governing their application. It should also be recognized that biotechnology can be used to improve animal welfare. For example, in the dairy industry, bull calves are unwanted and often suffer from inadequate care in auction yards, with a very high rate of mortality. Gender selection by semen processing will soon allow dairy operators to produce only heifer calves, eliminating one of the animal welfare concerns of the dairy industry (although there would be surplus heifers to be disposed of).

When we gripe about the animal rights activists and the urban public that doesn't understand agriculture, the importance of what the public thinks about livestock production practices to our future might be put into perspective by contemplating the following sentence (*The Ecologist,* 25:180, 1995): "The fact that corporations and governments feel compelled to spend billions of dollars every year manipulating the public is a perverse tribute to the ability of ordinary people to influence and change their society around them" (p. 180). These ethical and philosophical issues are not going to go away!

## GMO Crops and Livestock Production

The term "GMO crops" refers to genetically modified organisms (GMO), and in particular, transgenic plants. Examples include crops re-

sistant to the herbicide Round-up, and plants that contain genes from the bacterium *Bacillus thuringiensis* (Bt). Production of and trade in so-called GMO crops, such as corn and soybeans, have become very controversial. A high proportion of U.S. corn and soybean production now is GMO. Increasingly, major trading countries (EU, Japan) are refusing to import U.S. GMO grains, leading to trade disputes and involvement of such organizations as the **World Trade Organization (WTO).** The European position, based on the **Precautionary Principle,** is that we can't yet be certain that there are no unanticipated health risks associated with GMO crops. The U.S. position is that the Europeans are simply using concerns about GMO crops as a means of blocking U.S. grain exports for their own competitive advantage. It seems fairly obvious that if the rest of the world does not want GMO grains, it is a bit arrogant for the United States to continue to produce them for export purposes.

An even more murky area concerns whether or not animal products from livestock-fed GMO grains constitute a health hazard. It is difficult to imagine a scenario in which this would be the case. If there are any health hazards from GMO crops, which seems unlikely, feeding these grains to livestock would serve as a buffer against direct human exposure.

It appears that the major legitimate concerns with GMO crops are social and philosophical. A few companies, such as Monsanto, appear to be gaining control of the global food system. Many people are opposed to this concentration of wealth and power.

## Biotechnology's Great Yellow Hope

The so-called **GMO crops** have generated great controversy in the developed nations (North America, Europe, Japan, etc.). A common claim of anti-GMO activists is that this technology benefits only the rich, specifically the handful of companies that own the technology. The biotechnology industry has come up with a public relations gold mine: transgenic rice that contains β-carotene. Through a complicated process, including the inclusion of a daffodil gene, transgenic rice that synthesizes β-carotene was developed (Ye et al., 2000). The biotech industry is making extensive use of this development, by portraying biotechnology as the answer to many of the "world food problems." Millions of poor people in developing countries have suffered blindness as a result of vitamin A deficiency. The transgenic "golden rice" has the potential of preventing this deficiency. Critics of GMO technology point out that if we really wanted to eliminate vitamin A-deficiency blindness, it could be done with five cents worth of vitamin A per child per year.

Ironically, the development of "golden rice" was accomplished by research supported by public funds, including the Rockefeller Foundation (Potrykus, 2001), and the process was not patented. The technology is available to plant breeders everywhere. Potrykus (2001), in whose laboratory the rice was developed, believes that the opposition to golden rice is politically motivated in support of an anti-GMO agenda.

Similar technology has been used to produce oil crops enriched in vitamin E (Dellapenna, 1999). Canola oil enriched in vitamin E was produced by introducing a gene from a weed, *Arabidopsis*, into canola, producing vitamin E-enriched canola oil. Production of GMO crops with elevated nutrient levels has been termed **nutritional genomics** (Dellapenna, 1999).

One major type of GMO crop contains a natural insecticide derived from the bacterium *Bacillus thuringiensis* (Bt). Crops containing Bt genes produce a protein, Cry 9C, which is toxic to many insects. One of the concerns with Bt crops is that they may have effects on non-target insects, such as the **Monarch butterfly.** Despite a flurry of sensationalized reports suggesting that Bt corn is toxic to Monarchs, it appears that the effects on Monarch butterfly populations are negligible (Sears et al., 2001).

One advantage of Bt grain crops is that they may reduce the levels of mycotoxins in grains. In corn, for example, fungal invasion occurs following damage to the grain by the corn borer. A major reason for developing Bt corn was to reduce damage from the European corn borer.

Munkvold et al. (1999) found that the level of a mycotoxin, fumonisin, was lower in Bt than in conventional corn. Thus genetic engineering of corn for insect resistance may enhance its safety for human and animal consumption.

**Norman Borlaug,** who was awarded the Nobel peace prize in 1970 for his role in developing the "Green Revolution," is an outspoken advocate for biotechnology in agriculture and GMO crops. He refers to the "threat of anti-science zealotry" in a recent article (Borlaug, 2000), in which he concludes:

> Extreme environmental elitists seem to be doing everything they can to derail scientific progress. Small, well-financed groups are threatening the development and application of new technology, whether it is developed from biotechnology or from more conventional methods of agricultural science. (p. 490)

Borlaug has founded his own university, Norman Borlaug University (*www.nbulearn.com*). It is a for-profit, Internet-based knowledge company that delivers just-in-time learning to and about agriculture and the food system. Presumably environmentalists need not apply!

## ETHICS AND NATURAL RESOURCE ISSUES

Aldo Leopold (1949) coined the term **the land ethic.** In essence, it deals with our relation to land and to plants, animals and other organisms that live upon it. Land is not simply property to do with as we wish "because it's ours." Leopold states, "A land ethic changes the role of *Homo sapiens* from conqueror of the land-community to plain member and citizen of it. It implies respect for his fellow-members, and also respect for the community as such" (p. 240). It is apparent that we have a long way to go in developing a land ethic. However, we are making progress, as indicated by the whole development of the environmental movement and the concern of young people for the health of the planet. These are encouraging signs.[3]

In the United States, natural resource utilization is increasingly contentious, because many people are concerned about effects of resource extraction on the environment. Two of the major natural resource battlegrounds have been public rangelands and public forests, both collectively owned by citizens of the nation. Until fairly recently, ranchers and loggers had *carte blanche* use of these resources. The former head of the U.S. Forest Service, Jack Ward Thomas, is a coauthor of an excellent paper on ethical considerations in natural resource management (Cornett and Thomas, 1996). Some of the issues raised by these authors will be briefly discussed. Some of the quotations are from sources cited by Cornett and Thomas.

## Passion

The expression of passion has been felt to be inappropriate in natural resource professionals. Range managers and foresters were expected to be conservative and staid scientists, dealing with science and not making value judgments. As a result, the environmental movement stole the show.

"Value and emotion are inseparable. Any time we are dealing with people's values, we are faced with strong emotion; and whenever we are confronted with strong emotions, we can be sure that something of value is at stake. There is simply no way to avoid emotions when making important resource management decisions."

## Vision

"Moving forward may be difficult for those whose belief system and personal identity are totally invested in the old paradigm; they perceive no reason to change. . . . A pro-

---

[3] In my opinion, every student majoring in Agriculture should be required to read Leopold's book *A Sand County Almanac*.

fession can move forward only to the extent that individuals within the profession develop new philosophies . . ."

## Ethical Choices

"There are no black-and-white ethical decisions; ethics is about a realm of greyness, of complexity, and of questions that are difficult to answer."

## Integrity

[I]ntegrity is defined as uncompromised values. Professional integrity aligns passion, vision and action:

"[I]ntegrity is the result of harmony between what one thinks, says, and does and what one really feels—the motive in the deepest recesses of one's heart."

Aldo Leopold put it this way: "A thing is right when it tends to preserve the integrity, stability, and beauty of the community. It is wrong when it tends otherwise."

## To the Twenty-First Century

"[W]e stand on a very short bridge called the present. As the people who stand on this bridge at this particular time in history, we must make the best decisions to move most effectively from the past to the future."

"We will not survive the 21st century with the 20th century's ethics."

"It requires courage to be ethical. Professionalism is achieved through integrity, by having courage and humility to make decisions that maintain our values uncompromised."

"Most of us were carefully trained (and perhaps educated) in academic and organizational institutions geared towards disengaging our hearts and fully engaging our brains. The complexities we face today, and in the decades to come, demand both our hearts and our brains be engaged and smoothly meshing."

Not surprisingly, Jack Ward Thomas's period of leadership of the U.S. Forest Service

**FIGURE 10-17** Many farmers and ranchers in the western United States are highly suspicious of "the government," which they perceive as trying to steal their rights and property, through such mechanisms as the Endangered Species Act.

was brief. It is hoped that his period of influence will not be brief. Perhaps eventually, we can catch up with Aldo Leopold.

A currently contentious area in natural resource use is the conflict between public and private rights. In the western United States, **private property rights** is a rallying cry. Many farmers and ranchers are highly suspicious of what they perceive as government intrusion on their property and rights (Fig. 10-17). Public policy and the social contract are ethical issues in this debate. At what point does the public good overtake personal rights or ethics? Some issues of this type would include the right to use DDT, the right to grow a particular crop (e.g., tobacco, hemp), the right to cut down a tree with a bald eagle's nest, the right to allow erosion to erode your farm's soil down to bedrock, or the right to subdivide the farm into quarter-acre ranchettes. Ethical issues and public policy dealing with issues such as these are discussed at length by Thompson et al. (1994).

In natural resource conflicts, all sides tend to demand the application of "good science" and reject "junk science." A simple, cynical but true, definition of these terms is that **good science** is what supports my posi-

tion and **junk science** supports your position. This has been very evident in the western states with grazing issues and salmon recovery efforts. Science is unlikely to solve any natural resource disputes (Provenza, 2000; Thomas and Burchfield, 2000). These issues are disputes because there is a clash of values. Management decisions are value-driven and are ultimately made by elected officials responding to dominant societal values. The diminishing numbers of farmers and ranchers more or less inevitably means that their values won't prevail.

# GENDER ISSUES IN ANIMAL SCIENCE/VETERINARY MEDICINE

The demographics of animal science and veterinary medicine students have changed markedly over the past few years. At Oregon State University, for example, over 80 percent of the incoming students are female, with non-farm backgrounds, with primary interests in horses and companion animals, and who aspire to become veterinarians (Cheeke, 1999). From conversations with colleagues at other institutions, I perceive this to be a general pattern. Not so long ago, animal science students were overwhelmingly of the male gender, came from a farm background, and intended "to go back to the farm." The same pattern is seen in schools of veterinary medicine. These changing demographics hold considerable implications for these professions.

## *Gender Issues, Teaching, and Curricula*

Both animal science and veterinary medicine are, at many universities, experiencing the phenomenon of student enrollment being primarily female. Miller (1998) refers to this as "the feminization of the veterinary profession" (p. 340). She predicted that by 2004, 50 percent of the veterinarians in the United

**TABLE 10-3**  Geographical distribution of male and female veterinarians in the United States, 1998
(Adapted from Wise and Adams, 1999.)

| Region | Number | | Proportion | |
|---|---|---|---|---|
| | Female | Male | Female (%) | Male (%) |
| New England | 1,466 | 1,695 | 46.4 | 53.6 |
| Mass.* | 669 | 629 | 51.5 | 48.5 |
| N.H.* | 173 | 173 | 50.0 | 50.0 |
| Middle Atlantic | 2,610 | 3,920 | 40.0 | 60.0 |
| East North Central | 3,597 | 6,151 | 36.9 | 63.1 |
| West North Central | 1,760 | 4,661 | 27.4 | 72.6 |
| South Atlantic | 4,447 | 6,799 | 39.5 | 60.5 |
| East South Central | 1,125 | 2,520 | 30.9 | 69.1 |
| West South Central | 1,994 | 4,695 | 29.8 | 70.2 |
| Mountain | 1,754 | 3,192 | 35.5 | 64.5 |
| Pacific | 3,415 | 4,944 | 40.9 | 59.1 |
| Possessions | 42 | 141 | 23.0 | 77.0 |
| Total U.S. | 22,210 | 38,718 | 36.5 | 63.5 |

*States where women comprise 50% or more of veterinarians.

States will be women. This has already occurred in New Hampshire and Massachusetts where the proportion of female veterinarians equals or exceeds 50 percent (Wise and Adams, 1999). Their report also shows that from 1990 to 1998, the number of male veterinarians in the United States increased by 6.9 percent, whereas for females, the increase was 78.2 percent. The report of Wise and Adams (1999) gives a complete accounting of male and female veterinarians for all U.S. states and regions. An abbreviated version is given in Table 10-3.

The increasing "feminization" of animal science and veterinary medicine has implications that hit at the cores of these professions. The positions of leadership in these professions, such as university faculty members, and society officers (American Society of Animal Science, American Dairy Science Association, Poultry Science Association, American Veterinary Medical Association) are primarily men. Schillo (1998), in a very perceptive review article, discussed the issue of increased participa-

tion of women in Animal Science. His viewpoint is that the animal science community has traditionally embraced methods and outlooks that reflect values consistent with masculine views and experiences, and he claims that "most of the studies reported in the *Journal of Animal Science* assume an industrialized, capitalistic society based on economic growth and competition. The type of agricultural research system that corresponds to these conditions is one that values control and economic efficiency more than it values rural communities or sustaining the well-being of farm families and ecosystems. Efficiency is the dominant value of the economically privileged men who have controlled agriculture since the scientific revolution" (p. 2766). Schillo (1998) contends that animal scientists attempt to socialize female students to acquire male traits of aggression, competitiveness, and dominance, perpetuating professional behaviors that have got us to where we are now, which is an animal agriculture dominated by the industrial model. The great influx of female students into animal science offers the potential for a redirection of the discipline, embracing other values in animal production besides economic efficiency. Schillo (1998) concludes that "the Animal Science community could more effectively cope with issues if it would develop a social climate that encourages individuals with diverse perspectives to express their views in their work" (p. 2769).

Schillo's article is titled "Toward a Pluralistic Animal Science: Postliberal Feminist Perspectives." That is a challenging and contentious title for many animal scientists to digest and accept. "Pluralistic" implies a diversity of people and ideas. "Postliberal" literally means "after liberal." Schillo (1998) presents a brief review of liberalism. **Liberalism** provides a system of beliefs about the world, along with a social structure that is consonant with those beliefs. The United States was founded on a framework of liberal ideology, including the entitlement to certain rights. The liberal approach to gender issues is the basic premise that "all persons are created equal"; women, as autonomous, rational agents, have the same natural rights as men. Therefore, they should be afforded the same opportunities for goods, power, and freedom as men. These views are reflected in the analyses of three female animal scientists, Glenn (1996), Harlander (1996), and Pell (1996), in their review papers from a symposium "Trails to Success for Women in Animal and Dairy Sciences—Mentoring as a Stimulus for Success." According to Schillo (1998), these papers

> assert that employers have an obligation to place women on an equal footing with men. They advocate the use of mentors to clarify the rules for success so that everyone—regardless of gender, race, and so on—has an equal chance to excel. Glenn (1996) asserts that 'selection and advancement (in the animal science profession) must be based on qualifications and merit, not on a quota.' In other words, success should be based on traits assumed to be common to all good animal scientists, not particularized physical and (or) social traits, such as sex or ethnicity. Harlander (1996) echoes the view that em-

bodiment is irrelevant in the assessment of animal scientists. She states, The behavior that characterizes successful men must be encouraged in women. (p. 2764)

The postliberal feminist position departs from the preceding perspective in several areas, including mentoring and the issue of gender equality. Basically, **postliberal feminism** asserts that men and women are not equal, but different, and recognizes that there are characteristics typical of each gender that are different but equally valuable. Schillo (1998) argues that animal science is dominated by an androcentric (male-centered) ideology:

> In most professions, merit is determined by the extent to which an individual conforms to particular behavioral standards. Close examination of academic standards for merit evaluation, promotion, and tenure would reveal that the most valued personality traits are aggressiveness (how active one is in pursuing recognition), competitiveness (the extent to which one seeks to outperform his or her peers), and dominance (how much influence one exerts). The emphasis on these traits perpetuates certain professional behaviors: for example, addressing global rather than local issues, arranging laboratories hierarchically rather than collectively[4], conducting research secretively rather than openly, working independently rather than collaboratively, and interacting with colleagues in an adversarial rather than collegial manner. The professional reward structure is consistent with the liberal ideal of an autonomous individual and is more compatible with the characteristic lives of men rather than those of women. Whereas men frequently pursue competitive careers independently away from home, women have traditionally maintained families and interacted within networks of women as a means of enhancing their well-being.

---

[4] A tendency among many scientists these days is to use letterhead stating: "From the Laboratory of John L. Jones" (for example), with a great emphasis on "my lab."

Continuing, Schillo notes that "men commonly approach ethical questions by deferring to universal principles of justice. In contrast, women tend to be more aware of the contexts in which moral issues arise and focus on how their decisions might affect relationships. They also rely more on a process of communication to resolve such issues. Presumably, these differences in approaches reflect the different structures of typical male and female lives" (p. 2765). Schillo goes on to state: "Had women been dominant in this profession (animal science), the values underlying professional reward systems might be very different from the ones that currently prevail. Rather than promoting behaviors associated with competitiveness, aggressiveness, and dominance, reward systems might promote cooperating, caring, nurturing, interacting, and communicating. Thus, the practices and traits thought to characterize successful animal scientists might be very different from the currently adopted ones." In essence, the postliberal feminist position is that men and women really are different, and it is a disservice to women (and the profession) to mentor them to become more like men, in order to succeed in their careers. Or, as Schillo (1998) puts it: "The underlying premise of such programs (mentoring programs) is that female proteges require guidance in how to adopt the professional behaviors that garner success. . . This approach presupposes that the standards of the animal science profession are beyond reproach. But what if they are not? In particular, what if they are gender-biased?" (pp. 2763–2764)

For example, the animal science community is heading down a road leading to livestock production as an industrialized, corporate activity (see Chapter 8). This is a classic androcentric (male) approach of dominance, emphasis on economic efficiency, and disregard for animal and employee health and welfare. Increasing numbers of female students are coming to animal science because they love animals and want to work with horses and companion animals. Many of these students exemplifly characteristics of gentleness, nurturing, and concern for animal well-being. These are traits that are often disregarded, undervalued, or disdained by the animal science heirarchy. A more pluralistic animal science would welcome and embrace these diverse perspectives.

In this discussion, I have relied heavily on the paper of Schillo (1998) for several reasons. I think his paper should be read by all animal scientists, but it will likely be read by very few. How many animal scientists are likely to give a second glance at a title "Toward a Pluralistic Animal Science: Postliberal Feminist Perspectives"? He has stated the issues more clearly than I could, so I have quoted him extensively. Finally, I hope that my use of his work will help disseminate his ideas to a wider audience. In this regard, I also recommend his paper, "Teaching Animal Science: Education or Indoctrination?" (Schillo, 1997).

Increased intensification and concentration of animal production inevitably reduce the numbers of animal scientists needed (Cheeke, 1999). As discussed in Chapter 8, the intensification of the poultry industry was accompanied by a linear reduction in departments of poultry science. The intensive, vertically integrated corporate enterprises now controlling poultry production need fewer poultry scientists than formerly was the case. Does this fate of diminishing numbers await departments of animal science as animal agriculture industrializes? As I previously discussed (Cheeke, 1999), "There is an urgent need to develop new opportunities for young people who 'love animals,' especially horses and dogs, and who, therefore, may not be attracted to 'factory farming' of animals. These are not mutually exclusive, but I perceive that people who 'love animals' are not usually excited about intensive animal production" (p. 2037). Many universities are enlarging their equine programs and developing courses in companion animals in response to this need. Ironically, **equine programs,** which were once disparaged by animal scientists, may turn out to be our saviors. Besides the obvious effect of equine programs attracting large numbers of female students, there is the advantage of horses and companion animals being "non-globalized." With the industrial-ization and globalization of animal agricul-ture (Chapter 8), there may be a shift of animal production from the United States to other countries (e.g., Brazil, Argentina, Canada, Thailand, Russia, etc.). It is possible to raise pigs anywhere in the world and ship pork to the United States (or beef, or chicken meat). Horses and companion animals are "home-based." An American teenager is not going to go to Argentina to ride her horse, or to Brazil to pet her dog. These animals, and all the ancillary activities—stables, tack, dog sup-plies, veterinary services—are done in the neighborhood. As horses and companion an-imals become increasingly significant in the leisure activities of Americans, the demand for people with skills in horse and companion an-imal management, nutrition, behavior, and so on will grow. If animal scientists see the writ-ing on the wall in time, these people will be trained in animal science departments. As the veterinary profession becomes increasingly dominated by horse and companion animal specialists, the pre-veterinary programs in de-partments of animal science should increasingly emphasize these species, with less emphasis on livestock. This might be difficult for the "old guard" to accept and embrace! Exotic and zoo animal management and nutrition likely will become more common components of animal science curricula as well. For example, people leading programs for captive breeding of endan-gered species should be well grounded in ani-mal nutrition, reproductive physiology, and animal breeding.

Increased proportions of female students in animal science classes may necessitate changes in **instructional methodology.** Demonstrations of or participation in such techniques as castration, dehorning, slaugh-tering, and so on may have to be approached differently or perhaps eliminated from the curriculum. Language used in the classroom may have to be "toned-down" or made "kinder and gentler." Male professors may need assistance in finding ways of better con-necting with female students. For example, I have been recently described in student evalu-ations as being "bad tempered and rude," a charge that I find difficult to really believe. Perhaps I am perceived that way because my classroom style has not changed since the days when my classrooms held almost exclusively male students.

## The Leaky Pipeline

Pell (1996) used the term "the leaky pipeline" to describe the fate of women scien-tists in academia. This refers to the phenome-non that while many girls enter the academic pipeline in elementary school, few come out the other end as scientists. She identified several periods when the risk of leakage is particularly high: (1) early childhood, (2) adolescence, (3) sophomore year of college, and (4) the later part of graduate school and the job entry period. Low self-esteem is a concern, even at the graduate school level, and is referred to as **"the imposter phenomenon."** Pell (1996) de-scribes it this way: "Despite their selection for very competitive positions, the victims of the imposter phenomenon ascribe their success to luck, hard work, being in the right place at the right time, knowing the right people, and interpersonal skills instead of to ability or competence" (p. 2846).

Pell (1996) describes some of the unique problems of young female faculty members:

> Too often, young female/minority fac-ulty eat lunch alone in their offices while some of their peers eat together. This exclu-sion usually is unintentional and arises from the incorrect assumption that every-one feels welcome to participate in the brown bag lunch. . . . Being "one of a kind" or even "two of a kind" in a department can be a lonely and demanding experience. Fe-male undergraduate and graduate students often have unrealistically high expectations of the change that will occur due to the ini-tial hiring of female faculty. The first female faculty member in a department must face these unachievable demands with limited support, a justified fear of antagonizing senior faculty members, and without same-sex consultants for assistance. (p. 2847)

The leaky pipeline often dead-ends at a **glass ceiling** (Harlander, 1996). In departments of animal science, a major contributing factor is that there is a slow turnover rate of faculty positions and administrators. The typical career is 30 years or more. Thus while the graduate level of the pipeline is quite well supplied with female students now, academic positions are slow to come available. Even after obtaining an academic position, women usually take a longer time to receive promotion. Clearly, there are still hurdles to be overcome. Valian (1999) discusses these issues in the book *Why So Slow? The Advancement of Women*. In a review of the book titled "The Boys Don't Cheer," Zuk (2000) cites a poignant example from the book:

> My favorite example is from a fourth grade girl asked why the sexes didn't play soccer together at recess. Valian reports that "one girl who did play occasionally said: The boys never ask us to play. Then when we do play, only boys are chosen to be captains. And girls don't get the ball passed to them very often, and when a girl scores a goal, the boys don't cheer." This quote, with terms suitably transposed could describe adult professional life as well as fourth-grade recess. (p. 1022)

## Gender Issues in Veterinary Medicine

As previously discussed, the veterinary profession is becoming increasingly female-dominated (although it has a ways to go before it is truly female dominated). Miller (1998) reviewed the "feminization" of the veterinary profession and its likely future effects. Miller (1998) predicted that by 2004, 50 percent of U.S. veterinarians will be women. Why has there been a surge in entry of women into veterinary medicine? Miller (1998) discusses what she calls "the simplistic explanations"

and "the real reasons." The simplistic reasons are the following:

1. Gender discrimination in applications to veterinary schools has been eliminated; the proportion of women admitted approximately equals the proportion of women applicants.

2. New medicines and technologic innovations may have helped overcome physical limitations that women may have.

3. The desirability of the profession for women has increased because of the caring image the profession has projected to the public through popular books and TV programs.

4. Another reasons for the gender shift, according to Miller (1998), is that employers might prefer to hire women instead of men. "Employers of veterinarians might embrace stereotypes that cause them to believe they should prefer hiring women instead of men. For example, they may believe a woman will express more empathy towards clients (e.g., when dealing with grief resulting from the loss of a pet). Furthermore, an employer might believe that they can hire a woman for less money than they can a man" (p. 341).

According to Miller (1998), "I believe the main reason is most likely the relative decline in attractiveness of the veterinary profession for male applicants, coupled with the more simplistic explanations listed previously" (p. 341). As professions become female-dominated, salaries tend to flatten out or decline, making the field less attractive to men. Rather than applying to veterinary school, more men are focusing on medical schools, and if they don't get in, they go into an entirely different field than medicine (human or veterinary). The high cost of a veterinary education, with diminishing salary prospects, tends to make veterinary medicine less attractive to men than it once was. New emphasis on small animals rather than livestock may also be a factor. The rise in the number of employed veterinarians and the decline in the proportion of practice owners during the past two decades would tend to decrease autonomy and job security and might make the profession less attractive to men (Miller, 1998). Growth of corporate-owned veterinary practices is a relatively recent phenomenon, which happened after the proportion of women in veterinary schools had already started to increase. Corporate practice might offer special advantages to women, who often complete their veterinary education in their prime reproductive years. Thus, part-time work or predictable hours, both more common with corporate employment, may be attractive to some women. Miller (1998) discusses some other possibilities. "Influx of women into a profession will tend to reduce its prestige as a male occupation and exacerbate its feminization" (p. 342). "Women (as a whole) are more willing to put up with circumstances that men (as a whole) are less willing to tolerate" (p. 342). It is possible that women have different goals and objectives than men in choosing veterinary medicine as their profession.

It is well documented that female veterinarians earn less than do their male counterparts (Miller, 1998; Brown and Silverman, 1999; Slater and Slater, 2000). Slater and Slater (2000) point out that the veterinary profession has moved from a tradition of exclusion on the basis of gender to one of inclusion more quickly than any of the traditional professions. Despite this, male graduates usually receive more employment offers, larger salaries, and larger benefit packages than do their female counterparts. Slater and Slater (2000) also point out that "Women experience different vulnerabilities during late-night emergencies than men do, and women who are attempting to conceive or are pregnant must avoid exposure to certain animal diseases and other mutagens. Female practitioners looking to open their own practices often find that lending institutions tend to be much more chary of female borrowers" (p. 474). Studies from other countries show similar trends. In Australia, "female veterinary practitioners are less likely to own practices, and more likely to earn low incomes

than males" (Heath and Niethe, 2001, p. 546). They conclude, "The evidence points to a lower interest by women than men in the business aspects of veterinary practice" (p. 546). Similar findings were seen in Canada (Shaw, 2001).

In a report from Canada titled "Is the Veterinary Profession Losing Its Way?" Nielsen (2001) notes the following specific signs that the profession is losing its way:

> The profession is disinterested in vertebrate health problems and issues associated with environmental degradation; is failing to sustain its academic and research base; and is suffering serious erosion or stagnation of its laboratory infrastructure; fails to provide graduates with sufficiently high entry level competence to practice modern herd health, public health, and ecosystem health; has very low participation rates in the clinical specialty boards and colleges; has an increasingly disproportionate share of its members occupied in small animal and equine practice (too few in other branches of the profession); continues to have a low level of participation in public health, despite lip-service to its importance; and has leadership that is preoccupied with shorter-term economic issues. (p. 439)

The leadership of veterinary organizations is still primarily men. Perhaps as women work their way up the ladder of professional hierarchies, they can engage these challenges more effectively than men have.

## The Sexual Politics of Meat

I have found over the years that some ideas strike me as ludicrous when I first hear about them, but after a period of time they start to make a bit of sense. So it is with the book *The Sexual Politics of Meat. A Feminist-Vegetarian Critical Theory* (Adams, 1996). The book is similar to many philosophical treatises I have read: it is difficult to read. Philosophers can't seem to say what's on their minds in clear, concise English! I'll try to tease out the meat of this book (an incredibly bad pun!).

In his book *Beyond Beef*, Rifkin (1992) has a chapter, "Meat and Gender Hierarchies," in which the basic thrust is that meat-eating cultures tend to be patriarchal (male dominated), with meat, and especially beef, enjoying high status: "Red meat and beef are especially prized because of the qualities ascribed to them. It has long been held in myth and tradition that the blood flowing through red meat 'confers strength, aggression, passion and sexuality, all virtues coveted among beef eating people'... Blood conjures up notions of aggression and violence—valued emotions among warriors, sportsmen and lovers.[5] Soldiers have always been favored with beef before battle. So too have athletes entering the arena... Red meats, especially beef, have been associated with maleness and male qualities while the 'bloodless' white meats have been associated with femaleness and feminine qualities... The identification of raw meat with power, male dominance, and privilege is among the oldest and most archaic symbols still visible in contemporary civilization" (p. 239, 240, 244). Adams (1996—first published in 1990) supplies the philosophical underpinnings of these ideas.

Male domination of women is compared to male domination and killing of animals. There are many cultural references to "women as meat" (see Fig. 10-18). Adams (1996) views **vegetarianism** as a step in the overthrow of male oppression of women, and views male disdain for vegetarianism as part of the power play. Thus vegetarianism is linked to feminism. The power structure labels feminist concerns with rape, domestic violence, and pornography as hysterical, and the vegetarian's emphasis on death of animals as emotional. According to Adams (1996), "Meat eating is the re-inscription of male power at every meal... If meat is a symbol of male

**FIGURE 10-18** Women as meat. This poster is a mimic of the common posters showing cuts of meat animals.

dominance, then the presence of meat proclaims the disempowering of women" (p. 187). Thus, "Vegetarianism does more than rebuke a meat-eating society; it rebukes a *patriarchal* society, since we have seen meat eating is associated with male power. Colonialist British (male) Beefeaters[6] are not viewed positively if you do not approve of eating beef, male control, or colonialism" (p. 178). Thus Adams (1996) asserts that there is a moral imperative for people, especially feminists, to be vegetarians (**ethical vegetarianism**).

George (2000), in the book *Animal, Vegetable or Woman? A Feminist Critique of Ethical Vegetarianism*, takes issue with Adams' views, by way of postliberal feminist philosophy. In essence, George (2000) argues that since current nutritional requirement values have been

---

[5] As discussed in Chapter 2, red meat is red because the muscles contain the pigment myoglobin. It is not red because it contains blood.

---

[6] Beefeaters are a British military regiment.

determined largely with adult males, who have lower nutritional requirements than women and children, a vegetarian diet imposes extra nutritional burdens on women. Additionally, vegetarianism imposes extra challenges on people, especially women, in non-industrialized countries. According to George (2000): "Looking at the nutritional evidence, the best candidates for vegetarianism and veganism, in particular, are young, adult, healthy males living in industrialized countries. They do not have protein, vitamin or mineral stresses from feeding a rapidly growing fetus or a nursing infant, nor are they unduly stressed by their own growth requirements, as most of their growth is accomplished. Males have generally larger skeletons and higher iron levels than females and are at much less risk of anemia in adolescence and adulthood and osteoporosis in late middle and old age" (p. 105). Thus "ethical vegetarianism" is unethical! George (2000) concludes: "But I do think feminists must stop preaching the vegetarian life as a moral imperative. Vegetarianism is not morally required. It is an aesthetic choice that may be personally satisfying and healthful. To argue otherwise is divisive and self-defeating" (p. 169).

It is well acknowledged that there appears to be a link between cruelty to animals and violence against humans. Violence is usually associated with the need of an individual to exercise power and control over other living things. Some feminists argue that the dominant role of men in "violence against animals" is accompanied by acceptance of violence against women (Adams, 1996). Adams states: "Women are allied with animals because they too are objects of use and possession" (p. 168). In contemporary U.S. society, women as "objects of use and possession" is often exemplified by the behaviors of professional athletes (Ortiz, 2001), who, perhaps not coincidentally, are generally consumers of large amounts of red meat, especially beef. Some feminists also link domination of animals by men, including ultimately butchering and eating them, with male domination, subjugation, and sexual assault of women (Adams, 1996).

There is an ancient association linking violence with blood in meat. According to Shapiro (2000):

> Moses had previously taught the people a method of slaughtering animals. In Leviticus III:17 and VII:26, it is prohibited to consume flesh containing blood. The purpose was to tame man's instincts toward violence by weaning him from blood and implanting within him a distaste—a horror—of all bloodshed. The kosher method of slaughter causes the maximum effusion of blood, with the remaining blood being extracted by means of the washing and salting of the meat. (p. 29)

## Gender Issues and Production Agriculture

American agriculture, especially animal agriculture, tends to be dominated by men. This is particularly true of "factory farming." It is also true of the "family farm," which typically is patriarchal in nature. The dominant individual is usually the male farmer. The American family farm tends to be an isolated unit, physically distant from the neighboring farms, with the farm buildings located at the end of a lane. In contrast, in many European countries, the farmers and their families live in villages, maintaining an active social life with their neighbors, including sharing meals with other families in restaurants and cafes. The isolation of the American farm family can leave other family members under the thumb of a dominant, autocratic, and insensitive male.

Poultry production and dairying used to be the province of the "farm wife." Women had their own layer flocks to produce eggs for sale or barter and looked after and milked the family dairy cow. In developing countries, these activities are still usually performed by women. The development of industrialized agriculture has led to both poultry and dairy production becoming male-dominated, and animal-, human-, and environment-exploitive industries.

The plight of the American farm family has been described in the novel *A Thousand Acres*, by Jane Smiley (1991). A perceptive review of the book is provided by Bender (1998), entitled "What Is So Disturbing about Jane Smiley's *A Thousand Acres*?" As Bender (1998) states: "Much is troublesome in Jane Smiley's award winning novel. Subjects include loneliness, despair, domination, resentment, intimidation, incest, repression, anger, violence, ingratitude, selfishness, humiliation, among others. Especially startling is the setting—an American family farm—where we might expect more wholesome interaction among characters" (p. 153). The book is a critical appraisal of the popular assumption that family farming and sustainable agriculture are mutually reinforcing (Bender, 1998). The book provides caricatures of stereotypical farm families—where the dominant male is "Daddy" to his adult children, who are kept in line and allowed little true responsibility and decision making. Among the consequences is the maintenance of a static and environmentally unprogressive farming system. Introduction of new ideas by young family members is discouraged. These characteristics, exemplified by dominance and aggression, are typically androcentric, as described earlier in this chapter in the review of the paper by Schillo (1998). Animal agriculture, both academically (animal scientists) and on the farm, "values control and economic efficiency more than it values rural communities or sustaining the well-being of farm families and ecosystems" (Schillo, 1998, p. 2766).

The introduction of more feminine viewpoints into agricultural production could result in more environmentally sensitive and humane farming systems. Wells and Gradwell (2001) describe the role of gender in **Community Supported Agriculture (CSA)**. The CSA concept originated in Japan, when housewives, concerned with an increase in imported food and the loss of farmers and farmland, asked local farmers to grow vegetables and fruit directly for them. Farmers

agreed, if the women would pay "up-front." CSA has been introduced into the United States. Community members who participate in CSA commit in the spring to a full-season purchase of food, providing the money up-front. Farmers then know that their yearly expenses and income have been provided for, and they can concentrate on producing food. The members know where and how their food is grown and share a connection to the land and the farmers. In many cases, livestock have been incorporated into CSA programs. Wells and Gradwell (2001) reviewed CSA programs in Iowa. In Iowa, about two-thirds of the CSA growers are women, whereas regular commercial agriculture is dominated by men. Wells and Gradwell (2001) conclude: "Conventional agriculture is indeed a system socially coded as masculine. A more holistic system, one coded as feminine, would place value on cooperation, social relationships and connection, making a difference, future generations, nonhuman nature, and community . . . The emergence of CSA in Iowa and elsewhere signals a possible renewal of a smaller-scale, people-focused, nature-friendly, and community-based agriculture. The blurring of divisions between male/female and gardening/farming holds promise" (p. 118).

# OPPORTUNITIES ARISING FROM SOCIETAL CONCERNS

Some of the negative public perceptions of livestock production and meat that have been described in this chapter and others can offer opportunities to innovative people. Public aversion to "factory farming" and intensive livestock production presents marketing opportunities for products produced in a more "consumer-friendly" manner. The term **green pig** is used to describe a system of pork production that is "animal welfare and environmentally-friendly." This production system could include keeping sows loose-housed or outdoors, no early weaning, no castration or tail docking, straw-bedded pens for growing-finishing pigs, generous space allowance, no antibiotics or growth promoters as feed additives, welfare-considerate handling and loading facilities, and so on. Eggs produced by free-range hens or hens kept in more natural housing than battery cages are increasingly sought after by consumers. **Organic meat** produced by livestock not administered growth promotants and other "chemicals" is another opportunity, as is grass-finished beef with no cereal grains used. Many efforts in this direction have already been made. In Europe, supermarkets often offer a choice of several egg types, such as conventional, free-range, organic, and so on. These alternative products are by their nature unsuited for "factory farming," so these specialty market niches are likely to be filled by small-scale entrepreneurs. The limiting factor, of course, is economics. Any alternative to industrial production will by definition result in a more expensive product. Whether the public concerns are sufficiently strong to overcome the cost differential remains to be seen. Efforts to introduce "green" products into U.S. supermarkets have generally failed, after an initial flurry of interest.

Producing "pharmaceuticals" using genetically engineered animals might be another niche for small farmers. These animals might benefit from the close control and TLC (tender loving care) that individual farmers, especially women, could provide, and produce products of high economic value sufficient to sustain a family farm. However, the numbers of animals needed for producing pharmaceuticals is very small (Cunningham, 1999). For example, less than 10 animals are needed to meet the projected world requirements for some products (Table 10-4).

# PERCEPTIONS OF AGRICULTURAL AND ANIMAL RESEARCH

Another area of societal concern impacting the animal industry and the field of animal science is the public and peer perception of the quality and direction of **agricultural research.** In brief, agricultural and animal science research is often viewed by scientists in other fields as superficial, "barn-yard, feed'em and weigh'em research," and far-removed from the cutting edge of new developments in biology. From the other end of the spectrum, farmers, ranchers, and extension agents often view agricultural research as ivory-tower basic research without any practical application. The agricultural scientists in the middle often attempt to meet the criticisms coming from both sides and generally are unsuccessful in placating either group. A new factor has recently entered the equation: the mad scientist or Frankenstein factor. Many of the new developments in biotechnology involve agriculture.

**TABLE 10-4** Number of transgenic animals needed to produce the estimated world requirements for different pharmaceutical products (Adapted form Cunningham, 1999.)

| Product | Number of Animals Required | | |
| --- | --- | --- | --- |
| | Cattle | Sheep | Goats |
| Blood Coagulating Factor VIII | 1 | 1 | 1 |
| Blood Coagulating Factor IX | 1 | 1 | 1 |
| Glucocerebrosidase | 2 | 3 | 6 |
| Protein C | 2 | 3 | 6 |
| Antithrombin III | 3 | 6 | 12 |
| Fibrinogen | 17 | 45 | 83 |
| Human Serum Albumin | 35,000 | 93,000 | 175,000 |

These include transgenic plants and animals, genetically engineered organisms that are herbicide or pesticide resistant, cloning animals, and so on. Agriculture is on the cutting edge, along with medical science, of biotechnology and its impact on society. Many people are uneasy about these developments.

In the early 1970s, the agricultural research establishment of the United States was rocked by the publication of the book *Hard Tomatoes, Hard Times* (Hightower, 1973). This book was an intensive critique of the land grant college-agricultural experiment station complex of the U.S.[7] As stated in the preface of the book, "The message of the report is that the tax-paid, land grant complex has come to serve the elite of private, corporate interests in rural America, while ignoring those who have the most urgent needs and the most legitimate claims for assistance" (p. i). Those with the greatest need for assistance were identified as family farmers and farm laborers such as migrant workers. The title *Hard Tomatoes, Hard Times* refers to the genetic modification of tomatoes by agricultural scientists at the University of California so that the tomatoes could be picked mechanically. These "hard tomatoes" were reputed by Hightower to be inferior in eating quality, while "hard times" refers to economic hardships of farmworkers displaced by mechanical harvesting of crops. The essence of the dismay of Hightower (1973) about this development is that this research,

---

[7] In 1862, the U.S. Congress passed the Morrill Land Grant Act, providing an endowment of public land (or its monetary equivalent) to each state that would earmark the income from this endowment for the establishment and support of agricultural (A) and mechanical (M) colleges, for example, Texas A and M University. In 1887, the Hatch Act was passed, mandating that the land grant institutions conduct scientific research to improve the welfare of farmers and consumers. Hatch Act funds are a major source of support for the Agricultural Experiment Stations, which are now an integral part of land grant universities. Most professors at land grant universities have joint appointments, with teaching supported by state funds and research by federal (Hatch) funds.

paid for by the public, benefits only the corporate farmers who grow North America's tomato crop in California and Florida. Mechanical harvesting has displaced farmworkers, without any consideration of the social and moral costs of discarding these workers as throw-away, expendable "in-puts." It was claimed that the consumer suffers also, by being offered a tasteless, inferior tomato, and the only beneficiaries of the research are subsidized corporate farms. For example, Hightower (1973) writes, "The experiment stations at land grant colleges exist today as tax-paid clinics for agribusiness corporations, while others who need publicly supported research either benefit only incidentally, are not served at all, or actually are being harmed by land grant research. Land grant researchers are preoccupied with machinery, chemicals, systems and other gadgetry designed to assist agribusiness and to eliminate the human element from farming. As farms collapse, small town businesses follow, and small towns begin to die. Rural America is changing radically, irrevocably and for the worse. All America is affected. It is not demonstrated anywhere that the public wanted or expected these changes to come from its tax investment in mechanization research" (p. 44). An alternative point of view is that the hand harvesting of crops is hard, dirty work and that mechanical harvesting frees people from this drudgery to pursue a more rewarding life. The crux of Hightower's position is that our society has failed to assist people in making this transition. At one time, cotton was picked by hand, but now is harvested mechanically. America is still paying high social costs for failing to recognize that the farm workers who picked cotton were not expendable in-puts to be discarded and put out of mind as soon as machines were developed to replace them. The bill for these social costs will continue to be paid by many future generations of Americans.

Animal research was not immune to criticism in *Hard Tomatoes, Hard Times*, as revealed in the following quotations. Regarding chicken, "It is factory raised, indoor and immobile, chemically fed; and while I think the birds reach a kind of forced, physiological maturity, the only flavor they have is what will be absorbed from the cardboard and plastic wrappings they're presented in. Perhaps to compensate for the taste that they have taken out of meat, land grant scientists have added a cabinet-full of drugs. DES is a product of land-grant college research, and its wide use is a product of land-grant college promotion. DES is an example of land grant research at its worst—it is at once a service to industry, and a disservice to consumers. The advantage is all on one side—agribusiness, millions; folks, zero. It is an outrageous allocation of public resources" (pp. 83–84). Getting back to chickens: "Here is a disgusting example of all the problems we face in automated agriculture—poultry production. Chicken feed contains antibiotics to prevent disease that might be produced by crowding and stress. It also contains tranquilizers to prevent the upset chickens from eating each other. The chickens are also sprayed with pesticides. Other drugs, including arsenic and nitrofurans, are fed to them to increase their rate of weight gain. Before distribution, they are bathed in tetracycline or sorbic acid to extend their shelf life. They are lastly colored with a yellow additive to give their flesh an appetizing tint[8]. We do not want our chickens produced in this way at this cost."

Finally, these comments in *Hard Tomatoes, Hard Times*, about land grant university professors: "Behind these hallowed walls lies an almost unbelievable chaos. You learn that the term colleague means the S.O.B. down the hall, that the caste system is the accepted way of life, that tenure, paycheck, and career are the most important things in their life, that they are the most rested group you will ever meet, that they are without responsibility as they are 'experimental,' which in turn excuses them from being objective"[9] (p. 1).

---

[8] In fact, the yellow color is due to natural pigments (xanthophylls) included in the diets of poultry (see Chapter 5).

[9] For an amusing fictional look at land-grant universities and their professors, the book *Moo*, by Jane Smiley is a good read.

Needless to say, *Hard Tomatoes, Hard Times* caused a bit of a stir on land grant college campuses. However, over 30 years later, the issues raised are still with us and, in many respects, are intensifying. I am personally guilty of accepting several free breakfasts offered to faculty and students at my university to hear presentations on bST and what we university people should be doing to steam-roller the "gadfly activists" into submission. The breakfasts and presentations are, of course, sponsored by companies that hope to market bST. How cozy should universities and multinational drug companies be? Whom should publicly funded agricultural research serve?

The land grant system, which has served agricultural science and agriculture in the United States for over 100 years, is in need of reestablishment of a mission for the twenty-first century. In the late nineteenth and for most of the twentieth century, the land grant system was effective in teaching farmers new technology. Westendorf et al. (1995) reviewed the land grant mission and suggested that numerous modifications of the objectives and missions are needed if the land grant system is to retain its relevance to producers and consumers and to survive in the current American political climate. Various politicians have been advocating the dismantling of federally funded agricultural research. Agricultural scientists must face up to the need for restructuring or eliminating programs, development of regionalized programs, sharing resources between institutions, and focusing resources to strengths (Westendorf et al., 1995). We can no longer be all things to all people. The following quotation, cited by Vavra (1996) from the Western Council of Administrative Heads of Colleges of Agriculture in a statement on public land management in the West, is of relevance to the entire land grant mission:

> If Land Grant Universities are successful in identifying what society expects from its public lands and applies the land grant model to new issues, they can be as effective today as they were in changing America from a country of rural poor to

the well fed, industrial power of the 20th century. If they ignore public concern for new issues and listen only to their traditional client groups, they will find themselves increasingly at odds with the people Land Grant Universities are supposed to serve. If they refuse to acknowledge or are insensitive to societal changes, they will become irrelevant. (p. 1420)

Meyer (1993) suggested that land grant universities "are baffled by efforts to identify and address the challenges they confront" (p. 881). He recommends a change from emphasizing agrarian and production interests to an increased emphasis on the entire food chain, including food quality, environmental concerns and the rural-urban interfaces. Attention should focus on the whole topic of food from production through consumption, and the offering of undergraduate curricula that attract students from both urban and rural backgrounds. Whiting (1996), in a review of land grant institutions, likewise concluded "the land-grant institutions are confused about what they are" (p. 1755). He suggests a return to the "time-honored, original mission of serving people" (p. 1754). A return from what? A return to serving people rather than the shareholders of transnational food corporations is one example. However, an apparent trend is the formation of alliances between agribusinesses and land grant universities. The University of Arkansas has established the Poultry Science Center of Excellence, which is heavily funded by several Arkansas-based poultry companies. Cargill has formed a Higher Education Initiative designed to promote strategic business partnerships with several U.S. universities (*Feedstuffs*, Sept. 22, 1997, p. 15). The Cargill initiative will fund faculty internships and send students overseas to learn more about international agribusiness. The long-term effects of such arrangements on academic freedom, objectivity of research, and so on raise some concerns. What, if anything, have universities given up

when they have accepted these financial windfalls from multinational agribusinesses? An interesting account of how transnational corporations such as Cargill are moving to control world food production and distribution is provided by Kneen (1995a) in an article "The Invisible Grant. Cargill and Its Transnational Strategies" and in a book with the same title (Kneen, 1995b). Influencing university programs and the education of students appears to be another means that these corporations are using to control every step in the food production process. Should universities be willing participants in this? Should they negotiate with several companies to get the best deal?

Further discussion of the challenges facing animal agriculture and land grant institutions are provided by Braund (1995), Kunkel and Hagevoort (1994), and Zimbelman et al. (1995). The general thrust of these papers is that the status quo is unacceptable; without major changes in research, extension, and teaching, departments of animal science will become irrelevant to society's needs and will meet the natural fate of irrelevance.

## PERCEPTIONS OF AGRICULTURAL AND ANIMAL SCIENCE TEACHING

Departments of animal science (formerly animal husbandry) have traditionally emphasized the techniques of livestock production in their undergraduate curricula. Until fairly recently, most students had a farm background. Students have, for as long as I can remember, complained that the courses weren't practical enough, and they wanted more "hands-on" experience. A livestock judging team was highly desired. Farmers and ranchers, many of them alumni of the departments in question, complained about the curriculum: it wasn't practical enough, I didn't learn anything at college that was of any use when I went back to the farm, and so on. Colleges of Agriculture receive a lot of flak about their teaching programs, as well as their research.

Undergraduate education in agriculture was the subject of a major National Research Council symposium (NRC, 1992). The proceedings should be read by those with a particular interest in the future of teaching and curricula in agricultural science. Interestingly, the consensus of the symposium speakers and participants was overwhelmingly negative towards current curricula, but for entirely different reasons than those mentioned earlier. The bottom line is that Colleges of Agriculture are in danger of becoming totally irrelevant to the needs of society. In general, agricultural students are deficient in many of the skills that will be of paramount importance in the twenty-first century, including biological science, biotechnology, chemistry, and computer technology. Very few employers will be concerned whether or not animal science graduates are good livestock judges or have good sheep handling skills. The students that will be needed most are those most likely to think globally, to act creatively, to value diversity, and, above all, to be able to think. The emphasis will be on the educated person, rather than on the person trained in specific skills. The world is changing and developing so rapidly that training in a particular area quickly becomes obsolete. The educated person with a good foundation in science, with highly developed intellectual capacity and thinking abilities, never becomes obsolete. Curricula in animal science will have a greater environmental and ecological orientation, with a holistic approach to the management of natural resources.

Traditionally, animal science curricula have emphasized the presentation of scientific facts that are to be accepted by students without critical evaluation. Schillo (1997) is highly critical of this approach, claiming that it impairs the intellectual and ethical devel-opment of animal science students. Our goal in agricultural science education should be to teach individuals to think independently in an analytical and critical way (Schillo, 1997). Similar thoughts are expressed by Barkley (1995), who points out that agriculture today is surrounded by an enormous number of issues that require critical thought and evaluation. This is the environment that most of our graduates will be working in. A survey of graduates of Kansas State University College of Agriculture graduates revealed that over 95 percent of the respondents agreed or strongly agreed that communication, people skills, and problem solving were "important to me in my current position" (Barkley, 1995). An issue is a source of disagreement and conflict; otherwise it wouldn't be an issue. A contentious issue is presumably one that involves even more conflict than a mere issue! Animal agriculture is faced with many contentious issues. We've been blindsided by many issues—we haven't seen them coming. Animal science students need to acquire many facts, but they also need to learn how to use them in **conflict resolution**.

These changes, which in my view are desirable and inevitable, will be hard for the agricultural community to accept. In part, this is a conflict between value systems. The agrarian belief system, which has been the foundation of the land grant college system since its founding, is discussed further in Chapter 11. The inevitable changes in orientation of agricultural research and teaching, with a new global and environmental emphasis and a decrease in emphasis of traditional production agriculture, will likely be resisted by the agricultural community. It is hoped that university administrators will be up to the task of explaining to their constituency (stakeholders is the current buzz word) that they are not abandoning university programs in agriculture, but making them better.

# STUDY QUESTIONS

1. How would you define bioethics?

2. Is the raising, killing, and eating of other sentient animals by humans an ethical or moral issue, or both?

3. What is sentience?

4. Suppose you have a calf with a fatal illness, such as BVD (bovine viral diarrhea). The animal doesn't look too bad yet, but you are certain that it will die in a few months. Which of the following scenarios would you think would be the best way of dealing with the situation—(a) take it as soon as possible to a livestock auction and sell it, (b) let it die naturally, (c) kill it humanely, (d) take it to a veterinarian for possible treatment or to be euthanized?

5. Is it ethical to slaughter animals in view of other animals, or does it not matter because the others are just going to die anyway? Can the other animals (e.g., cattle) perceive impending death—do they really know what's about to happen to them? Even if they do, does it matter?

6. Why is a human life more valuable than that of a chimpanzee? Which life is more valuable: that of a chimpanzee or a cow? Why?

7. What is an ethologist?

8. It is obvious that animals feel pain when being dehorned, branded, or castrated. Does it matter if they experience pain? What alternatives are there to reduce or eliminate this type of painful experience?

9. What is an aviary system for housing laying hens?

10. What is the difference between the animal welfare movement and the animal rights movement? What is an animal rights extremist?

11. Mickey Mouse, Donald Duck, Peter Rabbit, and Bambi are examples of anthropomorphism. What is the role of anthropomorphism in animal welfare and animal rights debates?

12. Do animals have any rights at all? If so, what are they?

13. Do farm animals differ in intelligence? If so, does it matter? Explain.

14. When ducks are raised for meat, should they have access to water for swimming? Defend your answer.

15. Do you eat meat? If so, would you eat horse, dog, cat, rabbit, guinea pig, or rat meat, assuming that the animals had been raised under sanitary conditions? If not, why not? (These animals are all used as food sources in various cultures.)

16. Some people use the term "other animals" instead of "animals." What reasons might they have for doing this?

17. Assume you are a vegetarian. Is it ethical for you to eat a transgenic tomato that contains genes from cows? Explain.

18. What is the Frankenstein syndrome?

19. Is genetic engineering morally wrong? If so, should it be stopped? If so, by whom?

20. Suppose you are a farmer. On your farm there is a grove of oak trees; the trees are about 250 years in age. The grove occupies about one acre of land. You decide you want to increase the size of your tomato patch and will cut down the oak trees for firewood. The neighbors hear about your plans, and protestors chain themselves to some of your trees. You regard this as an invasion of your personal property—you have the right to do whatever you want with these trees because you own them and pay the property taxes. Discuss.

21. Compare liberal and postliberal positions on gender equality.

22. What accounts for the rapid increase in the number of female veterinarians?

23. Is "ethical vegetarianism" an oxymoron?

24. What is meant by the term "green pig"?

25. Are you a student at a land grant institution? Is the land grant concept obsolete? Explain.

26. Many universities are experiencing budgetary difficulties. One way of improving their financial situation is to form partnerships or arrangements with transnational corporations, such as the Cargill Higher Education Initiative. Discuss the advantages and disadvantages of this type of arrangement for the companies and for the university faculty and students.

27. Discuss the contention of Davis (2001) that a vegan diet is unethical because it results in the death of large numbers of "animals of the field."

28. If blind chickens are better and more efficient egg layers than normal chickens, do you think that the poultry industry, if left to follow efficiency and economics, would blind female chicks at hatching with a hot iron?

NIELSEN, N. O. 2001. Is the veterinary profession losing its way? Can. Vet. J. 41:439–445.

OELLER, P. W., L. MIN-WONG, L. P. TAYLOR, D. A. PIKE, and A. THEOLOGIS. 1991. Reversible inhibition of tomato fruit senescence by antisense RNA. *Science* 254:437–439.

ORTIZ, S. M. 2001. When sport heroes stumble: Stress and coping responses to extramarital relationships among wives of professional athletes. Annual Meeting, American Sociological Association, August 18–21, 2001, Anaheim, California.

PELL, A. N. 1996. Fixing the leaky pipeline: Women scientists in academia. *J. Anim. Sci.* 74:2843–2848.

POTRYKUS, I. 2001. Golden rice and beyond. *Plant Physiology* 125:1157–1161.

PROVENZA, F. D. 2000. Science, myth, and the management of natural resources. *Rangelands* 22:33–36.

PURSEL, V. G., C. A. PINKERT, K. F. MILLER, D. J. BOLT, R. G. CAMPBELL, R. D. PALMITER, R. L. BRIN-STER, and R. E. HAMMER. 1989. Genetic engineering of livestock. *Science* 244:1281–1288.

RALPHS, M. H., and J. D. OLSEN. 1992. Comparison of larkspur alkaloid extract and lithium chloride in maintaining cattle aversion to larkspur in the field. *J. Anim. Sci.* 70:1116–1120.

REGAN, T. 1983. *The case for animal rights.* University of California Press, Berkeley, CA.

REGOSIN, J. V., and M. FRANKEL. 2000. Conservation biology and western religious teachings. *Conserv. Biol.* 14:322–324.

RICHARDSON, N. J. 1994. UK consumer perceptions of meat. *Proc. Nutr. Soc.* 53:281–287.

ROBINSON, J. J., and T. G. MCEVOY. 1993. Biotechnology—The possibilities. *Anim. Prod.* 57:335–352.

ROLLIN, B. E. 1990. Animal welfare, animal rights and agriculture. *J. Anim. Sci.* 68:3456–3461.

ROLLIN, B. E. 1991. An ethicist's commentary on the case of the sow with a broken leg waiting to farrow. *Can. Vet. J.* 32:584–585.

ROLLIN, B. E. 1995a. *Farm animal welfare. Social, bioethical and research issues.* Iowa State University Press, Ames, Iowa.

ROLLIN, B. E. 1995b. *The Frankenstein syndrome. Ethical and social issues in the genetic engineering of animals.* Cambridge University Press, Cambridge, UK.

ROLLIN, B. E. 1996. Bad ethics, good ethics and the genetic engineering of animals in agriculture. *J. Anim. Sci.* 74:535–541.

ROLLIN, B. E. 2000. Equine welfare and emerging social ethics. *JAVMA* 216:1234–1237.

RUSHEN, J. 1991. Problems associated with the interpretation of physiological data in the assessment of animal welfare. *Appl. Anim. Behav. Sci.* 28:381–386.

RUSHEN, J., and A. M. B. DE PASSILLE. 1998. Behaviour, welfare and productivity of dairy cattle. *Can. J. Anim. Sci.* 78(Suppl.):3–21.

RUSZLER, P. L. 1998. Health and husbandry considerations of induced molting. *Poult. Sci.* 77:1789–1793.

SANDOE, P., B. L. NIELSEN, L. G. CHRISTENSEN, and P. SORENSEN. 1999. Staying good while playing God—The ethics of breeding farm animals. *Animal Welfare* 8:313–328.

SCHILLO, K. K. 1997. Teaching animal science: Education or indoctrination? *J. Anim. Sci.* 75:950–953.

SCHILLO, K. K. 1998. Toward a pluralistic animal science: Postliberal feminist perspectives. *J. Anim. Sci.* 76:2763–2770.

SCHNEIDER, B. J. 2001. A study in animal ethics in New Brunswick. *Can. Vet. J.* 42:540–547.

SCHONHOLTZ, C. M. 2000. Animals in rodeo—A closer look. *JAVMA* 216:1246–1249.

SEARS, M. K., R. L. HELLMICH, D. E. STANLEY-HORN, K. S. OBERHAUSER, J. M. PLEASANTS, H. R. MATTILA, B. D. SIEGFRIED, and G. P. DIVELY. 2001. Impact of *Bt* corn pollen on monarch butterfly populations: A risk assessment. *PNAS* 98:11937–11942.

SHAPIRO, L. 1999. *Applied animal ethics.* Delmar Thomson Learning, Albany, New York.

SHAW, D. 2001. Gender and the profession—An issue or not? President's message. *Can. Vet. J.* 42:12–13.

SINGER, P. 1975. *Animal liberation.* Avon Books, New York, (2nd Edition, 1995).

SINGER, P. 1979. *Practical ethics.* Cambridge University Press, New York, (2nd Edition, 1993).

SLATER, M. R., and M. SLATER. 2000. Women in veterinary medicine. *JAVMA* 217:472–476.

SMILEY, J. 1991. *A thousand acres.* Ballantine Books, Random House, Inc., New York.

SMILEY, J. 1995. *Moo.* Ballantine Books, Random House, Inc., New York.

SWANSON, J. C. 1995. Farm animal well-being and intensive production systems. *J. Anim. Sci.* 73:2744–2751.

TABE, L. M., T. WARDLEY-RICHARDSON, A. CERIOTTI, A. ARYAN, W. MCNABB, A. MOORE, and T. J. V. HIGGINS. 1995. A biotechnological approach to improving the nutritive value of alfalfa. *J. Anim. Sci.* 73:2752–2759.

TAUSON, R. 1998. Health and production in improved cage designs. *Poult. Sci.* 77:1820–1827.

THOMAS, J. W., and J. BURCHFIELD. 2000. Science, politics, and land management. *Rangelands* 22:45–48.

THOMPSON, P. B. 1997. Ethics and the genetic engineering of food animals. *J. Agric. Environ. Ethics* 10:1–23.

THOMPSON, P. B., R. J. MATTHEWS, and E. O. VAN RAVENSWAAY. 1994. *Ethics, public policy, and agriculture.* Macmillan Publishing Co., New York.

VALIAN, V. 1999. *Why so slow? The advancement of women.* The MIT Press, Cambridge.

VAVRA, M. 1996. Sustainability of animal production systems and an ecological perspective. *J. Anim. Sci.* 74:1418–1423.

VELANDER, W. H., H. LUBON, and W. N. DROHAN. 1997. Transgenic livestock as drug factories. *Sci. Amer.* 276(1):70–74.

WALL, H., R. TAUSON, and K. ELWINGER. 2002. Effect of nest design, passages, and hybrid on use of nest and production performance of layers in furnished cages. *Poult. Sci.* 81:333–339.

WARD, K. A., and C. D. NANCARROW. 1991. The genetic engineering of production traits in domestic animals. *Experientia* 47:913–922.

WEBSTER, A. B. 2000. Behavior of White Leghorn laying hens after withdrawal of feed. *Poult. Sci.* 80:192–200.

WEBSTER, A. J. F. 1994. Meat and right: The ethical dilemma. *Proc. Nutr. Soc.* 53:263–270.

WEBSTER, A. J. F. 2001. Farm animal welfare: The five freedoms and the free market. *Vet. J.* 161:229–237.

WEGNER, R-M. 1990. Poultry welfare—Problems and research to solve them. *World's Poult. Sci. J.* 46:19–30.

WELLS, B. L., and S. GRADWELL. 2001. Gender and resource management: Community supported agriculture as caring-practice. *Agriculture and Human Values* 18:107–119.

WESTENDORF, M. L., R. G. ZIMBELMAN, and C. E. PRAY. 1995. Science and agriculture policy at Land-Grant Institutions. *J. Anim. Sci.* 73:1628–1638.

WHITING, L. R. 1996. Challenges that face colleges of agriculture. *J. Dairy Sci.* 79:1754–1759.

WIEPKEMA, P. R., W. G. P. SCHOUTEN, and P. KOENE. 1993. Biological aspects of animal welfare: New perspectives. *J. Agric. Environ. Ethics* 6(Suppl. 2):93–103.

WILLS, P. R. 1996. Transgenic animals and prion diseases: Hypotheses, risks, regulations and policies. *N.Z. Vet. J.* 44:33–36.

WILMUT, I., A. E. SCHNIEKE, J. MCWHIR, A. J. KIND, and K. H. S. CAMPBELL. 1997. Viable offspring derived from fetal and adult mammalian cells. *Nature* 385:810–813.

WILMUT, I., A. L. ARCHIBALD, M. MCCLENAGHAN, J. P. SIMONS, C. B. A. WHITELAW, and A. J. CLARK. 1991. Production of pharmaceutical proteins in milk. *Experientia* 47:905–912.

WISE, J. K., and C.-L. ADAMS. 1999. Geographic distribution of female and male veterinarians in the United States, 1998. *JAVMA* 215:478–480.

YE, X., S. AL-BABILI, A. KLOTI, J. ZHANG, P. LUCCA, P. BEYER, and I. POTRYKUS. 2000. Engineering the provitamin A (β-carotene) biosynthetic pathway into (carotenoid-free) rice endosperm. *Science* 287:303–305.

ZIMBELMAN, R. G., D. K. WAGGONER, and D. B. WAGGONER. 1995. Changing paradigms in animal agriculture: Addressing animal issues in Washington by a professional society. *J. Anim. Sci.* 73:3178–3181.

ZUK, M. 2000. Review of: Valian, V. 1999. Why so slow? The advancement of women. The MIT Press, Cambridge, MA. *BioScience* 50:1021–1022.

# CHAPTER 11

# Livestock Integration into Sustainable Resource Utilization

**CHAPTER OBJECTIVES**

1. To introduce the concept of the biosphere and its regulation by biological processes.

2. To discuss the concept that agriculture and livestock production should be integrated into the biosphere as part of natural processes. This concept is the basis of the currently popular topics of sustainable agriculture and holistic resource management.

3. To present the idea that while a "new agriculture" is needed, giving greater attention to environmental and ecological concerns, this should be regarded as an opportunity rather than a threat. Changes forced by societal concerns have generally been improvements.

The objective of this chapter is to integrate the preceding chapters into a philosophical discussion of livestock production and the global environment. While some of the discussion may seem quite removed from livestock production, I hope my efforts to demonstrate its relevance will make some sense. The concept of Earth as a self-regulating biosphere is the framework upon which sustainable resource utilization rests.

It is now increasingly being recognized that our planet bears much resemblance to a living organism; in fact, it satisfies virtually all the definitions of life with the exception that it does not reproduce. This is the concept of the **biosphere**; that the planet is a self-sustaining system maintained by biological processes. It employs homeostatic mechanisms to modulate change. **Homeostasis** in biology refers to the constancy of the internal environment. For example, animals employ homeostatic mechanisms (e.g., eating, secretion of insulin, adrenalin, and glucagon, etc.) to seek to maintain a constant blood sugar level. When the level is changed, for example by fasting or eating, homeostatic changes in hormone secretion set into motion biochemical reactions to return the blood sugar to normal. What is normal is very characteristic for living tissues, and much different for what would occur if homeostatic mechanisms didn't exist. The concentrations of most chemicals in living tissues are far from equilibrium values. Consider that the concentration of chloride ions in hydrochloric acid in gastric juice in the parietal cells of the stomach lining is about a million times higher than in the plasma. Such concentration gradients far from equilibrium conditions are characteristic of life, and life processes to a considerable ex-

tent involve homeostatic mechanisms to maintain these chemically unnatural concentrations. Similarly, the composition of the atmosphere, oceans, and soil of the Earth are much different than would be the case for a lifeless planet; this unique composition of the biosphere is maintained by living, interconnected processes. These processes are now being impacted in a major way by human activity (Chapin et al., 1997; Matson et al., 1997; Vitousek et al., 1997). Most of these human impacts are negative, in terms of maintaining an environment consistent with a high quality of life.

The belief in the similarity of Earth to a living entity has been raised to a quasi-religious level by the scientist-philosopher James Lovelock who refers to the Earth biosphere as **Gaia**. This term is derived from the Greek Earth goddess Gaia or Ge (from which geology and ge-

## BIOSPHERE I AND II

Biosphere I is the planet Earth, on which we dwell. Obviously, Earth is a sphere, and its unique properties for life are in part sustained by biological processes (which we seem to be doing our best to disrupt): hence the biosphere. Biosphere II is the world's largest greenhouse, in Oracle, Arizona (Fig. 11-1). It was originally constructed as a prototype of a Mars space station with a closed ecological system that could maintain equilibrium and sustain life support for humans for a long period of time. Its construction was paid for by a wealthy individual. Biosphere II came under criticism by scientists for its lack of scientific rigor, and well-publicized goofs. For example, there was difficulty in growing enough food for a small group of people who

**FIGURE 11-1** Biosphere II, located in Oracle, Arizona. The world's largest greenhouse, it is a primitive attempt to build a self-sustaining, closed ecological system.

volunteered to live in it for several months, and there was a secret attempt to inject oxygen to correct an unexpected oxygen depletion. In 1995, management of Biosphere II was taken over by Columbia University, which is seeking to establish the complex as a state-of-the-art research and educational facility for environmental sciences.

The facility contains several ecosystems, including a rain forest, desert, savanna, marsh, ocean and an agricultural unit. Problems in getting Biosphere II to function as a self-regulating system help to demonstrate the complexity and exquisite functioning of Biosphere I and illustrate that when we disrupt these processes on our planet, we may be in for some unexpected, bad consequences. (A report on some of the soils research in Biosphere II is given by Torbert and Johnson, 2001.)

ography derive their names). In brief, the Gaia concept is that the composition of the planet's life support systems, the atmosphere, soil and oceans, is established, maintained, and controlled by biological systems responding to feedback mechanisms. According to Lovelock (1987), without life on Earth, the atmosphere would consist of 99 percent carbon dioxide and 1 percent argon. The actual composition, with 0.03 percent carbon dioxide, 78 percent nitrogen and 21 percent oxygen, can be maintained only by the presence of life. On a lifeless Earth, the oceans would contain 13 percent salt, rather than 3.5 percent. It is now being recognized that the mineral contents of the soil and oceans are regulated by **biogeological cycles**. For example, iodine and sulfur are conveyed from the oceans to the land by the volatile compounds methyl iodide and dimethyl sulfide produced by marine phytoplankton (Watson and Liss, 1998). Excess carbon dioxide in the atmosphere is converted by plankton growth in the oceans to calcium carbonate, which is deposited on the ocean floor (e.g., the origin of the white cliffs of Dover [Fig. 11-2] and other limestone deposits). The oceans are a reservoir of stored gases that are mobilized or deposited as needed by biological action to maintain a stable atmosphere and marine environment. The salinity of the sea has been stable for hundreds of millions of years (Lovelock, 1987) and is regulated by the formation of salt deposits. A silicon cycle is regulated by diatoms in the oceans. Marine cyanobacteria such as *Trichodesmium* fix nitrogen and evolve oxygen and play a major role in the global nitrogen cycle (Capone et al., 1997). Spawning salmon transport marine nutrients hundreds and even thousands of miles inland (see Chapter 7). On an extremely long-term geologic time basis, these cycles are connected with the exchange of minerals from the Earth's core to the mantle. Tectonic plate movements result in the Earth's mantle (cool outer layer floating on the hot core) being folded into the molten core, while volcanoes bring magma from the core to the mantle. An integral part of this

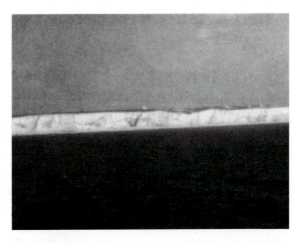

**FIGURE 11-2** The famous white cliffs of Dover. Limestone deposits such as these are a means of sequestering carbon dioxide in the form of calcium carbonate. A mist of insoluble calcium carbonate, formed by plankton at the surface, continually rains down to the ocean floor. The formation of these deposits is regulated biologically to modulate atmospheric carbon dioxide levels (Watson and Liss, 1998).

mineral cycling is biological activity. After molten lava cools, it is colonized by lichens, and the lengthy process begins of converting rock into soil by the biological activities (enzymatic action, acid secretion, rock splitting by roots, freezing and thawing of water in cracks made by roots, etc.) of lichens, algae, bacteria, fungi, and, ultimately, higher plants and animals. Thus on a time scale of hundreds of millions of years, minerals are cycled from the Earth's surface to the molten core and back to the surface. Biological homeostatic mechanisms modulate the changes in climate and soil and ocean composition caused by the volcanic eruptions. There is even a biologic involvement in geologic events deep in the Earth. There are **deep-living bacteria** in rocks thousands of meters beneath the Earth's surface (Gold, 1992). They dissolve rock (on an immensely slow time scale) by producing organic acids from their metabolism of carbon sources such as petroleum and coal deposits. Their activity creates chambers and channels in which groundwater and petroleum deposits occur. These deep-living microbes could have entered their harsh environment through shallow aquifers and percolating groundwater or may have been entombed with other biological material when the rock was deposited tens or hundreds of millions of years ago (Appenzeller, 1992). Deep-living microbes may be the product of millions of years of evolution deep in rock in the Earth's crust and contribute to biogeochemical and water cycles of the planet. It is possible that other planets such as Mars have similar life forms.

The concentration of greenhouse gases in the atmosphere is currently increasing, primarily because of human activities (see Chapter 6). According to Gaia theory, this will lead to increased plankton growth in the oceans, homeostatically lowering atmospheric carbon dioxide. Climate phenomena such as El Niño may have a role in modulating atmospheric conditions. During an **El Niño** event, natural warming of many areas increases plant growth in forests and grasslands, causing the removal of more $CO_2$ from the atmosphere. It needs to be appreciated that many of the Earth's homeostatic mechanisms operate on geologic time. Ice ages, global warming, glaciation, and so on occur in cycles involving tens of thousands of years. This is why the human impact on the Earth's environment now is so critical; "Mother Nature" or "Gaia" will respond to our disruptions with homeostatic reactions, but the adjustment may take several thousand years. **Mass extinctions** associated with climate change, such as the extinction of the dinosaurs, have occurred numerous times in Earth's history. Global climate changes have been common. Massive amounts of carbon dioxide have been taken out of the atmosphere in the past, to form coal and petroleum deposits. We are busy putting this carbon dioxide back into the atmosphere, in addition to the nitrous oxide arising from chemical fertilizer (Chapter 6). Nature will respond and adjust to our actions. The critical question is whether or not we can adjust. Since it is estimated that about 99 percent of all species of life that evolved during Earth's history have become extinct, the long-term future of humanity is by no means assured.

# MONOCULTURE AND THE LOSS OF BIODIVERSITY

**Monoculture** is a term originally referring to the widespread production of a single crop such as maize (yellow corn) in the U.S. corn belt or cotton in the southern United States. Increasingly, it has come to imply the production of a genetically uniform crop, such as hybrid corn. The advantages are economic, relating mainly to high yield. The potential disadvantage is that large areas of a genetically uniform crop could lead to devastating plant disease epidemics. At one time, farmers in North America and Europe saved their own seed for next year's crop. Thousands of different varieties of common crops existed, and hundreds of livestock breeds were developed. As intensification of agriculture occurred, many of these varieties and breeds fell by the wayside, and for most major crops in temperate countries there now exist only a few varieties. The same trend is occurring in the tropics. Thousands of varieties of common crops such as beans, maize, and rice exist, with each locality and in many cases each farmer having a locally adapted variety selected over many generations. One of the concerns about the **green revolution,** which involved the introduction of high-yielding varieties of rice (Fig. 11-3) and wheat and new techniques to maximize their yield, is that these locally adapted varieties are rapidly becoming extinct. The significance of this loss of genetic diversity is that some of these varieties may have resistance to pests and diseases or possess other valuable traits that may be useful in the future. Many efforts are being made to preserve genetic diversity of tropical crops, including the construction and maintenance of gene banks where the varieties are maintained by tissue culture techniques. Also of concern is evidence that yield increases obtained in the green revolution may decline over time (Matson et al., 1997).

In the case of livestock, there has been a similar loss of genetic diversity (NRC, 1993). Breeds of livestock were developed by selection under local conditions (Fig. 11-4). In the developed nations, local breeds have been

**FIGURE 11-3** The "green revolution" involved the development of high-yielding strains of rice, which are dependent on inputs of fertilizer and pesticides. Critics have claimed that the green revolution is not sustainable because of its dependence on expensive and non-renewable resources. (Courtesy of Robert A. Swick.)

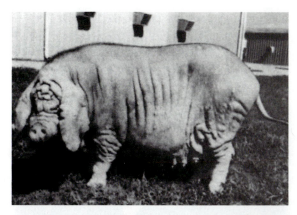

**FIGURE 11-4** Chinese breeds of pigs such as the Meishan (*top*) have various desirable traits, including high reproductive capacity and the ability to survive on non-concentrate diets, which may be valuable in swine breeding programs, despite the apparent lack of potential as judged by the animals' appearance. Similarly, indigenous African cattle breeds, such as the Ngumi (*bottom*) are much better adapted to African climates, diseases, and feed resources than are imported European breeds. (Top photo courtesy of L. D. Young.)

pushed aside by a few highly productive breeds. The dairy industry is now dominated by the Holstein-Freisan cow, and to a lesser degree, the Hereford breed has dominated beef production. Numerous organizations, both public and private, have been formed to preserve the less common breeds of livestock. Examples are the American Livestock Breeds Conservancy (USA)[1] and the Rare Breeds Survival Trust (UK). The Food and Agriculture Organization (FAO) of the United Nations has established a Global Animal Genetic Data Bank in Germany, regional animal gene banks in Africa, Asia, and Latin America, and a World Watch List and Early Warning System to identify **endangered breeds** of livestock (Hodges, 1991). Various biotechnological techniques can be and are used for germ plasm preservation. They include the preservation of sequences of catalogued DNA, genome mapping, collection and storage of semen and embryos, embryo cloning, and the preservation of live animals of rare breeds. An example of the potential of cloning was illus-

trated by Wells et al. (1998). A unique population of cattle, adapted to the harsh sub-Antarctic conditions on Enderby Island, the Enderby Island breed, was reduced to a single cow (Fig. 11-5). There were frozen semen samples from nine deceased bulls. Cells from the surviving cow were used via adult somatic cell nuclear transfer to produce a number of clones. Several live calves have been produced. Preservation of somatic cells from rare and endangered

[1] The American Livestock Breeds Conservancy, P.O. Box 477, Pittsboro, NC, 27312.

large areas of Africa because of the tsetse fly infestation has been a significant factor in preservation of forests and wildlife by limiting human encroachment (Matzke, 1983). The extensive use of artificial insemination (AI) in the dairy industry has led to a reduction in genetic variability. In some ways, the use of AI and embryo transfer are analogous to monoculture production of hybrid corn. In each case, highly productive, genetically uniform germplasm is sought. If you have a highly productive Holstein cow, why not clone her and have hundreds of identical highly productive cows? The use of AI in the dairy industry in the United States has dramatically reduced the number of bulls used. One Holstein bull, Mandigo, has produced 1 million units of semen and has sired over 45,000 calves. Limiting the genetic diversity could create havoc through expression of genetic diseases that are expressed in the homozygous state. For example, a fatal immune deficiency disease of Holsteins has been traced to a genetic disorder in a bull widely used in the 1950–1960 era (Husten, 1992). The disorder increased in incidence in the 1990s, because it takes several generations of inbreeding for the incidence of the homozygous state to become apparent.

**Losses of biodiversity** in wild populations of plants and animals, as well as species extinction, are major concerns with the increasing human impact on the environment. The loss of tropical rain forests is an example, resulting in species extinction, as well as in fragmentation of ecosystems and reduction of populations to remnants. Genetic diversity in a population is necessary for a species to be able to adapt to changing environmental conditions. Human impacts are resulting in a rapid acceleration of environmental changes (e.g., global warming) while at the same time diminishing the ability of many species to respond to these changes by reducing their genetic variation. The subject of biodiversity has been dealt with in detail by Wilson (1988).

# FORWARD TO THE PAST: SUSTAINABLE OR ALTERNATIVE AGRICULTURE

**Alternative agriculture** is defined by NRC (1989) as any system of food or fiber production that systematically pursues the following goals, as compared to conventional agriculture:

- more thorough incorporation of natural processes such as nutrient cycles, nitrogen fixation, and pest-predator relationships into agricultural production
- reduction in the use of inputs with the greatest potential to harm the environment or the health of farmers or consumers
- greater use of the biological and genetic potential of a diversity of plant and animal species
- long-term sustainability of agricultural techniques
- emphasis on conservation of soil, water, energy, and biological resources

To this list I would add the following:

- upholding of agrarian values. Alternative agriculture implies an emphasis on the family farm and traditional agrarian ethics. Most people interested in alternative agriculture would probably agree that a corporate farm owned and operated by industrialists should not be considered as alternative agriculture, even if some of the preceding criteria were met.

Other characteristics of sustainable agriculture (Fig. 11-7) noted by Lasley et al. (1993) include preservation of farm traditions and farm culture, cooperation rather than competition among farmers, and maintenance of the integrity of small communities. Lasley et al. (1993) characterize industrial agriculture as "motivated by self-

**FIGURE 11-7** Sustainable agriculture involves erosion-minimizing tillage and cultivation, trees and wildlife, and crop rotations with forages grown for livestock production. (Courtesy of D. Castaldo.)

interest, and lack of interest in farm traditions and rural culture. In this vein, small communities are not viewed as central to agricultural production. Farm work is viewed as drudgery and thus should be replaced by machinery and technology. In the industrial model, the emphasis is on speed, quantity and profit" (p. 134). It is noteworthy that one of the common measures of agricultural development is the rate of exodus of farmers; the conventional view is the faster, the better. It is also worth noting that most agriculture has not been sustainable. Great civilizations have risen on the strength of their agriculture and subsequently collapsed because their farming methods eroded or destroyed the natural resource base (Horrigan et al., 2002).

Some examples of techniques that characterize alternative agriculture are the following:

- crop rotations to reduce weed, disease, insect and other pest problems, to increase soil nitrogen and reduce chemical fertilizer use, and to reduce soil erosion
- integrated pest management (IPM) techniques such as crop rotations, pest and weather monitoring, use of resistant cultivars and biological control agents to control pests, with the overall objective of reducing pesticide use
- animal production systems that emphasize disease prevention through health maintenance, reducing the need for antibiotics and other drugs

The terms "sustainable" and "alternative agriculture" imply that current agricultural techniques may not be optimal for the long term. The implication is that maximum yields obtained with high external inputs of chemical fertilizer, petroleum energy in the form of fuel and agricultural chemicals, and so on may not be consistent with long-term sustainability of the agroecological system. Common perceptions are that modern agriculture is oriented towards corporate interests and incentives rather than rural families and traditional values. Further, it is perceived to have negative effects on the environment (e.g., soil erosion, groundwater pollution), produces unsafe foods with feed additives, pesticide residues, and so on and (in the United States) is driven by government policies that favor corporate farming over traditional farming. Opponents of "agribusiness" often claim that alternative agriculture could compete against corporate farming "on a level playing field." For example, the NRC (1989) report on alternative agriculture concluded:

A wide range of federal (U.S.) policies, including commodity programs, trade policy, research and extension programs, food grading and cosmetic standards, pesticide regulation, water quality and supply policies, and tax policy, significantly influence farmers' choices of agricultural practices. As a whole, federal policies work against environmentally benign practices and the adoption of alternative agricultural systems, particularly those involving crop rotations, certain soil conservation practices, reductions in pesticide use, and increased use of biological and cultural means of pest control. These policies have generally made a plentiful food supply a higher priority than protection of the resource base.

A key to high yield, sustainable agriculture is the substitution of knowledge for current large-scale inputs of chemicals and energy. The "information age" with its use of computer technology and instant communications has been a crucial development. With weather satellites, detailed soil mapping, global positioning satellites, and so on, **precision farming** has become a reality. Fertilizing, pest control, and other procedures can be accurately pinpointed to precise sites on individual fields on individual farms, using the **Global Positioning System (GPS)**. The GPS is a network of 24 satellites, which can provide precise locations on earth within a meter. This "new agriculture" can build on natural processes and work with them, rather than attempting to subdue nature with a heavy hand.

Heitschmidt et al. (1996) examined the long-term sustainability of animal agriculture in an ecological context. They define an **ecosystem** as an assembly of organisms and their associated chemical and physical environment, which can include anything from a goldfish bowl to a ranch to the entire world; it can be anything for which we have defined boundaries. Ecosystems contain abiotic (nonliving) and biotic components. In agricultural ecosystems, **abiotic components** include such things as climate, atmosphere, and soil. **Biotic components** include producers, consumers, and decomposers. **Producers** capture solar energy. They include crops and forages. **Consumers** are organisms that obtain their energy from eating other organisms; they include herbivores and carnivores. **Decomposers** are microorganisms such as bacteria and fungi that decompose the wastes of consumers to recycle them back into the food chain. **Food chains** are energy-processing pathways that determine the energy flow through an ecosystem. Regulation of energy flow is governed by the **First and Second Laws of Thermodynamics**. The First Law of Thermodynamics essentially states that energy can be transformed from one form to another (chemical, mechanical, electrical) but cannot be created or destroyed. The Second Law states that no transformation between energy forms is 100 percent efficient. In ecosystems, about 90 percent of the energy is lost each time the energy is transferred from one organism to another (i.e., from one trophic level to another). For example, when rumen protozoa eat rumen bacteria, there is a loss of most of the energy that was potentially available to the cow from digesting the bacteria. When humans consume chicken meat or pork produced by feeding corn-soy diets, we lose about 90 percent of the energy that would have been available if we had eaten the grain directly, without going through another trophic level. This is the concept of **eating low on the food chain**. The inefficiency of feeding high quality foods to livestock and poultry is a major, and legitimate, criticism of modern animal production. By contrast, the village chickens or pigs in many tropical countries feed almost exclusively on materials not directly useable or desired by humans (table scraps, insects, small wild mammals, etc.).

Heitschmidt and coworkers (1996) are at the Fort Keogh Livestock and Range Research Laboratory at Miles City, Montana. The activities of this USDA facility involve beef cattle research. This (to me, anyway), makes it even more noteworthy that Heitschmidt et al. (1996) conclude: "Results pointedly reveal the high level of dependency of the U.S. beef cattle industry on fossil fuels. These findings in turn bring into question the ecological and economic risks associated with the current technology driving North American animal agriculture" (p. 1395). They indicate that "the fundamental characteristic of sustainable animal agriculture systems must be that animals act as 'energy brokers'; that is, they convert low-quality human feedstuff (e.g., corn stalks, spoiled grains, waste products, etc.) into high-quality human feedstuff for their consumption (e.g., meat, milk, eggs, etc.)." Oltjen and Beckett (1996) point out that calculations of efficiency of conversion of feedstuffs into animal products, particularly for ruminants, should be based on **human-consumable feedstuffs** only. For example, if cattle are fed on cornstalks with a supplement of grain, the efficiency of conversion should be based on the grain only, because the cornstalks are not consumable by humans. When calculated on this basis, dairy production efficiency can exceed 100 percent (Oltjen and Beckett, 1996). The efficiency of beef production is largely dependent on the feedlot phase. It is possible to produce beef entirely from products inedible to humans. At the present time, it is uneconomical to do so in North America, because of the abundant supplies of grain and other high quality feedstuffs.

As reviewed by Vavra (1996), it is well known by scientists, economists, and informed people in general that our society is using natural resources at an unsustainable rate, ensuring that future generations will inherit a much depleted and degraded resource base. He suggests that sustainability may be a goal, like universal liberty and equality, that may not be reached but is a direction which guides constructive change. In some respects, livestock producers may be in front of animal scientists on these issues. Vavra (1996), who is at the Eastern Oregon Agricultural Research Center, describes the input of livestock producers (ranchers) on the Center's advisory committee: "Their suggestions for research do not place a high priority on increased weaning weights, heifer development, or growth-stimulating hormones. Rather, their plea has been for research directed at the environmental effects of livestock production, because that is perceived as the biggest threat to their future" (p. 1421). In rangeland and grazing issues, at least, the lack of interest and involvement of animal scientists has given (by default) the decision making to scientists in other disciplines (e.g., plant ecology, forestry, fisheries) who may have little knowledge of, or interest in, livestock production. According to Vavra (1996), on public lands in the West, new grazing systems are being implemented by the Forest Service and Bureau of Land Management, and "to my knowledge, neither federal agency employs even one animal scientist" (p. 1421).

The rise of industrialized agriculture has been accompanied by its critics, some of whom appear to have been vindicated decades later. In the 1930s and 1940s, a retired USDA agronomist, **Edward Faulkner,** proposed that one of the major causes of soil erosion, the 1930s Dust Bowl, and the need for high levels of fertilization of crops was the use of the moldboard plow (Faulkner, 1943). Faulkner mused on the common observation that weeds and other vegetation growing in fencerows and other undisturbed areas were often more vigorous than the adjacent crops, where the land had been plowed and cultivated, fertilized, and subjected to other agricultural practices. Faulkner recognized that in natural environments, the soil has a definite biological profile. Organic debris at the surface is broken down by bacteria and fungi and gradually incorporated into the upper layers of soil by the action of insects and earthworms. The soil develops **tilth,** with an open, friable structure, highly water absorbent and resistant to erosion. The friable, non-compacted soil favors abundant growth of feeder roots and rootlets and provides a supply of oxygen and nitrogen in the upper soil layers for nitrogen-fixing bacteria and algae to flourish, including the *Rhizobia* contained in root nodules of legumes. Decomposing organic matter produces organic acids that release minerals bound to rock and clay particles, providing plants with mineral nutrients. Under these natural conditions, **Mother Nature** (for want of a better term[2]) is more productive than modern farmers. Indeed, native vegetation,

---

[2] The term "Mother Nature" is of course widely used in everyday speech but may have special relevance in our relationship with the biosphere and the concept of Gaia, as discussed earlier in this chapter. There is also some relevance of referring to nature in the feminine nurturing sense, as opposed to the masculine, domination-of-nature sense.

including forests, accumulates more biomass than do agricultural crops, produced with intensive inputs. For example, according to Vitousek et al. (1986),

> average NPP (net primary production) in agricultural systems is lower than in the natural systems they replace, largely because most plants in agricultural systems are annuals. Longer-lived or perennial crops, multiple cropping, nutrient subsidies (fertilizers), and especially irrigation can offset or reverse this difference in some settings, but **traditional agriculture almost always produces less than natural systems** [emphasis mine].

Webb et al. (1983), in an extensive review of natural ecosystems, also documented the high productivity of nature as compared with modern agriculturalists. **Net primary production** is the amount of photosynthetically trapped energy that is produced by plants after subtracting the energy used in plant respiration. Perhaps it may be humbling to agriculturalists to ponder on the fact that with our big tractors and equipment, fertilizer, herbicides and pesticides, and other inputs of modern farming, our crops are producing less carbohydrate than a wild, natural forest on the other side of the fence, often growing on poorer soil. Of course we can't eat wood, but maybe we can, by close observation, learn something about growing crops from "Mother Nature."

Faulkner (1943) in his classic *Plowman's Folly* closely observed the natural formation of soil and concluded that the moldboard plow (which lifts up a layer of soil and turns it upside down) was in direct conflict with the natural processes. The tilth is destroyed, and the upper layer of organic matter is buried to form a layer that inhibits the natural capillary movement of water in soil. The plow exerts tremendous pressure, which creates an impervious layer (hardpan) at the bottom of the furrow. This hardpan layer restricts movement of soil water and plant roots. Agricultural engineers attempted to solve this problem by developing subsoilers, which in essence are longer plows. They break up the

hardpan but may create a new layer of compaction further down. Little (1987) described the action of the plow on the soil in these terms:

> All of this (the biological soil profile) is violently disrupted by a single pass of the moldboard plow. The only human-scale analogy that comes to mind is the awesome destruction that obtains from the geological shift of great tectonic plates, producing an equivalently violent wrenching of the earth that ruptures highways, disconnects gas and water mains, topples bridges, and buries neighborhoods in rubble. Just as the whole infrastructure of the human community can be disrupted by such gigantic seismic events, so the smaller-scale, but no less important, infrastructure of the soil community is catastrophically damaged and deranged by the tearing, slicing action of the plow, which not only cuts through the soil but lifts it, turns it, and dumps it upside down.

Faulkner's ideas were not well accepted by many of his contemporaries. However, over the intervening years, agriculture has gradually moved in the direction Faulkner suggested. Various types of **conservation tillage** (no-till) have been developed to minimize mechanical disruption of the soil (Fig. 11-8). In some no-till systems, cultivation is replaced with herbicides. A new crop is simply planted in the stubble left by last year's crop, and weeds are controlled chemically. This **chemical tillage** is criticized by some as increasing the use of and dependence on herbicides. Conservation tillage has increased rapidly in the grain-producing areas of the United States (Hendrix et al., 1986), resulting in lower fuel costs and improved soil and water conservation. Hendrix et al. (1986) compared the effects of plowing and conservation tillage on biological processes in the soil. No-till systems increase the role of fungi relative to bacteria as the primary decomposers of organic matter, whereas plowing favors disturbance-adapted bacteria with high metabolic rates that cause more rapid breakdown of organic matter and greater mobility

**FIGURE 11-8** Soybeans planted in wheat stubble, using a no-till or chemical tillage system. Weeds are controlled by herbicides rather than by cultivation. No-till or conservation tillage reduces soil erosion and preserves the soil profile. (Courtesy of T. McCabe and USDA Soil Conservation Service.)

(and thus potential leaching) of minerals. Thus Edward Faulkner seems to have been posthumously vindicated.

In 1962, **Rachel Carson** in her book *Silent Spring* ushered in the ongoing era of concerns about agricultural chemicals. While probably overdramatizing the situation ("Silent spring" refers to the silence of the countryside because of the loss of most of the songbirds to spraying of DDT, 2,4-D, and other agricultural chemicals), Rachel Carson's book brought worldwide attention to possible negative effects of the use of pesticides and herbicides and may have provided much of the impetus for the development of more environmentally friendly chemicals. Birds at the top of the food chain, such as raptors (hawks, eagles, owls, etc.) were especially affected by chemicals that bioaccumulate. Populations of these and other birds have recovered or are recovering following the banning of DDT in the United States. The "silent spring" without songbirds predicted by Rachel Carson is of concern now, not because of agricultural chemicals, but because of habitat loss. Many of the **neotropical songbirds** that migrate to North America in the summer are becoming scarce, because of the loss of their winter habitat through tropical deforesta-

tion in Central and South America (Tangley, 1996). Marra et al. (1998) observed that the quality of the tropical winter habitat of the migratory American redstart, a songbird, influenced its breeding success in the North American breeding grounds the following summer. The conversion of tropical rain forest to cattle pasture (Chapter 6) may thus affect songbirds in Iowa, for example. Global resource use is like pushing on a balloon—push on it somewhere and it will bulge out somewhere else! Clear-cutting Costa Rica can affect songbirds in Iowa, while monocultures of corn in Iowa cause soil erosion and elevated nitrogen levels in the Mississippi and Missouri Rivers that disrupt ecological function in the waters of estuaries in Louisiana and in the Caribbean Sea (Fig. 11-9).

Another person ahead of his time was **Robert Rodale**, the founder of the magazine *Organic Gardening and Farming*. Many of the concepts behind "organic" methods are sound. The importance of organic matter in maintaining soil structure and fertility is well-known. The emphasis on avoiding agricultural chemical use in organic methods cannot be criticized on biological grounds. It is not likely that anyone has suffered from a deficiency of herbicide or pesticide residues. Likely there will be a gradual fusion of organic methods with conventional agriculture, increased conservation tillage, integrated pest management, and more attention to soil microbial processes.

Western civilization for at least the last several hundred years has been exploitive of the natural environment, particularly in the New World. The Earth's riches seemed inexhaustible. Forests were felled, rivers dammed, mountains mined, rangelands overgrazed, and fisheries depleted. Humans have a long history of overexploitation of natural resources on a non-sustainable basis. Our ethic has been one of dominion over nature and subduing the wilderness (Table 11-1). The wilderness has been an enemy to be conquered. Instead of viewing ourselves as part of a sustainable ecosystem, we have viewed the natural world as a commodity for our use.

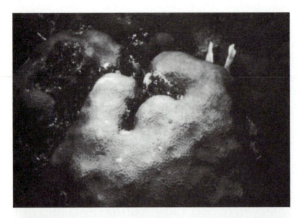

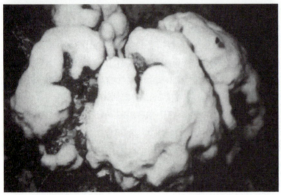

**FIGURE 11-9** (*Top*): A normal coral reef in the Caribbean Sea. Coral consists of a symbiosis between a small animal, and blue-green algae. (*Bottom*): A coral reef showing bleaching due to death of the algal component, believed to be caused by nitrogen and phosphorus pollution. When this bleaching occurs, the coral dies. Intensive livestock production and grain monoculture in the Midwest could be linked with the health of the Caribbean Sea, via transport of nutrients down the Mississippi River. (Courtesy of Raymond L. Hayes.)

We look at resources in terms of "what use is it?"[3] This ethic is under increasing challenge.

---

[3] Aldo Leopold (1949) had this answer to the question "What use is it?":

The last word in ignorance is the man who says of an animal or plant: "What good is it?" If the land mechanism as a whole is good, then every part is good, whether we understand it or not. If the biota, in the course of aeons, has built something we like but do not understand, then who but a fool would discard seemingly useless parts? To keep every cog and wheel is the first precaution of intelligent tinkering. (p. 190)

**TABLE 11-1**  Adverse environmental impacts of human action (From Cohen, 1997.)

| |
| --- |
| Loss of topsoil |
| Desertification |
| Deforestation |
| Poisoning of drinking water |
| Oceanic pollution |
| Shrinking wetlands |
| Overgrazing |
| Species loss (loss of biodiversity) |
| Loss of wilderness |
| Shortage of firewood and other fuel |
| Siltation of rivers and estuaries |
| Urbanization (human encroachment) of arable land |
| Declining water tables |
| Erosion of atmospheric ozone |
| Global warming |
| Rising sea levels |
| Depletion of mineral resources |
| Nuclear wastes |
| Acid rain |

In the United States, business as usual will not continue. In the Pacific Northwest, where I reside, one of the world's greatest remaining areas of temperate rain forests is a battleground of loggers and environmentalists. Virtually everyone must be aware of the controversies surrounding the loss of tropical rain forests. How can the world's rapidly increasing human population be sustained without creating an environmental wasteland?

In the Pacific Northwest, a concept of **new forestry** is being promoted (Norse, 1990). The present system of clear-cutting forests (Fig. 11-10), which is basically a slash and burn system, might be replaced with selective harvesting and multiple use of forestland for many purposes, including timber, recreation, medicinal plants, grazing, watershed management, fisheries, Native American religious rites,

**FIGURE 11-10** The old forestry. (*Top*): In the U.S. Pacific Northwest, traditional logging has involved clear-cutting all trees and replanting a monoculture. The so-called new forestry is intended to be more environmentally friendly, with selective cutting and maintenance of biodiversity. The one tree remaining in the photo meets present guidelines for leaving a few trees uncut. With the surrounding trees that once formed an unbroken canopy removed, it will likely snap in two with the next wind storm. This particular scene is next to a major highway in western Oregon. The public, inflamed by such scenes, is demanding the adoption of less destructive and exploitive activities in forestry and agriculture. (*Bottom*): Trees in an old-growth temperate rain forest in British Columbia are several hundred to over 1,000 years in age. These forests were clearcut in a few decades of the twentieth century, in an arrogant display of resource plundering. Why is our generation so important and our needs so great that we think we don't need to save any of these resources for future generations?

and a reservoir of plant and animal biodiversity. This contrasts with the current system of clear-cutting ancient forests and replacing them with monoculture plantations of genetically uniform Douglas fir. Clear-cutting a forest is the easiest and "most efficient" way of harvesting timber, and the most profitable system for the multinational timber companies. Perhaps it is not the most sustainable and the best system for the long-term health of the planet. Are today's broiler houses, managed to be the "most efficient" and most profitable for multinational poultry integrators, the agricultural equivalent of clear-cutting?[4] Similarly, drift nets on the high seas are an "efficient" means of catching fish, leading to rapid overfishing and depletion of fish stocks. Industrial agriculture, forestry, and fishing do not have good environmental track records.

Of what value is **biodiversity**? The subject has been dealt with extensively by Wilson (1988). While many reasons can be advanced to support efforts to preserve species and maintain biodiversity, there is always the practical matter that it may benefit us. For example, the Pacific yew is a scrubby, undistinguished looking tree in the Pacific Northwest forests that was considered worthless. It is now of

---

[4] In 1996, heavy winter storms hit the Pacific Northwest. The following is from a Corvallis *Gazette-Times* (Jan. 30, 1997) article:

> On February 6, 1996, a massive storm slammed into 150 years of road building, clear-cutting and development in the Pacific Northwest. In four days, 11 inches of rain tore tons of mud, rocks and debris from exposed hillsides and roads. The lethal mix roared from steep canyons . . . And when it was over, our belief that we could forever bend natural processes to our own desires collided head-on with reality.

Are these comments relevant to monocultures of corn and soybeans in the Midwest? Is the health of the Gulf of Mexico dependent upon farming practices in Iowa? Is there a message for industrial agriculture from industrial forestry?

great interest as the source of the anticancer drug taxol (Rowinsky et al., 1990). Another example is the thermophilic (heat-loving) bacteria that live in hot water in Yellowstone National Park (YNP). Some of these microbes can even grow in boiling water. Recently, microbes that live in high-temperature environments have been found to have numerous industrial applications. Significant efforts are underway to identify potentially useful microbes present in thermal areas of YNP (Chester, 1996). Commercial use of these biological resources raises new questions, including the important aspect of financial compensation. Should national parks, private landowners, developing countries, and so on receive compensation for genetic resources? **Thermophilic bacteria** are an example of the value of protecting biodiversity—their potential commercial value (and even their very existence) could not possibly have been foreseen when YNP was established. The ultimate question concerning biodiversity is whether we have the moral right to decide which creatures, the end-products of 3.5 billion years of evolution, should be exterminated for our convenience.

Upon reflection, we can readily appreciate that humans greatly value biodiversity. We go to great lengths to develop new varieties of flowers and trees and to landscape our homes with a wide diversity of plants. We have pets and companion animals, in which we value diversity in appearance and behavior. We appear to have a great affinity for other biological organisms. Wilson (1984) has even coined the term **biophilia** to mean a deep-seated, instinctual desire of humans to interact with other organisms. The development of many different breeds of domesticated animals and many varieties of plants, and even the concept of having ornamental plants and pets, reflects our desire to be in intimate contact with other forms of life. Our fascination with the possibility of extraterrestrial life, and the enthusiasm for the search for life on Mars, is a manifestation of biophilia. Thus our current agricultural infatuation with crop monocultures and genetically uniform factory-farmed animals seems inconsistent with our basic na-

ture. In fact, there seems to be an innate desire of people to have domestic animals, accounting for the huge numbers of pets and companion animals kept in our society. One of the negative aspects of factory farming is that it squeezes people out of the opportunity to raise animals (except as a grower, for whom all substantive decisions are made by the integrator). Perhaps the appeal of new or exotic animals, such as ostrich, emu, rabbits, yaks, bison, llamas, and so on, can be accounted for by biophilia. Many people want to raise animals, and if they are squeezed out of conventional animal agriculture by industrialization, they are attracted to exotic animals for which they perceive opportunities for enjoyment and profit.

In spite of our biophilic tendencies, loss of biodiversity due to human activity is nothing new. In contrast to the popular perception of "the noble savage," pre-industrial humans have not shown any evidence of innate conservation ethics. The arrival of humans in previously human-free lands has invariably precipitated an **extinction crisis**. The arrival of humans in North America via the Siberian-Alaska land-bridge has been linked to the disappearance of many of the large land mammals then present in North America (mammoths, mastodons, camels, rhinoceros, sabre-toothed cats, etc.) (see Aboriginal overkill, Chapter 7, p. 213). In fact, on all continents, and especially the Americas, Australia, and northern Eurasia, hunting by humans probably played a major role in the late Pleistocene extinctions of large mammals (Martin and Klein, 1984). The arrival of Polynesians on the islands of Oceania (southern Pacific), such as New Zealand, western Polynesia and Micronesia, led to the rapid extinction of many bird species (Steadman, 1995). Using Stone age technology, Polynesians exterminated over 2,000 bird species, or about 15 percent of the world total (Pimm et al., 1995). Human involvement in extinctions is not new; what is new is that we have the technology and knowledge to stop doing it. The issue is whether or not we have the will to sacrifice short-term economic gain for long-term environmental sustainability. Our history, unfortunately, does not provide grounds for much optimism.

# MAKING THE DESERTS BLOOM

Agriculturalists in the twentieth century and the early years of the twenty-first have had a deep-seated ethic that their role is to make two blades of grass grow where one grew before and make the deserts bloom, maximizing the production of human food as efficiently as possible. The **agrarian belief system** is based on the primary importance of agriculture to society, both socially and economically. Agrarians tend to view farming as the most honorable and virtuous way of life and often are a bit condescending towards city dwellers (a view that is reciprocated by many urbanites, often referred to in polite contemporary terms as the rural-urban divide or in less polite terms as hayseeds and manure kickers vs. city slickers). Economically, agrarians consider the production of primary products through agriculture, forestry, and mining as the only source of true wealth (see Chapter 7). In the United States, the agrarian ideology is largely based on the beliefs and writings of **Thomas Jefferson,** who believed that the nation should rest on a foundation of yeoman farmers. The family farm is viewed as the apex of traditional values (Wunderlich, 2000).

Many, if not most, of the independent farmers and ranchers in the United States have an agrarian belief system. They firmly believe that what they are doing agriculturally represents the best and wisest use of the land and its resources. Land is for growing crops or grazing livestock to produce food for humans. Forests are for the purpose of providing us with lumber and paper. Farmers and ranchers are the original environmentalists, in the agrarian view, because they care about the land and keep it productive. "Every day is Earth Day for loggers" proclaims a bumper sticker on a log truck leaving an Oregon clearcut, hauling out a load of 600 year old Douglas fir trees to be made into plywood. It is a deeply held agrarian view that primary production (farming, ranching, mining) is always the best use of the land, and anyone not in agreement with this view is a member of a suspicious "special interest group."

It is also a deeply held agrarian view that farmers and ranchers have been good stewards of the land, and only they are capable of truly understanding what it takes to provide good **stewardship**. They are particularly disdainful of the views of "city slickers" on natural resource use. Matthews (1992) provides a sobering view of the history of American agriculture:

> Even the briefest survey of our nation's history reveals that the reality does not accord with the agrarian myth of good stewardship. Whether it is a matter of the eroded badlands of Oklahoma, the Dust Bowl era immortalized in John Steinbeck's *The Grapes of Wrath*, the massive destruction of hundreds of thousands of acres of wetlands in the San Joaquin Valley of California, the depletion of the Ogallala aquifer of the High Plains,[5] the fertilizer-induced eutrophication of the Chesapeake Bay, or the salination and siltation of the Colorado River in Arizona, the conclusion is inescapable: U.S. farmers have not been good stewards of the land. (p. 173)

My intention in providing this quotation is not for the purpose of making us feel bad about being connected with agriculture, but rather to make the point that everything is not rainbows and roses, and maybe some of the flak that industrial agriculture is receiving is deserved and self-inflicted. When it comes to the current mood of society, many agriculturalists "still don't get it."

Agrarian beliefs are on a collision course with the ideology of the larger society. Mainstream America is concerned with preservation of wildlife, wetlands, rain forests (both in the tropics and in the Pacific Northwest), and other natural ecosystems. Increasingly, people are asking "why should we make the deserts bloom?" by irrigation and agricultural development. They believe the deserts to be ecosystems to be preserved, whereas agrarians regard deserts as wasteland. I encountered an example

---

[5] The depletion of the Ogallala aquifer is described by J. Opie, 2000. *Ogallala: Water for a Dry Land.* University of Nebraska Press, Lincoln.

of this conflict in belief systems when I heard a student talking about a trip across the United States. He was impressed by the Grand Canyon in Arizona but thought it too bad that it wasted so much land that couldn't be used productively for farming. People are now thinking globally. Under the agrarian belief system, livestock are servants of humans. They are beasts of burden and sources of food and clothing. The implication that animals might have "rights," including the right not to be raised in confinement on a factory farm, is in direct conflict with the agrarian belief that the role of these animals is to serve us. The faster and more efficiently they meet our needs, the better.

Current livestock production issues of concern to society (Chapters 6–10) have developed because urban society has moved away from its agrarian roots. People no longer put agriculture and forestry above all else. The list of issues seems endless: animal rights, reintroduction of wolves into the continental United States, predator control, public land grazing, riparian zones, wetlands, feed additives, logging, clear-cutting, modification of plants and animals by genetic engineering, bovine somatotropin, food safety, farm subsidies, free trade, concern for the natural ecosystems, and the list goes on. Farmers and ranchers are developing a siege mentality. They perceive themselves as under attack from all sides.

Where do we go from here? The "preservationists, radical environmentalists, food faddists, tree huggers, animal rights activists, consumer advocates, gadfly activists," and all other types of activists, advocates, and special interest groups—will they go away? Not very likely! What is a **special interest group?** To a rancher, it's an environmentalist; to an environmentalist, it's a rancher! I heard a livestock producer say this about environmentalists: "We'll fight 'em 'til hell freezes over, and then we'll fight 'em on the ice." Great rhetoric, but out of touch with reality!

Although it is possible to meet in the middle, there are extremists on both sides who are incapable of compromise. The majority of people, however, are capable of resolving conflicts. This process of conflict resolution begins

with all participants identifying their goals and concerns. One of the benefits of **holistic resource management** (see Chapter 7) is that it forces people to come to grips with their belief systems and goals. An example of a system in which this process has worked is the Oregon Watershed Improvement Coalition. Environmentalists, ranchers, hunters, sports and commercial fisheries groups, American Indians, and others with an interest in the future of rangelands have formed a coalition to, in a spirit of congeniality, work toward a solution to problems of rangeland degradation. Each group analyzes the situation within its own belief system and arrives at the "truth." Thus, each group may define the truth differently and feel equally passionately that it is the only correct truth. Rather than withdrawing at this point to shout at each other, the next step is to negotiate and compromise to reach a broad consensus that everyone can live with. It can be done!

A **new agrarianism** appears to be developing (Wunderlich, 2000), probably beginning with the publication of *Silent Spring* by Rachel Carson. New agrarianism has a strong environmental component, supporting the preservation of open space by resisting urban sprawl and congestion. It is supported by values and beliefs about the wholesomeness of farming and local commerce, and a suspicion of corporate agriculture and globalization. Livestock producers should be able to tap into this trend. Ranching maintains a cultural heritage and open space (greenspace). In the new agrarianism, environmentalists and livestock producers are natural allies.

# INTEGRATION OF CROPS AND LIVESTOCK INTO SUSTAINABLE PRODUCTION SYSTEMS

A couple of examples will be given of how sustainable systems integrating crop and livestock production could be developed. There is interest in the development of **perennial grains**. Wheat, for example, was developed by our ancestors from wheatgrasses (*Agropyron* spp.). Wheat is an annual; for wheat production, the land is cultivated and the wheat seeded each year. Soil erosion, particularly on sloped land, is high with annual grains such as wheat. Wheat is closely related to the perennial wheatgrasses of North America and can be hybridized with them to produce perennial wheat (Fig. 11-11). Several research groups are attempting to develop perennial wheat by crossing wheat with intermediate wheatgrass (*Agropyron intermedium*). Advantages of **perennial wheat** would be a dramatic reduction in soil erosion, improved soil texture and fertility, and an integration of livestock grazing into grain production systems. Wheat readily crosses with wild grasses (Mujeeb-Kazi and Asiedu, 1995). For example, *Triticum agropyrotriticum* is a synthetic species arising from a cross of wheat with quackgrass and may have potential as a perennial grain. *Tritordeum* is a cross between durum wheat and wild barley (Martin et al., 1999). According to plant breeders, it is virtually impossible for perennial grains to yield as much grain as annual crops, because a portion of the photosynthate is used to build and support a perennial root system. However, in the dichotomy of maximum yield vs. sustainability and minimization of external inputs, there may be advantages for perennial grain production. Perennial grain grasses could be grown in prairielike style, with nitrogen provided by associated leguminous plants (Piper, 1998).

Eastern gamagrass (*Tripsacum dactyloides*) is a native forage grass of North American Great Plains. It can be crossed with corn, and USDA researchers are attempting to develop a perennial grain crop from gamagrass with the selective transfer of desirable genes from corn (Hardin, 1994). Perennial grains could lead to a food production system with the advantages of the original grassland ecosystems, including the integration of grazing herbivores. Utilization of new techniques in genetic engineering have potential in accelerating the

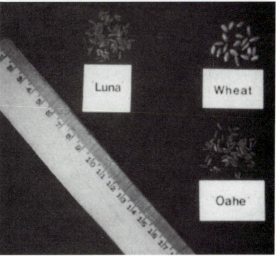

**FIGURE 11-11** (*Top*): Experimental perennial wheat plants produced by crossing wheat with perennial wheatgrasses. (*Bottom*): Seeds of perennial wheat (Luna, Oahe) compared with regular wheat. (Courtesy of Peggy Wagoner.)

development of perennial grain crops (Wagoner, 1990). Wes Jackson, of the Land Institute in Salina, Kansas, has been a leader in efforts to develop a sustainable farming system based on perennial grains and other perennial plants (Jackson, 1985).

Until the early part of the twentieth century, much of the transportation and agricultural production systems were powered by plant biomass via draft animals, mainly horses and mules. Vogel (1996) has proposed that plant biomass might once again be one of our major sources of fuel, by conversion of biomass to **ethanol**. Recent advances in genetic engineering have allowed development of microbes that can efficiently ferment plant cell walls with the production of ethanol. Vogel (1996) proposes the use of perennial grasses for biomass production, which would have favorable agronomic effects such as reduced soil erosion. Integrated systems involving perennial grasses for production of food grain, liquid fuel (ethanol), electricity (wind turbines; see Fig. 7-20), and livestock grazing could become a sustainable reality in the twenty-first century. Development of such systems will require a change from our fixation with maximizing short term yields at high input costs of chemicals and fossil fuels, and achieving a greater appreciation for long-term ecosystem health.

As part of the development of a perennial prairielike mosaic of crop plants based on perennial grasses, **perennial legumes** could also be introduced as sources of nitrogen fertilization. Ideally, rhizomatous legumes such as "spreading" alfalfa and clover cultivars that have rhizomes (creeping alfalfa) or stolons (some clovers) would be advantageous and resistant to rodent (gopher, vole) damage. Cropping systems that do not require annual soil preparation and planting could offer major savings in fossil fuel costs, preserve the land base, and reintegrate livestock production into cropping systems. Advances in biotechnology should facilitate development of perennial grasses with the desired characteristics for simultaneous food, fuel, and forage production.

# FOOD TECHNOLOGY: ANOTHER CHALLENGE TO THE LIVESTOCK INDUSTRY

Livestock production is a link in the food chain. Any product that can be produced from plants without going through the animal step will be produced more efficiently. Margarine is cheaper than butter, because it is more efficient to take photosynthate as vegetable oil and make margarine than it is to have cows eat the photosynthate in grass to make butter fat. Similarly, synthetic cheese and ice cream can be made using vegetable oils rather than butterfat. It is almost inevitable that advances in food science and technology will result in facsimiles of meat and other animal products becoming more palatable, more like the real thing, and less expensive. A pig farmer may prefer to be stricken dead than to eat imitation bacon. However, to the other $99^+$ percent of the population, if it tastes like bacon, looks like bacon (or looks better, without the extra fat), and is cheaper than bacon, whether it comes from pigs or soybeans is largely irrelevant. For many people, a plant origin would even be preferred. The utilization of plant protein sources by humans will be improved by food technologists. For example, an enzyme product to facilitate the digestion of complex carbohydrates in beans is available. This product allows people to eat beans without enduring the famous gas attacks. This will likely increase the importance of beans as a protein source in the human diet. Thus there is likely to be a continued erosion of the market for animal products because of the availability of acceptable, but less expensive, plant-based substitutes. How should the animal industries respond to this challenge? Will the future history of meat be like the past history of wool; will meat become largely replaced by substitutes? What will be the place for livestock production in the "new agriculture"?

# THE FUTURE OF LIVESTOCK PRODUCTION

I see several possible scenarios for the future of livestock production. I will present them as a basis for discussion, without suggesting that I favor one over another.

(a) Intensive, large-scale, technology-based livestock production

(b) Smaller-scale, alternative agriculture with livestock-crop integration

(c) Combination of A and B

In the developed countries, intensive livestock production will likely continue to dominate, unless blocked by legislative action. Intensive production can produce uniform, high quality meat, milk, and eggs more efficiently and at lower cost to the consumer than less intensive systems. The beef industry will be forced to become more intensive and consolidated in order to be competitive with poultry and pork production. The poultry and swine industries will likely become even more intensive and dominated by a handful of very large corporations. In the United States, the poultry industry is already dominated and controlled by a few large corporations, some of which are moving into industrial-scale swine production as well (Chapter 8).

In the United States, the most likely way that this development will be reversed or slowed is if animal rights groups and groups opposed to intensive agriculture are successful in getting laws passed that make it difficult or impossible for factory farming of animals to continue. In my opinion, this probably won't happen, except perhaps in isolated instances. People are not likely to vote to increase the cost of their food, just as they are not likely to vote for increased taxes. Self-interest comes ahead of idealism for most people. Another factor that could cause the demise of intensive animal production would be the depletion of

fossil fuels without the development of new sources of cheap energy.

The populations of most developed countries have stabilized, and there is excess food production capacity. There will probably be a trend to the permanent withdrawal of agricultural land from production and its return to forest or grassland. This will have beneficial effects on wildlife and natural ecosystems, while (it is hoped) providing better economic returns for the remaining agricultural producers.

Agricultural practices and animal production will vary considerably among countries. In the former Soviet Union and Eastern Europe, livestock production units of state farms were huge and very inefficient. These are being broken up into small, privately owned farms. Eventually, consolidation of these farms into larger, more economic units will occur, as has taken place in North America. One of the recurring themes in Latin America has been the breakup of large farms under land reform programs to distribute land to peasant farmers. This policy has largely been a failure. I have seen formerly very productive large farms in Chile that were seized by the Communist government of Chile in the 1970s and distributed to the former farmworkers. In a few years, the farms were abandoned, the peasants moved to the cities, and formerly productive farms were turned into wastelands. Mexico has recently reversed its policy of land distribution to peasants and is encouraging large commercial farms to develop. Increased productivity has been the result. In parts of Western Europe, farmers are "guardians of the countryside." In England, for example, hedgerows and stone walls are protected by law, to preserve the character of the English countryside. Switzerland similarly has legislation to protect its small alpine dairy farms and subsidizes small farmers so they can earn an adequate income. As societies become mature, these policies become important to the populace. Thus in many countries, small farms, including livestock production units, may be preserved through economic incentives via tax laws and subsidies to preserve the character of the countryside.

Perhaps subsidized public land grazing fees and other interventions could help preserve ranching in the North American West, and cowboys and ranchers may eventually be seen as guardians of the countryside. In fact, conservation groups such as **The Nature Conservancy** are beginning to recognize that helping ranchers to survive may be in the best interests of the environment. The American West is rapidly becoming urbanized (Fig. 11-12). Prime farm- and ranchland and wildlife habitat are being "developed" at an astounding rate, as anyone who has looked out the airplane window while flying to Denver can appreciate. Wildlife are especially vulnerable to this human invasion. Prime winter habitat of elk, deer, and other large herbivores is also prime land for "development" (a one-word oxymoron!). It is hoped that the American public will begin to realize before it is too late that the countryside they cherish can best be preserved by preserving farming and ranching.

In the tropical developing countries, livestock production tends to be small-scale and well integrated with crop production. This is likely to continue. Agroforestry will become increasingly common, and livestock will be an important component of agro-

**FIGURE 11-12** Invasion of farmland and wildlife habitat by suburbia is a major threat to agricultural production and the preservation of wilderness.

forestry systems. With the rapidly increasing human populations of developing countries, food production is of paramount concern. Concepts such as animal rights may not be received very favorably under such circumstances, particularly in countries where human rights are lacking.

My experience in tropical countries has been to observe a high degree of importance of livestock raising. People in cities keep chickens, goats, or pigs. Millions of draft animals are in use, including donkeys, horses, oxen, and buffalo. Feeding and management systems have evolved over hundreds or thousands of years that very effectively integrate animals into crop and vegetable production. Of course, this was also true of the farms of Europe and early America. However, these self-sufficient small farms in the developed countries have largely been replaced by more intensive production, which has been much more efficient and productive in terms of food production and use of labor. Critics maintain that this is largely based on cheap fossil fuel energy inputs, which are not sustainable on the long term. It is likely that small subsistence farming with animals as key components will continue in most developing countries. Human populations in the tropics continue to increase rapidly, and cities are expanding beyond the capacity of the infrastructures to provide food, water, and sewage disposal. Rural subsistence farming will be essential for many people for survival.

Many lesser-developed countries are now well on their way to becoming economically developed, such as China, countries of southeast Asia, and several Latin American countries. As per capita incomes rise, so does the demand for meat and other animal products. As discussed in Chapters 3 and 8, this demand will be met by industrialized animal production, particularly of poultry. Corporate poultry and swine producers from Europe and North America are rapidly moving into these emerging markets. At present, people in developing countries are not very receptive to such concepts as animal rights and are anxious to enjoy the "good life" that Europeans,

Japanese, and North Americans have. In many lesser-developed countries, the dichotomy of livestock production will remain for many years: high technology, intensive animal production providing products for urban dwellers, while in the countryside, millions of people will continue to engage in subsistence agriculture for their survival.

# OUR ACHILLES' HEEL— THE POPULATION BOMB

The world's human population increased dramatically in the twentieth century, increasing at an exponential rate (Fig. 11-13). This growth rate is clearly not sustainable, and many efforts to control population growth have been made. In the developed or industrialized countries, population stabilization has largely been achieved, whereas in the tropical developing countries, huge increases in population are projected before stabilization may occur. These population increases account for the "feeding the world" mentality that has led to the promotion of agricultural development at all cost. In the 1970s and early 1980s, farm-

ers in the United States were encouraged to plant crops "fencerow to fencerow" and many people believed that mass starvation was just around the corner. However, human population growth is not usually constrained by limitations of food. In modern times, populations have grown fastest in countries with the least food and slowest in countries where food is abundant (Cohen, 1997).

Obviously, there are clear limits to the number of people that the biosphere can accommodate (Cohen, 1995). The upper limit of human population is probably a great deal higher than the number that is consistent with a high quality of life. Limiting the rate of population increase, and achieving a stable world population, should be our ultimate aim. Sustainable agriculture and sustainable resource utilization imply a stabilized human population. The world's present population is expected to double by 2050 (Kendall and Pimentel, 1994). This would imply at least a doubling of present food production, just to meet current standards of nutrition. Millions of people now suffer from inadequate food or nutrient intake. The challenge of meeting the global food needs of the future is formidable and could well necessitate a diversion of grains from livestock use to direct human consumption. Global (human) carrying capacity involves both absolute numbers and quality of life. Increasing affluence reduces the number of people that can be supported by a given resource base (Barrett and Odum, 2000). For example, one American may consume 100 or more times the amount of energy and resources as a citizen of a developing country.

The 1994 genocide of 500,000 people in Rwanda is a textbook example of political problems exacerbated by overpopulation (Uvin, 1996). Rwanda is the most densely populated country in Africa. Resource scarcity played a key role in precipitating the violence, which in turn led to major ecological damage. According to Uvin (1996): "Rwanda's ecological problems cannot be addressed without resolving its political problems and vice versa. Any solution that addresses only one aspect of the dynamic is bound to fail" (p. 15). When

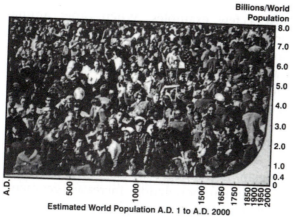

**FIGURE 11-13** More people are alive today than all of the people who ever lived prior to 1900! The world had an astounding increase in population in the twentieth century.

$1.0 billion. Others making the list were Donald John Tyson of Tyson Foods at $1.2 billion and those of the Cargill fortune (a total of $7.875 billion for seven members of the Cargill family).

While Donald Tyson of Tyson Foods joined the *Forbes'* 400 list, the average hourly earnings of poultry processing plant workers in the United States in 1995 was $5.27 (Schrimper, 1997). Trickle-down economics isn't working well for the people who actually do the work in the poultry industry—the growers and processing plant workers. Instead of a trickle-down, there's a cascade-up to a single-profit center!

The success of the industrial age has given human society a new challenge. We have the industrial capacity to meet everyone's basic needs. However, once people's basic needs are satisfied, **quality of life** becomes of concern. Unfortunately, consumerism and acquisition of expensive toys do not improve quality of life. Absence of wealth can cause misery, but having it is no guarantee of happiness. The effect of wealth on happiness or contentment was assessed by Myers and Diener (1995). Assessment of happiness, using standard social science methodology, showed an interesting relationship (Fig. 11-14). Increased wealth has not increased happiness. Compared with Americans in 1957, people today have twice as many cars, plus microwave ovens, color TVs, VCRs, air conditioning, answering machines, and $12 billion worth of brandname athletic shoes (Myers and Diener, 1995). Judging by rates of depression, violent crime, divorce rate, teen suicide rate, young people "unable to find themselves," and so on, Americans are richer but not happy campers.

Lane (2000), in the book *The Loss of Happiness in Market Economies*, discussed in detail the ironic situation that has occurred over the years since the Second World War. The overwhelming economic successes of the market economies have not brought the beneficiaries of these successes (the labor forces of developed economies) the emotional satisfaction being sought. As people have strived to "get ahead," they are less and less happy to have ar-

rived. All evidence suggests that a happy family life is the single most important factor in achieving personal satisfaction. Lane (2000) suggested that the decline in birth rates in most of the developed economies might reflect hormonal disturbances related to stress responses associated with the postindustrial economy. The capitalistic market-driven economic road we are on is not leading us to the personal happiness pot-of-gold. Perhaps this can be related to animal agriculture by considering the personal fulfillment possibilities open to a worker on a "factory farm," as compared with those achievable as a family farmer (in an economic atmosphere that made it possible for a family farmer to earn a living). The industrialization of animal agriculture, producing cheap food, may have come at the expense of quality of life.

In many respects, it is easier to live with adversity rather than with affluence. Humans have not been very successful in learning how to develop sustainable self-regulating societies that don't depend on continual growth. Civilizations have come and gone. Like a biological organism, they may experience a period of rapid, robust growth, which eventually levels off and then declines. Development of a great civilization is accompanied by rapid population growth, which outstrips the resource base and crashes. We need go no farther than the U.S. Southwest or Central America to observe the relics of past, failed civilizations (e.g., Maya, Inca, and Aztec), let alone the Egyptians, Greeks, and Romans. Medieval Europe was saved from the Dark Ages by the discovery of the "New World," allowing a 500 year period of unrestrained growth, which must come to an end soon. Unless future Mars expeditions provide more promising evidence than that seen to date, there are no "new worlds" as safety valves.

The world is in a transition to a new paradigm (to use a popular buzz word). Personal fulfillment and **quality of life** (freedom from war, clean air and water, safe and abundant food, good health care, personal safety, high literacy, meaningful or intellectually satisfying work, just wages, etc.) may replace the pursuit of money as goals of life. Global capitalism, chasing ever cheaper sources of labor, cannot

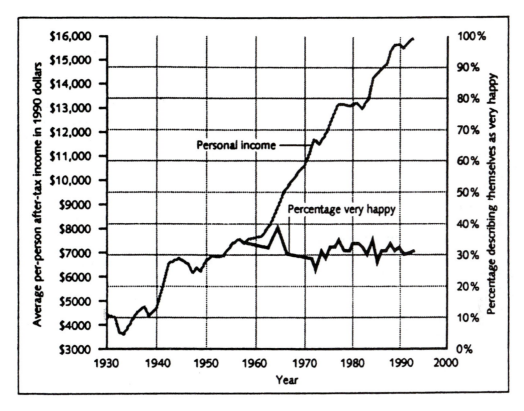

**FIGURE 11-14** Inflation-adjusted income versus happiness in the United States. Increased wealth has apparently not made us more happy. Perhaps the old saw "Money isn't everything but it sure beats what's second" isn't true after all! (From Myers and Diener, 1995; courtesy of *Psychological Science*.)

satisfy quality of life issues. As Greider (1997) has summarized, once people are freed from scarcity, they are faced with a profound reckoning with self-discovery. "Who am I? What is my life for? If people are no longer bound to endless toil, what is their real work and purpose here on earth? If survival and accumulation are no longer the challenge, where does one find meaning and pleasure? As wealthy people can attest, these are the hardest questions" (p. 440).

So what's this got to do with animal agriculture? First, the direction in which animal production is heading—globalized, industrial production seeking the lowest labor costs, maximum short-term economic efficiency, and the concentration of wealth in the corporate headquarters—is an example of the classic industrial process that is clearly unsustainable and exploitative, creating a great "single-profit center" supported by an army of poorly paid indentured workers (growers and processing

plant workers). On the other hand, maintaining a component of "farming as a way of life" in agriculture could help satisfy the basic desires of people to be independent, to like and be interested in what they are doing, and to feel satisfied with their quality of life. Raising livestock as an independent farmer or rancher achieves these goals for many people, whereas doing the "grunt work" on an industrialized, automated factory farm may not.

How will animal scientists and land grant universities be involved in societal changes? Will we continue to worship efficiency and technological wizardry that benefit the corporate few, or will we help to lead the way to what President George H. Bush called "a kinder and gentler society?" The following comments by Uhl et al. (1996) are relevant:

The job of running the American university is generally left to managers—people who, by and large, think about the "bottom line" in terms of enrollment, grant income, endowments, alumni rolls, and investments and not necessarily in terms of citizenship, global responsibility, moral fiber, and ecological literacy. (p. 1310)

At a time when all the world looks to the United States as a role model, they see a country drunk on consumption, a country whose people are often apathetic, imagining that citizenship requires little or nothing of them. If U.S. universities are to truly fulfill their responsibilities of education, research and leadership, they must (1) begin producing students who are ecologically literate and who understand both the demands and power of an active citizenry; (2) take a proactive role in convincing public and private sector funders to support research on pressing sustainability-linked issues; and (3) operate according to sustainable practices. They must, in short, take courageous steps to prepare students for a world founded on sustainable practices and not on the impossible dream of perpetual growth. (p. 1311)

Current emphasis on corporate "clients and customers" by land grant universities, the formation of official corporate linkages (e.g., The Cargill Higher Education Initiative) with universities, and grant-seeking "feeding frenzies" by university faculty indicate that the "bottom line" still wags the dog. In a follow-up article to Uhl et al. (1996), Uhl and Anderson (2001) suggested that sustainability should become a central organizing focus for higher education, instead of "mindlessly promoting economic globalization and homogenization of culture" (p. 41). Universities could be the catalyst in creating a new generation of socially and ecologically responsible citizens.

Another implication of globalization of food production is the loss of biological and cultural diversity and a global homogenization of culture and values. In political terms, the concentration of economic power in fewer and larger transnational corporations blurs the concepts of nationalism and nation-states. Is that good or bad? Ironically, in the United

States, it is probably true that the most ardent proponents of global capitalism and "the market" are also among the most ardent flag wavers and patriots.

Lacy (1994) discusses the effects of science, technology and capitalism on the cultural values of agriculture. Until recently, production of and eating food had sacred connotations,

with rites and rituals surrounding food and agriculture in each culture. Great festivals marked the spring sowing and the fall harvest. Animals were slaughtered according to certain carefully regulated practices. Bread was broken only after certain prayers were performed . . . Today the link between seeds and fertility and the earth has been broken. Mechanization of food production, processing, and consumption has removed the mystery of food and agriculture . . . The food production process has been standardized so that food products have become indistinguishable from other products sold in our consumer society. Perhaps this is nowhere more evident than in the fast food industry where the very uniformity, sterile environment for dining and the speed at which the food is consumed all convert food into merely an instrument, a tool and a resource for our survival. (p. 6)

Concentration of germplasm into a few crops and livestock breeds, patented and owned by a few corporate giants, so that worldwide we all get the same food, prepared the same way, with high economic efficiency (e.g., ADM-self-styled supermarket to the world) is the logical conclusion of global capitalism in agriculture. This is where we are heading. Are we sure that's where we really want to go?

There have been several well-publicized efforts around the world to fight back against global cultural homogenization. A French sheep farmer, Jose Bove, led the French Farmers Union in an attack on a McDonald's restaurant in France. His views are publicized in the book *The World is Not For Sale. Farmers Against Junk Food* (Bove and Dufour, 2001). An activist in India, Vandana Shiva, has campaigned against corporate control of the world's food supply, and published the book *Stolen Harvest. The Hijacking of the Global Food Supply* (Shiva, 2000). The book *Fast Food Nation*, by Schlosser (2001), is a critique of the fast-food industry. A "Slow Food Movement" has emerged; its objective is to encourage the preservation of traditional cultural dietary and eating habits, in opposition to the McDonaldization of global cuisine.

The **World Trade Organization (WTO)** was formed to rationalize rules for global trade. It has served as a lightening rod for opponents of globalization, culminating in the "Battle in Seattle" at the 1999 WTO meeting in Seattle. Opponents of the WTO are not opposed to world trade, but are opposed to global trade regulations that appear to allow transnational corporations to gain control of the world's economy for the benefit of the wealthy at the expense of the world's poor. According to David Gergen, editor of *U.S. News and World Report* (*USNWR* May 14, 2001), half the people on Earth today live on less than $2 per day. The three richest people on Earth have wealth that exceeds the combined gross domestic product of the 43 poorest nations. The "Third World" deserves to be more than a place to mass produce cheap food, clothing, and electronics for the United States, Europe, and Japan.

## SOME PERSONAL THOUGHTS

My objective in this book has been to provide a balanced discussion of the importance of livestock production to society and to consider some of the major concerns and issues that have been raised about the animal industries. Livestock producers, animal scientists, and others connected to animal production cannot ignore these issues, and to be effective in responding to them, they must have some appreciation of the rationale behind society's concerns. This is especially important for young people beginning careers in agriculture. I hope this book will provide a framework for these discussions. I cannot claim to have provided the truth, because the "truth" is in the eye of the beholder.

I grew up on a farm in Canada and possessed a full-fledged agrarian belief system. I believed that land was meant to be used for farming, and if not suitable for farming, then it should be used for logging. The population of Canada when I was a youth was 12–15 million people. Like many Canadians, I fervently believed that Canada needed 75–100 million people, as soon as possible, to provide a larger internal market for manufactured goods. Then we could cease being a branch-plant economy and take our rightful place as a major industrial power. I fantasized about developing the north, with mining, agriculture, and pulp production, and developing large northern cities. I liked to think that Prince George, B.C., would become the Chicago of Canada. I also fantasized about agricultural development in the tropics. I mentally developed a huge machine that would be driven through the Amazon jungles, chopping up the forest and converting the jungle to farmland. I thought the Fraser River in British Columbia should have numerous hydroelectric dams in the Fraser Canyon, to produce electricity for industrial development. I couldn't understand the objections of biologists to building dams on the Fraser. So what if the salmon runs became extinct? Electricity for economic development was more important, and there were plenty of salmon in other rivers.[6]

---

[6] Ironically, serious consideration is now being given to the removal of dams, such as those in the Columbia River system (*Science* 284: 574, 1999). New technology makes it possible to harness the energy of rivers for hydroelectric power without the need for huge reservoirs. Of further irony is that while developed countries such as the United States are considering dam removal, China is moving full-speed on damming the gorges of the Yangtze River, which will have devastating ecological and cultural consequences. A special issue of BioScience (vol. 52, No. 8, 2002) was devoted to dam removal and river restoration.

Growth, development, and subduing the wilderness to put it to human use formed the core of my belief system.

I like to think that somewhere along the way my thinking matured into a deeper understanding of the biological and ecological processes of our planet. This has led me to recognize the need for sustainable activities, rather than exploitation. What do we want Earth to be like 100 years from now? 1,000 years from now? What would our planet be like now if people had thought this way 100 years or 1,000 years ago?

I recognize that in this chapter I seem to have drifted some distance from a discussion of the role of livestock production in human welfare. What I hope comes through is that for livestock production to continue to be important in human welfare, greater emphasis must be placed on integrating animal production into sustainable agroecology. Society will insist on it, with legislation dealing with animal welfare, waste disposal, air and water pollution, grazing on rangelands and riparian zones, predator and pest control, use of agricultural chemicals and feed additives, industrialized farming of animals, transgenic animals, and so on. Those in the animal industries can face these perceived threats in an adversarial way, or they can react positively. At present, as I view these controversies in the western United States, I perceive that the "don't give an inch" mentality is alive and well, but if it persists, the livestock industries will go down the tube. To react positively will require a change in traditional agrarian values to include a greater appreciation of the global environment. With current technology, people can survive and live well without consuming meat or using animal products, and they have every right to do so. In fact, increasing numbers of people are voluntarily choosing such a lifestyle and rigorously avoid the use of any animal-derived products. A primary motivation for making this dietary choice is their objection to current systems of intensive animal production and the desire to avoid financially supporting such systems. People interested in careers involving production of

domestic animals, either as farmers or as scientists, must face up to and respond to these issues. If we are not up to this challenge, livestock production will, in my opinion, not survive as an important industry. I hope that this book will help young people meet the challenges facing the livestock industries and that the importance of domestic animals in our civilization will continue. Finally, I hope that animal scientists and livestock producers will not look at this book and think "with friends like this, who needs enemies?"

Agriculture can make contributions to human welfare beyond simply producing food. These functions are known as **ecosystem services** and include regulation of atmospheric gases, flood control, water storage, soil formation, waste treatment and pollution control, wildlife habitat, recreational opportunities, cultural and aesthetic aspects (e.g., the view), and so on. These ecological and cultural services are generally provided free by agriculture. Current economic analyses do not consider these contributions (Costanza et al., 1997). Pimentel et al. (1997) have calculated that the ecosystem services attributable to biodiversity are worth $319 billion per year, about 5 percent of Gross Domestic Product. For the world, benefits are estimated to be $2,928 billion per year, or about 11 percent of the total world economy. Perhaps in the future, farmers can be compensated for ecosystem services. How much is a nice view worth? Who pays for the view, and how do they pay? Farmers and ranchers should somehow be compensated for these contributions; such compensation would increase the economic viability of sustainable production systems.

Nielsen (1992) summed up the relevance of sustainable agriculture to the livestock producer in this way: "Presently farmers are viewed as producers of food. If sustainable agriculture is to become a reality, farmers must be regarded as managers of healthy ecounits that have as only one of their objectives, the production of food" (p. 25). I would only add that farmers and ranchers should begin to regard themselves in this way. Their goal should shift from maximizing yields of crops and livestock to optimizing the production of human food using techniques that minimize or prevent environmental damage and maximize quality of life. Where possible, as on rangelands, for example, food production should be integrated optimally with other human uses such as recreation and with enhancement of the environment for native plants and animals. Such changes will occur only when environmentally friendly farming is also the most profitable system. Society brings about such changes for the common good by tax incentives, subsidies, and penalties that lead people to change by the carrot approach and to minimize forced changes by the stick approach. Agricultural interests should participate fully in developing guidelines for the "new agriculture," rather than fighting the inevitable. Farmers and ranchers should become true ecosystem managers and validate the common claim that "farmers and ranchers are the original environmentalists."

# STUDY QUESTIONS

1. How does life on Earth contribute to maintaining the conditions necessary for continued life on Earth?

2. Rivers have been carrying dissolved mineral salts into the oceans for eons. Why don't the oceans get increasingly salty, until they're all like the Dead Sea or Great Salt Lake?

3. What is the Gaia concept? Does it seem reasonable? Explain.

4. What is the value, if any, of conserving rare breeds of livestock? What is the relationship, if any, between rare breed associations and the concept of biophilia?

5. With biotechnology techniques such as cloning, it may be possible to have "monocultures" of genetically uniform animals. What are the potential advantages and disadvantages of this?

6. What relationship is there, if any, between the philosophical bases of vegetarianism and the Second Law of Thermodynamics?

7. Discuss the use of the concept of human-consumable feedstuffs in evaluating the competition between humans and livestock for food resources.

8. What has been the contribution of Edward Faulkner to the development of sustainable agriculture?

9. In her book *Silent Spring*, Rachel Carson speculated that songbirds would become scarce or extinct because of pesticides. Has this happened? What are now the greatest threats to songbirds in North America?

10. How can farming practices in the midwest influence the survival of turtles and shrimp in the Gulf of Mexico?

11. What is the real agenda of the environmental movement?

12. Do you think that perennial grain crops will become a significant part of agriculture? Explain.

13. Do you think it is possible or probable that "synthetic meat" will be developed from soybeans and other plant products, which will be accepted by people as a substitute for meat? Assume that food scientists are successful in developing "synthetic meat" that looks, tastes, and feels like the real thing. Would this be a good or bad thing?

14. Describe what you think farming and ranching will be like in North America 50 years from now.

15. On a strictly mathematical basis, it is obvious that the human population of the world cannot increase indefinitely. What do you see as the most likely scenarios for the regulation of global human population?

16. Do you think farmers and ranchers should be compensated for ecosystem services? Do you believe in subsidies for agriculture? Discuss.

# REFERENCES

APPENZELLER, T. 1992. Deep-living microbes mount a relentless attack on rock. *Science* 258:222.

BARRETT, G. W., and E. P. ODUM. 2000. The twenty-first century: The world at carrying capacity. *BioScience* 50:363–368.

BOVE, J., and F. DUFOUR. 2001. *The world is not for sale. Farmers against junk food.* Verso (New Left Books), New York.

CAPONE, D. G., J. P. ZEHR, H. W. PAERL, B. BERGMAN, and E. J. CARPENTER. 1997. *Trichodesmium,* a globally significant marine cyanobacterium. *Science* 276:1221–1229.

CARSON, R. 1962. *Silent spring.* Houghton Mifflin Co., Boston.

CHAPIN, F. S., III, B. H. WALKER, R. J. HOBBS, D. U. HOOPER, J. H. LAWTON, O. E. SALA, and D. TILMAN. 1997. Biotic control over the functioning of ecosystems. *Science* 277:500–504.

CHESTER, C. C. 1996. Yellowstone's resources. *Environment* 38(8):11–15,34–36.

COHEN, J. E. 1995. Population growth and earth's human carrying capacity. *Science* 269:341–346.

COHEN, J. E. 1997. Population, economics, environment and culture: An introduction to human carrying capacity. *J. Appl. Ecol.* 34:1325–1333.

COSTANZA, R., R. D'ARGE, R. DE GROOT, S. FARBER, M. GRASSO, B. HANNON, K. LIMBUR, S. NAEEM, R. V. O'NEILL, J. PARUELO, R. G. RASKIN, P. SUTTON, and M. VAN DEN BELT. 1997. The value of the world's ecosystem services and natural capital. *Nature* 387:253–260.

FAULKNER, E. H. 1943. *Plowman's folly.* University of Oklahoma Press and Grosset and Dunlap, New York.

GIBBS, M. J., J. S. ARMSTRONG, and A. J. GIBBS. 2001. Recombination in the hemagglutinin gene of the 1918 "Spanish Flu." *Science* 293:1842–1845.

GOLD, T. 1992. The deep, hot biosphere. *Proc. Nat. Acad. Sci.* USA. 89:6045–6049.

GREIDER, W. 1997. *One world, ready or not. The manic logic of global capitalism.* Simon and Schuster, New York.

HARDIN, B. 1994. Corn's comeback cousin. *Agr. Res.* 42(4):12–15.

**HEITSCHMIDT, R. K., R. E. SHORT, and E. E. GRINGS.** 1996. Ecosystems, sustainability and animal agriculture. *J. Anim. Sci.* 74:1395–1405.

**HENDRIX, P. F., R. W. PARMELEE, D. A. CROSSLEY, JR., D. C. COLEMAN, E. P. ODUM, and P. M. GROFFMAN.** 1986. Detritus food webs in conventional and no-tillage agroecosystems. *BioScience* 36:374–380.

**HERN, W. M.** 1993. Is human culture carcinogenic for uncontrolled population growth and ecological destruction? *BioScience* 43:768–773.

**HODGES, J.** 1991. Sustainable development of animal genetic resources. *World Anim. Rev.* 68(3):2–10.

**HORRIGAN, L., R. S. LAWRENCE, and P. WALKER.** 2002. How sustainable agriculture can address the environmental and human health harms of industrial agriculture. *Environ. Health Perspectives* 110:445–456.

**HUSTEN, L.** 1992. The legacy of Ivanhoe. *Discover* 13(5):22–23.

**JACKSON, W.** 1985. *New roots for agriculture.* University of Nebraska Press, Lincoln.

**KENDALL, H. W., and D. PIMENTEL.** 1994. Constraints on the expansion of the global food supply. *Ambio* 23:198–205.

**KOLATA, G.** 1999. *Flu. The story of the great influenza pandemic of 1918 and the search for the virus that caused it.* Farrar, Straus and Giroux, New York.

**LACY, W. B.** 1994. Biodiversity, cultural diversity, and food equity. *Agric. Human Values* 11(1):3–9.

**LANE, R. E.** 2000. *The loss of happiness in market democracies.* Yale University Press, New Haven.

**LASLEY, P., E. HOIBERG, and G. BULTENA.** 1993. Is sustainable agriculture an elixir for rural communities? *Am. J. Alternative Agric.* 8(3):133–139.

**LAVER, G., and E. GARMAN.** 2001. The origin and control of pandemic influenza. *Science* 293:1776–1777.

**LAWRENCE, A. (ED.)** 1990. *Genetic conservation of domestic livestock.* CAB International, University of Arizona Press, Tucson.

**LEOPOLD, A.** 1949. *A Sand County almanac.* Oxford University Press.

**LITTLE, C. E.** 1987. *Green fields forever. The conservation tillage revolution in America.* Island Press, Washington, D.C.

**LOI, P., G. PTAK, B. BARBONI, J. FULKA, JR., P. CAPPAI, and M. CLINTON.** 2001. Genetic rescue of an endangered mammal by cross-species nuclear transfer using post-mortem somatic cells. *Nature Biotechnology* 19:962–964.

**LOVELOCK, J.** 1987. *Gaia.* Oxford University Press.

**MACLACHLAN, G. K., and W. S. JOHNSTON.** 1982. Copper poisoning in sheep from North Ronaldsay maintained on a diet of terrestrial herbage. *Vet. Rec.* 111:299–301.

**MARRA, P. P., K. A. HOBSON, and R. T. HOLMES.** 1998. Linking winter and summer events in a migratory bird by using stable-carbon isotopes. *Science* 282:1884–1886.

**MARTIN, A., J. B. ALVAREZ, L. M. MARTIN, F. BARRO, and J. BALLESTEROS.** 1990. The development of tritordeum: A novel cereal for food processing. *J. Cereal Science* 30:85–95.

**MARTIN, P. S., and R. G. KLEIN. (EDS).** 1984. *Quaternary extinctions.* University of Arizona Press, Tucson.

**MATSON, P. A., W. J. PARTON, A. G. POWER, and M. J. SWIFT.** 1997. Agricultural intensification and ecosystem properties. *Science* 277:504–509.

**MATTHEWS, R. J.** 1992. Designing an environmentally responsible undergraduate curriculum. pp. 173–187 In: *Agriculture and the undergraduate.* National Academy Press, Washington, D.C.

**MATZKE, G.** 1983. A reassessment of the expected development consequences of tsetse control efforts in Africa. *Soc. Sci. Med.* 17:531–37.

**MUJEEB-KAZI, A., and R. ASIEDU.** 1995. The potential of wide hybridization in wheat improvement. *Annals of Biology* 11:1–15.

**MYERS, D. G., and E. DIENER.** 1995. Who is happy? *Psychol. Sci.* 6:10–19.

NATIONAL RESEARCH COUNCIL. 1989. *Alternative agriculture.* National Academy Press, Washington, D.C.

NATIONAL RESEARCH COUNCIL. 1993. *Managing global genetic resources: Livestock.* National Academy Press, Washington, D.C.

NIELSEN, N. O. 1992. Ecosystem health and veterinary medicine. *Can. Vet. J.* 33:23–26.

NORSE, E. A. 1990. *Ancient forests of the Pacific Northwest.* Island Press, Washington, D.C.

NOTTER, D. R. 1999. The importance of genetic diversity in livestock populations of the future. *J. Anim. Sci.* 77:61–69.

OLTJEN, J. W., and J. L. BECKETT. 1996. Role of ruminant livestock in sustainable agricultural systems. *J. Anim. Sci.* 74:1406–1409.

OPIE, J. 2000. *Ogallala: Water for a dry land.* 2nd ed., University of Nebraska Press, Lincoln and London.

PIMENTEL, D., C. WILSON, C. MCCULLUM, R. HUANG, P. DWEN, J. FLACK, Q. TRAN, T. SALTMAN, and B. CLIFF. 1997. Economic and environmental benefits of biodiversity. *BioScience* 47:747–757.

PIMM, S. L., G. J. RUSSELL, J. L. GITTLEMAN, and T. M. BROOKS. 1995. The future of biodiversity. *Science* 269:347–350.

PIPER, J. K. 1998. Growth and seed yield of three perennial grains within monocultures and mixed stands. *Agric. Ecosyst. Environ.* 68:1–11.

PRESTON, R. 1994. *The hot zone.* Random House, New York.

ROWINSKY, E. K., L. A. CAZENAVE, and R. C. DONEHOWER. 1990. Taxol: A novel investigational antimicrotubule agent. *J. Natl. Cancer Inst.* 82:1247–1259.

SCHLOSSER, E. 2001. *Fast food nation. The dark side of the all-American meal.* Houghton Mifflin Co., Boston.

SCHRIMPER, R. A. 1997. U.S. poultry processing employment and hourly earnings. *J. Appl. Poult. Res.* 6:81–89.

SHIVA, V. 2000. *Stolen harvest. The hijacking of the global food supply.* South End Press, Cambridge.

STEADMAN, D. W. 1995. Prehistoric extinctions of Pacific island birds: Biodiversity meets zooarchaeology. *Science* 267:1123–1131.

TANGLEY, L. 1996. The case of the missing migrants. *Science* 274:1299–1300.

TAWAH, C. L., J. E. O. REGE, and G. S. ABOAGYE. 1997. A close look at a rare African breed—The Kuri cattle of Lake Chad Basin: Origin, distribution, production and adaptive characteristics. *S. Afr. J. Anim. Sci.* 27:31–40.

TORBERT, H. A., and H. B. JOHNSON. 2001. Soil of the intensive agriculture biome of Biosphere 2. *J. Soil Water Conserv.* 56:4–11.

UHL, C., and A. ANDERSON. 2001. Green destiny: Universities leading the way to a sustainable future. *BioScience* 51:36–42.

UHL, C., D. KULAKOWSKI, J. GERWING, M. BROWN, and M. COCHRANE. 1996. Sustainability: A touchstone concept for university operations, education, and research. *Conserv. Biol.* 10:1308–1311.

UVIN, P. 1996. Tragedy in Rwanda: The political ecology of conflict. *Environment* 38(3):7–15,29.

VAVRA, M. 1996. Sustainability of animal production systems: an ecological perspective. *J. Anim. Sci.* 74:1418–1423.

VISSCHER, P. M., D. SMITH, S. J. G. HALL, and J. A. WILLIAMS. 2001. A viable herd of genetically uniform cattle. *Nature* 409:303.

VITOUSEK, P. M., P. R. EHRLICH, A. H. EHRLICH, and P. A. MATSON. 1986. Human appropriation of the products of photosynthesis. *BioScience* 36:368–373.

VITOUSEK, P. M., H. A. MOONEY, J. LUBCHENCO, and J. M. MELILLO. 1997. Human domination of earth's ecosystems. *Science* 277:494–499.

VOGEL, K. P. 1996. Energy production from forages (or American agriculture-back to the future). *J. Soil Water Conserv.* 51:137–139.

WAGONER, P. 1990. Perennial grain development: Past efforts and potential for the future. *Crit. Rev. Plant Sci.* 9:381–408.

WATSON, A. J., and P. S. LISS. 1998. Marine biological controls on climate via the carbon and sulphur geochemical cycles. *Phil. Trans. R. Soc. Lond.* B 353:41–51.

WEBB, W. L., W. K. LAUENROTH, S. R. SZAREK, and R. S. KINERSON. 1983. Primary production and abiotic controls in forests, grasslands, and desert ecosystems in the United States. *Ecology* 64:134–151.

WELLS, D. N., P. M. MISICA, H. R. TERVIT, and W. H. VIVANCO. 1998. Adult somatic cell nuclear transfer is used to preserve the last surviving cow of the Enderby Island cattle breed. *Reprod. Fertil. Dev.* 10:369–378.

WILSON, E. O. 1984. *Biophilia.* Harvard University Press, Cambridge.

WILSON, E. O. (ED.) 1988. *Biodiversity.* National Academy Press, Washington, D.C.

WUNDERLICH, G. 2000. Review essay. Hues of American agrarianism. *Agriculture and Human Values* 17:191–197.

YOUNG, L. D. 1992. Effects of Duroc, Meishan, Fengjing, and Minzhu boars on productivity of mates and growth of first-cross progeny. *J. Anim. Sci.* 70:2020–2029.

# Index